T0317734

Systems Engineering

Systems Engineering

A 21st Century Systems Methodology

Derek K. Hitchins
FIET, FRAeS, FCMI, INCOSE Fellow

John Wiley & Sons, Ltd

To my beloved wife,
without whom . . . very little.

Contents

2   ADVANCES IN SYSTEMS SCIENCE   31

3   ADVANCES IN SYSTEMS THINKING   63

CASE G: DEFENSE PROCUREMENT IN THE 21ST CENTURY   **447**

18   SYSTEMS ENGINEERING: INTELLIGENT SYSTEMS   **459**

CASE H: GLOBAL WARMING, CLIMATE CHANGE AND ENERGY   **477**

Foreword

Derek Hitchins is truly a long term pioneer in systems engineering. He had a variety of experiences, initially serving in the Royal Air Force until his retirement after 22 years service. He subsequently held positions in the public and private sector in a variety of positions including serving as the UK Technical Director for the NATO Air Command and Control System (ACCS), and in two leading systems engineering companies in the UK as Marketing Director, Business Development Director and Technical Director. He first became an academic in 1988.

He was the inaugural president of the UK Chapter of the International Council on Systems Engineering (INCOSE), and also the inaugural chairman of the Institution of Electrical Engineers (IEE) Professional Group on Systems Engineering. He has also been a member of the UK Defense Scientific Advisory Board.

His current research is into system engineering on a broad scale, including: system thinking, system requirements, social psychology and anthropology, command and control system design, and world-class systems engineering. He published his first book titled *Putting Systems to Work* in 1992, and a second book titled *Advanced Systems Thinking, Engineering and Management* in 2003. He has also completed an on-line electronic book titled *Getting to Grips with Complexity* which examines complexity, what is it, how it comes about, and how we can exploit. This, and a description of his other works can be found at http://www.hitchins.net/SysBooks.html.

He has accomplished much that should support establishment of systems engineering as the dominant paradigm for managing complexity in industry. His work has done much to develop a large scope view of systems engineering comprising product, project, business, industry and socio-economic levels. His apparent objective in this is to support systems engineering as the route to simultaneous effectiveness, efficiency and quality in industry, government, and education.

This 400 page new work describes this image of systems engineering. It is comprised of 3 major parts and 18 chapters within these parts. There are:

I   SYSTEMS – ADVANCES IN SYSTEMS SCIENCE AND THINKING (1 Systems Philosophy, 2 Advances in Systems Science, 3 Advances in Systems thinking, 4 Systems Engineering Philosophy, 5 System Models)

II   SYSTEMS METHODOLOGY (6 Overview of Systems Methodology, 7 Addressing complex issues and problems, 8 Exploring the Solution Space, 9 Focusing Solution System Purpose, 10 Architecting/Designing system Solutions, 11 Optimize Solution System Design, 12 Create and Prove Solution Systems, 13 Systems Methodology–Elaborated, Setting the Systems Methodology to Work)

III   SYSTEMS METHODOLOGY AND SYSTEMS ENGINEERING (15, Systems Engineering –
      The Real Deal, 16 Systems Creation: Hand of Purpose, Root of Emergence, 17 System of
      Systems Engineering Principles and Practices, 18 Systems Engineering: Intelligent Systems)

In addition, at appropriate places in the text, 8 pragmatic case studies, based strongly on the systems
engineering experiences of the author, are used to good advantage. These are:

A   Japanese Lean Volume Supply Systems
B   Practice Intervention
C   Total Weapon System Concept
D   Architecting a Defense Capability
E   Police Command and Control System
F   Fighter Avionics System Design
G   Defense Procurement in the 21st Century
H   Global Warming, Climate Change and Energy

This book is about the ways to address and resolve problems, from small scale to those of very large
scale and scope. This work *Systems Engineering: A 21st Century Systems Methodology* addresses
a large variety of problems, and discusses how they might be solved both in theory and in practice.
The wide variety of case study presentations illustrate both how issues have been approached in
the past and how they might be addressed more effectively in the future.

The author, Derek Hitchins, is familiar with the evolution of systems engineering from its
initiation around a half century ago to the present time and has made a number of definitive
contributions to system engineering methodology. He demonstrates this knowledge well in this
work. In particular, the book presents a systems methodology that can in principle be employed
when confronting a very large set of issues and synthesizing potentially appropriate resolution for
them. That one single methodology can address systems of all kinds from small technological
systems to global socioeconomic systems involving humans, technologies, and organizations may
seem unlikely. Such a methodological process has been the goal of systems engineers - thinkers,
analysts, architects, integrators, and designers - for several decades. The author sets forth the
claim that this has only now become possible as a result of new work. Through use of this
approach, he suggests that it should be possible to prove and potentially disprove the acceptability,
suitability, viability and optimality of potential resolution to a variety of complex contemporary
issues.

There can be no question but what systems engineering pioneer Derek Hitchins has produced
a valuable work. It is steeped in its discussions of the early works of other pioneer systems
engineers. It is steeped in its knowledge of recent contributions to systems engineering
methodology. It is steeped in its synthesis of these into new methodological processes for
systems engineering that are original with Professor Hitchins. Thus, it is a distinct pleasure to
welcome this book by Professor Hitchins into the Wiley Series on Systems Engineering and
Management.

Andrew P. Sage
Editor, Wiley Series on Systems Engineering and Management
21 June 2007

Preface

Systems engineering has been recognized as a discipline for over half a century. It emerged from the study of whole systems and of *gestalt* that started in the first half of the 20th century, and was greatly accelerated by World War II, particularly by the advent of operational research, mathematical modeling and computer-based simulation. Some whole systems exhibited properties that were not exclusively evident in any of their parts, and it was found that these emergent properties, as they were called, could be synthesized by engaging the right system parts in the right way to create a unified whole that was potentially greater than the sum of its parts. Moreover, this seemed to be true for all kinds of systems.

This was, and is, more an organismic view than the mechanistic metaphor adopted by many engineers. Looking at whole systems in this way served, *inter alia*, to reduce perceived complexity. In particular, systems were seen as part of some greater 'whole,' open to, and interacting dynamically with, other systems within the environment of that greater whole — as an organ interacts with other organs within a body. Regarding the world in this way became known as 'the systems approach,' characterized by addressing whole problems and synthesizing whole solutions, principally to overcome perceived shortfalls in contemporary, piecemeal Cartesian reductionist practices in government, defense, and aerospace engineering.

In the second half of the 20th Century, systems engineering — a practical application of the systems approach — was instrumental in, and further developed during, NASA's iconic Apollo program, and was widely used in such major defense programs as Polaris, Vanguard, Aegis, Strategic Defense Initiative (SDI), and many more, together with wide application in the developing nuclear power industries on both sides of the Atlantic.

For Apollo, systems engineering had clear goals: the limited rocket payload had to deliver a complete system for going to the Moon and back. The whole system had to comprise a variety of astronauts and technological subsystems operating in close harmony; these had to be organized, arranged, interconnected, modified, etc., so that they fitted within the volume, shape and mass limits, yet operated together as a *unified whole* and exhibiting requisite emergent properties. Achieving this involved continual compromise, test, training, revaluation, and compromise again, to eventually produce an optimum (best) solution, one that would do the seemingly impossible job. The process, notion, and achievement exemplified all that was best in systems engineering. So successful was the enterprise that today some are incredulous, preferring instead to believe in conspiracies to deceive the public and the opposition during this Cold War era.

Europeans who had contributed particularly to the Apollo program returned to their respective countries, taking with them the concepts, processes, methods and practices that they had learned. So, systems engineering was widely adopted, notably by aerospace and defense organizations operating under the NATO umbrella.

The approach adopted by the military, however, changed: it became aimed, not so much towards achieving the unprecedented, innovative and emergent — such as Apollo — as it was aimed at meeting the requirements of military and government customers in terms of timescales, budgets and life cycles. Military engineering staffs pursued a linear mechanistic business-management approach, assuming and stating standards and requirements, decomposing those requirements into discrete functions and subfunctions for the solution system to perform. Then there was 'specialty engineering,' including Life Cycle Cost, Supportability, Reliability, Maintainability, Human Engineering, Safety, Electromagnetic Compatibility, Testability, Software, Producibility and Manufacturability, Value Analysis and Design to Cost. Life Cycle Cost, for example, was considered under such headings as:

- predicted costs for basic engineering;
- test and evaluation;
- experimental tooling;
- manufacturing and quality engineering;
- recurring production costs;
- nonrecurring production costs;
- logistics and maintenance support;
- operational costs and disposal costs.

Each of these cost headings was further broken down into many subheadings, in an attempt to cover all conceivable eventualities. The practices created such complexity and complication that the net result became referred to as 'paper engineering;' that is, the filling of forms and the ticking of boxes. Ironically, the fundamental concept of 'systems' as a way of managing complexity had been turned on its head by this complicated parody of the original. And the notion of 'system' in the defense engineering management context seemed to refer more to the thoroughness and comprehensiveness of the reductionist engineering management approach, than to any sense, in system terms, of the whole being greater than the sum of the parts.

US aerospace companies continued to design and engineer excellent aircraft, ships, tanks, etc. Defense engineering management, despite its undoubted high cost overhead, did not appear to detract from the factory design/manufacture of the products; on the other hand, it undoubtedly helped in the provision of through-life support facilities for operational systems.

Not all areas of defense were amenable to this reductionist approach: command and control (C^2) was one such, which, together with non-defense air traffic management, police, fire and ambulance services, government services and many more, constitute a class of systems dubbed IDA (information–decision–action) systems. IDA systems are sociotechnical systems: teams of people undertake tasks, using technological support facilities for acquiring, handling, storing and presenting information, and for supporting decision-making. IDA systems can be highly technological; on the other hand, they can be virtually technology free, depending solely on humans, their intellects and ability to communicate. An individual is, *de facto*, an IDA system in his or her own right.

Typically, teams collect intelligence, assess situations, identify threats and opportunities, develop strategies and plans, decide courses of action — often based on incomplete, even incorrect, information — and implement their plans. The various parts of the IDA system have to act as a unified whole, exhibiting emergent properties, including responsiveness, timeliness, integrity, decisiveness and strategic/tactical flair. Systems engineering continued to develop in the conception, design, development, implementation, work-up and evolution of such nonlinear sociotechnical IDA systems.

It was, however, systems engineering more of the 'whole exceeds the sum of the parts' nature than of the 'systemic engineering management' variety adopted in the US for defense engineering.

Meanwhile, the 1990s saw a realization in the West that defense 'systemic engineering management' version of systems engineering was unsuccessful, even counterproductive. Japanese approaches to procurement and manufacturing were providing a powerful counter-example of how to do these things much more efficiently and effectively; Japanese methods were 'joined up,' consensual and synthetic, rather than piecemeal, authoritarian and reductionist. Recognizing the inevitable, the contemporary US administration led the way by discarding their military standards and systemic engineering management practices, seeking instead to adopt Japanese methods, styles and even culture in their revised approaches to defense procurement.

Damage had been done, however: the reputation of systems engineering was tarnished. Redundant US DoD military standards and practices had imprinted a persistent legacy of people, trained and experienced in the DoD practices, who still believed that those methods and practices were sound. Even the core ideas of what systems engineering was about had been subverted. Instead of being associated with innovation, creativity, managing complexity, excellence and integrity, it had become associated with complication, 'gold-plating,' introspection and the engineering philosophy of 'giving the customer what he/she *wants*,' as opposed to the original: solving the customer's problem and providing the whole of what the customer *needs*.

The demand for world-class systems engineering persisted, however, and many realized that there had to be a way of creating better systems in all areas and walks of life. The 'whole systems approach' was the only rational answer when viewing complex issues and problems; piecemeal practices clearly did not work, more often than not exacerbating issues. Would-be systems engineering practitioners were uncertain as to how to proceed; few could recall the ideas, the vitality, the enthusiasm, the processes and the methods of earlier times. Whereas previously these had been handed down typically within major aerospace companies by successive generations of dedicated systems engineers, now employees might spend as little as three or four years in any one job. The legacy was frittered away.

There had been a vibrant pool of classic systems engineering know-how in the so-called 'systems houses.' These were companies, notably in the USA and Europe, which undertook the task of solving customers' problems objectively by conceiving, designing and providing whole-system bespoke, or 'turnkey,' solutions. In general, they were not manufacturing companies — to manufacture would inevitably prejudice objectivity — but instead they either contracted engineering companies to make parts to specification, or selected suitable existing parts that were available in the marketplace and interfaced/integrated them so as to synthesize the whole solution. Systems houses valued their integrity as well as objectivity, and they were generally creative and innovative, but it was not a high-profit business, principally because they did not manufacture and sell hardware. Similarly, there was limited profit in IDA systems, with minimal hardware, some software development and some training of customer's personnel.

The 1990s saw most systems houses driven out of the systems engineering business by the large aerospace companies, who offered to do the work of the systems houses, particularly concept, feasibility and project definition studies, often for nothing — an offer cash-strapped governments found too tempting to forego. Unfortunately, some aerospace companies were sometimes less than creative, and their solutions were invariably comprised of products from their own product range — not the objective, innovative solution that government was seeking. Further, they had little to offer in relation to IDA systems. This episode illustrated yet another example of the so-called Law of Unintended Consequences, which so often seems to associate with piecemeal initiatives by disjointed government.

Nothing if not resourceful, engineers in industry started to reinvent systems engineering. Instead of working at 'whole system' level as had the originators, engineers employed their reductionist-engineering practices on parts of whole systems. Such linear practices assumed, fundamentally, that the whole is equal to the sum of the parts: no more, no less. So: create the right parts to perform the right functions; integrate/interface/join them to fit into some conceptual architecture; and the result is a product as required by a customer. It all seemed seductively simple and straightforward.

Of course, this did not work when there were humans in the system, as they were inconveniently variable and unpredictable. So, this engineers' version of systems engineering — sometimes referred to, confusingly, as 'the engineering of systems' — did not address such human activity systems, but concentrated on the creation of mechanical, electrical, electronic and electro-optical artifacts. Emergent properties did not exist; or if they did, they were incidental, and probably undesirable. IDA systems were unsuitable subjects, owing to their people content. Similarly, businesses and enterprises were inappropriate subjects for the 'engineering of systems.'

Degree courses in engineering appeared with the sexy term 'systems' added, to turn, e.g., electrical/electronic engineering into electrical/electronic systems engineering, aeronautical systems engineering, mechanical systems engineering, communications systems engineering, and so on, without there being any significant different in course material compared with the straight engineering courses. Similarly, aerospace and engineering companies added 'systems' into their titles, to 'add luster to their cluster,' as the contemporary saying had it, but without any significant change in principles, procedures or practices. And the wider application of systems engineering to sociotechnical and socioeconomic systems languished, at least in the West.

The notion 'system of systems' sprang into being to meet this perceived shortfall. The 'system of systems' (SoS) concept is still not entirely mature, but it seems to refer to the bringing together in some way of a number of extant, independent enterprises or businesses, and referring to the association as a system. Creating a system by integrating a number of extant subsystems has been practiced for many decades. An avionics system in a modern aircraft, for instance, can be created by purchasing and integrating several extant systems: primary and secondary radar systems; automatic flight control systems; attitude sensing and control systems; flight instrumentation systems; communications systems; navigation systems; etc, etc. Is an avionics system a 'system of systems'? It seems not. The jury seems to be still sitting on just what a SoS might be, and if it even qualifies as being a system at all, as opposed to an association, a collection, a family. . . , etc., of systems. Not to be thwarted by such niceties, academics advertised a new subject to be learned, promulgated and practiced: System of Systems Engineering (SoSE.)

Meanwhile, the Japanese global lean industrial supply systems continued to sweep the world, notably in the production of automobiles/motor vehicles. This was (sociotechnical) systems engineering resurgent, but in a different guise, and of a different culture: it was — and is — taking the West's largely reductionist manufacturing industries to the proverbial cleaners.

Throughout this period of change, researchers have been continuing to seek 'systems enlightenment.' So-called soft systems have emerged as a way of addressing complex and fuzzy problems, i.e., those where the objectives may be uncertain. Soft methods are aimed particularly at Human Activity Systems (HAS), and have an underlying concept and theory of systems with which the originators of systems engineering would have been entirely comfortable.

And, it also has to be said that there was a small body of practitioners and researchers who kept faith with the original systems approach to synthesizing all kinds of systems. Their continuing researches have identified new methods and techniques, and have underpinned them with systems science. Today, there are new ideas and new ways of conducting systems engineering that are not only scientifically sound, but also for which there is great need both on local and global scales.

World *Problematique* is a concept created by the Club of Rome to describe the set of the crucial problems facing humanity: environmental, political, cultural, social, economic, technological and psychological. The heart of the World *Problematique* lies in the mutual interdependence of these problems, and in the long time delay between action/cause on the one hand and reaction/effect, often counterintuitive effect, on the other.

This book is about ways and means of addressing and solving problems, from the small-but-complex, perhaps even to those of the World *Problematique*. *Systems Engineering: A 21st Century Systems Methodology* addresses all kinds of problems, how they might be solved in theory, and how the solutions can be manifested in practice. It also presents a variety of case studies showing how different issues have been tackled in the recent past and how they might be addressed even more effectively in the future.

The book introduces a comprehensive systems methodology (SM) that can, in principle, be employed when tackling any issue or problem and creating a solution to solve it. That one, single methodology can address systems of all kinds from small, technological to global socioeconomic may seem unlikely. Such an SM has been the goal of systems thinkers, analysts, architects and designers for decades, however: it is only now becoming possible as a result of new tools, methods, science and ideas. The SM employs both established and new systems-scientific methods, with provability/falsifiability in mind throughout; it should be possible both to prove and disprove the acceptability, suitability, viability and optimality of potential solutions to a complex problem.

The SM is not a fixed-for-all-time entity. It is a morphing, evolving framework for the generation and management of information, independent of problem, solution, context, or environment, all of which are brought to the methodology by practitioners and proponents seeking to find answers to complex problems. The SM can be adapted, evolved and employed in the exploration and solution of problems of all kinds, at all levels, in any environment and on all scales, in the service of humankind and of our common environment. The Companion website for the book is http://www.wiley.com/go/systemsengineering.

Derek Hitchins
April, 2007

Part I
Systems: Advances in Systems Science and Thinking

Part I

1

Systems Philosophy

Observe how system into system runs
what other planets circle other suns

Alexander Pope, 1688–1744

Emerging Systems Movement

The industrial revolution started in Britain in the late eighteenth, early nineteenth century, spreading rapidly through Europe and the northern US. Two centuries later, the industrial revolution is still sweeping around the world, with populous nations such as China and India emerging as major economic players on the World Scene, in swift succession to Japan and other South East Asian nations. So significant has the industrial revolution been that it has been likened to the previous 'ages' in human social development: Stone Age, Bronze Age, Iron Age, and now, perhaps, the Machine Age? And like those successive Ages, the advances in tools, methods and artifacts have fuelled a social revolution that has not always enhanced the human condition and quality of life.

The industrial revolution emerged from the European Renaissance, and owes much to the iconic figure of René Descartes, of '*cogito ergo sum*' fame. His name is enshrined in the term Cartesian Reductionism — the concept he espoused, of breaking down big things into smaller, and hence more understandable things. If it was possible to understand and explain the smaller parts, then the whole could also be explained by bringing together the various part-explanations.

The ideas and practices of Cartesian Reductionism pervade our everyday lives, our buildings and engineered artifacts, even the way we are taught. Whenever we list, prioritize, disassemble, disaggregate, decompose, etc., we pay implicit homage to René Descartes and Cartesian Reduction. And, it has a great track record. Many of our greatest achievements have relied on reductionism.

At the start of the twentieth century, scientists began to notice that not everything was amenable to the reductionist approach. Some things, systems, seemed to function and operate only as wholes. They certainly might have discernable parts, but the parts did not explain the whole. Or, looked

Systems Engineering: A 21st Century Systems Methodology   Derek K. Hitchins
© 2007 John Wiley & Sons, Ltd

at in another light, the parts seemed to be mutually interdependent and to adapt to each other, such that looking at any part on its own was not only impractical, but also irrational. Such things, systems, were dubbed 'complex,' or 'complexes.'

Systems were defined: a system is an organized or complex whole: an assemblage or combination of things or parts forming a complex or unitary whole (Kast and Rosenzweig, 1970). The definition may not, in retrospect, offer much insight into the nature of systems, but it did serve to indicate, on the one hand, the essential integrity of systems, and on the other the basis for the integration of scientific knowledge across a broad spectrum.

A movement grew, notably among scientists of all persuasions outside of the 'hard' sciences, that came to be referred to as the systems movement, that regarded the whole system from a functionalist perspective (q.v.). Wholes were fairly evident: a whole person, a whole personality, a whole team, a whole organization, a whole army, a whole economy, or a whole nation. And the natural world seemed intent on creating wholes: whole atoms and molecules; whole animals and plants; whole civilizations; whole planets; and so on. True, many of these wholes subsequently decayed, but they were replaced by other wholes....

To understand any part of these wholes, it was seen as vital to view the part operating in concert with, and continually adapting to, the other interacting parts making up the whole; a part could not rationally considered out of context, excised, sans interactions. To understand the operation of the heart, it was vital to see it functioning in a live body, responding to stimuli from other bodily systems, adapting to demands from the whole body, growing, aging, and so on.

This systems approach, as it came to be called, proved highly successful, and it was widely adopted in organizational and management research, economics, psychology, anthropology, sociology, political science, geography, jurisprudence, linguistics and many more. So much so, that a new age was declared: the Systems Age. In this bright new age, dynamic Systems Age thinking was compared with static Machine Age thinking (Ackoff, 1981, see Table 1.1):

A classic example concerned a mythical Martian arriving on Earth and finding a mechanical clock. He might open the clock, take it apart, reassemble it and hence find out how it worked. However, he would gain no idea of the concept of time in this process.

Table 1.1   Machine Age vs System Age paradigms (Ackoff, 1981).

Machine Age procedure	Systems Age procedure
■ Decompose that which is to be explained (decomposition) ■ Explain the behavior or properties of the contained parts separately ■ Aggregate these explanations into an explanation of the whole	■ Identify a containing system of which the thing to be explained is part ■ Explain the behavior or properties of the containing whole ■ Then explain the behavior of the thing to be explained in terms of its roles and functions within its containing whole.
Machine Age analysis	Systems Age synthesis
■ Analysis focuses on structure; it reveals how things work ■ Analysis yields knowledge ■ Analysis enables description ■ Analysis looks into things	■ Synthesis focuses on function; it reveals why things operate as they do ■ Synthesis yields understanding ■ Synthesis enables explanation ■ Synthesis looks out of things

Or, consider a proposed new university department, dedicated perhaps to research into medieval French literature. Its containing system, the university as a whole, is dedicated to research, to developing new knowledge, and to the dissemination of that new knowledge both to students and for the enlightenment of all. It may be funded, or it may have to engage in research and in teaching for reward. So, the proposed new department has to be seen in the context of the whole. In what degree will it contribute to the research goals of the university, over and above those already targeted by existing departments? How will it contribute to the dissemination of knowledge? How will it be funded and/or earn reward? In what degree will it interact synergistically with existing departments, so as to enhance the capability of the university as a whole? By thinking about the proposed new department in terms of its contribution to the whole, it becomes possible to understand the essential role, purpose and value of the new department as it will help to achieve the goals and objectives of its containing system, the university.

Systems Age thinking spread even into systems design: parts were no longer to be designed like static pieces of a jigsaw puzzle; instead they were to be designed to fit, and even adapt to, each other so as to work together harmoniously as well as efficiently and effectively. Since some designs and combinations of parts could prove more effective than others, the idea of 'best,' or 'optimum' design solution emerged in parallel.

The Nature of Systems

Systems may be real, tangible wholes, or they may be concepts. They are comprised of parts, which may be arranged in some way. A fundamental idea about systems is that they possess some degree of order, i.e., that there is discernable pattern or configuration, which leads to notions of structure and architecture. Yet many systems are thought of as 'doing' things, i.e., functioning in some way — which is suggestive of action, activity, process, etc. One of the simplest, yet curiously compelling, definitions of a system is a 'dent in the fabric of entropy.'

'Systems' must, in truth, be one of the most overworked words in the English language:

- Washing powder manufacturers promote their 'washing system'.
- Gamblers frequenting the Las Vegas casinos may have their pet 'gambling system'.
- Some industries declare themselves to be 'systems companies'.
- In Linnaeus' seminal work, nature was divided into three kingdoms: mineral, vegetable and animal. Linnaeus used five ranks: class, order, genus, species, and variety. (Or, more generally: kingdom, phylum, class, order, family, genus, species and form.) This is a biological 'classification system'.
- People who try to do things in an unorthodox way are described as 'trying to buck the system,' where the system seems to be 'the way things are,' or — for the more paranoid — the way 'they' make things happen. Whoever 'they' are . . . .
- At least one aerospace company declares that systems are everything on an aircraft other than airframe and engines, so includes avionics systems, cabin-entertainment systems, coffee- and team-making systems, air conditioning systems, toilet flushing systems, luggage loading and unloading systems, etc., etc., all of which are needed for the transport aircraft to be considered a 'whole.'
- Some engineers working in, e.g., aerospace companies declare themselves to be systems engineers; in the light of the previous bullet, that title does not seem to be too specific in its meaning and implications.

Despite this overtaxing of the word, 'system' retains its sense of order and completeness. Systems have been around since before the solar system. Artificial systems have clearly existed for millennia before we humans thought about them.

The ancient Egyptians had a system for building the Great Pyramid in only twenty years using only some 4000 men. How do we know? Well, by deduction — without a high degree of purposeful effort from organized teams of dedicated people, so much work could not have been achieved with such precision in so short a time to a clear and evident design plan. Consider: the only technology they had was the rope. The lever had not been invented, nor had the wheel. And, on the motivation front, the builders were not paid in money, because that had not been invented in 2650BC, either. So, there had to be some kind of system for enthusing and motivating the builders to work in such hot, desert conditions year in, year out . . . there was organization, motivation, competition, reward, supply chains of food and beer (water was too tainted to drink), leadership, control, direction, supervision . . . .

Civilizations are systems, too, of course — large, diffuse ones perhaps, with uncertain boundaries, and apparently with a limited lifetime, since they all inevitably pass away — *sic transit gloria mundi*, and all that!

Causality and Teleology

Cracks in the bastions of Cartesian Reduction appeared early in the twentieth century. Teleology — purposeful, goal-seeking behavior — was a problem. A goal-seeking system responds differently to events until it produces a particular state or outcome: the system has a choice of behavior.

In a mechanistic world, the idea was to decompose parts to find more basic components, with which to explain how things worked. Decomposing a goal-seeking system failed to reveal any component as the root of the goal-seeking behavior. Yet, such purposeful behavior was all around in organisms. Causality proved similarly problematic. Causality was unidirectional. One gene corresponded to one deficiency in an organism, for instance: one bacterium caused one disease. Only, it didn't . . . at least, not always.

The notion of individual units acting on their own in unidirectional causality proved inadequate to explain observed phenomena. Evidently, groups of things worked together in some way, so that the behavior of the whole could exhibit purpose, could be multi-causal . . .

Consider, too, DNA — the so-called blueprint of life. DNA is a complex woven, folded molecule, comprised of nucleotides. Despite its complexity, it is nonetheless an inert molecule. There is nothing in the DNA that indicates whether the 'owner' is alive or dead; the DNA of a 100-million-year-old extinct dinosaur is indistinguishable in this respect from that of a live modern human. To some, the notion that the very essence of life is itself lifeless is curious, even bizarre . . . which came first, the organism or the DNA?

Then there are soldier ants, with enormous colonies made up from hundreds of thousands of ants. How can the whole colony act as one in its pursuit of food, its setting up of a bivouac within a net of living, interconnected ants, its overpowering of prey many times the size and power of any ant, its ability to cross rivers . . . yet each ant is blind and can carry out only a very few functions? There seems to be only one way to understand the army of soldier ants — as a whole, as some kind of extended, unified organism.

Animals in groups presented real problems, too. How did hives of honeybees know when to swarm? How do flocks of birds and shoals of fish wheel, whirl and dance as a single entity?

Humans in groups have been the source of commentary for many years before systems ideas emerged. *E pluribus unum*, (out of the many, one) was the motto on the first Great Seal of the US, which signified the coming together of the 13 states: it a system concept of wholeness. Similarly, there is the well-known Aesop fable: the Bundle of Sticks. A man had sons who were always quarreling. One day he brought a faggot of sticks and asked each of the sons if he could break the faggot. None could. He then untied the bundle and gave individual sticks to each brother, who broke each stick easily. The moral was evident: united we stand: divided we fall.

Benjamin Franklin is similarly reputed to have remarked to John Hancock, at the Signing of the Declaration of Independence, 4th July 1776: 'We must indeed all hang together, or, most assuredly we shall all hang separately.' The words are different; the sentiment remains the same. All these ideas are expressions of the power of cooperation, coordination and complementation that seem to underlie the concept of the system as a whole being in some way greater than the sum of its parts.

Emergence

Cartesian Reductionism could not explain why some wholes possess capabilities, have properties, and behave in ways that were not evident from examination of their parts in isolation. This observation was labeled 'emergence,' and some wholes were observed to possess or exhibit properties, capabilities and behaviors not exclusively attributable to any of their rationally separable parts.

It was evident, for example, that the human brain was made from many different neurons, each of which was of itself relatively simple, being able to adopt a very few discrete states. Yet, somehow, the combined effect of all these simple, interconnected, interactive neurons was to create self-awareness, which astonished — and still astonishes — any scientist who cared to think about it. How could that be?

There were many examples of this initially mysterious emergence, once people began to look. How could bringing together two odorless gases, nitrogen and hydrogen, result in ammonia, with its pungent odor? How could a film, made up as it was of a series of still frames, present apparent motion to a cinema audience?

Life and the Second Law

The Second Law of Thermodynamics is a central tenet of classical physics. It can be stated in a number of different ways:

- The processes most likely to occur in an isolated system are those in which the entropy either increases or remains constant; or,
- the entropy of an isolated system not at equilibrium will tend to increase over time, approaching a maximum value.

Whichever way it might be stated, the conclusion may be drawn that entropy (disorder) will pervade all systems eventually. Using the Second Law as a guide, astronomers have concluded that the Sun will cease to shine in another five billion years or so, and the Universe will eventually become cold and inert. Current research suggests they may not be right, but the principles expressed by the Second Law seem sound enough for all that.

Only . . . life and living things seem to confound the Second Law. Living things create order in their being, in their growth, in their ability to organize, categorize, form groups, etc., etc. Similarly, civilizations seem to confound the Second Law, and on a smaller scale so do companies, organizations and societies. Indeed life is itself an apparent anomaly, inconsistent with the Second Law.

Of course, the Second Law is not really being confounded — it refers specifically to an isolated system. Life does not occur as isolated systems. For a system to be isolated, it has to be alone and receive no inputs. Living things — plants, animals, organizations and civilizations — receive inflows and emit outflows. So, an animal ingests food, negative entropy, and converts the food into parts that it can use to replace and rebuild its own internal parts. Hence, by ingesting negative entropy, the entropy of the whole can reduce, rather than increase.

Entropy and Work in Human Organizations

A more pragmatic view of entropy and the Second Law indicates that the ability of a system to do external work reduces with increasing entropy. This indication is compatible with the observation that disorganized companies do not produce much output, even though the people inside the organization might be working feverishly; streamlining the organization can improve its performance.

Government departments are often organized using committee structures. There are high-level committees, and then a number of intermediate-level committees with perhaps the committee chairmen also sitting as delegates on the higher-level committee. Lower-level committees are set up similarly, providing delegates to intermediate level committees. Additionally, there are cross-connections, with delegates and representatives operating laterally and diagonally across the committee structure. Such committee structures are found in such contentious areas as defense ministries and departments, where difficult choices have to made about costly defense projects such as new fleets of ships, aircraft, tanks, missiles,etc.

It has been observed, somewhat cynically no doubt, that the committee structure has two characteristics: (a) it ensures that no one person can be made responsible for any mistakes; (b) the net outcome from so much work done by such committee structures is to decide whether, or not, to implement some proposed major project. Logically — perhaps even rationally — the same decision that took perhaps years of reporting, vigorous debate and heated counter-debate, could be taken instantaneously on the turn of a coin. For, after all, the final decision is simply to proceed, or not to proceed with the project. The high-entropy system, which may expend copious amounts of internal energy operating and maintaining itself, is capable of very little external work. And, it really would be cynical, would it not, to suggest that, on the evidence, spinning a coin might have a better chance of success in choosing the right project . . . .

Of course, the many participants in such complex committee structures will argue that it is necessary to give everyone their say, that each advocate or opponent has a valid contribution to make, that it is important to give everyone a voice . . . and so on. There is copious evidence to show, however, that the ability of any group of people to make a cogent decision is inverse to the number of people involved in the process; beyond a handful, perhaps, five or six only, it becomes increasingly difficult and eventually impossible to make any sensible, consensus-based decision in any reasonable time.

Entropy Cycling

While the Second Law suggests that entropy inexorably increases over time, this is not what we observe with natural and many artificial systems. Instead we see systems being created, growing, becoming more complex perhaps, and then collapsing only to be supplanted by newer or resurgent systems.

The weather is a global system, always in a dynamic state, as the atmosphere is churned by the rotation of the earth and differential heating from the Sun at poles and equator, and by day and night. We see periods of calm interspersed with periods of turbulence, dry with wet, hot with cold, in a never-ending panorama. The entropy of the weather system seems to both reduce, as clear patterns of anticyclones, cyclones and ridges form, and then to increase again as the various patterns disperse — only to reform again, although never in quite the same way.

Life appears to be an example of entropic cycling, too. In the natural world, plants grow, are eaten, so animals survive and grow, then age and die. Animal and plant remains decay into nutrients for more plants, and the cycle continues — seemingly forever.

In the real world, it seems that we are not dealing with closed systems, with increasing entropy, but with open systems and entropic cycling, which we may perceive as birth–death–birth cycles.

General Systems Theory and Open Systems

In the middle of the 20th century, an attempt was made to develop a General Systems Theory. A Society for General Systems Research was founded in 1954, with the following principal contributors:

- Ludwig von Bertalanffy, a biologist
- Kenneth Boulding, an economist
- Ralph Gerard, a physiologist
- A. Rapoport, a biomathematician

Their General Systems Theory had four aims (Checkland, 1981):

1. to investigate the isomorphs of concepts, laws and models in various fields, and to help in useful transfers from one field to another;
2. to encourage the development of adequate theoretical models in areas that lack them;
3. to eliminate the duplication of efforts in different fields;
4. to promote the unity of science through improving communications between specialists.

The general theory envisaged by the four contributors has not really emerged, although their work has been highly influential. Some specialists, in particular, were scathing — but then, that was to be expected. Much of the work was mathematical in nature and has not stood the test of time so well as some of the many models that were created. Today, systems people use many of these models unaware of their origins . . . .

Boulding's Classification of Systems

Kenneth Boulding attempted one of the more difficult tasks — a coherent classification of systems. In Table 1.2, the first three levels are appropriate to physics, astronomy and the related 'hard' sciences; at these levels, a system is essentially regarded as closed, having no significant contact with its

Table 1.2   Boulding's classification of systems (von Bertalanffy, 1968).

Level	Characteristics	Example
Structures	Static	Bridges
Clockworks	Predetermined motion	Solar system
Controls	Closed-loop control	Thermostat
Open	Self-maintaining	Biological cells
Lower organisms	Growth, reproduction	Plants
Animals	Brain, learning	Birds
Man	Knowledge, symbolism	Humans
Social	Communication, value	Families
Transcendental	Unknowable	God

environment, and not adapting to that environment. The next three levels, four to six, are the concern of biologists, botanists and zoologists, and address open systems, those that are in contact with, and adaptive to, their environment. Levels seven to nine, also open systems, concern themselves with human and social systems, the arts, humanities and religion. Then, of course, there are the sciences that cross the layers: sociology, sociobiology, ecology, anthropology, psychology, organization, linguistics, and many more.

Categorizing systems is notoriously difficult. Even providing examples may be fraught. For instance, there are many systems engineers who deny that the solar system is a system at all, because a system has to be 'manmade and purposeful' (*sic.*) Boulding's Classification has stood the test of time, and it certainly provides a basis for discussion and much head scratching.

Parallels and Isomorphisms

Bertalanffy suggested that the many and various fields of modern science exhibit a continual evolution towards a parallelism of ideas, which allows for the identification of isomorphisms within and between the various separate scientific disciplines. Such isomorphisms offer the basis to formulate and develop principles that are applicable to systems in general. 'So has arisen systems theory — the attempt to develop the scientific principles to aid us in our struggle with dynamic systems with highly interactive parts.' (Ashby, 1964)

Isomorphisms had been observed increasingly between the various hard scientific disciplines. An elementary example concerns simple harmonic motion (SHM) and resonance. Physicists described SHM mathematically, observing that the acceleration of a particle obeying SHM was directly proportional to its distance from the point of zero displacement, and directed towards that point. They talked of resonant columns of air vibrating longitudinally, plates vibrating when a violin bow was pulled across the edge, and so on.

Electronics scientists and engineers regarded resonance somewhat differently, as caused by the exchange of energy between a capacitor and an inductor. Such resonance could be induced, usefully, by broadcasting a collection of radio waves of different wavelengths, so that the capacitor and inductor combination resonated only at the wavelength of a particular component of the broadcast spectrum, so enabling a radio to be tuned to a particular broadcast.

Mechanical engineers thought about vibration in much the same way, except that it was generally undesirable, and structures could be developed which did not vibrate at certain frequencies.

A notable failure in this respect was the catastrophe of the Tacoma Narrows suspension bridge, which was driven into resonance by a strong wind; so violent did the resonance become that the bridge collapsed into the river below.

Despite the many and various ways of looking at SHM and resonance, the underlying mechanisms were mathematically the same; and there were many other 'behavioral isomorphs,' as these phenomena came to be known. Behavior parallels (and convergent evolution) were observed in animals, even of quite different species. It became apparent, too, that some complex systems might be described by their behavior without necessarily having to delve into the depths of the systems design, structure, processes, etc. Here, then, was a potential way of managing complexity . . . .

The Concept of the Open System

Bertalanffy's concept of an open system was a masterstroke; clearing the way for a singular theory of systems that could apply to soft, open systems, as well as closed, hard systems. An open system is in contact with its environment, receives inflows and emits outflows. It is able to adapt to its environment, yet may retain a steady state. Essentially, therefore, it is internally dynamic, which suggests that stability is conceptually different from that proposed for closed systems in physics.

The following equations (Bertalanffy, 1950) show some general systems mathematics which may seem rather obvious, once stated, but which show the developing nature of systems theory at the time. Equation (1.1) states, in essence, if the sum of all the elements arriving at a system equals the sum of all the elements leaving the system, then the system will neither increase nor decrease.

$$\frac{\partial Q_i}{\partial t} = T_i + P_i \tag{1.1}$$

where: $Q_i =$ is a measure of the ith element of a system; $T_i =$ the velocity of transport of Q_i at that point in space; $P_i =$ the rate of production or destruction of Q_i at a certain point in space.

A system defined by Equation (1.1) may have three types of solution: first there may be an unlimited growth in the system, Q; second, a time-independent state may be reached; and third, there may be periodic solutions.

In the case where a time-independent solution is reached:

$$T_i + P_i = 0 \tag{1.2}$$

In these two simple equation can be seen both the conservation laws of physics and the open systems (dynamic) stability of organisms. To an electronics engineer, for example, these equations may be redolent of Kirchoff's Laws used in circuit analysis, concerning the conservation of energy and the conservation of charge. These laws simply state that the algebraic sum of the currents flowing into a node is zero — total charge in, equals total charge out — and that the algebraic sum of the voltage 'drops' around a closed circuit is also zero.

However, it was perhaps more the general ideas and models emanating from GST that most influenced the fledgling disciplines of systems science and systems engineering.

Understanding Open System Behavior

The following nine characteristics seemed to define all open systems (after Katz and Kahn, 1966):

1. Importation of energy. Open systems import some form of energy from the external environment. This is evident for open systems since they invoke dynamic interactions between the parts, which therefore do work and expend energy. Less obviously, perhaps, this is also true for the personality, which is dependent on external stimulus.
2. The throughput. Open systems transform the energy that is available to them.
3. The output. Open systems export some product into the environment, whether it is the invention of an inquiring mind, or a bridge constructed by an engineering firm. At a more basic level, open systems export waste which has arisen as the result of metabolism, i.e., as the result of transforming energy.
4. Systems as cycles of events. Energy exchanges with the external environment and within the open system itself have a cyclic nature. Breathing, heartbeat, peristalsis, the hunger–eating–digestion–metabolism–excretion cycle, etc., are well known examples from the body. Consider, too, a manufacturing organization. It takes in parts from suppliers, as needed, machines, assembles, etc., in a series of processes, sells the product for money, uses the money to buy in new parts, and to sustain, enhance and perhaps expand the internal machinery, processes. Like the body, the whole is characterized by a continuing series of interconnected (peristaltic?) processes.
5. Negative entropy. To survive, open systems must 'ingest' negative entropy. This may come in the form of food, new staff, new organization, even new concepts and beliefs . . . .
6. Information input, negative feedback, and the coding/categorization process. All open systems receive information, which they categorize both to reduce perceived complexity, and to identify that which is relevant. Open systems generally receive negative feedback, of a form that allows them to correct deviations from some course. (Research subsequent to this work suggests that control in many systems does not depend primarily on negative feedback, but instead on the opposition of forces, as in agonistic and antagonistic muscles and processes. Negative feedback, where it occurs, may act as incremental control.)
7. The steady-state and dynamic homeostasis. Importing energy to arrest increases in entropy can result in a steady-state condition, or quasi-stability, which may be dynamic in nature. Body temperature is a good example. Le Chatelier's Principle (q.v.) can be seen in operation: changing any element within the open system causes other elements to rearrange themselves so as to oppose the change, and to restore the body as near to its previous state as possible.
8. Differentiation. Open systems move in the direction of differentiation and elaboration.
9. Equifinality. Open systems can reach the same final state from differing initial conditions by a variety of paths. (Von Bertalanffy suggested the principle of equifinality; it has not emerged as a universal truth, however.)

Since whole humans were whole systems too, it was possible to regard humans individually and in social groups and societies as exhibiting behavior. Comparing human behavior with the nine characteristics above, humans evidently exhibited additional characteristics, which were the source of intense study. Freud and Jung were foremost in this field of endeavor. It also became evident with research that groups of humans did not behave as individuals. Jung observed, for instance:

It is a notorious fact that the morality of society as a whole is in inverse ratio to its size; for, the greater the aggregation of individuals, the more the individual factors are blotted out, and with them morality, which rests entirely on the moral sense of the individual and the freedom necessary for this. Hence every man is, in a certain sense, unconsciously a worse man when he is in society than when acting alone, for he is carried by society and to that extent relieved of his individual responsibility. Any large company composed of wholly admirable persons has the morality and intelligence of an unwieldy, stupid and violent animal. The bigger the organization, the more unavoidable is its immorality and blind stupidity . . . the greatest infamy on the part of . . . a man's . . . group will not disturb him so long as his fellows steadfastly believe in the exalted morality of their social organization. (Jung, 1917)

As this extract shows, Jung was acutely aware of the ways in which groups of people could behave, and that such group behavior could be, was likely to be, significantly different from the behavior of individuals. We recognize in his words the ideas of 'group think,' mob behavior, 'risky shift,' and of the typical excuse of the hard-nosed businessman: 'business is business; no room for sentiment in business.' Jung's insights make it evident that there was rather more to understanding social systems than had been thought previously, and in particular that, while a social system might be in equilibrium or homeostasis, it could at the same time evolve and adapt when stimulated.

Jung introduced the notion of types: introvert and extrovert, which have become everyday terms. He identified the introverted thinking type and the introverted feeling type, together considered introverted rational types. He also identified the introverted sensation type and the introverted intuition type, together the introverted irrational types. Extrovert types were similarly characterized, creating a total of some sixteen types in all. His seminal work was taken up by others and can be seen today in psychological tests such as the Myers–Briggs Type Indicator (Briggs, 1990), much used in industry as part of recruit selection testing.

Both Freud and Jung might be described as early systems thinkers: in their different ways they were trying to understand the whole human intellect, and both saw the need to view the parts only in the context of the whole. Each developed models of the intellect, Freud with his ego, super ego and id, Jung with his conscious, subconscious, collective unconscious, archetypes, etc. While the work of both emphasized the individual, Jung in particular seemed to recognize that the social group might be considered to have a collective intellect, too.

Developing systems theory encompassed and incorporated the ideas of psychology, group psychology, social anthropology, etc., and deployed them into organizational and management theory, to address the behavior of whole social systems, of whole socioeconomic systems, and whole sociotechnical systems. This last was of particular importance, since organizations, businesses and industries were generally sociotechnical systems, with their people-content forming teams, groups and divisions, and using machines which may also form social entities, such as distributed computer systems, sequential processing machines, etc. It had become evident, notably with the advent of mass production and its adverse effects on many of the factory workers, that technological devices interacted with their human operators/users, each affecting the behavior of the other, such that the conception and design of either the human team or the technology on its own was unlikely to afford success; instead, the whole had to be conceived and designed as one.

Gestalt and Holism

Aristotle said:

> The whole is more than the sum of its parts.
> The part is more than the fraction of the whole.

> Composition Laws (Hall, 1989)

The world, it seemed, would not be ready for such a profound systems concept for a further 2000 years; not, that is until Gestalt – a German word with no clear English translation, but meaning something like 'form,' or 'shape.' The Gestalt movement started early in the twentieth century: Gestalt psychology was launched in 1912 by Max Wertheimer, who published a paper on the visual illusion of movement created by presenting a series of still photographs of a galloping horse. The central tenet of Gestalt thinking was that the whole was greater than the sum of the parts.

A more current view might be that a Gestalt entity is a physical, biological, psychological, or symbolic configuration or pattern of elements, so unified as a whole that its properties cannot be derived from a simple summation of its parts. From this perspective, the whole is different from, not necessarily greater than, the sum of the parts. . . .

The Gestalt notion is contained within contemporary ideas of holism:

- Holism: the theory that the fundamental principle of the universe is the creation of wholes, i.e., complete and self-contained systems from the atom and the cell by evolution to the most complex forms of life and mind;
- Holism: the theory that a complex entity, system, etc., is more than merely the sum of its parts (*Chambers Dictionary*).

Gestalt has left a legacy, often overlooked, but nonetheless deeply embedded in today's systems thinking. Contemporary systems engineering, for instance, seems to owe more in practice to Gestalt than to operations research, since ideas of holism and emergence are firmly embedded, whereas mathematical optimization might be proposed by academics, but seems to be of little interest to engineers. Without optimization, however, requisite emergent properties may not be fully exhibited . . . .

Indivisibility

Whole system behavior was seen as indivisible; it was possible to observe and understand the behavior of any part only in the active, dynamic context of the whole and, therefore, the other parts too. It was not possible to observe and understand the part in isolation.

This was, of course, the antithesis of Cartesian Reductionism. Checkland put it as follows:

> Some view a system as being like a bag of pool balls: you can put your hand in the bag, pull out a ball, observe it, understand it, replace it in the bag, and nothing has changed. That is a reductionist view. On the other hand, a system may be seen as more like a privet hedge: if you try to remove a complete branch, you will have to rip it out from the complex growth, damaging both the branch and the hedge in the process; and, you will never be able to replace it. That is a complex, open systems view.

Interaction dynamics

It became clear with time and research that emergence was not really mysterious. As the example above of the mixing of hydrogen and nitrogen gases showed, emergence was not confined to complex systems either. Emergence comes about from the dynamic interactions between the various parts of a whole, or system.

For the two gases, the interaction concerns itself with outer electrons orbiting the ammonia molecule, and as chemists had known for some time, it was the pattern of electrons orbiting in the outer shells that gave a substance many of its chemical and physical properties.

The appearance of motion from a series of cine-stills is caused by interactions between the eyes and the brain, specifically between the retinal image and the visual cortex. The retina forms an image in such a way that it takes time for the image to form, and time to decay and disappear, while being overwritten by a succeeding visual image. The visual cortex has evolved the ability to integrate, or smooth out, this otherwise 'jerky' sequence that we would 'see.' It is also true that the optic nerve, which employs a sophisticated digital code, lacks the capacity to send to the brain sufficient information for us to 'see' the full visual images that we appear to see.

The problem is alleviated in part by having only the central point of vision transmitted in detail; so-called foveal vision. Even this is insufficient, apparently, and the brain also possesses the remarkable facility for 'filling in' parts of the scene from memory. The subjective result is that we observe a seamless, jerk-free moving picture. However, the processing takes time: it has been estimated, for instance, that a top-class tennis player has already received and played the ball by the time his/her brain forms an image of the ball traversing the net. So, while he/she may think they are watching the ball, in truth they have anticipated its trajectory and base their reactions on this anticipation and on years of training and practice. Now, that is interaction dynamics and emergent behavior!

Emergent behavior is not always beneficial. It appears that part of the human brain is dedicated to identifying pattern and sequence. We become so good at observing and detecting pattern and sequence that we can sometimes 'see' what is not there. Police witnesses have been found to subconsciously create sequences of what happened, and in what order, during the commission of some crime or the build-up to some incident. Different, unconnected witnesses may tell the same story, but it may not be so. Our brains have the ability to 'fill in the blanks' in a supposed sequence of events or activities, to the point where we will be absolutely convinced that our version of events is correct. (This phenomenon has been observed since the advent of time-stamped CCTV, which has allowed investigators to build an unassailable timeline of events, against which to test observers' reports.)

Stability and Steady-state

Stability in closed systems is a well-understood phenomenon. Generally, stability is indicated by a minimization of potential energy. Stability in open systems is more complex. Many open systems rest at relatively high levels of potential energy — a horse sleeps standing up, for instance, a candle flame stabilizes with the flame at a particular length, and a factory stabilizes when it has significant throughput. The reason for the difference, of course, is because open systems maintain themselves through internal processes and interactions, and these do work and expend energy.

The weight of a person is a balance between energy intake, internal energy transformation, energy expenditure, work and disposal of waste substances. If the energy intake, in the form of food, exceeds the rate of energy expenditure and waste disposal, then the body will convert some of the excess energy into fat, or stored energy, and weight will increase. This is a simple, obvious example of Equation (1.1) in action.

So, stability (homeostasis) in open systems is more a matter of dynamic balance than of minimum energy. An alternative, and perhaps more fruitful, way to look at open system stability is as the

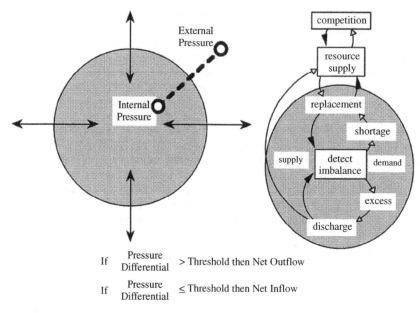

If $\dfrac{\text{Pressure}}{\text{Differential}}$ > Threshold then Net Outflow

If $\dfrac{\text{Pressure}}{\text{Differential}}$ ≤ Threshold then Net Inflow

Flow rates = f{Threshold ~ Pressure differential}

Figure 1.1　Homeostatic concepts. At left, a cell maintains its volume by osmosis, with outflows and inflows through the cell membrane being 'driven' by the pressure difference between the cell and its environment. Analogously, at right, a commercial enterprise senses the difference between supply and demand (the analog of pressure differential) and uses that difference to vary inflow and outflow rates of resources: manpower, machinery, materials, money and information.

maintenance of a steady state: see Figure 1.1. Such maintenance will, in general, be dynamic, varying about some mean. If open systems are regarded as cycles of events, then this notion of steady state starts to make more sense. Cyclic events for the human body would include sensing hunger, eating and digestion including peristalsis – the waves of involuntary muscle contractions which break food up into boluses, and transport food, waste and matter through, and out from, the alimentary canal.

A person at rest, under anesthetic perhaps, appears to be motionless and inactive. If it were possible to open up the flesh, however, we would see inside a ferment of activity as the various subsystems and their organs interact dynamically and rhythmically. These cycles of events are analogous to the flow of materials through a manufacturing organization, payments through an accounting system, students through a university, and designs through a systems design team . . . and the feet of a swan swimming, apparently serenely, upstream against a strong current.

The Systems Approach

The systems approach, in which it became the practice to understand the part only in the context of the whole, interacting with, and adapting to, its environment, entered into almost every sphere of scientific endeavor, including the social sciences, and the life sciences — it had its roots, of course, in biology where there was no alternative. Management and organization sciences, in particular, adopted the systems approach.

It came to the attention of the engineering fraternities, too, who had been experiencing difficulties in applying their engineering practices to systems that included people — sociotechnical systems. The systems approach offered ways of understanding and addressing human activity systems and sociotechnical systems that had not previously existed.

The systems approach was seen as a way of addressing complex problems and issues. Ackoff (1981) suggested that there were three ways in which problems could be addressed:

1. Problems could be *resolved*. To resolve a problem is to find an answer that is 'good enough,' one which satisfices.
2. Problems could be *dissolved*. To dissolve a problem is to change the situation in some way such that the problem disappears, to 'move the goalposts.'
3. Problems could be *solved*. To solve a problem is find the correct answer, as in solving an equation.

In general, most people resolved problems, often by dealing more with the symptoms than by getting to the roots of the problem: sometimes they had to make decisions in absence of full knowledge. Satisficing was not seen as bad, more pragmatic. Sometimes satisficing resulted in more knowledge about the real problem, enabling further satisficing and more knowledge, so homing in on a complete solution to a problem.

Some people, however, were good at dissolving problems, making them go away. Politicians are often thought of as working in this way, and it can prove smarter, less confrontational and less expensive than other methods. However, it also can come at a delayed or hidden cost, as when the UK attempted to appease Hitler before World War II. Appeasement, in that context, was an attempt to move the goalposts, and it did not work. In the TV film, *Star Trek II: the Wrath of Khan*, the iconic Captain Kirk told of an examination he and his colleagues had faced at Starfleet Academy; the examination, concerned with the space vessel Kobyashi Maru, was known to have no solution. Kirk managed to surreptitiously reprogram the exercise computer simulation so that he could pass; that is, he moved the goalposts. Hence, in some circles, Kobyashi Maru is synonymous with dissolving the problem.

Some systems engineers chose the third route — they sought the best solution, the optimum, to a complex problem by so balancing the interacting components and coupled processes of a complex solution system that it gave the best results in its environment. This was a management task and, potentially at least, a mathematical task . . . as well as requiring understanding of just how emergent properties, capabilities and behaviors could be synthesized and realized.

Systems Thinking

Systems thinking is thinking, scientifically, about phenomena, events, situations, etc., from a systems perspective, i.e., using systems methods, systems theory and systems tools. Systems thinking, then, looks at wholes, and at parts of wholes in the context of their respective whole. It looks at wholes as open systems, interacting with other systems in their environment. Instead of thinking in the abstract sense, systems thinking has developed into dynamic modeling of open systems, often using smart simulation programs (Roberts *et al.*, 1983, Richmond, 1992).

Because systems ideas are applicable to all kinds of systems, and are hence not limited by particular physical/structural/procedural constraints, systems thinking has evolved as modeling, particularly, the *behavior* of systems. This offers the opportunity to take maximum advantage of behavioral isomorphs. It also affords the ability to manage complexity, so that highly complex phenomena, situations, organizations, etc., may be modeled with some degree of confidence.

Like any form of computer simulation, behavior modeling is not infallible. This issue is alleviated, in some degree, by the way in which behavior modeling and systems thinking are used. In general, these methods are not used to provide specific numerical answers to complex mathematical problems. Instead, they are used to model the interactions between various systems-of-interest to explore likely outcomes from such interactions in some future environment. The models assume, too, that each system and its interactions affect other systems in the model and surrounding the model, so that the whole behaves dynamically and generally nonlinearly — like the real world.

In this way, models are used as experimental laboratories, to explore what might happen in some future situation, to explore the 'what ifs'... to see if there are likely to be any counterintuitive effects (Forrester, 1971) from unexpected interactions.

Systems thinking has been fuelled and enabled by the development of systems dynamics tools, of which there are many now available on the market. One such is STELLA™, which stands for: Systems Thinking and Experimental Learning Laboratory Approach: the title typifies the approach.

Functionalism and the Organismic Analogy

Functionalism is the principal theoretical perspective in sociology and many other social sciences. It is balanced on twin pillars: the potential application of the scientific method to the social world; and, development of an analogy between the individual organism and society. The concept of considering society as a system, and therefore subject to scientific investigation, might be viewed as a functionalist concept.

The analogy between the organism and society points functionalists to the bases for a social system to exist. A functionalist view would consider that every society must undertake, perform, or be capable of, certain basic functions for that society to survive and flourish, just as the organs of the body perform functions, which are necessary for the body's survival.

For a society to exist, for example, it must be able to overcome shocks and traumas and yet return to a state of normality, homeostasis or equilibrium; for the various elements of a society to function harmoniously, there will have to rules of behavior, codes of conduct, etc., which encourage cooperative interaction, and which discourage conflict. There will need to be organs of society that deal with members who do not observe the rules, such as the ability to impose sanctions on the one hand, and the provision of schools, colleges, etc., on the other hand.

The organismic analogy

The organismic analogy (Rapoport and Horvath, 1959) became central to the systems approach. The analogy does not state that all open systems, such as civilizations, societies, industries, enterprises, etc., are organisms. It does state, however, that in many ways they *behave* as organisms, in that they have a life cycle: a conception, a beginning, a development, maturity, and eventually a collapse. Bertalanffy noted that the collapse was often spectacular, or implosive, as for instance in the collapse of the Soviet Union, or the sudden collapse of seemingly invulnerable industries.

The organismic analogy suggested that, in viewing a system as made up from interacting, adapting subsystems, it was necessary to see these subsystems as complementary in some way, and as performing quite different functions on behalf of the whole system. As in society, there was a need to be able to impose sanctions on individuals who misbehaved, so in the human body there was a need for an immune system to suppress pathogens, and so in modern radar there was a need to detect and isolate defects and failures, and to undertake repairs.

The last example shows that systems thinking could be applied equally to technological systems, and that in so doing the technological system could be perceived differently: not so much as a collection of boxes; more as a synthesis of subsystems each of which performed different functions that interacted with those of other subsystems and which contributed to the performance, effectiveness, efficiency, stability and longevity of the whole. Viewing technological systems in this fashion was radically different from the engineer's view of technological systems and, in particular, allowed and encouraged the human users, operators, maintainers, etc., to be part of the whole system — which became a sociotechnical system in consequence.

The machine metaphor

It had been the practice to consider organisms as essentially biological machines. For the human body, the heart was a pump, the brain was a processor, the eyes were a camera, the kidneys were a filtration unit, and so on.

Bertalanffy (1968) had identified a number of difficulties with this metaphor, which he believed rendered it invalid. First, there was the problem for the origin of the machine. Whereas Descartes depended upon his Blind Watchmaker to build his 'biological machine,' machines did not create themselves in nature.

Second was the problem of regulation. While it was true that machines could regulate themselves, could a brain or an embryo be programmed for regulation after disturbances of an indefinite, possibly immense, number? It seemed not.

Third was the continuous exchange of components. Life, von Bertalanffy proposed, is a machine, composed of fuel, spending itself continually, yet maintaining itself. This creates a paradox: a machine-like structure of the organism cannot be the ultimate reason for the order of life processes because the machine itself is maintained in an ordered flow of processes. The order, then, must lie in the overall process itself.

Although engineers and some scientists still cling to the machine metaphor, the organismic analogy, or organic metaphor, has been widely recognized and employed, and can be seen at work in many aspects of systems engineering.

Mechanistic Control Concepts

Cybernetics offers a control view of the world that is neither particularly mechanistic nor organismic. The cybernetic model involves some input 'signal' which is amplified to drive a mechanism. Information from the mechanism output is fed back to, and differenced with, the input signal. In this way, the actual output is driven to meet the desired output as determined by the input signal.

The cybernetic model can, in principle, be used not only for machinery, but also for control of many different kinds. Management methods have included cybernetic concepts for the control of workers, although it generally proved unacceptable to the workforce, and hence to the management, since the workforce would be alienated by, and invariably rebel against, close control.

Organismic Control Concepts

Control in organisms generally works in two ways. The basic control strategy seems to be to form two opposing influences, such that that which is to be controlled finds itself balanced between the two. Limbs in the body are moved under the control of opposing sets of muscles: agonistic and antagonistic. Growing embryonic buds are directed in their growth pattern by chemical or nutrient gradients. Levels of glucose in the blood are regulated by the metabolism of glycogen on the one hand and by insulin on the other; and so on. Le Chatelier's Principle applies:

> When a constraint is applied to a physical system in equilibrium then, so far as it can, the system will adjust itself so as to oppose the constraint.

In so doing, the physical system seeks to restore equilibrium. Le Chatelier's Principle was first expressed in relation to chemistry and the behavior of interacting solutions, but it clearly has wider implications, addressing physical systems in general, as well as organismic systems. The principle evidently applies to social systems, for example. Note however, that Le Chatelier's Principle says nothing about the manner, direction, linearity or speed, etc., of this 'adjustment.'

Feedback regulation can also be found in organisms, where it may act incrementally, as fine control superimposed on the first method of control, i.e., through balancing influences. A well-known example of regulation is that of homeothermy, the maintenance of constant body temperature. For humans, the understanding is that as environmental temperature rises, there comes a point at which the body starts to perspire; the perspiration evaporates into the surrounding atmosphere, taking with it the latent heat of vaporization, and so cooling the skin and, possibly, reducing the rate of perspiration.

This well-known explanation falls short of explaining how the body maintains temperature when the environmental temperature falls. To appreciate this, it is important to understand that, as a warm-blooded, or endothermic, animal, humans metabolize food to provide energy, some of which inevitably emerges in the form of heat. This would raise the core body temperature, except for the conduction and convection of heat from the core of the body towards the surface, where it is lost to the environment. There is, therefore, a continual flow of heat energy from the core to the outside world in the steady state.

Should the external temperature fall, the consequent fall in skin temperature encourages the surface capillaries to close up, diverting blood flow away from the skin surface. The outer layers of skin, devoid of blood, insulate the core of the body, reducing the flow of heat energy so that the core body temperature remains effectively constant. While body temperature is higher than that of the surrounding environment, core body temperature is regulated in a manner analogous to that of a bath, in which the taps are running to raise the level of water, with the outlet left unplugged. As the water level rises, so does the head of water over the outlet, so the rate of outflow increases. If the bath is deep enough, a point of equilibrium will be reached at which the inflow from the taps equals the outflow through the plughole. This equilibrium, like that of core body temperature, is regulated without feedback.

From this perspective, the body's sweat evaporation-cooling mechanism specifically addresses the situation where the environment is warmer, such that heat loss from blood circulating near the skin's surface is insufficient to accommodate core body metabolic heat generation. (Sweating is not the only solution; dogs, for example, do not sweat, but pant instead.)

This greatly simplified description of homeothermy in humans illustrates that control and regulation in organisms and analogous open systems is not primarily of the cybernetic, feedback variety, but more generally of the opposing influences variety, with opposing influences acting differentially on coupled, cyclic processes.

Basic Percepts, Concepts and Precepts

Emergence and hierarchy

Whole systems exhibit emergent properties, where the whole is greater than the sum of the parts. If a subsystem could be a whole system, then it, too, may exhibit emergent properties. So, a hierarchical structure of systems within systems within systems is perceivable, in which levels of hierarchy are determined by emergence at each level.

This is distinct from the more general use of the term 'hierarchy,' as for instance in management structures or military rank structures. In systems, where a number of complementary systems come together and interact such that they form a coherent whole with emergent properties, then a level of hierarchy is established. Should that coherent whole interact with a number of complementary wholes, and should they then form a set with emergent properties, then a 'higher' level of hierarchy would be identified, and so on.

This concept is consistent with both the natural and the human activity worlds. Subsystems in the body interact to synthesize a whole human with emergence. Several humans can come together to form a team or a family, and the team/family can exhibit emergent properties: a team/family is, in this respect, different from a group of people, because the members of a team/family interact with each other. Such cooperative and coordinated interactions 'bind' the members of the team/family into a coherent whole that is capable of more than the sum of the separate individuals. Several teams can come together to synthesize a department with emergent properties, and so on.

The concept works for technological systems, too. The complementary parts of radar system might be: a transmitter subsystem, a receiver subsystem, an antenna subsystem, a power supply subsystem, an intra-communication subsystem, an operator/user control and display subsystem, and an operator or user. Without the user, the radar would, to use the Zen metaphor, make the 'sound of one hand clapping.' Individually and separately, none of subsystems can detect, locate or track a target at distance: brought together to interact in just the right ways, the whole radar system has these emergent capabilities; moreover, the nature and degree of interactions results in different capabilities.

By seeing the whole radar within its containing system, i.e., one hierarchy level higher, it is possible to establish just what the radar systems emergent properties should be. For instance, if the radar were to be one of a set of radars forming a sensor network to detect ships crossing some boundary, then there would be a potential risk of each radar transmitter interfering with the receivers of other radars within line of sight. This would mean either restrictions to the transmitter power, or the use of spread spectrum transmissions and receiver correlation, or simply, overall network frequency management, such that adjacent transmitters always worked on different frequencies. Only by looking at the next hierarchy level up, is it possible to determine just what the current whole systems emergent properties should be . . . .

(And, of course, a true systems engineer might question whether radar was the right solution to the problem in the first place — hence: 'just what is the problem, and might there be other, better solutions?' has become the byword of the professional systems engineer.)

Systems as comprised of interacting parts, themselves systems

Systems can be envisaged as made up from parts, themselves having the characteristics of systems, where the parts mutually interact and adapt to each other 'symbiotically.' So, the human body is made up of sensor systems, a cardiovascular system, a pulmonary system, and so on. A company might be made up from a production department, a research department, an administrative department, a sales and marketing department, and so on. In each case, these rationally separable parts cannot exist in isolation, are mutually interdependent, and are definable and comprehensible only in the context of the whole system, and of their interactions with other departments.

The associated concept of containment indicates that a whole system may 'contain' a number of complementary, cooperative, coordinated, interacting parts, themselves systems, or subsystems. Each subsystem could, in its turn, contain a number of mutually interacting sub-subsystems, and so on down to elementary particle level.

Interacting subsystems would, generally, mutually adapt to the inflow that each gave the other. Human activity systems do this all the time, of course, but it is possible to create technological parts that do not adapt to inflows — they process inflows, of course, but need not, in any way adapt to them. This led some to believe that subsystems could be created as building blocks, and could be mixed and matched. However, it is rare indeed that a piece of technology forms a subsystem; in the more general case, a subsystem would be found to be a person, or perhaps a group of people, using either one-on-one tools or machines, or perhaps using a distributed machine, such that the subsystem was in reality sociotechnical, not technological. Sociotechnical subsystems interact and adapt, because of the human element at their focus.

A classic example of this is the ubiquitous fighter aircraft. Without its human crew, it is a lifeless composite of metal, plastic, fuel, oils, gases, lubricants and chemicals. With its human crew it can take off, fly, detect, locate, identify, choose to intercept, intercept, warn, deter, return, land and report. Or, it can change its mind and . . . intercept, kill, return, land and report. So, all the higher functions, the decision-making, the prosecution of mission, the achievement of purpose, are in the hands of the human crew, but are enabled and supported by the live, active, interactive technology. It is the complete fighter, operational with its crew, that is a subsystem in an air defense system – not the sleek-but-inert fighter, leaking quietly in a hangar. The properties, capabilities and behaviors of the fighter emerge only when it is crewed by experienced experts: in that state, and only in that state, of wholeness, it is adaptable and flexible.

Variety in whole systems

Since the parts of a whole must complement each other, interact, cooperate and coordinate, it follows that the parts must be different from each other. Variety is therefore an important ingredient of whole systems: variety in subsystems, variety in interactions, variety in processes, etc. It is possible, in principle, to identify the minimum variety required of any whole, open system such that it can achieve homeostasis. This notion of variety suggests that the whole is made up from a necessary and sufficient set with minimum variety, and tends to emphasize the coupled process view of open system stability/homeostasis. That there is a minimum, or requisite, variety within an open system suggests that there might be adequate variety and even excess variety, too . . . .

Potential synthesis of open systems with desired emergent properties: systems engineering

With the (limited) understanding of emergence came the notion that it should be possible to synthesize systems with desired emergent properties: this was the driver for systems engineering. There are two convergent models of open system synthesis: in one model, the whole system is synthesized by bringing together suitable subsystems and causing the subsystems to interact; in the second model, the whole open system is seen as being formed from coupled processes; drawing in resources and energy, operating upon the resources, expending energy internally, pushing out product and waste, repairing and defending, etc., all of which is necessary for homeostasis. The two models converge where the various interlinked processes from inflow to outflow are executed by suitable subsystems. Processes operate, as it were, 'across' the hierarchy to create emergence.

Putting the two models together, there is an evident relationship between emergence, complexity and highly dynamic interaction between the parts of a whole system. The key to achieving desired emergence is to identify the essential pattern of coupled processes in system design that would result in emergence, and to ensure in any subsequent creative activities that the links in the pattern are not disturbed or distorted. This realization militates against the employment of Cartesian Reductionist practices during the creative phases of systems engineering, where such practices would address parts of the whole system in isolation, would treat such parts as independent, and would disregard the essential coupling and interaction pathways and routes through the design and operation of the whole.

So, a central tenet of systems and of systems engineering was identified as synthesis, as opposed to Cartesian Reductionism. However, it was not, and is not, entirely clear how to go about synthesizing highly interactive, adaptive systems. Nor is it clear what approach should be used. There seem to be two schools:

1. *Hard Systems School.* To create a new system that can be introduced into some problematic situation to neutralize/solve the problem, or . . .
2. *Soft Systems School.* To look for the symptoms of dysfunction in existing, dysfunctional systems, and to seek to repair, diminish, or work around, the dysfunctions so as to suppress the symptoms. The result is not so much a new system, as one that has been 'mended,' 'repaired,' 'enhanced,' 'improved,' etc.

The first school was characterized by the concept of 'hard' system solutions, i.e., solutions that had a clear singular purpose and which could be 'manufactured' *in vitro*, as it were, delivered and set to work. While recognizing the importance of interaction and process, this school emphasized functional, structural and architectural aspects of potential solutions systems, and that such systems would be delivered as a packaged, whole, 'turnkey' solution. Solution systems were conceived, created, delivered, supported, and eventually replaced after they had effectively 'died.'

The second school, on the other hand, applied itself *in vivo*, investigating problematic situations hands-on, seeking to understand the nature of the dysfunctions, and to recommend palliatives that would improve the situation. This second school viewed systems as complex, highly interactive and adaptive, and without any singular clear purpose. Such systems were described as 'messy,' or 'soft,' and were not necessarily amenable to revolution, more likely perhaps, evolution.

Soft systems approaches included 'action research,' in which systems consultants immersed themselves in the dysfunctional system, looking for causes of dysfunction and ways that might repair damaged or missing processes, interactions, etc. Often the systems concerned were large organizations which had formed internal disjoints, omitted necessary internal communications and interactions, or were even at war between internal parts. It was not for some external consultant to redesign the organization; rather, the external consultant might be able to recommend ways to ameliorate the issue or problematic situation. Solutions were conceived, proposed, adopted (or not), results assessed. There was no sense in which the 'solution system' had a life cycle, as such.

Both schools considered that they were systems engineering. And both were undoubtedly correct, although neither recognized the claims of the other. However, a more abstracted view might suggest that, rather than soft and hard as opposites, it was possible to perceive a spectrum from soft to hard, from higher entropy to lower entropy. Problem situations may start out as messy, unclear, lacking objectives, etc., but solutions may include not only repair of dysfunction by reconnection and adjustment, but might also include repair of dysfunctions by introducing new (e.g., sociotechnical) solution system as interacting components of some whole. The sense of this viewpoint becomes clear when the solution system turns out to be a modification, an upgrade, a retraining program, a streamlined process, a reconfiguration, etc.

Problem space and solution space

Systems engineering, both schools, were/are in the business of solving problems. Indeed, it was generally necessary to identify the seat of the real problem, since customers were often too close to the problem to see clearly. The concept of a 'problem space' emerged, a model of the wider system surrounding and encompassing the perceived problem, or at least the symptoms that led to a belief that a problem existed.

A problem space might be a diagrammatic 'universe of discourse' showing participating parties, interactions, coupled processes, disjoints, conflicts, omissions, etc. Of itself, it offered no solution, but sought to encompass and highlight the extent of the situation that engendered symptoms of dysfunction.

Consequent upon the concept of problem space, there emerged the concept of 'solution space.' A solution space showed the 'universe of discourse' in which the solution system would find itself, interacting with surrounding systems and the environment, and potentially adapting to both.

For soft systems, the solution space was similar to the problem space, in that the same participants might be present, but now there would be a changed pattern of interactions and coupled processes, such that the previous symptoms of dysfunction no longer appeared, homeostasis was restored, and so on. Soft systems approaches were seen as potentially progressive, in that making changes would alter the situation, perhaps making some aspects better, but potentially introducing unforeseen problems. So, there might be a number of successive 'solutions,' each taking account of the results from predecessors. This is satisficing, homing in on the ideal solution to a complex problem, but always mindful that such problems are dynamic and constantly changing: so, sometimes it would be impossible to keep up with the dynamics of the (complex) problem.

For hard systems, the solution space was also similar to the problem space except for the addition of the newly created hard solution system, which was seen in its 'operational' environment, interacting with other systems in that environment, all mutually adapting to each other. This was seen as something of a 'big bang' approach, one in which problems were hopefully eliminated at a single stroke, i.e., this is a form of solving. In practice, it was found that, in the time taken to create the big bang hard solution, the problem that it was designed to solve may have morphed, or even disappeared, such that the solution system was less than ideal for the new situation at delivery.

Further, it proved difficult to anticipate how future systems in the solution space would interact with, and adapt to, the newcomer . . . .

Hard systems engineering was attractive to engineering and aerospace organizations and to the military. It promised creative, innovative ready-made, turnkey solutions to problems. There was and is a tendency to exclude, submerge, or overlook the role of the human in the solution system where possible, perhaps because the human operator, user, supervisor, etc., is not possessed of a clearly understood transfer function.

Much of hard systems engineering took to creating complex, technological and expensive solution systems, comprised largely of interacting artifacts, with humans seen almost as 'plug-ins,' i.e., outside of 'the system,' and acting as an operator or user. Lip service was paid to the users with the development of MMI, the man–machine interface, HFE, human factors engineering, and ergonomics. For systems such as command and control, emergency services, air traffic management, etc., in which teams of people worked in harmony, using technological support systems, hard systems engineering proved less than satisfactory, so for these kinds of sociotechnical solution system, systems engineering developed along a pathway intermediate between the two schools, soft and hard. The variability and adaptability inherent in people was reduced by training them to perform particular tasks in particular ways, and by forming them into teams, such that the behavior of the team could be sensibly described and relied upon.

Soft systems, on the other hand, was attractive to management and organizational interests, and so found its home more in academia and in consultancy organizations, some of which applied the basic ideas with varying success, while others developed patent snake oil medicines for improving corporate health.

Evolving adaptive systems

Natural systems are able to adapt to changing situations, changing environments, changing climate, etc., provided the changes are neither too rapid nor too extreme. Human activity systems also display a degree of adaptability, and the ability to evolve. If adaptive systems could be designed and created, they would be able to track changing problem situations, operate successfully in changing environments and situations, and hence offer increased utility and longevity.

There are, potentially, a number of ways in which adaptive systems might be synthesized. If it is possible to perceive a problem space, understand the nature of a problem within that space, conceive, test, evaluate and the create/introduce a remedy into the problem space that neutralizes the problem, then it should feasible to incorporate all those facilities and capabilities into a complex system, such that the system can sense problems, can solve/resolve them, and can reconfigure itself, reconstruct itself, etc., continually so as to be at its most effective in addressing the contemporary problem. We might call such a system intelligent, since it would be able to change its behavior according to situation and experience.

An approach that offers promise is one based on so-called genetic methods. In this, some future system is generated using so-called genes: for a military system, there might be a gene that 'codes for' a particular kind of tank, another for a truck, another for a command and control system, another for a communication link, an infantry platoon, and so on. Using such genes it is possible to synthesize a force, with capability, within a computer simulation, so that there might be, say, genes coding for platoons, for tanks, and so on. The simulation would also contain representations of the environment — in this case, territory, perhaps — and of course an enemy.

The simulation would proceed by causing the two opposing forces to engage, perhaps as intelligent cellular automata, moving over, and making use of, the terrain, with force elements on both sides being threatened and damaged as the engagement proceeded until there was a victor. The simulation would then be run repeatedly, each time with a different pattern of genes, corresponding to a different force mix. Again, the combat results would be recorded. By choosing the various combinations of genes as a random pattern, a wide variety of potential force structures and tactics could be simulated, and the results for each structure/tactic combination recorded. Eventually, the 'best' combination would present itself. This combination would then be used as the 'seed' for a further round of simulated combats, i.e., as the initial force configuration, disposition, resources, communications, vehicle speeds, etc. In this way it is possible to progressively generate a force structure that should be able to successfully combat the chosen enemy on the chosen terrain to best effect.

If it were now supposed that this genetic design capability was 'built-in' to a combat force, such that the force could run simulations and determine ideal configurations and combat tactics before engagements, then the force would become an adaptive system. If this could be achieved for a fighting force, then it could be achieved, too, for a business, an enterprise or an industry.

Readers will have seen the flaw in the above argument, i.e., that the force structure would have been tuned to a particular enemy over a particular terrain. Trying out the evolved fighting force against different foes over different terrains, and so engendering a broader capability, would ameliorate this concern. In addition, the design of the fighting force might be frozen, and the tables turned, such that the enemy force is the one being evolved, using the same genetic approach. This should result in a more capable enemy force. The tables could then be turned yet again, and the friendly fighting force evolved once more to combat and overcome this new, improved enemy, in a kind of auto-ratcheting capability design system (Hitchins, 2003).

Such methods for developing sophisticated system designs are not without their flaws, but they afford great potential if employed with care.

Self-organized criticality

Some systems drive themselves to a critical state, one in which they are on the edge of collapse or disaster, but in which they seem rarely to go across the edge, and if they do they possess the ability to withdraw to safety and regroup.

As an example, consider ancient Egypt. During the Old Kingdom, the River Nile inundated its banks annually, providing rich black silt for farming. Food was plentiful and the population prospered and grew, until a point was reached at which the food supply was just able to meet the needs of the population. This was a state of self-organized criticality: if the inundation of the Nile should be either significantly greater or smaller than usual, the crop yields for that year would reduce and famine would ensue. If, as a result of famine, some of the population died, then the situation would repair itself — for a while, that is, until the population rose again. So, the population teetered 'on the edge of chaos,' as the situation is popularly described.

Such situations are common, and were epitomized by the sandpile experiment (Bak and Chen, 1991), conducted as part of an investigation into the frequency–size pattern of earthquakes. In the experiment, grains of sand were dropped on to a horizontal, circular plate positioned on a chemical balance. Successive grains formed a pile, which rose into a cone, until a point was reached at which adding another grain caused an avalanche to occur. As further grains were added, the cone grew again until further avalanches occurred. So, in the steady state, the height of the cone had a critical value; as it grew above the critical value, avalanches occurred which could bring it below the critical value, whereupon it would rise again to and above that critical value. In the jargon of chaos theory, the critical value was a simple attractor. In systems-speak, homeostasis was established.

The use of the chemical balance made it possible to calculate how many grains fell during each avalanche; some avalanches were comprised of only a few grains, others of many grains. As expected, there were more small avalanches and fewer large avalanches. Surprisingly, there was a clear mathematical relationship between the number of grains of sand in an avalanche, and the frequency of such avalanches. The relationship formed a power curve, of the form:

$$y = ax^b \tag{1.3}$$

Weak chaos

So, on a graph plot of log (grains of sand per avalanche) against log (frequency of avalanches with that number of grains) the result is a straight line. This phenomenon is sometimes called 'weak chaos,' and we are familiar with it in connection with earthquakes, where the Richter scale for the magnitude of earthquakes is logarithmic, such that a magnitude six earthquake is ten times more powerful than a magnitude five, while a magnitude seven earthquake is ten times more powerful than a magnitude six earthquake, and so on. It is found that the frequency of magnitude five earthquakes is greater than that of the more severe magnitude six earthquake, and again that the frequency of these is greater than the magnitude seven earthquakes.

This unexpected mathematical relationship was found to apply in many different, unrelated spheres: stock exchange price movements; distances between cars on a busy motorway; noise in an electrical conductor; meteors entering the Earth's atmosphere; deaths in wars; and many, many more. As the last instance shows, this rule applied both to natural and to human systems. It appears to be that rare thing — a universal rule that applies to whole systems, and is not explained by looking at parts of the whole.

System precepts

From the foregoing fundamental precepts, or tenets, emerge as central to sound systems thinking, systems practice, and systems engineering:

1. *Holism.* A system is a whole. An open system is a whole. The whole is different from, and may be greater than, the sum of its rationally separable parts.
2. *Organicism.* A whole (system) may be an organism, or may be analogous to an organism, in that the many interacting parts behave as a unified whole. The rationally separable parts exist in virtual symbiosis, each depending upon, and being defined by, the sum of the other interacting parts
3. *Synthesis.* It is possible to form a whole from open, interacting parts such that the whole may exhibit desired, or requisite, emergent properties, capabilities and behaviors. This is a functionalist viewpoint, or *Weltanschauung*, and is the *raison d'être* of systems engineering
4. *Variety.* The parts of a subsystem are complementary: they cooperate and coordinate their various actions. The parts are therefore mutually different, i.e., they exhibit variety, and so too must their interactions and interconnections to complement each other. There is a minimum variety of parts for any system to exist and continue to exist.
5. *Emergence.* Emergent properties, capabilities and behaviors derive from interactions between the parts, and are traceable therefore principally to coupled processes, rather than to structure. Emergence arises only when the parts of a whole interact, or conversely when the coupled

processes flowing through the system are active. An open system, therefore, is only a whole while it is both complete and internally dynamic.

6. *System*. A system is an open set of complementary interacting parts, with properties, capabilities and behaviors emerging, both from the parts and from their interactions, to synthesize a unified whole. The definition encompasses the first five precepts: holism, organicism, synthesis, variety and emergence.

7. *Homeostasis*. Stability in open systems occurs at high, rather than low, energy. Stability is a dynamic steady state, brought about when inflows balance outflows. Homeostasis is necessary in a variety of parameters in complex systems, including energy, resources, waste, material inflow, product outflows, and many, many more.

8. *Viability*. The viability of open systems depends in part on achieving and maintaining homeostasis, but also on their ability to neutralize threats from without and within, to adapt and change with circumstance, and to maintain synergy — cooperation and coordination between the parts, to act as a unified whole in achieving some desired external effect.

9. *Purpose*. Manmade systems are viewed as having purpose, one perhaps that they were designed to achieve, or one that they have adopted. It is 'helpful' to consider the human element of systems as having purpose, or intent, and to consider the technological/artifact element of systems as serving that human purpose. So, the parts within a whole may be purposive, i.e., an observer might attribute purpose to them. The parts contribute to the objectives and purpose of the whole, but the purpose of the parts need not aggregate to the purpose of the whole.

10. *Behavior*. Open systems exhibit behavior, i.e., they respond to stimulus. Since open systems are interconnected and interact with other open systems, they are constantly stimulated and exhibiting behavior. Where the behavior of a system is consistent and predictable, the system may be usefully described by its behavior, so diminishing the need to describe the internals of the system; this is a direct means of reducing perceived complexity. Intelligence is marked by the ability of a system to change its behavior according to situation, e.g., it may not respond to the same stimulus in the same way every time, as would a machine or a simple organism.

11. *Isomorphs*. Different systems may exhibit the same behavior, i.e., they respond identically to the same stimulus, although they may be comprised of different parts. Clearly true of many physical systems, it may also be true of some natural systems at some times.

12. *Ideals — the Ideal System*. A concept exists in systems thinking and in systems engineering of the ideal system: it is the best that planners and designers can conceptualize (Hall, 1989). The ideal system can become a yardstick against which to compare options and alternatives, or against which to measure that which is realized.

13. *Values*. Value in artificial systems is often related to utility; the more useful a system, the more it is valued. The value of a subsystem or part of a whole may be judged in the degree with which it contributes to its containing system's objectives in concert with the other subsystems and parts. The value of a subsystem or part may also be judged by the degree in which it complements the other subsystems, particularly where, without such complementation, the other subsystems and the whole would not exist.

From the foregoing, we observe, engineers notwithstanding, that the solar system is indeed a system, organisms are systems, a complete set of ideas can be a system, a series of strategically placed stepping-stones across a river can be a system. We can also see that it is possible to create artificial, or human activity systems in which the whole is greater than the sum of the parts, which has important implications:

- systems can be created that exceed the capability implicit in their technology, or implicit in the sum capabilities of individuals, or both;
- systems can be created that proffer greater value than the sum cost of their parts would indicate;
- systems can be created that may proffer the desired value at less than the sum cost of their separate parts would indicate.

Not everything is a system, however, since not everything is a whole. As we have seen, a fighter aircraft sans crew is not a whole. Similarly, software sans processor is not a whole, simply a set of instructions; a computer without its software is similarly not a whole, simply a machine without instructions. A marriage is not a whole, by definition, unless it joins together complementary man and woman into a single union.

Assignment

Soft and hard systems schools both purport to explore problems and afford system solutions. Compare and contrast their approaches and propose in which kind of situation and problem these might be most appropriate. What do you think might be the flaws, limitations and constraints of each method in addressing complex issues and problems, and how might these be ameliorated?

2
Advances in Systems Science

In everything that relates to science, I am a whole
Encyclopaedia behind the rest of the world

Charles Lamb, 1775–1834,
The Old and the New Schoolmaster

System Theory, System Science

Systems science was recognized from the 1960s, and has been described both as the science of complex systems and as the science of wholes; that is, the science of how wholes form, how they stabilize, how they behave, how they function, how they are structured, how they remain viable, how they decay, fall apart, reconfigure, become moribund, etc.

Systems theory and science address self-organizing systems, autopoietic (self-reproducing) systems, multi-agent systems, closed and open systems, feedback systems, and many more. Systems science investigates how systems behave as a whole, without necessarily having to explain behavior in reductionist terms, e.g., by basing it on the behavior of the rationally separable parts.

Bak and Chen's sandpile experiment (described in Chapter 1) typifies the issue. If the sandpile is regarded, using a physical model, as a pile of tiny spheres placed upon each other to form a cone, then the weakly chaotic, dynamic behavior of the whole will not be observed in such a static model. If the human brain is regarded as an accumulation of neurons, then no basis for the emergence of intelligence and self-awareness will be evident in the model. If the army of soldier ants is regarded as an accumulation of ants, then no basis for the rich behavior of the whole (army) may be detected.

It may be possible to find explanations for whole system behavior at the level of the constituent parts. Weak chaos is exhibited in situations where a flow or stream in a channel is inhibited, such that there is a build-up of particles/force elements into a line or queue, which then releases, only for another build-up to accumulate. This offers an explanation for earthquakes where two tectonic plates rub past each other; snags occur, building up tensions and increasing forces until the snag eventually gives way, with the resulting earthquake, tremors and release of tension. Some snags 'rub off;' for others, there is a greater build-up of tension, and occasionally there is a major snag, and a major earthquake as the snag releases, the plates start to slide relative to each other, and perhaps grate roughly together, giving aftershocks.

Systems Engineering: A 21st Century Systems Methodology   Derek K. Hitchins
© 2007 John Wiley & Sons, Ltd

Similarly, when electrons flow in a conductor, some electrons can become impeded by the ions forming the metal lattice structure of the conductor. Temporary queues of electrons can build up behind an ion, only to release and flow on. This gives rise to noise in the conductor; the so-called $1/f$ noise to indicate that noise amplitude varies inversely with frequency.

Entering a school just as the signal is given for the children to go home is analogous to the electrons in the conductor. Children rush towards the exit, heads down or looking and chatting sideways, until one of them find the path blocked by an adult going in the opposite direction. The child cannot proceed as there are children streaming homewards to left and right. Another child finds their way blocked by the first child, and so on, until the mini-queue of children builds and disperses into the streams to either side; and, the process repeats.

It is possible, too, to find explanations at the level of the individual of how a flock of starlings returning from a city center to roost can present such dazzling displays of coordinated behavior: they swoop, wheel, change shape, divide, re-form, yet all the time behave like a single organism before diving unexpectedly into the darkness beneath to their covert roosting site. What triggers the starlings to come together, and to form this 'super organism' is not clear, although the motive for the flock, once formed, may well be protection against predation. Shoals of fish behave similarly in the face of predation. Goldfish, and other animals, display a slightly different pattern of behavior, in which the shoal moves around with individuals moving seemingly at random as they search for food, but periodically all sinking to the bottom in synchronism, usually pointing in the same direction, apparently for a rest. Ant colonies also exhibit rhythm, coming to a rest every 28 minutes; individual ants do not behave in this way — only when many ants are interacting does such behavior of the whole colony emerge.

Systems science purports to address all kinds of wholes. As von Bertalanffy demonstrated, systems theory is founded in well-established laws and phenomena. So, the queues of electrons and of children in the examples of weak chaos would all, of necessity, conform to the conservation laws and to the rules of queues. Systems science not only includes the physical sciences, but also the life sciences; it is, therefore, an inclusive science, investigating the natural, social and physical domains — wherever, systems are to be found, natural, and artificial.

Conservation Laws and Transport Phenomena

The physical conservation laws are significant to the understanding of systems behavior and to the development of system theory, owing to the high degree of interactivity within and between complex systems. The high interactivity between parts of a complex system is implicated in the generation of, and variability in, emergent properties, capabilities and behaviors. In physics, the continuous random motion of particles gives rise to the net macroscopic transport of matter by molecular diffusion; of energy, by thermal conduction; and of momentum by viscosity.

Dynamic simulation models developed as the basis for systems thinking are generally designed so that conservation is 'built-in' to the program. As a simple example, see Figure 2.1, which illustrates homeostasis and conservation of matter using the analogy of a bath with taps left on and plug left out. In the figure, which uses the simple STELLA™ notation, there is a header tank, which receives water in spurts (inflow) from an external pumped source (not shown). The header feeds water out to the bath, which has a drain arranged such that the rate of outflow is proportional to the level of water in the bath.

Graph 2.1 shows the result of running the simulation: the zigzag line shows the level of water in the header tank as the water spurts in and flows out. The rising line shows the level in the bath,

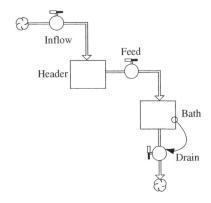

Figure 2.1   Conservation of matter — the Bath analogy.

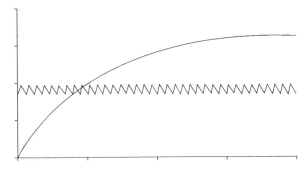

Graph 2.1   Homeostasis — the Bath analogy.

rising to reach a steady level. This is analogous to the manner in which homeostasis develops in, say, the human body. Food is taken in periodically, at mealtimes and during snacks. Peaks and troughs of energy in the body are smoothed out by converting some of the food intake into glycogen, largely in the liver, to be released over time and as needed, so that the level of, e.g., glucose in the blood remains within sensible limits. The body mass, as shown by the other line on the graph, rises to some mean level, after which it will vary about the mean with the varying balance of food intake, work done and energy outflow.

So, the header tank is crudely analogous to the action of the liver, in that it smoothes out peaks and troughs in resource and energy inflows, and the bath is analogous to the body of the open system, stabilizing at high energy levels, rather than low, and with stability being a homeostatic balance between opposing influences, inflow and outflow, i.e., without regulation via feedback.

Queuing Phenomena

The formation, behavior and dispersion of queues can prove invaluable in understanding behavior of whole systems; particularly where intrachanges between parts of the system, or interchanges between systems are comprised of discrete entities. Queues might be formed of people in a supermarket, signals in a communication channel, assemblies in an assembly plant, patients in doctors' surgery . . . the behavior of the queue transcends the nature of the discrete elements forming the queue. When people form into parallel queues, in particular, they are likely to leave the system if the queue is too long (customer impatience), or jump between queues if they observe that another queue is moving faster than theirs. Queues can form in parallel, in series, or in any combination of parallel and serial.

Queuing theory and queuing simulation, of which a simple example is shown in Figure 2.2, are invaluable tools for understanding the behavior of complex systems. The model could represent parts being inspected (Service A) in a factory prior to assembly (Service B), or, triage arrangements in a hospital emergency room, or any number of situations and processes.

In this instance, it represents part of a recruiting office. Service Channel A might be the process of applicants providing their background and experience to an interviewer — technicians in Q1 and operators in Q2. Those who forget to bring their résumés might find themselves having to rejoin Q1 through Re-queue, after having retrieved or reconstructed the missing document. Service Channel B might be the final interview, with those applicants who failed returning to the general population . . . . The time taken to pass though each service channel would not be fixed; rather it would be distributed in some way. So, the time taken by each new recruit to pass through the whole process might vary significantly, as shown in the figure.

By simulating the end-to-end process, including the insertion of distribution patterns for arrivals and service times, it is possible to develop a statistical 'behavior profile' for the queuing system, indicating the mean time taken for an operator or a technician to go through the system, with and without the feedback loops, and with differing presumptions. This is systems thinking, and although the example might concern a recruiting office, the technique and tools are widely applicable. Note, that the model of Figure 2.2 is a 'closed' systems model, as opposed to a model of a phenomenon, i.e., there are no outflows or inflows to or from any unspecified external parties as there were in Figure 2.1. Note, too that the model shows feedback control, with the accumulated number of

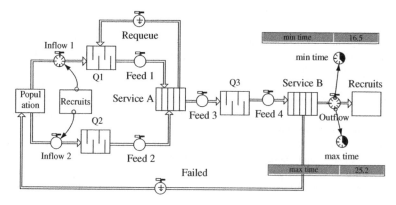

Figure 2.2 Model of parallel and single queues. Q1 and Q2 are queues of discrete entities forming at Service Channel A (a conveyor in the STELLA™ notation, analogous to a factory conveyer belt.) On passing through this channel, the entities form/join a third queue, Q3, for Service Channel B. Some entities are sent back from service Channel A to rejoin the back of Q1. Some entities are sent back from Service Channel B to rejoin Q2. The resulting flow of entities through the system over time can be calculated, but it is simpler and more effective to simulate the behavior of the queuing system, as in this example.

recruits being fed back to close the taps, Inflow 1 and Inflow 2, when sufficient recruits have been interviewed.

Queuing is ubiquitous. It occurs wherever there are flows of entities, whatever those entities might be. We can say with certainty, for example, that the 2.5-tonne stones being laid in ancient Egypt on the Great Pyramid of Khufu, at a rate of one every two minutes, will have formed queues en route from the quarry to the pyramid, with more queues forming as stones were dragged, raised, or 'magically elevated' up its sides. We can even work out the likely statistical patterns, lengths and queuing times, and be positive about our estimates. Queuing behavior applies equally to people in a supermarket; automobile parts in a global lean-volume manufacturing and supply chain, and digital words in a data stream.

Chaotic Phenomena

Lepidoptera Lorenzii?

Edward Lorenz is credited with the initial identification of what turned out to be chaotic behavior and strange attractors.

At the time, he was studying the weather using a fairly primitive computer, by today's standards. Line printouts took a long time, so he would suspend a run and record the readouts to only three decimal places, although the computer stored to six decimal places — he didn't think it important. Whenever he restarted, he found that he didn't get exactly the same results. They were tantalizingly close, and the same graphical sequences seemed to emerge, but they were never precisely the same. The model replicates his three original equations:

$$dx/dt = -10x + 10y \tag{2.1}$$

$$dy/dt = 28x - y - xz \tag{2.2}$$

$$dz/dt = -8z/3 - xy \tag{2.3}$$

Graph 2.2 shows a phase-plane chart of x vs y, which illustrates the now-famous Lorenz, or strange, attractor, looking suspiciously like a butterfly. It illustrates how the weather progresses: the graph space might be described as 'climate, ' in that the instantaneous weather as represented by each dot can vary extensively, but always stays within overall bounds, which correspond to the limits of climate. No matter how extreme the weather might be, it is always 'attracted' back to this curious pattern. Each dot (plot) is a small progression from the previous dot: the weather is not random. Were it so, each dot would owe nothing to any other dot, and each dot could appear anywhere within and without the climate envelope of the graph.

Generating chaos

This is true for chaotic phenomena in general; they are bounded, and while each event in a chaotic series of events may not be simply predictable, the overall pattern of events is bounded. There is, moreover, an underlying order to the seeming disorder. Chaotic phenomena exhibit 'no-go' zones: areas/spaces in their phase-space where they will never go. So, chaotic phenomena are, it seems, more amenable in some respects than random phenomena.

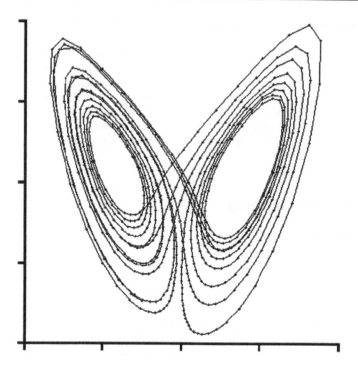

Graph 2.2   Lorenz's butterfly, or strange attractor. STELLA™ simulation.

Chaos is associated with events or processes where the output from some process becomes the input to the next, same process repeated. For example, if a video camera is set up on a tripod, pointing directly at a TV screen, and if the camera output is shown on the screen, then the camera is taking a picture of its own output. If the process is started off in the dark, nothing happens. However, if a match is struck then the camera picks up the light from the match, shows it on the TV, and the TV picture is seen by the camera . . . remove the match and the process is now self-sustaining. Waves of complex patterns sweep across the TV screen as each 'circuit' of the repeated process experiences the nonlinearity in the amplifiers, displays, camera, etc. The displayed patterns are 2-D chaotic, ever changing, never repeating, but nonetheless similar over time.

It is this repeated process, coupled with nonlinearity, and continuous/continual flow, that seems to characterize the generation of chaos. Chaos can be associated with linear behavior, too: Hyperion, one of Saturn's moons, rotates linearly about its planet, but at the same time it tumbles chaotically about its axis, so that its direction of pointing is unpredictable.

Figure 2.3 shows how easy it is to generate chaotic behavior. The model could not be simpler, consisting as it does of only three simple reservoirs, systems A, B and C. The chaotic behavior emerges from the nature of the interchanges, which are discrete rather than analogue, see Graph 2.4.

How could such behavior arise in the real world: three prisoners locked up in a tiny cell; three divisions in an organization coordinating and cooperating their activities; three stars tumbling about each other in close proximity under their mutual gravitational attraction? As this simple model shows, complex behavior can potentially emerge from the simplest of open interacting

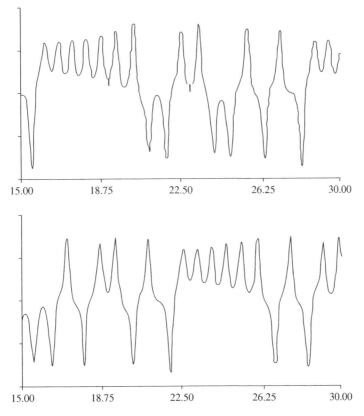

Graph 2.3 Lorenz weather simulation patterns. Note that the graphs run from 15 to 30 time units; from zero to 15, the two graph runs appeared identical, although the upper graph started with a value of $x = 1$, while the lower started with $x = 1.0001$, a difference of only 0.01%. Note, too, the repetition of patterns, hinting that the patterns might coincide — but they never do. (STELLA™ simulation)

systems. Mathematically, perhaps the most complex entity is the Mandelbrot Set, named for Benoit Mandelbrot, which derives from iteration of the simplest of equations:

$$z^2 + C = z$$

The recognition of chaos and its sources might have been expected to cause a major stir amongst the ranks of both physicists and engineers. Generally, this does not seem to have been the case, however. Other than being in denial, one reason may be that chaos, while easy enough to observe in simulations and stellar motion, may nonetheless be difficult to observe in human activities. It has long been suspected, for instance, that projects experience chaotic behavior, particularly as they fall behind schedule and the various personnel seek to coordinate and cooperate more intensively in an effort to recoup lost time. Although coupling may be increased during this frantic effort, proving that chaos is implicated has not been possible to date.

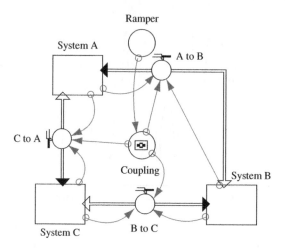

Figure 2.3 Simple, coupled reservoir model, representing three systems, A, B and C, mutually interacting. The degree of coupling between the three can be altered, i.e., how much of the contents of each reservoir can be interchanged at any moment. (STELLA simulation model).

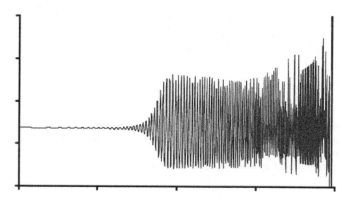

Graph 2.4 Stability–chaos–breakdown. The graph shows the behavior of the model in Figure 2.3 as the coupling is progressively increased. At first, there is little change, then chaos ensues; finally, at the end the whole goes unstable. (STELLA™ simulation).

Another reason may be that chaos in everyday experience may not be so neatly expressed as the deterministic chaos of the simulation model or the weather model. Everyday chaos may be 'noisy;' i.e., parameters that would contribute to a chaotic model are constantly changing and adapting, so that any chaotic behavior which may emerge does not appear to fit neatly into anticipated patterns.

Despite these concerns, chaos and chaotic effects are real enough and are undoubtedly present in, and contributing to, uncertainties in human affairs. Instead of being concerned with avoiding the presumed harmful effect of chaos, scientists and engineers are now moving in the direction of harnessing the potential value of chaos in the design of military tactics, security coding devices, communication systems, graphical design programs and many more.

Self-similarity

Investigations into chaotic phenomena observed that many systems contained replicas of themselves on smaller scales. This is self-similarity; for instance, our solar system has a number of planets orbiting the sun, while several of the planets themselves have orbiting satellites. In the natural world, self-similarity in plants has been widely observed, notably in the design of fern fronds, where each branch is of the same form, though smaller, than the whole frond. This continues, with each branch from the frond being of the same form again, but of course much smaller still. It is thought that this observation may cast light on the genetic mechanisms that govern the development of the plant from the spore. Self-similarity is evident in the human activities, too. An army might have three divisions. Each division might have three brigades. Each brigade might have three battalions. Each battalion might have three companies. (In human affairs, the fractal chain is not infinite.) Similarly, the corporation might have three groups. Each group might have three divisions. Each division might have three departments, and so on through sections and teams.

For self-similarity to be meaningful in human activity terms, each of the self-similar entities forms a unified whole: so, a company is a unified whole, so is a team, an army, a division or a platoon. This is not to suggest, however, that our human propensity to form self-similar organizations is associated with chaos; on the contrary, it seems more likely to be based on some innate desire to manage and control a limited number of subordinate groups. A manager is more likely to be comfortable managing three groups under his control than ten groups. In the military, the talk would be of 'span of control,' and the issue would be both of the number and of the diversity of units under direct control of one individual.

Fractals

Fractals are part of the chaos scene. Benoit Mandelbrot described fractals so: as you get closer to straight line, the smaller section you can see remains straight. As you get closer to an irregular coastline, the smaller section you can see remains irregular. Fractals are all those things that, as you get closer, remain the same, but that are not straight lines.

These are phenomena where the behavior may be described as being of the form $y = ax^b$ where b is a non-integer, so is 'fractional' — hence fractal. Many physical phenomena are linear ($y = mx + c$) as in distance traveled at constant velocity or square law ($y = ax^2$), as in distance fallen from rest under the influence of gravity. To observe phenomena with fractional indices was novel, but once observed, examples cropped up in unexpected places.

Graph 2.5 shows the reported crime statistics over some 25 years for a county in England. The annual figures show a seemingly erratic nature, allowing politicians of either flavor to successively deplore rising crime, or to state that their policies were clearly working, as witness the sudden dramatic fall in reported crime . . . .

The result of applying one method of identifying fractals is shown in Graph 2.6, and it tells a different story. The method used was to measure the 'bumpiness' of Graph 2.5; this involves using rulers of different length (hence y-axis, ruler length), and using the full length of each ruler to see how many lengths made up the overall length (hence x-axis, periphery). A small ruler would be able to go in and out of the various peaks and troughs, while a large ruler would miss them out. The straight-line 'goodness of fit' factor of 0.98 is sufficient to indicate that the reported crime statistics may be reasonably considered as fractal. This suggests that the reported crime statistics vary somewhat chaotically, without seeming influence from politicians or policemen, although

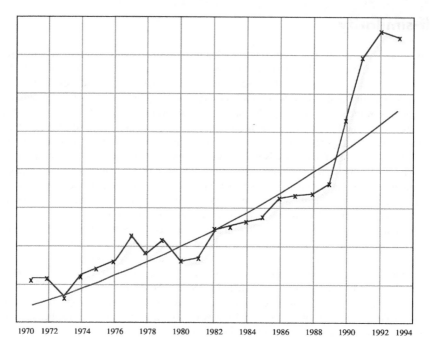

1970 1972   1974   1976   1978   1980   1982   1984   1986   1988   1990   1992 1994

Graph 2.5   Reported crime statistics in an English county. The graph variability suggests two separate features: 1. Exponential increase in crime over 23 years; 2. High degree of variation about the regression line.

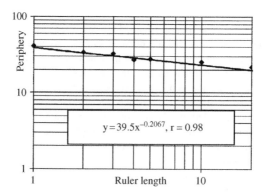

$$y = 39.5x^{-0.2067}, r = 0.98$$

Graph 2.6   Reported Crime statistics as a fractal.

there is clearly an underlying trend upwards for whatever reason. Moreover, using the notion of self-similarity, and noting that the county in question was comprised of a number of divisions, it is reasonable to suppose that each of the divisions had similarly fractal crime statistics — and so it proved.

There is a relationship between weak chaos and fractals, in that both correspond to a power law, i.e., $y = ax^b$; for fractals, the index b is a non-integer. The development of fractal ideas is

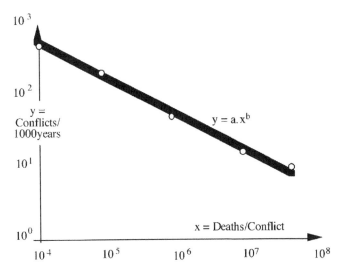

Graph 2.7 Deaths in war. from Lewis F. Richardson, the British meteorologist. (Richardson, 1960).

not really new, although those developing them may have been largely unaware of the underlying consistencies. One such was Lewis Richardson, and English meteorologist, who plotted deaths in war from 1820 to the end of World War II. When Richardson figures are plotted on a log–log scale, Graph 2.7, the result is a straight line with a negative, fractal index.

The similarity between Graphs 2.6 and 2.7 is striking; but perhaps the more surprising observation comes from considering the advances in warfare and the science of killing that had taken place over the period from 1820 to 1946. In 1820, rifles were fairly primitive, machine guns and motor vehicles were nonexistent, and air power was virtually nonexistent in and over the battlefield. By 1946 two atomic bombs had been dropped. Yet, in spite of tremendous advances in the science of warfare, the number of deaths in war conformed over the period to a straight-line graph. Recent work suggested that Richardson's work could be extended at the lower end to the level of violence between individuals, and the fractal relationship still held.

There is much more to the burgeoning science of chaos, and it is clearly of major significance to understanding systems behavior and in designing system solutions. Whereas current practices tend to avoid chaotic regimes, it seems likely in the future that it will be possible to make great use of chaos in the design, development and creation of robust, adaptable and self-sustaining systems.

Period doubling

Another phenomenon associated with chaos is period doubling. This can be observed in the dynamic behavior of many physical objects. A dripping tap will drip regularly: this was, after all, the basis of some ancient clocks. Open the tap slightly, and the drops will come out, not singly, but in groups, with a delay between the groups. Open the tap some more, and the groups have more drips, and the delay between them gets longer still. Finally, the stream becomes chaotic. This phenomenon came to be known as period doubling — the period between the groups of drips first doubled, then quadrupled, then octupled, and so on.

Period doubling occurs widely in dynamical systems. It is observable in electronic oscillators, where increasing feedback causes the period to first double, then quadruple, and then to go rapidly into chaos. In the medical field, researchers have observed the same phenomenon in people with heart disease; in this case the phenomenon was called 'electrical alternands,' and presented as alternate stronger and weaker beats, such that the period between strong beats doubled: this occurred prior to the onset of fibrillation, offering the prospect of being able to anticipate heart attacks for some patients.

Information: Conserved, or Non-conserved?

Systems are pools of information concerning their functions, interactions, processes, architectures, purpose, objectives, etc.; they may also receive information, interpret it, use their interpretation, and exchange information with other systems. Information is unlike matter or energy, seemingly, in that information can be given away, yet the supplier still has it: it appears not to be conserved.

This is a simplification, of course, since information is negative entropy; receiving information reduces uncertainty in the receiver. Since there is a relationship between entropy and energy, a system receiving negative entropy is, in principle, potentially capable of doing more work. For practical purposes in simulation and modeling, however, it is often not unreasonable to consider information as a nonconserved component.

Information of interest to viable systems comes in many forms. Information about the environment may be sensed at a distance using eyes and ears, radar and sonar, passive and active sensing, and so on. In each case, desired information has to be extracted from a 'cacophony' of background information, and interpreted in some way. Generally, a system acts upon its own interpretation of information received, rather than on some 'ground truth.'

So, people often see what they expect to see. Doppler radars see only things that move. Air traffic management (ATM) area radars around London were puzzled by concentric circles of echoes moving towards the center of London each morning and outwards to the suburbs and beyond each evening. Initially dubbed 'angels,' these echoes were caused by large flocks of starlings coming into London to feed, and returning home to roost at night.

Information can be inferred, too. Supposedly, the adult human does not receive enough visual information via the eyes and the optic nerves to enable successful driving at night with only car headlights by which to see. That we can drive at night indicates that we are inferring information about the whole scene in front of us from visual cues. This system can break down: drivers in flat featureless areas were found to drive straight into ditches at night. Upon investigation, it transpired that the road system wound its way around the edges of fields, while the telegraph poles had been set in straight lines, sometimes alongside the road, but then going straight across fields where the road deviated. Drivers were using the telegraph poles as a cue to the road ahead, and were crashing into the ditches, but only on moonless nights. Accidents were rare on clear nights with a full moon . . . .

Similarly, some commanders in battle seem able to make successful decisions that should have required intelligence about the enemy, even though that intelligence was missing. Investigation suggested that these commanders were empathic, could infer what the enemy commander would do in the situation facing him, and exploit that inferred knowledge as intelligence.

Systems Science as Natural Science and Social Science

Systems science incorporates the natural sciences, which form the basis for the applied sciences: physics, chemistry, biology, astronomy, earth sciences, etc. It also incorporates the social sciences, including: psychology, anthropology, sociology, jurisprudence, economics, linguistics, and many more. Systems science invokes the scientific method, involving the proposition of hypotheses to explain phenomena, predictions based on the hypotheses, experimental studies to test the hypotheses leading either to the rejection of the hypotheses or the formation of theory, so binding specific hypotheses into logically coherent wholes. Where physical experimentation would be impractical or unaffordable, system science must needs resort to experimental studies, using dynamic computer simulations as experimental test beds.

Action research, in which researchers involve themselves in the physical and social environment, may be considered as scientific experimentation where the presence of the researcher can be shown not to affect the outcome of the experiment. On the other hand, some situations preclude action research; it would be a brave, or stupid, researcher who would explore the psychology of a prison riot by asking the participants to pause in mid-riot while he investigated their feelings and motivation.... Similarly, it has proved necessary to simulate the behavior of dangerous systems/operations where wide-ranging practical experimentation would be too risky and/or expensive. Such measures are acceptable only where the simulations are rigorous; where rigor is difficult to come by, experiment-by-simulation must take a broader view, address a wider potential spectrum of outcomes, and consider experimental results with caution.

Psychology is of particular importance in the development of models of systems behavior, since nearly all problem and solution systems concern social or sociotechnical systems, and necessitate the representation of human behavior, either individually or in teams/groups/societies. Psychology differs from anthropology, economics, political science and sociology in seeking to capture explanatory generalizations about the mental function and overt behavior of individuals, rather than relying on field studies and historical methods.

Behavior

Behavior refers to the actions or reactions of a whole, usually in relation to its environment, or to some stimulus, e.g., behavior as stimulus–response. Behavior can be conscious or unconscious, voluntary or involuntary. The more complex a system, generally the more complex will be its behavior, although some simple systems can exhibit remarkably complex behavior. Also, the more complex a system, the greater is its propensity to learn and to adapt its behavior. Intelligent behavior is marked by differing responses to the same repeated stimulus: while a simple system will respond in the same machine-like way every time, an intelligent system will change its response, perhaps to evade the stimulus, or to investigate its source.

So, if a woman strikes a man in the face, he may not respond. If she strikes out a second time, he may duck, flinch or catch her hand to stave off further attack. On the other hand, if a man strikes a man in the face, the striker may expect an immediate parry, if not a return blow. Should he strike a second time, he is pretty well guaranteed a fight, or else his victim will retreat if able. Of course, those with a strong moral and ethical code might choose to 'turn the other cheek,' which will present the striker with a dilemma. . . hence the role of psychology as a predictor of behavior. In each instance above, the behavior as response to stimulus might be deemed intelligent, since the response differed for the same repeated stimulus. The difference in response between a man and a woman illustrates in a small way the impact of culture, mores and situation on behavior.

For social systems and sociotechnical systems, there is perhaps a greater emphasis on group psychology, so that sensible predictions can be made about the behavior of teams, platoons, companies, etc. While not a precise science, it is surprising to the uninitiated to discover just how accurately a psychologist can predict behavior of an individual, provided the psychologists is aware of the individual's background, culture, and particularly their recent experiences up to the moment of prediction.

Interpretation and categorization

Complex organisms display the ability to categorize information: it appears to be a fundamental ability. A newborn calf can immediately distinguish mother, teat and food from non-food. Psychologists suggest that categorization is an essential feature of the brain, enabling it to reduce an otherwise bewildering set of sensory stimuli to significantly fewer categories. In some animals, the sensory organs take part in the process of categorization: the horizontal pupils of some grazing animals enables them to detect vertical stems of grass with greater accuracy; the eyes of some animals are tuned to the infrared to enable them to detect warm-blooded prey; and so on.

Tacit knowledge

To facilitate categorization higher organisms, and some mechanisms, too, create 'libraries of tacit knowledge.' These are everyday items of knowledge, developed since infancy, such as grass is green; sky is blue; things fall to the ground when dropped. Brains draw upon this tacit knowledge to facilitate categorization and recognize stimuli.

World models and world views

Higher organisms, and humans in particular, create so-called world models of the environment in which they live and operate. A world model is a stored image of how the particular world works, or appears to work. So, helicopters hover with the rotating blades above the fuselage; were you to observe a helicopter without blades, or one with the blades underneath the fuselage, it would confound your world model, and you would be alerted by a mental disjoint. A world model is a mental model space, within which tacit knowledge determines the basic rules of form, function, behavior, color, etc.

Weltanschauung, or world view, is an allied concept, describing as it does the philosophy or viewpoint from which a person sees and justifies situations and events. One *Weltanschauung* might consider it good, right and proper to drop atomic bombs on a country with which one is at war, in the interests of saving many lives by shortening the conflict. Another *Weltanschauung* might consider that killing is fundamentally and morally wrong under any circumstances, and 'just because they are doing it to us does not justify us doing it to them.' Neither *Weltanschauung* is either right or wrong, and neither is logical or illogical – they are what they are: philosophical viewpoints. The implications of adopting and expressing such viewpoints might prove important, however.

Interpretation

When a stimulus is received, it is, in effect, compared with memories of previously received stimuli recorded as tacit knowledge and world models; the particular stimulus is thus recognized and its meaning and implications drawn from memory. This process is subject to error where, for example, the stimulus is similar to previous ones, but actually from a different source. Alternatively, a new, previously un-sensed stimulus may be received, which does not accord with tacit knowledge and world models. The brain may actually ignore it, since it has no way of even registering it; alternatively, it may choose to ally the stimulus with others that it does recognize. In this case, it may interpret the novel stimulus as being like, or even the same as, something familiar. On the other hand, it may decide something is amiss and investigate further. Or, there is a third option . . . .

Belief system

Higher organisms, especially man, develop a so-called belief system. This is a set of beliefs reinforced by culture, theology, experience and training, as to how the world works, cultural values, stereotypes, political viewpoints, etc. If a stimulus is received, it may be interpreted with the effective aid of the belief system, to be whatever the belief system might lead the recipient to rationalize. So, a naval radar operator is viewing a radar screen in friendly waters when a radar track appears on screen coming from the open sea toward the ship. The operator is expecting an aircraft, which launched an hour earlier with a faulty identification system, to return any minute; his belief system assures him that the incoming track is the anticipated 'unidentified friendly.' He may be right: but, on the other hand . . . .

A belief system need have no basis in reality, so long as it consistently provides adequate explanations. The Mayans believed, apparently, that sacrificing the blood of young men would ensure that the sun rose each morning, would ensure the rains, would prevent flooding; every time they sacrificed young men, sure enough, the sun rose, the rains came, etc., so reinforcing their belief . . . and if the rain failed to come, perhaps the gods thought the sacrifice too small, so sacrifice more blood, until the rains arrive?

Figure 2.4 shows conceptually how an individual's and a society's belief system might be brought together and might mutually sustain each other. The figure shows interlinked reinforcing loops. The top loop shows that an individual's belief system gives a believer a straightforward world model, so that he or she can find satisfactory explanations and interpretations of everyday events and situations. This reduces the individual's psychological uncertainty, so reinforcing his faith in his belief system.

Lower loops show the relationships between belief and society. A shared belief system is at the heart of a culture. The model indicates how shared beliefs sustain the belief system, promote social cohesion, and enable the growth of class and power structures.

Conflict between two groups, including war, may be characterized as a 'battle between belief systems.' Icons emerge strongly in such conflicts: they may be revered objects such as stones, writings, buildings, flags or badges; whatever they may be, they may symbolize the central core of the belief system. When people become icons, the real person may become obscured behind the projected iconic image or persona.

Organizations develop their own, in-house culture and belief system, too, which leads them to act and behave in ways that might not seem entirely rational to an outsider. Marketing campaigns represent a company, not so much as it is, but more as it would like to be; perhaps in an attempt to

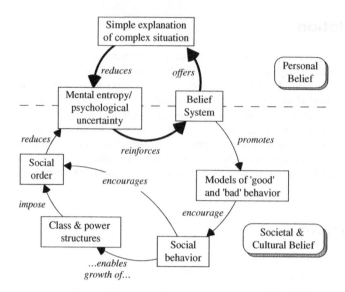

Figure 2.4   Sustaining belief systems.

influence the belief system of its own employees as much as those of customers. Call-answering services put customers on hold, repeating the mantra: 'we value your call;' while clearly evidencing, by not answering, that they do not value the call. Such companies hope (forlornly?) to influence the belief system of the caller into judging an inferior service to be acceptable.

Instinct and archetype

The way in which the human brain develops is not well understood; the most remarkable thing about it, perhaps, is that for so many people on the planet, the brain is a relatively stable, highly capable organ. The basis for this stability is hard to find; experiments with sensory deprivation show that we depend on sensory inputs, without which we start to hallucinate and lose our frame of reference.

It is observed, for instance, that young children experience the most horrific nightmares, in which they seem to see monsters and apparitions, which bear no resemblance to anything they have ever experienced in their short and sheltered lives. Where can such images come from? Are they, somehow, inherited? Are they some imprinted memory of more primitive, prehistoric existence? Do they emerge naturally as the structures and functions within the brain develop?

Various theories of how the human brain functions have been put forward, some such as those of Freud and Jung based on 'models of the mind.' Other approaches seek to unravel the brain's operation by examining its physical structure and observing which parts of the brain correspond to which activities. The well-known phrenological bust by L.N. Fowler was an early, discredited attempt to relate the bumps on the skull to human behavior. Since that time, the advent of sophisticated scanning machines has enabled researchers to observe how activities occur in the brain when the person being scanned is shown pictures, undertakes simple activities, etc; as a result, there is a much greater understanding of the complexities of the human brain, augmented in particular by studies of people who have had some impairment to the brain, such as the severing of the corpus callosum, the bundle of nerves joining the two halves of the brain.

Significant though such advances may be, they are insufficiently advanced to be of much use in comprehending the behavior of individuals, or groups of people. For this, we have to look to the work of Freud and Jung. (Jones and Wilson, 1987). Freud, the psychoanalyst, took a somewhat biomechanical view of the brain, identifying an id, ego and superego. Conflicts exist between id impulses, ego defenses and superego restrictions, with the ego mediating between 'primitive forces of the id' and the 'censoring, guilt-inducing power of the superego.'

Jung, on the other hand, created a richly populated, multilayered model for the mind, in which there were:

■ conscious
■ personal unconscious
■ collective unconscious; the deep species-wide layer of the psyche underlying the personal unconscious
■ archetypes; mythic images and motifs that go to makeup the collective unconscious
■ complex; a group of interrelated and emotionally charged ideas or images
■ individuation; the progressive emerging of the mature individual mind, coming to terms with thinking/feeling and sensing/intuition axes that determine the psychological types
■ extrovert/introvert and anima/animus (the woman inside every man and the man inside every woman)
■ active imagination; that which enables one to write or paint one's unconscious fantasies
■ synchronicity; meaningful coincidence of two causally unrelated events
■ self; the very center of one's being

Jung's model of analytical psychology is complex, even fanciful in some respects, but it was formed from years of research, and seems to offer explanations for some of the more bizarre behaviors which people display, particularly under times of emotion. Jung, too, considered the human mind as being part of some whole — the collective unconscious, which shaped and influenced humanity.

Another way of looking at his idea of collective unconscious, with its constituent mythic archetypes is to consider that, as humans, no matter what our individual differences and cultures, we all share some basic human instincts and patterns of behavior. Mothers will defend their children. Sons will rebel against their fathers. Teenage girls will be attracted to boyfriends that are unlike the girls' fathers. Men will engage in sex at the drop of a hat; women are less so inclined. Jung's mythic archetypes can be seen in a similar vein. The shepherd will protect his flock from wolves; as will the shepherd–king protect his subjects from marauders.

Figure 2.5 shows a diagrammatic representation of the developing psyche of the ancient Egyptians, as Jung might have seen it — both he and Freud were fascinated by ancient Egypt, with its overt psychological overtones. The figure shows the collective unconscious, which, in Jung's view, would underlie the personal unconscious and conscious minds of the people. Archetypes were in evident in abundance — the ancient Egyptians placed great emphasis on symbolism. The figure shows a few only:

■ The king as shepherd of his people. The king carried a flail and a *heqa*, a shepherd's crook, as symbols of his authority. Today, archbishops carry crosiers, for much the same reason.
■ The magus, or magician as high priest. Today we have people who can turn round ailing businesses, entrepreneurs, etc. Financial directors who can work on the books and turn an apparently ailing company into a seemingly-successful one overnight are modern magi.
■ The pharaoh as the self and individual of the people, an icon of the nation.
■ The divine king, leader and ruler ordained by divine right.

■   The healer, the creator, the ideal . . . .

■   Zoomorphs; gods who were men and women with animal heads, such as Anubis, the jackal-headed god of funerary rights. In the legends of the time, he helped Isis bind together the sixteen parts into which her brother/husband Osiris had been cut by his brother Seth, so creating the first mummy. Isis formed a phallus out of Nile mud, and conceived the child Horus, so identifying with both the virgin birth and the perfect mother and child image, iconic archetypes that were handed down to the early Coptic Christian church amongst others.

■   Duality, the propensity to see everything in terms of opposing pairs which must be reconciled. Today we have political left and right, for or against, black or white, man versus woman, good or bad, common or classy . . . duality is alive and well, as Jung would have expected.

According to Jung, these mythical archetypes form the basis for instinctive behavior. The myths may be just that — mythical — but these patterns of behavior, he suggests, are contained within us, at the level of the collective unconscious. In the normal course of events, we would remain unaware of these behavioral archetypes, but they can influence our behavior, and particularly they may behave the way in which someone feels he or she ought to behave.

So, when someone is put in charge of a group of people, he may feel that it his role to 'look after his new charges,' to guide them and protect them from others. On the other hand, he might decide that as the new boss he can do no wrong; in consequence, he lays down the law, enforces the rulebook, micro-manages, and brooks no criticism. In the first instance, we see the shepherd–king behavioral archetype; in the second, we see the king as divine ruler archetype. Similarly, the healer is someone who can seemingly make problems disappear, and so on.

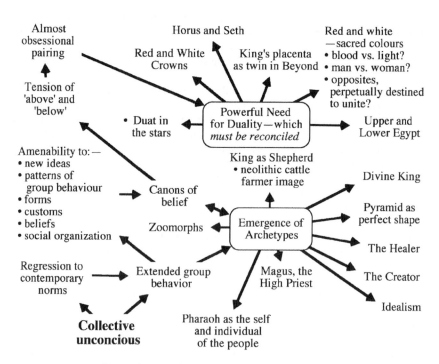

Figure 2.5   Jungian view of ancient Egyptian psyche.

Psychology forms a central thrust in the application of systems science, together with social anthropology.

Social and Cultural Anthropology

Anthropology is the study, particularly, of cultural and developmental aspects of humankind. In contrast to psychology, with its emphasis on theory, social and cultural anthropology are generally investigated though field studies, so that understanding may be gained about human behavior in context. Typical of the areas of interest might be social implications of technology; shamanism, new religious movements; political and economic changes; legal forms and institutions; cultural creativity; artifact-based theory; nationalism and the state; material culture; person and gender; relatedness and kinship; social development; hunters and gatherers; and many, many more.

Figure 2.6 illustrates how the anthropologist Desmond Morris proposes that humans developed their unique body hair pattern. The model suggests how man transitioned from tree-dwelling to hunter–gatherer, developed subcutaneous fat (uniquely among the great apes) to sustain him between meals, and shed much of his body hair to prevent overheating during the hunt; overheating would have been a threat with both subcutaneous fat and a covering of hair acting as insulators. The figure also illustrates the vital role of pair bonding between man and woman, such that the woman with child, who could not follow the hunt, was assured of food from the hunt. Compared with other mammals, the human female also evolved to make herself more frequently available to the male, who might return at any time from the hunt, leading to selective hair retention under the arms and

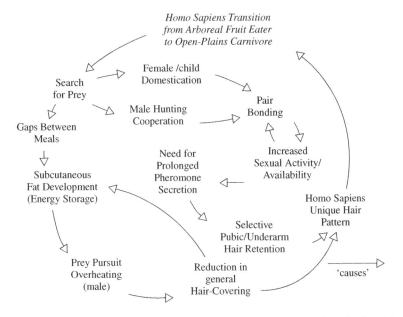

Figure 2.6 The Naked Ape. Causal loop model, showing how *homo sapiens* may have developed the pattern of body hair that is unique amongst mammals. (Hitchins, 1992)

around the genitals for the secretion of pheromones. All of which leads to the formation of the family as the instinctive social nucleus for humankind: families encouraged survival.

Man as the hunter–gatherer is evident in our everyday existences, if we care to look. The supermarket is laid out like so many hedgerows, and the most attractive pickings are just below eye-level, for both hedge and supermarket stall. The chairman of the company, who furiously seeks out new business, even though he is too old for such efforts and should have retired many years since, is acting out the role of the hunter pursuing the quarry.

Technology can also impact on this subliminal hunter–gatherer underlay. The effects of such technologies as television, the Internet and the mobile phone are undoubtedly great, but their full effect may be difficult to gauge in the short term.

The kinds of system of interest to both social scientists and systems engineers have major human content and context. Understanding and anticipating the behavior of people is often core to the achievement of successful outcomes; and, the behavior of people, individually and en masse, is evidently contextually, culturally, politically and economically influenced.

Social Capital

> Whereas physical capital refers to physical objects and human capital refers to the properties of individuals, social capital refers to connections among individuals — social networks and the norms of reciprocity and trustworthiness that arise from them. In that sense social capital is closely related to what some have called 'civic virtue.' The difference is that 'social capital' calls attention to the fact that civic virtue is most powerful when embedded in a sense network of reciprocal social relations. A society of many virtuous but isolated individuals is not necessarily rich in social capital. (Putnam, 2000)

Social capital is a relatively new expression of a concept that is well established — that the value of a social group is more than the sum of the individual parts. Social groups are cohesive, for instance, and that cohesiveness is related to the bonds that form, break and reform between individuals and groups within the society.

The notion of social capital is important in the context of systems science, since social capital moderates the desired effects of laws, changing economic circumstances, cultural upheavals, etc. At a trivial level, for instance, the rate at which a society may change, i.e., their resistance to change, is likely to be related to their social capital. Similarly, the ability of a society to recover from traumatic shock may relate to the degree of social capital. Policemen recognize the value of social capital, and find evidence of it, *inter alia*, in 'neighboring', i.e., the propensity of people to help each other out without external prompting, and in politeness on the roads such as letting people enter from side-roads into a stream of traffic.

Social capital is a whole system concept, and relates to the emergent properties, capabilities and behaviors of social wholes.

Social Genotype

We may be familiar with the concept of the organic genotype: the genetic makeup of an organism. We may think in terms of DNA, as being unique to an organism, with its molecular helix of coupled nucleotides. DNA, although not alive in any real sense, does seem to be eternal in that it is 'handed down' from organism to organism.

Societies may be thought of by analogy as having a social genotype. Instead of nucleotides and molecular bonds, there are social roles and interpersonal bonds. Like the organic genotype, the social genotype may be eternal; individuals may come and go, but the roles they adopt and act out remain largely unchanged, locked in by the relationship with other roles. So, a new manager coming into an organization will find himself interacting with subordinates, superiors and peers, all of whom will expect familiar behaviors from him. If he does not behave 'in role,' and conform to expected norms, he is likely to find himself unemployed — after a period of adjustment, of course, during which he will be expected to 'learn the ropes.'

In this manner, the behavior of a social group may persist from generation to generation, locked as it were into the role–relationship structure of the group or society. By further analogy, the genotype of the social group may interact with its environment to exhibit a social phenotype. Overt group behavior may change according to the environment in which the group finds itself.

To understand this, consider a group of families that have moved from, say, India to Indiana. In their new surroundings, they still form a cohesive group, maintaining their culture, their religion, their beliefs and their social practices. Some of these social practices may be alien to their host nation, such as arranged marriages for instance, and so may be concealed — or at least, not publicized. On the other hand, some of the members of the group may adopt western clothing, western customs and western practices when operating outside of the group in the wider community. On returning to the bosom of the family, they revert to the social and behavioral patterns of their social genotype.

The social genotype, then, contains social memory within its role–relationship structure. This is evident from the traditions, myths, legends and stories that are handed down within societies and cultures; but it also evident, more subtly, in the actual patterns of roles and relationships that, once formed and set, seem thereafter immutable.

It may be possible to bring together ideas of social capital with those of the social genotype. Moreover, for any society to persist, it has to exhibit homeostasis: there has to be a system in place for sustaining the society, for formulating social rules and for ensuring that such rules are observed. Similarly, there has to be a system for establishing and maintaining social assets, resources, foci for social interaction, etc. If these concepts are blended, a social genotype can be perceived in which there is a spectrum of archetypal roles and relationships that bind together to form an integral and essential part of an enduring social genotype with determined social capital.

For example, a society will endure only if it is able to manage antisocial elements from within and without. There will be a need for promulgation of the rules for acceptable and unacceptable social behavior: for the detection, location and apprehension of wrongdoers; for the imposition of sanctions to restrain, deter or reform offenders; and so forth. This appears to be true for any society or social group, including those forming companies, industries, enterprises, teams, brigades, squadrons, gangs, etc.

Managing Complexity

System theory affords the opportunity to manage complexity; that is, to effectively conceal perceived complexity so that our limited intellects can comprehend without being swamped by detail and variety. A system may be represented, not by its structure, function, form, etc., but by its emergent properties, capabilities and behaviors. If the system in question is dynamically stable, then so too will be its emergent properties, capabilities and behaviors, and these may be used to 'describe' the system without need to mention its structure, form and internal functions.

In the general case, and since our concern is exclusively with open systems, the properties, capabilities and behaviors in question will be those that emerge while the system is complete/whole, and dynamically interacting with other systems — this is the basis of the systems approach, in which we seek to understand things as dynamic parts of some dynamic whole, not in isolation. Of particular interest in managing complexity is the determination and manipulation of behavior, where behavior can mean both action and reaction to some stimulus.

Aggregation of emergent properties

It is possible and practicable to simulate the behavior of an open system. This may be done either by representing the object's interacting parts in detail, or by using some form of isomorphic behavior generator, such as a set of nonlinear difference equations, which exhibit the same behavior. In the second instance, the equations serve as a descriptor, or label, for the system and may be employed within simulation models in place of the system and its detailed internals.

If it is possible to represent the behavior of a system in this manner, then it should also be possible to represent the behavior of a second system; and, since such an open system may interact, it may also be possible to represent several systems simply by simulating their interconnected behaviors. This is reasonable within the notion of 'behavior,' since it is classically considered as a response to stimulus. When two or more systems are interconnected, the outflow from one is the inflow to another, and vice versa, so each stimulates the other(s), so preserving the context of 'behavior.'

Figure 2.7 shows the concept figuratively. It a shows a system/whole containing three parts, each represented by their (complementary) emergent properties, capabilities and behaviors, and all mutually interacting. They are also open to the 'outside world,' whence some of their interactions.

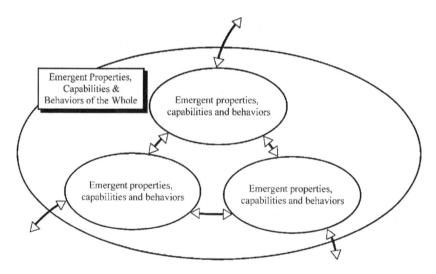

Figure 2.7 Synthesizing emergence. Emergence of the whole derives from the emergent properties of the parts and from their mutual interactions.

The emergent properties, capabilities and behaviors of the whole are derived exclusively from the emergent properties, capabilities and behaviors of the three contained, interacting parts — there is, essentially, no other source.

The figure, and the concept it represents, helps us to manage the complexity of the real world, which might otherwise overwhelm us. Suppose the three contained systems were a marketing department, a production department and a research and development (R&D) department, all working closely with each other in the conception, design, testing and manufacture of a new consumer widget, then it is to be supposed that the internal processes needed to make the whole function as needed might be legion. If we were to follow the activities of individuals, the story might become even more complex. But, if we can represent the departments as shown, we might find that we can use such a model as part of some wider investigation and experimental activity, without any further elaboration or detailing of the parts or the whole.

Anti-chaos

Instead of becoming progressively more disordered, as proposed by the Second Law, some systems exhibit 'anti-chaos,' that is, they develop order from disorder. This phenomenon has been observed in the natural world. The famous Belousov–Zhabotinsky experiment demonstrates that four substances (malonic acid, sodium bromate, sulfuric acid and ferroin), brought together in solution, result not in the normal chaotic mixing that might be expected, but instead form oscillating waves sweeping outwards across the surface with a set period. Instead of the disorder of chaos, this is anti-chaos: order out of disorder.

The Belousov–Zhabotinsky experiment may be a contrived example, but the individual organisms that group together to form slime moulds do so naturally: the phenomena described under the banner of anti-chaos are very real. The manner in which an enormous ant colony, or beehive can behave as a single, superorganism is also anti-chaotic. Human society en masse may exhibit anti-chaos, although we humans might be loath to accept such a notion, jarring as it would with our (illusory?) sense of independence and self-determination.

When a fertilized human egg, a zygote, starts to develop, it divides into cells; over a few weeks these cells differentiate into nerve cells, muscle cells, spleen cells, kidney cells, etc; there are some 250 different types of human cell. Each cell has the same DNA, of course, and the same number of genes — about 100 000 in the human genome. Differentiated cells have some of their genes active, some inactive: active cells make RNA and protein. And some genes can make other genes active or inactive, suggesting a potentially enormous complex of interactions.

Professor Stuart Kauffman of the Santa Fe Institute modeled the behavior of this complex, using an array of light bulbs as a simple analog of genes, with each light bulb being potentially on or off, as each gene might be active or inactive. The bulbs were interconnected, as the genes might be, by a web of connections forming a Boolean network, so each bulb could be switched on, or not, according to a set of rules. Rules might be 'this bulb will light if two other bulbs to which it is connected are lit.'

The array of lights could be set into dynamic interaction by starting from some state in which some bulbs were alight and then progressing through states to see what bulbs lit/went out as a result of the interconnection rules. The objective was to see if there were any stable conditions in which particular bulbs, either individually or in groups, stayed alight from state to state.

With a fully connected set of bulbs, i.e., each bulb connected to every other bulb, there were very few stable states. Paradoxically, when the Boolean network was simplified such that each bulb was connected to only two others, stable states did occur. This was true even if both the connections, and the rules, were allocated at random.

This was, Kauffman observed, a stunning result. For 100 000 human genes treated simply as fully connected on–off switches, the potential combinations would be an astronomical, unthinkable number: 1 followed by 30 000 zeros. Provided with a simple interconnection pattern, with very few interactions, the number of stable combinations reduces to some 300 — which is comparable with the number of different cells in the human body.

Essentially, the simple connection of gene/element/light bulb/etc., to only two others, boxes the potential variety from 100 000 interacting entities into just 300, or so, stable options. This may help to explain the ubiquity of life on Earth, and suggest that life may be expected throughout the Universe. Does it explain, too, how an ant colony, in which each ant obeys only a very few rules, and is in contact with only a few other ants at any time, can operate as a superorganism with a limited set of behaviors? Does it suggest, perhaps, how the neurons in the brain can give rise to thought, awareness and behavior? And, does it suggest that humans living in societies with neighbors, and working in companies, organizations with colleagues in departments, etc., may be governed by the same underlying universal rules of organization?

Systems Life Cycles and Entropic Cycling

Open systems are interconnected in a never-ending, n-dimensional space; truly has it been observed that all things are connected. For those seeking to understand systems behavior, this may present a problem: it would be daunting indeed to have to understand the universe of conceptual system space in order to understand local phenomena.

However, another view of this n-dimensional space sees it exhibiting regions and layers, not unlike a spring mattress that can be held up and shaken; waves of movement flow to and from across the steel-spring mattress, but their mean behavior is bounded in regions of the mattress, as waves of motion move back and forth.

With such a complex backcloth, understanding and predicting open systems behavior would be daunting were it not for the many and varied examples to be observed, particularly in the natural world where all systems are open, interacting, dynamic, etc. We see natural systems come and go, to be replaced by others, often the same, sometimes different, in a continual shimmer of life, death, rebirth, growth and decay.

Searching for a theory to support this observed behavior in both the manmade and the natural world results in a number of propositional principles which will be presented individually first, before being bound into a hypothesis or theory.

Principle of system reactions

If a set of interacting systems is in equilibrium and either a new system is introduced to the set or one of the systems or interconnections undergoes change then, as far as they are able, the other systems will rearrange themselves so as to move to a new equilibrium.

This principle is axiomatic for physical system, and is a restatement of Le Chatelier's Principle, expounded in 1888 in connection with chemical equilibrium. It is also reminiscent of Newton's Third Law of action and reaction. The principle is also applicable to social, political, economic, biological, stellar, or any other wholes/systems.

The principle is noteworthy, too, for what it does not say. It has nothing to say about the *manner* of the rearrangement, its speed, direction, linearity or otherwise. Moreover, in the phrase 'as far as they are able,' it incorporates a let-out term. It may alternatively be stated as '. . . will rearrange themselves as to oppose the change,' which amounts to the same thing as moving to new equilibrium.

A trivial example of the principle might concern a person whose weight had been stable for many years, until he gave up playing squash and rugby, only to find that he put on another thirty pounds within six months before his weight stabilized at this new level. A nontrivial example might concern the level of employment in a factory subsequent to a downturn in business, resulting in reduced throughput and return on sales. These in turn will result in reduced purchase of raw materials, and the need for fewer workers. Once redundant workers have left, the factory may return to equilibrium, but at reduced throughput and manpower levels.

Principle of system cohesion

Within a stable system set, the net cohesive and dispersive influences are in balance.

Again for physical systems, this principle may seem axiomatic, a restatement of Newton's Third Law. However, it is not so obvious in social systems. Consider the hymenoptera, and in particular the honeybees. Honeybees live in a hive with a single queen. The queen exudes a pheromone which is passed around all the bees in the hive, and which exerts a calming influence, such that all the bees get on with their allotted duties. As the hive grows, forager bees have to go further and further afield to find food, having scoured the immediate environs of the hive. At the same time, the hive numbers increase as new workers are hatched out.

A point is reached at which the amount of pheromone exuded by the queen has to be shared between so many bees that there is insufficient per bee to maintain her calming influence, and the time has come to swarm. A new queen takes off with a retinue to start up a new hive, and the process renews.

Not dissimilar processes can be seen in operation within human families and societies. Social bonds may hold the group together, while external influences tend to separate them. Equilibrium persists as long as these two are in balance . . . .

Principle of adaptation

For continued systems cohesion, the mean rate of systems adaptation must equal or exceed the mean rate of environmental adaptation.

This principle may seem rather obvious from the evolutionary biology perspective. It has been suggested that the dinosaurs died out some 64 million years ago, not owing to some meteor impact, but instead due to their inability to adapt quickly enough to falling levels of atmospheric oxygen in the atmosphere. After all, there were many creatures around before and after the dinosaur extinction that survived quite well, including sharks, turtles, crocodiles, beavers (!), and many others. Surely, a meteor strike would have wiped them out, too? No, an inability to adapt seems more likely . . . .

However, the principle applies to all interacting systems, not just prehistoric biological ones. Technological artifacts are a prime example, with manufacturers regularly updating their products, particularly brown goods and white goods, to keep pace with environmental change as well as fashion. For instance, impending shortages of potable water and fossil fuels are driving the evolution of improved reverse-osmosis desalination plants, and of lean-burn automobile engines, electric cars, hydrogen cars, and even cars that run on vegetable oils, rendering them CO_2 neutral — at least in the use of fuel, if not in their manufacture.

Principle of connected variety

Interacting systems stability increases with variety, and with the degree of connectivity of that variety within the environment.

This principle is less obvious. To get a feel for it, consider Figure 2.8, which represents some archetypal open subsystems within systems, with visible interconnected variety. The organic shapes may represent physical systems, social systems, ecological systems, economic systems, or whatever. As the outflow from any subsystem changes, its effects migrate throughout the system and some of them return to affect the initial system change.

The greater the variety of subsystems, the greater the prospect that the outflows from some of the set will match the inflow needs of others in the set — this is complementation. In the limit, there may be so much connected variety that it will prove impossible to invoke any change through the maze of cross- and feedback-connections. This would be a stodgy, overfull kind of stability, were it not for...

Principle of limited variety

Variety in interacting systems is limited by the available space and by the degree of differentiation.

The principle requires explanation of both 'space' and 'degree of differentiation.' Consider an organ pipe. Air in the pipe can vibrate in a variety of modes; the distance between the ends of the pipe,

Figure 2.8 Abstract opens systems (photograph and original painting by the author).

i.e, the available space, limits variety. The degree of differentiation is determined by the physics, which require each vibration mode to be an integer number of half wavelengths.

Variety appears to be limited in many spheres. There is a limited number of ethnic types on the planet: the terms caucasoid, mongoloid, negroid and australoid may be no longer in general use, but they illustrate the limits to variety in racial classification. Race is described today not in observable physical features, but rather in such genetic characteristics as blood groups and metabolic processes; again, blood groups are limited in variety.

Variety is limited in such things as varieties within a religion. Christianity may divide into Roman Catholic and protestant, and protestant may further divide into Methodist, Baptist, and so on, but the list of varieties is really quite short. Predators on the Serengeti Plain are limited in variety: there are lions, cheetahs, leopards. . . types of antelope, zebra, buffalo, etc., are similarly limited.

In general, even where there is seeming-profusion, there is a general limit to varieties within any category.

Principle of preferred patterns

The probability that interacting systems will adopt locally stable configurations increases both with the variety of systems and with their connectivity

As the weave of interactions between systems becomes more complex, it is increasingly likely that feedback mechanisms will arise, and that homeostatic balance will arise, perhaps acting through a series of interacting systems — see Figure 2.8, which illustrates the situation qualitatively.

In the figure, if the large object that looks like a chicken, on the left were squeezed rhythmically, and if all the various tubes were filled with fluids, then the tubes leaving the 'chicken' would pass pulses of fluid pressure to other entities in the figure. These would, in their turn, pass on the pressure pulses, and some of these would return as inflows to the 'chicken.' Looked at overall, we might reasonably expect waves of pulses going out, waves coming back in, and standing waves might be anticipated. Of course, there would be some absorption of the energy as the fluids passed through the pipes, inflated the various 'systems,' and so on, but it is reasonable to suggest that diminishing patterns of behavior would propagate outwards from the source. The greater the variety of systems, and the greater the web of interconnections, the more likely this would be to occur . . . .

Principle of cyclic progression (entropic cycling)

Interconnected systems driven by an external energy source will tend to a cyclic progression in which system variety is generated, dominance emerges and suppresses the variety, the dominant mode decays or collapses, and survivors emerge to regenerate variety.

This is entropic cycling, and it is observed in all complex systems. A classic example is to be seen in Yellowstone National Park (Romme and Despain, 1989). The subject of interest was the relatively rare occurrence of major fires in the forest. Fires were caused, generally, by lightning, which occurred every year, yet major fires were relatively rare, occurring about once every forty years in any given area of the Park. Why?

It transpired that the reason for the relative rarity of major fires was due to ecological succession. Each major fire created a space in which virtually nothing was growing. A few species were

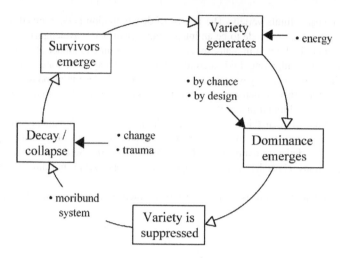

Figure 2.9   Cyclic progression — entropic cycling.

adapted, however, and these started to grow from corms. Birds and bats brought in seeds, and some of these took root where the conditions favored them.

The space encouraged species diversity, so a wide variety of soft-stemmed plants soon started to shoot up towards the sunlight, creating lush green undergrowth. Fast-growing, softwood trees started to grow, mature and then fall into the lush undergrowth, where they rotted down to provide nutrients for flora and fauna, and in particular for the slower growing hardwood trees that took many years to mature. As these hardwoods grew and matured, their canopies progressively shut out the sunlight and the ground level vegetation reduced. The hardwoods also drained the nutrients from the surrounding soil to fuel their growth, further diminishing the prospects for ground level vegetation. Finally, the hardwood trees lost branches, died, and fell on to the now tinder-dry forest floor where they presented an opportunity for the next lightning strike to start a major forest fire . . . .

The whole cycle, from fire to fire, took on average forty or more years. Before the forty-year period had passed, lightning strikes might set trees alight, but fires could not spread because of the damp undergrowth and the moisture retaining plants and softwoods.

And, the whole cycle presents an allegory for entropic cycling in civilizations, companies, enterprises, and complex systems in general — see Figure 2.9. Systems with variety can absorb/adapt to change, trauma and shock. Dominance tends to suppress variety. Once variety has been suppressed, the system becomes vulnerable — it lacks the ability to respond to changing circumstances, should they occur. A system which decays or collapses does not necessarily die, but may recover or re-form — not as the same system, but as something new, attuned to the contemporary context. And, what falls apart is the structure, the interaction between components — but not necessarily the components themselves.

System life cycles: the Unified Systems Hypothesis

The seven principles above have been presented independently; however, they can also be seen as complementary elements in a theory of systems, the Unified Systems Hypothesis, which offers a generic view of how systems synthesize, evolve, stabilize, decay and recycle in a never ending round of entropic cycling. An analogy is drawn between these processes and the human life cycle; a complex system is also said to exhibit a life cycle, in that it has a point of conception, periods of growth and change leading to maturity, followed by cessation of existence in its mature form.

Bringing the seven system principles together creates the causal loop model (CLM) of Figure 2.10. The figure offers insight into how systems evolve and change over time, and most usefully, what are the likely cause of downfall, and what might be done to prevent such downfall . . . .

To appreciate the model of Figure 2.10, start at energy in the top right-hand corner. Energy 'creates' variety, or rather it increases the space within which variety may manifest. We see this in all walks of everyday life: in the variety of housing in richer environs; in the variety of cars in richer cities; in the variety of jobs in cities as opposed to villages; in the variety of species in a tropical jungle compared with a tundra; and so on.

With variety generation, there is increased opportunity for varieties to interact and to react. Where some of that interaction is cooperative, symbiotic, and mutually sustaining, complementary sets of varieties may form. This may be seen as connected variety, which leads to stability owing both to the potential for homeostatic balance and constructive feedback. Complementary sets and connected variety are precursors of stability, or dynamic equilibrium as it might more properly be described in open system terms, i.e., the attainment of steady state. This is holism in principle and in practice, i.e., this is the formation of wholes. Note, there is nothing stated about the manner in which stability arises: it may be linear, chaotic, and even catastrophic . . . .

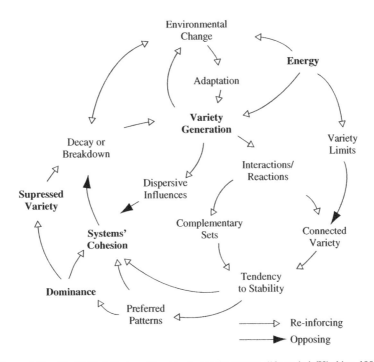

Figure 2.10 The Unified Systems Hypothesis and the system 'life cycle.' (Hitchins, 1992).

As an interacting web of systems forms, it adopts preferred patterns which, although exhibiting high energy, are generally in local energy wells; so, high energy, but not as high as they might otherwise be. And increase in energy would be required to move the system web from one preferred pattern to another; in other words, there is an energy barrier separating preferred patterns.

Preferred patterns encourage systems cohesion, the tendency of the various systems elements to cohere in some way, not necessarily in terms of physical space. This systems cohesion is challenged by dispersive influences, which may be varieties generated by energy and which did not connect to other varieties. In the human body, dispersive influences might be pathogens. In business, dispersive influences might be competition for skilled workers, increases in the base lending rate, and so on.

There is an observable tendency for systems with complementary varieties to encourage one of the varieties to become dominant, to be the leader, to overshadow, to set the rules and limits, etc. this may come about by chance, perhaps, but in the affairs of man it seems to occur as a by-product of male competition. Male-dominated societies require a strong leader. Curiously, female societies seem not to have the same need, although they, too, tend to establish a central coordinator or focus.

While there is no essential need for the dominant member to suppress variety, this seems to be the usual pattern. During times of hardship, in particular, a dominant leader may decide that 'too much variety within the system is superfluous, even wasteful,' and may eliminate some of it. Judging what is 'too much' variety is fraught with difficulty, so a blunt axe is generally employed. As with the forest in Yellowstone Park, a dominant element may simply absorb so much energy and nutrient that other varieties wither. In business, this is referred to by terms such as 'returning to core business,' as opposed to 'establishing a wide portfolio against the vicissitudes of trade variability' which occurs earlier in many large organization's strategic policy.

With a reduction in variety, the system may still appear robust to an external viewer — as the forest of hardwood trees would appear robust. However, the system is vulnerable: when the environment changes, it will lack the variety with which to adapt and respond. In this event, the system will decay or collapse, and its constituents may rejoin the pool of varieties generated by energy, so rejoining the entropic cycle at the start point.

System longevity: system decay

The continual entropic cycle described above is not the only outcome that the Unified Systems Hypothesis proposes. The organismic analogy observes that complex systems, like organisms, often end in spectacular collapse. In the real world, complex systems often seem to collapse in a spectacular implosion, witness the fall of the USSR, and of many major corporations over the years. The model of Figure 2.10 illustrates a mechanism for this explosive/implosive tendency. There is a positive feedback loop in the figure: dispersive influences — systems cohesion — decay or breakdown — variety generation — dispersive influences. In the case of the USSR, once one country had broken away, it joined the throng of international dispersive influences, raising the stakes and encouraging others to break free — with the ensuing domino fall.

However, there are other feedback loops within the figure, which indicate that things do not have to be that way. Consider the loop: variety generation — interactions — tendency to stability — systems cohesion — decay — variety generation. . . . That is a negative feedback, or control loop. It indicates that a system may continue unabated, provided that it does not allow its variety to be suppressed or, if it is occasionally suppressed, to continually renew it.

These notions suggest a rather different basis for the conduct of systems engineering, which in the view of some practitioners is about creating new technological artifacts as solutions to complex problems. Using ideas from the USH as guides, it may be that complex, open systems may be sustained, enhanced, or even destroyed by using connected variety, or the lack of it, as tool or weapon in the armory of the system designer/architect.

The USH is a useful contribution to systems science, in that it is scale independent, system type independent, understandable, and lends itself to simulation and modeling, and in the solving of real-world problems.

Summary

The chapter has looked briefly at systems science: a fuller treatment would have absorbed the whole book and more: the body of systems science knowledge is extensive. Only the main threads that will be of use later, during the development of the systems methodology, have been sensibly addressed. In particular, these addressed the fusion of the natural with the physical sciences. Systems and systems engineering must, of necessity, address all kinds of systems, including natural ones, artificial ones and those comprised wholly or largely of humans — the social, socioeconomic and sociotechnical systems: though, on reflection, a social system is really a natural system: as an apiary or formicarium, so a human society?

The chapter has cursorily addressed the principal physical laws, notably the conservation laws, and those referred to as transport phenomena. Information, it is suggested, is best viewed in pragmatic terms as a nonconserved quantity although energy and entropy considerations still apply as information is passed between systems. Queuing and the analysis of queues are ubiquitous in understanding systems, their interactions, and their ability — or otherwise — to maintain dynamic equilibrium.

Regulation is seen as being significantly less about cybernetics, i.e., negative feedback, and more about the evolution of homeostasis by the balancing of opposing influences. That this is the case arises principally because all systems of interest are open systems, with continuous/continual inflows and outflows; this places emphasis on regulation of the whole by balancing inflows and outflows, as opposed to negative feedback. Cybernetics is an interest within systems science, but its role in underpinning systems stability appears to be secondary.

The need to understand human behavior has raised the profile of social psychology and social anthropology, particularly, within applied systems science. In this, the role of systems engineering, sometimes seen as an adjunct to engineering, is seen in distinct form, since some engineers seem less than interested in psychology, at least in connection with the conception, design and creation of systems. A useful differentiator between engineers and systems engineers, might be that, while the former provides artifacts for people to use or operate, the latter designs systems with people as focal points within the system. Are people inside the sociotechnical system-to-be-created performing functions, or are they instead outside of the artifact-system to be created, as users and operators of such artifacts? That is an important question that will be explored in later chapters.

Also presented in this chapter are systems science models, representing hypotheses rather than established, fully fledged theories. The first of these, the social genotype, draws an analogy between the biological genotype and a social genotype in which, in place of DNA, there are roles and relationships that form and set over a period, to become established. Once set, occupants of each role may come and go, but the role and its relationship with other roles goes on. A newcomer is expected to interact with others in the adopted role through the interaction matrix, and his or her behavior is expected to conform to the established norms; else, the newcomer will be rejected. A group's social belief system is harbored in this role relationship network, with tales, myths and legends being handed on from generation to generation. This appears true of cultures, companies, families, industries, political parties, etc., etc.

A second systems science model is touched upon, briefly — anti-chaos, the appearance of order from disorder. There does appear to by some underlying organizing principle at work in the Universe, presently only dimly perceived and little understood. . . .

The third systems science model is the so-called Unified Systems Hypothesis (USH), which highlights universal systems principles and draws them together into a holistic model of system entropic cycling, that we often refer to as life-cycling, by analogy with our own, organic mortality. The USH suggests the basis for systems cohesion, systems demise, and hence for systems engineering of a rather different variety than that usually pursued. . . on the other hand, it also expresses the basis for holism, the seemingly universal principle of whole formation.

Assignment

You are an international systems consultant, and have been called in to review a government's policy about government computing and processing systems, of which there are very many. Some are concerned with social care and benefits, some with national identify and insurance, some with defense, some with taxation and treasury, and so on; the range is vast. Concern has been expressed about the ultimate vulnerability of the government's computing resource. There are different attitudes being expressed.

One view observes that these various computing systems, although each individually large, are not fully integrated; there is, for instance, no single network connecting, say, defense, with treasury, or social benefits with taxation. Moreover, none is connected to the Internet, and so there has been no need to introduce any anti-viral measures. To the best of their knowledge, no such infections have been experienced. . . .

A second view considers the possibility and risk of some systemic weakness that might affect all the computing systems at once, and which might then bring down the government. The risk, they say, is simply too great to do nothing about it.

You listen to both views and undertake to review their facilities and report back in one week. Using the Unified Systems Hypothesis as your guide, or otherwise, and without actually having to visit any of their sites, identify potential weaknesses in the government's computing systems and strategies, and suggest ways in which they might reduce their risk, if any.

3

Advances in Systems Thinking

Man is only a reed, the weakest thing in nature; but he is a thinking reed.

Blaise Pascal, 1623–1662

Scope, Limits and Values

Systems thinking is a developing discipline. It grew out of the systems approach, in which it was seen as necessary to view systems as open; to view them acting and interacting as wholes within their environment and context. Thus an open system, or whole, is observed, analyzed, understood, or posited, as an interactive part of something greater. The systems approach has brought considerable insight and benefit to those seeking enlightenment in nearly all fields of human endeavor. It owes some of its heritage, too, to operational research, which developed spectacularly during World War II.

Looking at an existing or putative system as an open, interactive part of some greater whole is not necessarily simple or straightforward. The system of interest (SOI) may interact contextually with its environment in various and complex ways, adapting as it does so. System thinking may be viewed as the process of envisioning these actions, interactions and adaptations. For some systems, and for some people, this is an exercise for the mind; systems thinking is cerebral. For others, the complexity may be too great to accommodate mentally, and there may be a need to resort to models. These may be of several types, including, but not limited to:

- Rich pictures — symbolic diagrams highlighting the essential features of some situation or problem space, looking particularly at paths of communication, lack of communication, areas of conflict, and so on. These are associated with so-called soft systems, but may be of much wider application.
- Causal or influence diagrams, usually formed from signed digraphs.
- Dynamic simulations of phenomena using one or other of the many systems dynamic simulation models. These encourage the exploration of the dynamic aspects of problem space, whereas the previous two bullets may describe the dynamics, but cannot simulate it — that must remain in the observer's imagination
- Dynamic models, not of phenomena directly, but of interacting open systems which exhibit properties of their interactions within the simulation

---

Systems Engineering: A 21st Century Systems Methodology   Derek K. Hitchins
© 2007 John Wiley & Sons, Ltd

Rich pictures

Figure 3.1 shows a simple example of one kind of rich picture (there are various formats, some including stick figure of people and sketches of factories, vehicles, counters, etc.).

Simple though it is, the rich picture has a strong message. We try to predict/predetermine the effectiveness of our artificial systems, i.e., the effect one system has upon another. If the rich picture is correct, however, once our systems start to interact with each other, they will be changed in the process as they adapt to inflows, so that effectiveness, instead of being some fixed measure, is largely an indeterminate variable with time.

You think not? Well, if Red and Blue were World War II naval destroyers, pounding each other with heavy guns, then it seems reasonable to suppose that each will experience damage, each will attempt to repair the damage, each will find its capability to respond impaired, and effectiveness will, indeed, turn out to be time variant. An alternative version of this 'rich picture' might show a sketch with two ships, shells flying from each to the other, and perhaps a stick figure of a man on deck with a speech bubble saying: 'How can I undertake damage repairs when the other side is still shelling us?' 'Meanwhile, another stick figure, perhaps the opposing captain, might be saying: 'I don't like this — perhaps we should withdraw and fight again another day.' Conventional rich pictures of this type often highlight personal reactions as a way of expressing conflict, or lack of essential information.

Yet another way of presenting a rich picture is the ubiquitous causal loop model, as in Figure 3.2. In this example, the rich picture highlights the factors leading to the collapse of the Old Kingdom of ancient Egypt. The example shows an early instance of counterintuitive behavior. Successive kings granted lands to nobles, who set up hereditary 'mini-dynasties' of their own, not only depending less on the favor of the king, but even possibly starting to challenge his divine authority. A succession of poor Nile inundations resulted in famine, internecine war between the mini-dynasties, civil war, and the breakdown of the Old Kingdom. (In later Kingdoms, land would be granted to nobles, but

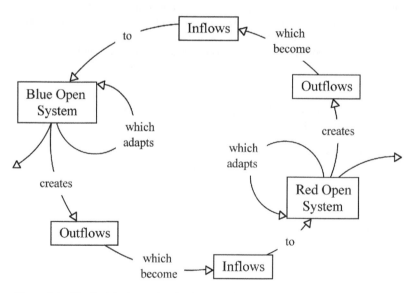

Figure 3.1 Simple rich picture showing how perceived effectiveness may vary with time.

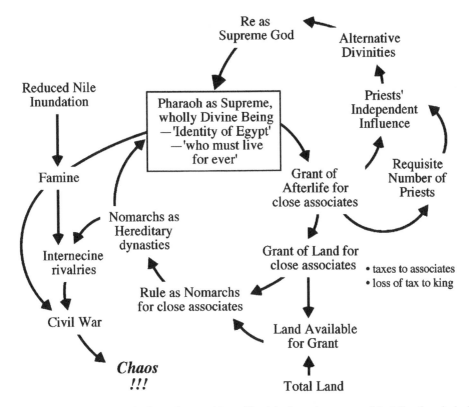

Figure 3.2   Rich picture in the form of a causal loop. The rich picture is a means of depicting the principal features, conflicts, etc., in a situation. In ancient Egypt, the king (pharaoh) reined supreme as a divine being. He granted land to nobles who became rulers of nomes (localities). These nomarchs established hereditary dynasties which fell into civil war when famine the land, leading to the collapse of the Old Kingdom.

only for the period of their life, so that they could not establish independent power bases up and down the country. . .).

Causality and causal loop models

A useful way to explore the behavior of systems is to trace and predict cause and effect. There are different ways of looking at causality — see Figure 3.3. One view tends to isolate one cause and one effect, so that there may be several causes and several consequent effects, but there is no association between them. This is a view seemingly taken by some politicians and government agents. For example, they might (intentionally?) see no relationship between raising the price of rail fares on commuter trains and the concomitant increase of road traffic congestion to and from the same city during morning and evening rush hours. . . .

A different view is often seen by managers and those seeking to impose control. This viewpoint sees a linear chain of causality, in which one cause creates an effect, which, in its turn, becomes a

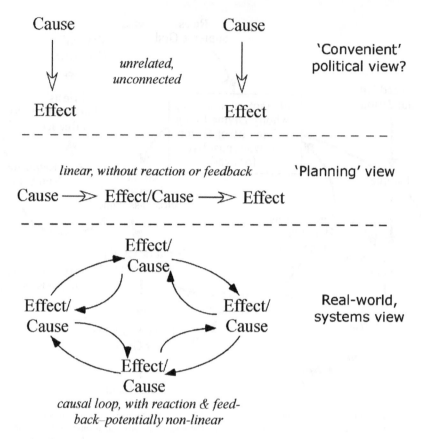

Figure 3.3 Views of causality. Some view cause and effect as singular, one-on-one; this is a reductionist view. Some consider that there are 'chains of causality,' where a cause produces an effect, which is seen as cause, so producing a further effect, and so on. In practice, however, causes not only produce effects, they also provoke reactions, or feedback, giving rise to the third viewpoint — the causal loop, nonlinear feedback viewpoint.

cause producing another effect, and so on. This is the viewpoint taken, for example, by planners, project managers, systems engineering managers, etc., when they formulate plans and when they seek to cost and implement those plans.

The third viewpoint is that every cause creates some reaction, which will either reinforce the cause (positive feedback), or counter the cause (negative feedback). At the most basic level, this is a reiteration of Newton's Third Law: 'action and reaction are equal and opposite.' However, while that can never be wrong, loops of causality may be more complex. Figure 3.4 shows an example of a causal loop model (CLM), which shows a simplistic view of population dynamics. Such a simple model would serve as a seed or nucleus on which to build a more sophisticated model showing, for instance, infant mortality, the effects of famine, nutrition and medical care, etc., as they affected births and deaths.

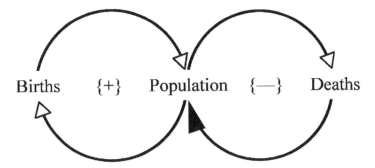

Figure 3.4 Population dynamics. The arrows are termed 'signed digraphs,' and indicate causality. Read open-headed arrows as 'increase(s);' read solid-head arrows as 'reduce(s).'

Note that the CLM in this instance is proposing causality in loop form, rather than illustrating any process. This is a classic open system concept, in that the population numbers are a balance between the opposing influence of births (increasing) and deaths (decreasing).

The CLM is a useful device:

■ It presents a viewpoint simply and clearly, such that an observer is in a position either to understand and to agree, or to challenge, disagree and suggest changes, corrections and improvements.

■ For this reason there may be 'good' CLMs and 'poor' CLMs. A good CLM would be one that is simple, clear and convincing, in that it appeals to reason and logic, and — by virtue of its emphasis on causality — addresses the core of the subject in hand.

■ CLMs are dynamic in nature; rather than presenting a static, structural view, the use of causality results in a dynamic view capturing the essence of change — in the case above, the dynamics of population.

■ CLMs provide a useful basis for dynamic simulation modeling; often, a good CLM can be transformed directly into a dynamic simulation, which can then examine those aspects that the CLM cannot address, such as rates of change. In the example, for instance, there might be an interest in assessing the likely change of population over the next fifty years. This cannot be sensibly assessed from the CLM, but a simulation based on the CLM may be more successful.

Influence diagrams also use signed digraphs. Influence diagrams introduce, in addition to chains of causality, 'influences:' an influence is a factor which may affect something, but which may not be considered directly, or uniquely causal. Influence diagrams may be more complex than causal loop models; causality is harder to justify than influence. CLMs tend therefore to be both sparser, and more pointed.

The CLM presentation, using loops of positive and negative feedback, is also used to present interactive processes. In such instances, they should, perhaps, be more properly called process loop models. Nevertheless, the CLM is valuable in exploring, understanding and anticipating complex problems and behavior.

Dynamic simulation of phenomena

It is possible to dynamically simulate phenomena, without necessarily representing whole systems. This is still generally referred to as systems thinking.

An example of this approach to systems thinking is shown in Figure 3.5. At the time of the Pyramids, *ca.*2600BC, the population of Egypt was about 1.5 million, and some 4000 men were directly employed on building the Great Pyramid of Khufu, ably supplied and supported by food, beer and other supplies from the whole of Egypt.

The figure, which uses the simple STELLA™ notation, shows the Nile producing the annual inundation or flood, on which the early civilization of ancient Egypt depended. The component representing the Nile is part of the weather model according to Edward Lorenz, shown in Graph 2.3 in Chapter 2. This model of chaotic weather has been used to 'drive' the population model, since variations in the annual flood affected the food supply each year.

The dynamic simulation uses the amplitude of the annual inundation to determine the amount of food grown each year. This in turn feeds the population, which grows exponentially under this 'regime of plenty,' until a point is reached, at which the population's needs match farming's capabilities. Thereafter, a poor inundation that results in little food will either slow population growth, or cause famine and major loss of life. This is known to have happened unpredictably in ancient Egypt. See Graph 3.1.

The simulation of Figure 3.5 is evidently the simulation of the phenomenon of self-organized criticality, in that the population was auto-expanding to a critical state, where it perched 'on the edge of chaos,' to use the popular expression. Graph 3.2, a phase-plane chart drawn from the same figures that generated Graph 3.1, highlights the situation. The phase-plane chart shows the variability in population, year-on-year, and, although complicated, the graph indicates that there is a central 'attractor,' i.e., the population may go above and below some level, but always returns from the extremes of low and high, to this intermediate level, as with Bak and Chen's sandpile. The situation was always self-correcting: famine reduced the population, restoring balance between food supplies and consumers; it was never possible for the situation to cross over to the 'region of chaos.'

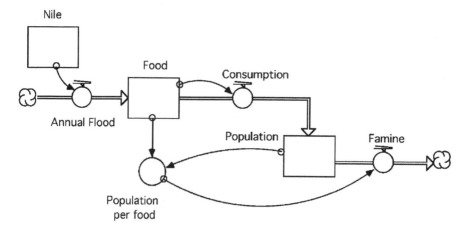

Figure 3.5 Dynamic simulation of the annual inundation of the Nile and its effect on population growth, *ca.* 5000 BPE.

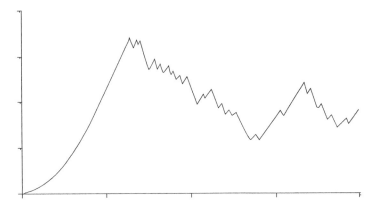

Graph 3.1 Population growth from the simulation of Figure 3.5. This is an example of self-organized criticality.

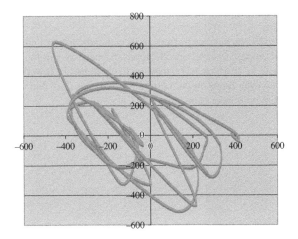

Graph 3.2 Phase-plane chart, drawn from Graph 3.1, showing the variability of the population, year-on-year over a 100-year period — the 'signature' of self-organized criticality?

Dynamic simulation may be conducted for almost any kind of phenomenon. Figure 3.6 shows a dynamic simulation of an archetypal systems engineering process, starting from requirements provided, presumably, by a customer, and ending with the solution system being commissioned. This particular simulation was occasioned to refute the claim by a tool vendor that, no matter how long was spent at the outset of such a process rectifying errors and omissions in customers' requirements, such efforts would not increase the overall project time. On the face of it, this was a nonsense claim. Unexpectedly, the simulation indicated that it was quite correct, however.

Time to clear defects found in requirements varies in practice. Simple, obvious errors are found quickly and easily. As work progresses, further errors prove more difficult to locate. The simulation assumes a square law time distribution: if it takes one unit of time to find 50% of errors, then it takes two units to find 60%, four units to find 70%, and so on up to 32 units of time to find

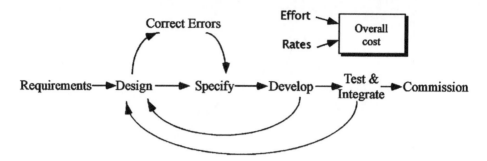

Figure 3.6　Model for dynamic simulation of a simple systems engineering process working from requirements at left to commissioning at right. The requirements contain errors and omissions that take time to correct, with some even going undetected until later in the process where they cause rework to be needed — indicated by the two right-to-left feedback loops.

100% of all defects. 100% may be impossible in practice, but it serves as a benchmark of 'best conceivable performance.'

As the simulation model shows, many errors are found and corrected during the system design phase, from which will emerge systems specifications. Some may slip thorough, however, to be found during development — at which stage, they are likely to invoke rework, shown in a feedback loop in the model. Some even more obscure errors may emerge only during test and integration, where they, too, are likely to invoke re-work.

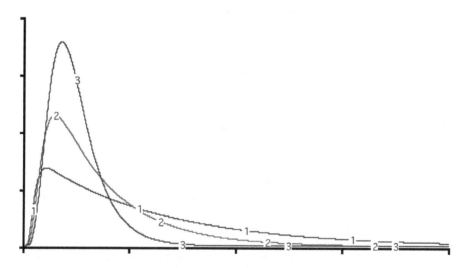

Graph 3.3　The graph shows the rate at which requirements are effectively commissioned. The x-axis is time, while the y-axis is effectively effort. Lines 1, 2 and 3 correspond to increasing design error detection rates Line 3, the 100% line, takes longer to climb due to the increased error detection time, but it peaks higher (implying a greater concentration of effort) and finishes earlier — i.e., completes the commissioning phase of the project. STELLA™ simulation of Figure 3.6.

The results from running the simulation are shown at Graph 3.3. These show that the overall time from 'requirements' to 'commissioning' is dominated by feedback — anything that can reduce, or eliminate feedback will have a beneficial effect on both time and cost. In this case, any time spent at the outset eradicating errors in requirements is well spent, reducing the overall time and cost out of all proportion to the time taken to eradicate errors.

Dynamic interactive systems simulation

There is a difference between dynamic simulation of phenomena, such as the requirements error correction simulation above, and a systems simulation. The latter requires that the essential characteristics of the systems approach be in evidence, i.e., the simulation must present the system of interest in its context of being open to, and adapting to, its environment, including other interacting systems. This accords with everyday common sense: we would not model the performance, say, of car driving through London without due consideration of the road, road surface, cross-wind, other traffic on the road, etc; were we to look at the car in isolation, or in a wind tunnel, we might achieve rather pointless results.

In the general case, there is an implication that we will need a model, both of a specific system of interest (SOI), and of its environment. This view of dynamic systems modeling is organismic: it works for flora, fauna, the body, and for any complex system. Figure 3.7 shows a notional system model in which the system of interest might be, say, the one on the left. The system on the right might represent the whole environment, of which the left-hand system is effectively part. So, in the case of the human body, the cardiovascular system can be seen as interacting with all the other systems within the body, all of which together constitute its environment. Similarly, the pulmonary system can be seen as interacting with all the other systems within the body, which together constitute its environment, including the cardiovascular system. So, perhaps, the SOI should be seen, not so much as interacting with its environment, but as being an intimate part of the environment along with all the other interacting systems. . . .

It is the essence of complex systems that there is a high degree of interaction, and of mutual adaptation, between the parts. Not to represent this in any systems simulation is to risk misconstruing systems behavior.

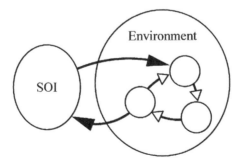

Figure 3.7 Systems modeling. Any system of interest to be examined, explored, understood, or modeled, can be sensibly addressed only when active, interactive with, and adapting to, its environment. For systems modeling, this indicates that any system model should appear only as a part of some larger, wider, containing, whole.

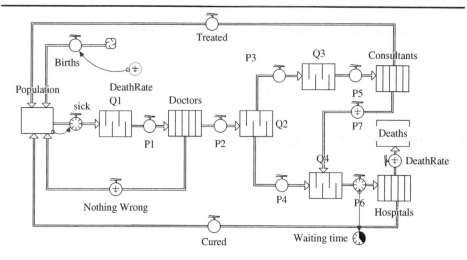

Figure 3.8 Closed loop system model with systems free to adapt to their environment.

Figure 3.8 illustrates the impact of such considerations in more practical terms. The simulation shows a simplified healthcare scenario: doctors treat patients from the general population; they hand some patients off to consultants, who assign some to hospital treatment. Patients either return to the general population in good health, or expire. The birth rate is coupled to the death rate, although the death rate is lower due to improved medical care; as a result, the population is increasing. Notice that there are no inflows and no outflows — this is a closed model of open systems: population, doctors, consultants, and hospitals. The doctors serve as a choke, or control system between the population, the consultants and the hospitals as systems. The inflow of 'sick' persons to the doctors, then, is from the population. But the outflow from the doctors also goes into the population, either directly or through the other systems/agencies. This is true for all four systems — each finds that the others constitute its environment.

This is an important feature of a system model — the system of interest appears as part of its whole, not in isolation. Moreover, the 'behavior' of the system model depends in part on its interaction with the other systems forming its 'environment;' so, those interactions must be in evidence, too, else neither the whole nor its parts will exhibit sensible behavior.

Graph 3.4 shows some results from running the simulation of the healthcare system of Figure 3.8. The x-axis is time: the y-axis is the number of patients seeing doctors at any time. The graph is limited in the y-direction by the number of doctors available, which is presumed constant in the model. Excursions below this upper limit indicate when the doctors are not fully occupied with patients. Note that this amount of 'less than full occupancy' is reducing with time; the doctors are becoming busier as time progresses. This is an inevitable result of improved healthcare, which leads to reduced death rate, to increasing population, to more patients to see the doctors. The price of success in healthcare, then, is an ever-increasing workload. . . .

Behavior modeling

Systems thinking is often coupled with behavior modeling. Behavior is the way something, a system, acts or reacts. As the simulations above have shown, it is reasonable and practicable to

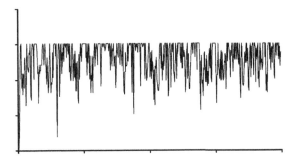

Graph 3.4 Graph from the dynamic simulation of Figure 3.8, showing the numbers of patients consulting doctors. Notice that there is a gradual, but perceptible, increase in the mean number of patients with time. Improved medical care prolongs life and so increases the population and the number of potential patients . . . .

represent and simulate the actions and reactions of systems without explicitly presenting their form or internal functions. This is behavior modeling. It relies for validity upon sensible understanding of the actions and reactions of the various parts within some whole. So, in the healthcare simulation example, how the various systems within the whole function, act, or behave, is well known; there is no need to represent the doctors' surgeries, the hospital wards and the nursing staff, etc. How the whole healthcare system behaves as a result of the interacting population, doctors, consultants and hospital behaviors is less well understood, and is the subject of the simulation.

As this example shows, then, it is possible to bring together the behaviors of a variety of systems making up some whole, to cause these behaviors to mutually interact in closed cause–effect loops, and so to 'observe' the evolving behavior of the whole. In the example, we may even see (simple) emergence in the population increase, brought about by improved healthcare. . . this is reasonable within the definition of emergence: properties, capabilities and behaviors of the whole that are not exclusively attributable to any of the rationally separable parts.

Systems Thinking and the Scientific Method

The scientific method experiments, formulates a theory about how something works, operates or behaves; uses the theory to make predictions about further working, operating or behavior; and, compares the predictions with reality. If the predictions do not match reality, then the theory is wrong. The scientific method can only disprove a theory; it cannot prove it, since it is generally impossible to experience all possible conditions and situations.

Systems engineering seeks to anticipate the future, in that it conceives and furnishes a solution (system) to some problem. The solution system will adapt to its environment (and it is in this *continually adapting condition* that it will be expected to meet its goals and objectives): systems engineering is expected to anticipate such adaptation and create solution systems accordingly. This usually invokes the use of some kind of dynamic simulation to accommodate the complexity, and to address future situations about which there may be only restricted information.

It is not possible to predict the future sensibly: certainly not beyond a near-term time horizon. Beyond that, it may be possible to predict trends, but little more. As we come to understand chaotic systems with their strange attractors better, we may be able to improve on longer-term projections. For the time being, however, systems engineering faces problems in prediction.

No matter how accurate the detail of simulation models may be, they offer no guarantee that the behavior they exhibit will match that of some future real world. This has led some systems engineers to design and build solution systems on the assumption that they will not adapt once in use and operation; and, it has led some systems engineers to abandon modeling and simulation in favor of a simple linear building-block approach to building rigid solutions, in which the linearity encourages them to believe that the solution will behave in its environment just as they designed it to behave in isolation. This is, in effect, abandoning science — and experience — for a more pragmatic approach of meeting the requirements of some customer by creating precisely what the customer asked for. . . which might be considered ideal in business terms — provided the customer does not seek guarantees of systems utility in operation.

A more reasoned approach is to make best endeavors to predict the behavior of the future solution system in its future environment, and to adjust the various (simulated) parts of the solution system so that the whole solution system exhibits the required effects, achieves the required goals and objectives, while interacting with, and adapting to, its future environment and interacting systems. Knowing that accurate prediction is not possible, the kind of simulation model that should be employed is one in which experiments could be carried out — a so-called learning laboratory (Richmond, 1992).

Using a learning laboratory enables the seeker after enlightenment to conduct experiments and to observe the outcome in the simulation. Seekers may observe counterintuitive effects (Forrester, 1971), which could be potentially problematic or advantageous. Since they are experimenting, such findings should serve as warnings or as windows of opportunities rather than precise results. But, not to experiment might be deemed as an oversight at least, and possibly irresponsible.

The simulation model of Figure 3.8 is a simple example of a learning laboratory. Using the model, it is possible to alter the numbers of doctors, consultants and hospitals; to consider what would happen if birth rates and death rates changed, and what effects that might have within the healthcare system; to estimate waiting times throughout for seeing a doctor, seeing a consultant, getting a hospital place, etc. By altering, e.g., the number of doctors working in parallel, it is possible to cut queues and speed up throughput — at a cost, which can also be estimated. In the model, the icon 'waiting time' calculates the mean time from an individual becoming sick, queuing for and being seen by a doctor, waiting for and receiving a consultant's appointment, and waiting for a hospital bed to become available. Waiting time is evidently an emergent feature of the whole.

System thinking in this manner can be reasonably deemed scientific, in so far as the simulation model used exhibits appropriate behaviors, and the experimentation is conducted rationally and logically. It also has the added bonus of falsifiability (Popper, 1972), i.e., a knowledgeable observer can see at once if the model is incorrect, or if the experiments being conducted are unsound, unfair, improper, give unfavorable results, or are insupportable in any real-life context. In systems thinking, it is possible to pursue the scientific method with the only proviso that experiments are conducted in simulation of the real world, since they cannot be conducted in a future, as yet nonexistent, world.

Representing and Modeling Systems

Systems may be represented in several different ways, according to one's understanding of what a system is, and what purpose the representation is to serve. Defining a system as a 'dent in the fabric of entropy,' suggests that bringing disordered entities into an ordered state would constitute forming a system, and indeed some engineers look at systems this way. There is no right or wrong in such definitions, but there is 'more useful' and 'less useful.' It is possible, but not useful, to view a brick wall as a system, but it is a static, closed view; notwithstanding, a fundamental aspect of any system is that it represents relative order, or reduced entropy.

Definitions of 'system' abound. Dictionaries offer 'A combination of related elements organized into a complex whole;' and, 'complex whole formed from related parts.' Wikipedia offers: 'System is an assemblage of elements comprising a whole with each element related to other elements. Any element which has no relationship with any other element of the system, cannot be a part of that system. A subsystem is then a set of elements which is a proper subset of the whole system.' None of these captures the essence of Gestalt, which is at the heart of systems and systems concepts, although they all recognize the concept of 'whole,' which by association with holism, should carry with it the implication of the whole being greater than the sum of the parts.

The Oxford American dictionary offers 'system:

1. *A system of canals*; STRUCTURE, organization, arrangement, complex, network, informal setup
2. *A system for regulating sales*. METHOD, methodology, technique, process, procedure, approach, practice, means, way, mode, framework, modus operandi, scheme, plan, policy, program, regimen, formula, routine
3. *There was no system in his work*. ORDER, method, orderliness, systematization, planning, logic, routine.
4. *Youngsters have no faith in the system*. THE ESTABLISHMENT the administration, the authorities, the powers that be, bureaucracy, officialdom, status quo.'

Little wonder that 'system' is an overworked word. Each of these four 'definitions' is more a list of examples of how the word is used and abused, than what a system is, or might be. Ideas of wholeness, synthesis, etc., are conspicuous by their absence.

A definition from INCOSE (International Council on Systems Engineering) offers:

'a system is an interacting combination of elements viewed with regard to function.'

A chemist might apply this definition to common salt, which is an interacting combination of the elements sodium and chlorine. Sodium, a soft, silvery alkali metal, and chlorine, a greenish yellow gas, combine to form the familiar translucent cubic crystals of common salt — a neat example of emergence. Salt occurs naturally, and just *is*, but people have put salt to many uses: preserving and flavoring food; as natron (a naturally occurring mix of sodium and potassium salts) for desiccating and preserving mummies in ancient Egypt; as freezing mixture, making ice cream and preserving food; and many, many more. This form of definition says less about 'system' *per se*, more about what a system is perceived as doing: no Gestalt, no whole, no synthesis. . . indeed, it is rather hard to determine just what this form of definition either means, or offers to those seeking enlightenment. In short, it does not appear to be particularly helpful.

INCOSE has another definition of system (INCOSE, 2006):

A system is a construct or collection of different elements that together produce results not obtainable by the elements alone. The elements, or parts, can include people, hardware, software, facilities, policies, and documents; that is, all things required to produce systems-level results. The results include system level qualities, properties, characteristics, functions, behavior and performance. The value added by the system as a whole, beyond that contributed independently by the parts, is primarily created by the relationship among the parts; that is, how they are interconnected (Rechtin, 2000).

All of which seeks to define by giving examples, and to describe emergence and hierarchy without mentioning either term. The core definition is given in the first sentence, while the remainder is descriptive narrative of parts, emergence and hierarchy. The final sentence just misses the point that the emergent properties, capabilities and behaviors are created, not by relationships and how parts are interconnected, but by the dynamic interactions between parts.

Overall, however, this definition contains useful notions of 'Gestalt,' 'wholeness,' synthesis,' 'organicism,' and 'value.' This last implies that a system can be measured, and that some value can be put upon it. Evidently, too, this definition of system is concerned with human activity in making, forming, or encouraging the formation of purposeful systems. It excludes, for instance, the solar system, but it includes social systems, human activity systems, sociotechnical systems, socio-economic systems, etc. The inclusion of 'hardware and software,' in place of, say, technology, suggests that this definition was brought about through some form of uncomfortable consensus, with some of the consensual members being engineers concerned with processing. . . the definition seems to be aimed at tangible solutions, and to exclude, for instance, a process as a system.

Previously, a system has been defined as:

> An open set of complementary, interacting parts with properties, capabilities and behaviors emerging both from the parts and from their interactions to synthesize a unified whole.

Such a definition encompasses most of the above definitions, and allows a process to be a system, and the solar system to be a system, too. Defining a system in this way indicates that a system could be represented as, or by:

1. An open set of complementary interacting parts, in the same way as the human body could be represented as a set of complementary organs, or organic subsystems.
2. A set of emergent of properties, capabilities and behaviors which together describe the system 'from the outside,' as it were, and with no indication of the systems internal form, functions or behaviors.
3. A unified whole, i.e., without describing either internals or behaviors, etc. This is more normal practice; we call an airplane 'an airplane,' without going into any detail about how it works, what size it is, how many subsystem it has, or whatever.

The three representations offer decreasing levels of detail. A key attribute of systems thinking, science and theory is the ability to represent a whole system using a simple label, such as its emergent properties. This is the key to managing complexity, by concealing it.

From the various definitions given above, a system may be a process, a procedure, a method, a arrangement, etc. What is less clear, is how one might represent a system diagrammatically, or in a model. Figure 3.9 illustrates the dilemma.

In the figure, the upper diagram might be called a structural viewpoint. It shows a bounded containing system, within which are three complementary, interacting subsystems within a bounded environment. One of the subsystems is shown as containing a further three complementary, interacting sub-subsystems. (In reality, it is likely that there would need to be more than three subsystems to furnish a full complement — three are shown for convenience of presentation.)

The top diagram is known, somewhat irreverently, as the 'poached egg' diagram, for obvious reasons. It indicates that an open, interacting system is comprised of open, interacting parts which are themselves systems, and that these subsystem parts are, in their turn, comprised of open interacting parts which are themselves systems, or sub-subsystems. This is a nesting model of systems within systems within systems, *ad infinitum*; like the proverbial fleas. In a real example, it may show

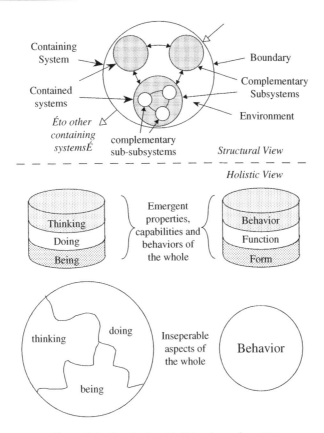

Figure 3.9 Structural and holistic views of a system.

considerable detail: it has the advantage of presenting order, structure and hierarchy to the viewer. It has nothing overt to offer with respect to emergence, unless it is understood that each system, and each system within a system is defined as a system by virtue of its emergent properties, capabilities and behaviors.

The lower diagrams of Figure 3.9 show a quite different view — they offer the viewer *only* the emergent properties, capabilities and behaviors of the whole, without internal detail, structure, etc. From one perspective, the emergent properties of a system may be seen as aspects of being, or existing, doing, and thinking, suggesting intelligence. An alternative view sees the emergent properties, capabilities and behaviors as comprised of form, or morphology, function and behavior; (some analysts propose technology, process and people for appropriate sociotechnical systems)

In each of the center diagrams of Figure 3.9, there is an implication, not so much of structure, but of substrates. For example, being or form is considered as a foundation: it is difficult to envisage process and function floating in space without supporting mechanism, structure, effectors, etc. Doing, or function, then exists on the substrate of physical entities; processes, as it were, travel over the surface of the substrate which offers pathways and energy to direct, enable and support the processes. Not all system have a third layer: thinking or behaving. Where they do, this may be thought of as sitting above the level of doing or function, in the context of responding to external stimulus.

Thinking and behaving systems may respond differently to the same, repeated stimulus according to experience, history and context. This implies some evaluative and decision-making process . . . .

Finally, in the lower figures, a purely holistic view is presented: on the left, the three aspects of being, doing and thinking are presented as irretrievable intermixed, but without any sense of structure. At the right, only behavior is shown. This is reasonable only where the dynamic interactive properties of some system are to be represented in open interaction with other systems, and where the behavior of the system to be represented can be described, and relied upon to remain dynamically stable for the duration of simulation time.

This is not quite as radical as it might seem. If the systems in question were, for example, trained and experienced soldiers either as a platoon or individually, then it might be quite reasonable to predict and rely upon their behavior in a given set of circumstances when they come into contact with some hostile opponent. We would not, generally, consider it necessary to know, in detail, about each soldier, his particular role in the platoon, his physical dimensions, what make of rifle he is carrying, etc. We would hope and expect that the platoon would act as a whole, would engage the hostile force — if appropriate — and would attempt to minimize own casualties. . . .

Similarly, if the system was a collision avoidance radar, part of an avionics system in a transport aircraft, we might have little interest in its construction, the many functions of which it was capable, etc., provided we could be confident of its behavior in detecting, and alerting aircrews to, relevant danger. In simulating air traffic management, therefore, the details of particular radar on one of many aircraft in the 'system' at any time are largely irrelevant; the behavior of that radar would be sufficient for purpose.

Nonlinear Systems Thinking

Many real world systems behave in a nonlinear fashion. Figure 3.10 presents a classic example, of interlinked rabbit and fox populations. Rabbit population increases with a plentiful food supply. A plentiful supply of rabbits 'feeds' the fox birth rate and fox population. Foxes hunt rabbits, increasing the rabbit death rate and reducing the rabbit population, allowing the grass and vegetation to recover. With less rabbits, there will be less food for foxes, the fox birth rate falls, predation rates fall, and the rabbit population will start to recover. And so the cycle continues.

This cycle may be formulated as a set of nonlinear, simultaneous difference equations, to which there is no fixed solution; instead, there are an infinite number of solutions as the two populations

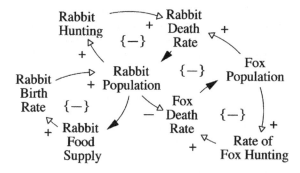

Figure 3.10   Rabbit and fox populations — nonlinear system behavior (Hitchins, 1992).

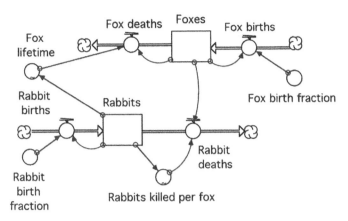

Figure 3.11 Fox and rabbit population simulation.

continuously interact. As Figure 3.10 indicates, it is possible to introduce foxhunting as yet another loop in the model.

Figure 3.11 shows a STELLA™ version of the causal loop model of Figure 3.10; the model has been simplified, and excludes both grass and foxhunting, concentrating instead on the interaction between the fox and rabbit populations. Each population is shown with its births and deaths respectively. Cross-linking between the two populations is shown in both directions: increasing rabbit population also increases fox lifetime, so reducing the fox death rate; and the more foxes, the greater the rabbit death rate.

The results of running this simple simulation are shown in Graph 3.5. As the graph shows, the two linked populations oscillate, some 90° out of phase, with the fox population (line 1 in the graph)

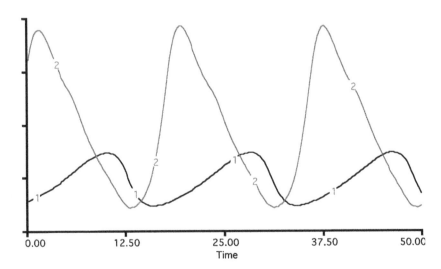

Graph 3.5 Graph showing oscillating fox–rabbit populations from the simulation of Figure 3.11.

lagging behind the rabbit population, line 2. The simultaneous, nonlinear difference equations governing the two oscillating populations are developed in STELLA™ within the dynamic simulation model.

Adding the effects of grass shortage and of foxhunting would be straightforward, but would, of course, result in a more complex graphical result. With such additions, the simulation model might form the basis for a learning laboratory to explore the effects off hunting on, say, vegetation and grassland, when hunting should best be undertaken to benefit the farmers who usually own the land being hunted upon, etc.

There is, then, no single solution to the question of population size; instead, there is an infinite set of solutions. This oscillating behavior is characteristic of much social, biological and physical behavior; such behavior might be viewed as a form of homeostasis, or dynamic equilibrium of, in this instance, the overall system of fox–rabbit interactions. The emergence of such behavior can be counterintuitive, however, which underscores the advantages of dynamic simulation as part of exploring and understanding the problem.

A further consideration stems from the observation that the nonlinear, indeterminate behavior shown in this instance and more widely occurs despite many of the elements in the overall system behaving linearly. This is something to be noted by engineers and other designers who seek to make 'their part of ship,' i.e., their part of some overall design and development strictly linear, thereby expecting the whole to be linear; the whole may well not behave linearly, even where some — many — of the parts do. A simple, biological example might consider a heart-transplant patient who has received a mechanical heart with a fixed rate and volume of pumping blood around the cardiovascular system. Would this steady rate of pumping, enable the patient to act and behave as a normal person with a normal heart? Or, would the linear heart pump impose limitations on the capabilities and behaviors of the patient? That bears some (systems) thinking about. . . .

Summary

Systems thinking is still developing as a discipline: it is a potentially broad subject since it may be possible to think in systems terms about almost anything. The term, systems thinking, is starting to imply some rigorous application of the systems approach, in which a part is considered, not in isolation, but in the context of its containing whole, such that it is open to, and adaptive to, inflows and interchanges with other parts in that containing whole. Systems thinking, the term, is also applied, perhaps with less rigor, to the dynamic simulation of phenomena.

Systems thinking often faces abstract, complex and even obscure issues and problems; methods have been developed to help cope with such vagaries, including so-called rich pictures associated with soft systems methodology (see page 192), causal loop modeling, N^2 charts (see page 370), and many more. Causal loop modeling seeks to identify causal loops, both reinforcing and reducing or opposing, which describe the active and reactive features in any situation, complex system, etc. N^2 charts show interacting entities and their interfaces, which can also be seen as active and reactive pathways between entities. All of these methods seek to capture the essential dynamics, conflicts, synergies, etc., as a basis for understanding the behavior both of the parts and of the whole.

Causal loop modeling (CLM) is notably useful as a precursor to dynamic simulation, as the CLM elements can be transposed almost directly into the model elements. Dynamic simulation similarly can be conducted by examining the dynamics of phenomena, but can be more rigorous when examining representations of whole system-parts interacting dynamically with other system-parts within a containing whole. In such full models, the system part of interest is free to act, react and adapt to the other parts in its 'environment,' and can be enabled to find its natural level, in balance and harmony with the other parts.

One lesson from chaos is that predictions are of limited value, particularly in complex problems and situations. Simulation models that purport to predict accurately may be unable to justify their predictions. This is true of systems thinking and systems thinking dynamic simulations, too. Such models are often about the future, attempting to understand what will happen as a result of some change, the introduction of a new organization, a new process or some new technology. So, are such models useless?

One way to address the problem is to employ the scientific method. This involves conducting experiments, learning, formulating a theory about the problem based on the results of those experiments, and making predictions about the outcome of the problem based on the theory. If the outcome fails to match the theory, then the theory is wrong. If the outcome matches the predictions, then the theory may not be wrong – but can never be proven right.

Dynamic systems simulation models can be employed sensibly as an environment within which experiments can be undertaken, alternative theories tried out, outcomes tested, etc. Such simulations can never prove that the outcome in the real world will match the outcome in the simulation. However, they can identify some theories as flawed, some as likely to be better than others, and so on. With time, experience and further research, systems thinking has the potential to make better predictions; that is, predictions that are wrong less often, and predictions that are less wide of the mark. That potential suggests that it is far better to think, in systems terms, rigorously and to our best ability, than not to think at all. Moreover, the tools and methods for systems thinking are now so freely available that not to think systems might be deemed irresponsible. . . .

Assignments

1. You are advising on the rules to be applied for foxhunting in a rural area of England. The area suffers from rabbits damaging farmers' crops. Farmers also complain that foxes, who would hopefully keep the rabbit numbers down, are often more interested in raiding hen-houses, catching young game birds, etc. Farmers also report periodic increases in the numbers of rabbits, in particular, and that the problem has got progressively worse since foxhunting with hounds was made illegal. You are required to judge and justify:

■   Whether foxhunting should be resumed, with or without hounds.
■   If fox hunting is to be resumed, at what point in the cycle of varying fox populations should the hunting be undertaken to offer maximum protection to the farmers' crops from rabbits.

2. A typical assembly plant takes in parts from suppliers, assembles them into subassemblies, assembles the subassemblies into assemblies, and ships these out to be sold. The money received from the sales goes to pay the workers, train the workers, equip the plant with effective machinery, purchase more parts from suppliers, etc., and pay profits to shareholders, if any. Present this overall process as a causal loop model (CLM), with view to developing a dynamic simulation of the whole system at some later time. Identify what value and insights the CLM provides, and also identify what aspects of the model are uncertain and would require the full dynamic simulation to clarify and expose.

3. A factory follows the practice of producing items, placing them in inventory store and allowing the consumption rate to reduce the inventory; this is called 'production push' since the rate of production bears no direct relationship to the rate of consumption. At the same time, the factory

has been attempting to increase the productivity of the workforce, introducing new machinery and improved training and practices. An unexpected side effect of this increased productivity is that the company has entered a cycle of alternatively hiring and firing workers, as the inventory builds up and reduces. So, both the number of employed workers and the inventory that they manufacture vary cyclically. Not surprisingly, the workers are unhappy with this 'sword of Damocles' hanging over them, and the factory management is unhappy at spending money on training employee only to fire them within weeks or months. Develop a CLM of the situation, explaining how the cyclic variation occurs. Outline a manufacturing and selling scheme that does not result in significant amounts of inventory and which, at the same time, doe not result in cyclic hiring and firing: develop an alternative CLM for this alternative scheme.

4
Systems Engineering Philosophy

All philosophies, if you ride them home, are nonsense, but some are greater nonsenses than others.

First Principles, Samuel Butler, 1912

Why Systems Engineering is Important

Systems engineering has been around for a long time — the Pyramids have been mentioned already, and there are even earlier examples. It may not have been recognized as such, as far as we know, if only because builders at that time did not differentiate between different disciplines; anyone could be an architect, in principle, if he had the will to learn and the talent to practice — and he would not need to be told that he had to do the whole job; that parts alone would not suffice. However, systems engineering really started to be recognized as a discipline during World War II.

There seem to have been several triggers: first, it was recognized that things that were connected could be usefully seen as wholes; even their connecting networks might be seen as entities. Second, it was recognized that systems made up from lots of interconnected parts could sometimes be made to work better, faster, more economically, etc; there was potential to make things more effective by treating them as a whole. Third, it became evident that engineers' methods for designing and controlling technological systems did not work when there were either people in the systems, or the systems were comprised of people. Fourth, systems were starting to become larger, more complicated and complex, with greater variety of parts, increased connectivity, etc., and there were mounting concerns that people knew neither how to design such complicated systems, nor how they might behave in practice. Fifth, it was becoming evident that, correctly synthesized, whole systems could exhibit emergent properties, capabilities and behaviors, affording more than the sum of the parts in terms of capability, performance and effectiveness.

A real stimulus for understanding systems and systems behavior is to be found, at least in part, in disasters. Arthur D Hall III (Hall, 1989) cites the chemical plant leakage in Bhopal in 1986; the explosion of the NASA Challenger space shuttle (1986) and the Apollo fire (1987); the sinking of the Titanic (1912); the nuclear explosion at Chernobyl (1986); and the disaster at the Three Mile Island power plant (1979). He could also have cited the problems facing humanity where some are over-farming land, creating unwanted mountains of food and drink, while others around the world

Systems Engineering: A 21st Century Systems Methodology   Derek K. Hitchins
© 2007 John Wiley & Sons, Ltd

starve; where nations continue to contribute to greenhouse gases in the atmosphere, unwilling or unable to prevent them from building to damage or destroy their economies and environment. He could have cited the emergence and spread of new diseases such as HIV/AIDS, CJD, MRSA, etc., and the reemergence of tuberculosis — not forgetting man's age-old nemesis, malaria.

Systems engineering is important, vital, because it offers the prospect of creating whole systems where man can live, prosper and flourish in harmony and balance with the rest of the environment, where man sees himself as part of, not separate from, that environment. It is that important: but it may not be that simple.

Early Examples Set the Style

Systems engineering is continually evolving around a basic set of tenets and principles. We have met some off the tenets already: synthesis, holism and organicism; principles will emerge later. Some of these ideas appear self-evident — or are they? If a system is large, diverse and complex, is it obvious that changing one part will also change the others, and not necessarily for the better; perhaps only in retrospect?

A retrospective look at the Battle of Britain, and in particular at the command and control system of No. 11 Group in southeast England in 1940 might illustrate the issue; this was before the term 'systems engineering' had been coined, supposedly in 1941.

Battle of Britain

The Luftwaffe on the other side of the English Channel outnumbered No. 11 Group of the Royal Air Force more than three to one. The RAF had an ace in the hole: a new ground radar, Chain Home, which was sited along the south coast of England, and which could detect German aircraft formating over France, preparatory to attack. This new device, about which the Luftwaffe was unaware, could provide essential early warning.

Time was still insufficient, however, to intercept intruders before they reached English shores, so the commander, Air Chief Marshal H.C.T. Dowding, set about 'tuning' the command and control (C^2) system, cutting out delays and speeding up processes. He provided direct telephone lines; he introduced codebooks so that raids could be described in one or two short phrases; he developed map tables with markers being moved in real time to show where friend and foe were located, and their direction of movement; he even redesigned the clock in the control centers with colored sectors such that the markers on the plotting tables could be seen as new, or older and less reliable. He had the pilots sitting beside their Hurricanes and Spitfires at forward dispersal airfields, with the electrical supplies plugged in, so that the fighters could take off at the drop of a hat. While they waited, they were able to listen to relayed intelligence broadcasts, so that they knew what to expect when they did get airborne. In short, he went around the whole command and control loop, tightening up procedures, cutting delays, until the whole system 'came up to speed:' which it eventually did; just.

This command and control system for the Battle of Britain set the pattern for subsequent UK Air Defence Systems, of which there have been several since. Each may have employed the latest, up-to-date technology, but the system design changed little: the original set the style and the standard of what was to become operational system engineering.

NASA's Apollo

If Dowding set the standard for command and control systems, then NASA set the gold standard for space travel in the 1960s. In the general case, every 'new' system is really a revision of some prior system. For instance, we have had air defense since we had hawks, bows and arrows, rocks and spears, shields and caves. . . . Rarely, very rarely, there is a genuine 'green-field site,' and the Apollo missions to the Moon were the first time that man stepped on to solid ground that was not Earth. It was also one of the first instances of the conscious application of systems engineering, and arguably still stands as its greatest achievement: so great that some today are unwilling to believe it ever really happened. . . .

The problem facing the NASA system designers was daunting. At the start, they had no clear strategy for how to get to the Moon, how to land, if there was any firm base on which to land, how to move about on the surface, how to get back and land safely back on Earth, etc. They had no initial idea of how many men should go — should it be one on his own, a pair, three, four. . . ? They did know that there was a limit on the payload that the Saturn V rocket could lift, and they knew that what must be lifted would have to include everything for the round trip.

They developed systems engineering principles and practices that have stood the test of time:

■ Systems engineering requires a clear singular objective or goal.
■ There should be a clear concept of operations (CONOPS) from start to finish of the mission.
■ There should be an overall system design that addresses the whole mission from start to finish. The full CONOPS should be demonstrably realized in the design, step by step. . . .
■ The overall system design may be partitioned into complementary, interacting subsystems. Each subsystem is a system in its own right, and has its own clear mission and concept of operations.
■ The overall system design addresses more than the spacecraft, its contents and the rocket; in addition, the whole design includes the ground control system, the telemetry and wireless systems, the ground maintenance systems, the crew training systems, etc., etc. The whole mission system is all of those, even though only part actively goes on the mission. The parts that do not go on the mission may take a very active part in the mission especially when things go wrong (e.g., Apollo 13)
■ Each subsystem should then have its own system designers and systems engineers who look at the whole subsystem in its containing system (the whole system) and its interactions with, and adaptations to, all the other subsystems (i.e., the systems approach). Subsystem design teams work hand-in-glove with the containing/whole system designers, and with their own subsystem engineers.
■ Each subsystem may be developed independently and in parallel with the others, provided that fit, form, function and interfaces are maintained throughout. Where any emerging deviations are unavoidable, whole system design may be revisited
■ Upon integration of the subsystems, the whole system should be subject to tests and trials, real and simulated, that expose it to extremes of environment and to hazards such as might be experienced during the mission. These would include full mission trials where recovery from defect was possible.

Working in such a clear environment made it easy for system designers to see that any change to any part might affect the overall size, volume, mass, shape, moment of inertia, etc. If such change was inevitable, then complete redesign may be needed, or at the very least a complete rebalance of the affected design parameters. And it was clear, too, that the whole depended on each and every part: a holistic view was essential, but without losing sight of the parts and their interactions.

Their approach was organismic, too, although the term was probably not in use then. The mission system had to operate as a unified whole, with each of the parts having high levels of interaction with the other parts, electrically, electronically, resource-wise, through the operators in mission control, and via the astronauts. The many parts had to adapt to each other, too, if only to keep within tight weight, volume and shape budgets, The end design, with each phase of the mission based around its own craft, and the mission phase craft fitting together like so many Russian dolls, is even reminiscent of an organism.

Is it 'Systems'?

There is an ongoing philosophical discussion about whether systems engineering is engineering or systems. The above two examples show, perhaps, the futility of such arguments. The example from the Battle of Britain was concerned, not with engineering, manufacture, or whatever — that had already been completed. Dowding was working at whole system level, trying to enhance the emergent properties of the whole: particularly, of timeliness. So, he was reducing delays that occurred due to organization, lack of direct communications, use of verbose speech, unclear mapping, fighters taking too long to take off, climb to height, etc., etc. This was really, nothing directly to do with technology — it was to do with process, and particularly speeding up the process. Technology was not the system of interest: the overall process was the system of interest. Some of the process-as-a-system was supported by technology: some was conducted by people (operators, controllers, support staff, aircrew), in the form of verbal communication, map creation, choice of tactics, etc.

Looking at NASA'a Apollo, there was plenty of technology about, so was systems engineering about the technology there? Not really. Again, there was a process, described effectively in the CONOPS (concept of operations). The mission and the astronauts had to pursue that process, step by step, supported and enable by the technology. Getting the CONOPS right was the first solid step on the road to success. What about the design of the tightly packed space launch vehicle? Surely, there was lots of technology in there? But again, the system design was less about technology, more about physics, balance, structure and flow. More fuel needed here, less mass permitted there. More volume here, need to rearrange there to make way.

So, it was a balancing act, rather than classic engineering. The balancing act could have been undertaken with mocked-up blocks of wood, metal, material — the problem would have been largely the same, and so would the solution. . . .

Is it Engineering?

At some point, the design of the whole system, and of the whole subsystem may give way to the need to actually change something — physically. If the something to be changed is a piece of technology, then engineering a new piece of technology, or re-engineering an existing one, may be involved. That this is done within the overall systems engineering process makes engineering either a part of systems engineering, or an adjunct to systems engineering.

However, in many cases, the 'something to be changed' may not be technology. It may be operator training, or a revised process, a reorganized team, a reconstituted communication channel, a new maintenance scheme, all of the above, etc., etc., none of which can be sensibly classed as engineering.

So, not engineering in any classic sense, then. On the other hand, the bulleted list of principles and practices that were devised and employed in both examples above can be applied to a device that is to form a largely technological solution to some problem. If this turns out to be engineering, it is different from classic engineering, which is linear and based, generally, on Cartesian Reductionism. Instead, it would observe holism, synthesis and organicism, and it would take note of the bulleted list above. But, would the outturn be the same? Probably not. . . .

Would you, then, call it engineering? Or, perhaps would it be more reasonable, since it would use the same principles and practices as any other system-to-be-changed, to call it systems engineering?

It is entirely possible, and practicable, to design and develop a technological system, part of some sociotechnical whole, using the systems approach, and the same systems methods and practices that would be used for any open system functioning in, and adapting to, its environment. To do this would be systems engineering. It is, on the other hand, possible to construct/build a technological system, that is to form part of some sociotechnical system, from piece-parts; this too would be systems engineering, in the sense that it would form part of the whole systems engineering process. However, this construction process need not apply the systems approach, nor employ systems methods and practices; classic, tried and trusted engineering practices may be appropriate.

So, it seems there may be two ways of designing and constructing a technological system:

- The classic engineering approach, based on reduction of the whole to recognizable parts, their manufacture and subsequent integration and test; there is an emphasis on functions, form, structure and physical architecture. Emergence may be neither recognized nor sought; instead the objective may to be make the whole equal the sum of the parts. . . .
- and the systems engineering approach, in which the technological system is regarded in its dynamic, open, interactive context, potentially adapting to other systems in its environment, and capable of exhibiting emergent properties, capabilities and behaviors. This approach emphasizes dynamic interactions between the parts, and traces threads of coupled interactions from external systems, through the internal parts and back out to the environment; hence the emphasis is on behavior, function, functional architecture, process and dynamics.

In simple systems, the results may be similar: in complex systems, different. In practice, too, there are other characteristic differences; one such is the concept of what constitutes a system. For many engineers (not all), a system is seen as a technological artifact, which a human may use or operate. For systems engineers, a system is more likely to include the human who is interacting with some artifact to extend and enhance his limited human powers and capabilities in some way.

So, a command and control system might be seen as a set of linked workstations with data communications to a central computing system with databases, or, teams of men working on intelligence, operations, logistics, etc., using networked processors in support. Either viewpoint appears to describe the same thing; both might wish to describe their view as that of systems engineering. The latter view, which focuses on human activities, is, however, the richer viewpoint, since it highlights purpose, function, behavior, adaptability, flexibility, diversity, capability, etc., of the whole, much of which is vested in the teams of people, even without their technological support — which serves to enhance, perhaps, but not replace.

Problem Solving, Resolving and Dissolving

Russel Ackoff suggested that there were three ways of addressing problems: they could be dissolved, resolved or solved (see page 17). To dissolve was to move the goalposts, or to change the situation such that the problem no longer existed. To resolve was to satisfice, to find a solution that was 'good enough.' To solve was to find a 'correct' answer, a best solution.

Systems engineering has been principally associated with the notion of solving complex problems. This is not to say that there is only one solution to a problem; in the real world, there may be many solutions. However, it is to say that some solutions are better than others in the sense that they will fulfill their mission more precisely, more effectively, and that they will satisfy the needs of those with the problem better, too, in terms of affordability, and of restoring harmony and balance where dysfunction and disorder previously existed.

So, the notion exists of a 'best' solution to a problem, where 'best' takes account of the environment, interaction with other elements, adaptation, stability/homeostasis, synergy, and both the 'value' and the 'cost' of the solution. It may be that this 'best' solution to a problem is unattainable, but it nonetheless serves as a benchmark against which to judge solution options.

There is no good reason why systems engineering should not be associated with dissolving and resolving problems, too. Often, dissolving a problem by changing the rules, or the situation, or even the viewpoint, requires an understanding of the whole, an overview if you like. Systems engineering is adept at developing such overviews, and, in searching for potential solutions to problems, should almost inevitably find ways of dissolving problems. Indeed, classically that has been one of the principle tasks for systems engineering: to find fundamentally different, innovative ways of addressing problems. In recent years, the practice has grown of some customers of deciding upon the solution that they want and presenting this *fait accompli* to systems engineering organizations. It is to be hoped that the customers have already examined ways in which they might dissolve or resolve their problem, and discarded them, before presenting their solution to be created. . . .

Be that as it may, systems engineering has traditionally been concerned with dissolving problems, too. Resolving problems may be different. If to resolve is to find a solution that is 'good enough,' then there is an implication that the solution found is less than 'best.' Satisficing in this way is indicative of working with lack of time, or lack of information, such that judgments have to be made. Systems engineering has been traditionally employed in this arena, too, although it may not be so obvious.

The Apollo missions were progressive, in the sense that each of the early missions went a step further than its predecessor, with none of them able to complete the full mission of getting a man to the Moon and back again. There were earth orbit missions. There was a figure-of-eight mission where a craft flew out to the moon and back in a figure-of-eight pattern, making a loop behind the Moon, but without landing. Then there was the first mission that landed, and subsequent missions that further explored the surface of the Moon using a vehicle; and so on. So, each of the successive missions 'took another bite out of the cherry,' until the problem was eventually resolved.

Looking at Apollo in this way suggests that satisficing (i.e., resolving the problem by producing a series of answers, each closer to the final answer than its predecessor), is systems engineering, too. And, after all, this is little different from prototyping or trialing of solutions to find out what works and what does not, what the environmental reactions are, how the solution will adapt, etc.

Systems Engineering: Definitions and Descriptions

So, systems engineering adopts the systems approach to solving, resolving and dissolving problems. Does that define systems engineering? Apparently not.

INCOSE, the International Council on Systems Engineering, defines as follows:

> *INCOSE A.* Systems engineering is an interdisciplinary approach and means to enable the realization of successful systems. It focuses on defining customer needs and required functionality early in the development cycle, documenting requirements, then proceeding with

design synthesis and system validation while considering the complete problem. . . . Systems Engineering integrates all the disciplines and specialty groups into a team effort forming a structured development process that proceeds from concept to production to operation. Systems Engineering considers both the business and the technical needs of all customers with the goal of providing a quality product that meets the user needs.

This definition is, perhaps, more about the 'how' than the 'what.' Stripping out from the statement INCOSE's view of their particular methods, INCOSE defines systems engineering as '. . . (a way of) realizing successful systems. . . .'

Walden University does not define: instead, it offers the following description of one of its courses:

The M.S. in Systems Engineering program prepares students to solve complex multidisciplinary problems in the real world, With an up-to-date understanding of the methods of analysis, synthesis and design, graduates of this program are equipped to be leaders in industries as diverse as aerospace, agriculture, manufacturing and medical device technology. . . .

Evidently, Walden University takes a broad view of systems engineering as complex problem solving in the real world.

Sheffield University offers the following description of their systems engineering course, within the Automatic Control and Systems Engineering Department:

Many systems in daily use are relatively complex in nature and include mechanical and electronic components, computer hardware and software and possibly additional sub-systems. Systems Engineers study the whole integrated system rather than one particular component within it. This involves modeling, design, analysis and implementation and in particular ensuring that such a system is designed so that all its components interact together in an efficient way to achieve specific and meaningful objectives.

Sheffield University looks upon systems engineering as the synthesis of a complex whole system. It concerns itself particularly with synergy.

Virginia University, Department of Systems and Information Engineering, offers the following introduction to their graduate systems engineering course:

An integrated introduction to systems methodology, design, and management. An overview of systems engineering as a professional and intellectual discipline, and its relation to other disciplines, such as operations research, management science, and economics. An introduction to selected techniques in systems and decision sciences, including mathematical modeling, decision analysis, risk analysis, and simulation modeling. Overview of contemporary topics relevant to systems engineering, such as reengineering and total quality management. . . .

Virginia is concerned with systems methodology, highlighting design and management, with professional and intellectual discipline, so with systems of all kinds.

INCOSE offers a second definition as follows:

INCOSE B. Systems Engineering is an engineering discipline whose responsibility is creating and executing an interdisciplinary process to ensure that the customer and stakeholder's needs are satisfied in a high quality, trustworthy, cost efficient and schedule compliant manner throughout a system's entire life cycle. This process is usually comprised of the following seven tasks: State the problem, Investigate alternatives, Model the system, Integrate, Launch the system, Assess performance, and Re-evaluate.

INCOSE's first definition (A, above) identified realization of a successful system, while this second definition identifies the creation and execution of a process as its goal: one identifies with the end, the other identifies with the means to that end.

Is it possible, reading through the various descriptions and definitions and elaborations, to detect a common thread running through them? Is there one systems engineering, or many? Or, is there, perhaps, a common systems engineering philosophy that guides each of the definitions and descriptions presented above?

Looking through the various offerings, certain features recur:

- **Wholes.** Systems engineering is about looking at wholes, understanding wholes and creating wholes.
- **Synthesis of the whole from complementary parts.** Repeated mention of 'interdisciplinary' or 'multidisciplinary,' indicates that parts differ from each other, but are encouraged to operate harmoniously — hence complementary. Also implicit in 'integrated.'
- Finding answers/solutions to 'whole problems:' this feature recurs explicitly or implicitly.
  - Addressing the whole problem is entirely rational, since it can be shown that addressing only part of a problem can result in counterintuitive responses, such that the whole problem becomes worse, rather than better.
  - Politicians, amongst others, often appear to misunderstand this, and address symptoms rather than the root problem: or, is this expediency?
- **Analysis,** in the sense of detailing constituent parts and how they relate to, and interact with, each other.
- **Design,** whence innovation.
- **Complexity** and its accommodation.
- **Discipline and science.**
- **Integrity:** quality, risk, etc.
- **Planning,** decision-making.

So, it seems that there may be one systems engineering philosophy underlying a seeming plethora of definitions and descriptions. The seeming differences emerge from situating the basic philosophy in different disciplines: situating systems engineering philosophy in an aerospace industry results in an apparently different definition of the discipline from situating it in, say, control engineering, agriculture, commerce, economics, social services, healthcare, etc. Despite these surface differences, then, there does appear to be a common notion of systems engineering, that it is a discipline, that it is about addressing whole complex problems and about finding complete, whole, solutions to those problems.

Is it, then, possible to provide a general definition of systems engineering, i.e., one that is not situated in some particular industry or sphere of activity? Lexicographers are good at addressing this kind of problem. Chambers dictionary defines architecture, a particularly complex subject with obvious parallels to systems engineering, succinctly as 'the art or science of creating buildings.' This avoids the trap of describing the 'how' of architecture, which would take many books and on which few pundits would agree.

The definition of architecture suggests a useful template for that of systems engineering: not how it does it, which would also take many books upon which pundits would not agree, but what systems engineering *is*.

Systems engineering is the art and science of creating whole solutions to complex problems.

A glance back at the definitions and descriptions above will indicate that this simple definition does indeed encompass all of them, although pundits within particular industries or disciplines may find it lacking specificity relevant to their application. But then, like architecture and the systems approach, systems engineering is a broad, philosophically based, universally applicable discipline, with its dedicated theory, science, tenets and principles. None of these need be changed by the environment in which systems engineering is applied, although the words used, the descriptions, the methods, etc., will properly and inevitably adapt to that environment.

Looking at this definition, and recalling the ongoing discussions about 'systems' or 'engineering,' there is a possibility that systems engineering might have something in common with, say, doctoring, or police detective work. Doctoring seeks to solve the problem of what is wrong within a highly complex system (the human body, or animal body for veterinarians) and to identify and apply a cure. Police detection seeks to find out what really happened, and to piece together a trail of evidence leading to the identification, prosecution and conviction of a perpetrator. Systems engineering similarly explores and unravels the real problem within some complex whole and seeks to conceive and manifest some way of 'curing' the dysfunctions: and like doctoring and detection, the methods and processes, can be applied to a wide variety of different problems, situations, types, categories, etc. The analogy suggests, too, that domain knowledge and experience will be important.

This catch-all definition of systems engineering combines art and science; art is very much in evidence during the innovative, creative stages of conceptualizing potential solutions to problems, while science, logic and rationale are in evidence in the approach taken to conceptualizing, to problem solving, to planning, to detailing, validating, etc. It has to be said, however, that many practicing systems engineers care little for art or science, recognizing neither in their everyday work. When painting the Sistine Chapel ceiling, Michelangelo might have been more concerned with adjacent colors running and brush strokes showing, than with the overall picture — which, after all, he could not see from close up.

The Real Objectives of Systems Engineering

Given this catch-all definition, it follows that the objective of systems engineering is to create a whole solution to some complex problem. This supposes that the problem can be identified, and that it is solvable (resolvable or dissolvable) — not all problems are. So, it seems that systems engineering has several objectives:

- to scope the problem space;
- to explore the problem space;
- to characterize the whole problem;
- to conceive potential remedies;
- to formulate and manifest the optimum solution to the whole problem, that is, the best solution achievable in the situation, constraints and circumstances;
- hence to solve, resolve or dissolve the whole problem

The notion of 'optimum' can be seen, for example, in the Apollo mission designs, where a balance had to be drawn such that all the various parts could be fitted within the volume, shape and mass constraints imposed by lifter capacity, aerodynamics, etc. The optimum solution was one in which all the essential parts were present, in the necessary amounts, and in the best configuration possible within the constraints. The optimum solution was, then, the one best able to achieve the mission.

Regarded from a different viewpoint, this means that an objective of systems engineering is to select, bring together, and cause to interact, parts such that the whole exhibits requisite emergent properties, capabilities and behaviors; methods for achieving this are core to the skill of systems engineering.

The real objectives of systems engineering can be seen in another sense, and at a higher level. Systems engineering seeks to create systems of many different kinds that work harmoniously together in the service of, and to the benefit of, mankind.

Strategies for Solving, Resolving or Dissolving the Problem

It is often considered that systems engineering solves problems by creating new systems, comprised largely of operators working with technology. To be sure, there are many such instances, but there are other ways of solving, resolving or dissolving problems. Some of these are shown diagrammatically in Figure 4.1.

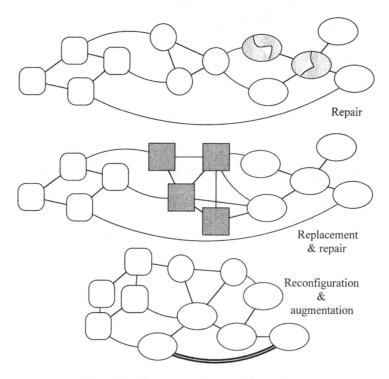

Figure 4.1 Notional strategies for solving problems.

Complex systems may exhibit dysfunctional behavior for a wide variety of reasons. One is that some parts are not working/operating, as they should, as they once did, or as the evolving problem situation requires that they should, e.g., to maintain stability, to meet demand. The simplest, straightforward solution is to repair the inadequate parts. This is shown conceptually in the upper diagram of Figure 4.1, where the shapes are suggestive of part of systems, while the lines are suggestive of paths of interaction between the parts. The parts may be tangible entities, they could equally be activities in a process, they could be people undertaking tasks in a shop, or whatever.

In the central diagram of the figure, the defective or inadequate parts have been repaired, but now the three linked circular parts have been replaced by four linked squares. This represents the strategy of solving the problem by introducing a new system. In general, introducing a new system amounts to replacing an existing one, although it may not always seem so to those creating the new system. So, the team of designers, technicians and engineers concerned with introducing a brand-new management information system (MIS) into a large corporation might not consciously realize that there had been a management information system in place since the corporation formed: it could not have formed or survived without one. It may have been the case that staff that used telephones, filled in ledgers, wrote reports, etc., constituted previous management information systems. The new, shiny MIS will most likely find itself competing with the old, 'handraulic' version, as the corporation executives cling to their experience of tried and trusted methods — and people.

Similarly, a new computer-based accounting system may be taking over from a room full of men with ledgers and quill pens. A new digital news service may be replacing an old analog one, which replaced teleprinters which replaced the old town crier. . . there is, in truth, very little that is new under the sun, and new systems are invariably replacing something that existed before, no matter how different, inadequate and antiquated it may have seemed.

A third strategy for solving system problems is to augment, i.e., to leave the existing system in place untouched, and still performing its role, doing its job. In some arenas, this augmenting might be referred to as 'jury-rigging,' while in others it would simply be increasing capacity by enhancing the degree of parallelism, perhaps. It is represented in the lowest diagram by the additional double interaction link.

A fourth strategy, also shown diagrammatically at the bottom of the figure, might be to reconfigure a problematic system, i.e., to rearrange entities and relationships, but without necessarily altering any of the entities and/or relationships per se. A topical example of this might be the rearrangement of parts in a depot so that picks could be undertaken with less effort and less distance traveled. (Picks refers to the process of collecting all the parts from various bins to fill some 'shopping list,' or order. According to how the bins are laid out, and what is stored in each location, any one pick might take more or less time. The problem, then, is to minimize the time for the 'average' pick, so making the whole process more efficient and timely.)

A second example might consider the practice of arranging all the workstations associated with a particular sequential process in a straight line, so presenting the operators with a clear view of the chain, and enabling them to see their role in the context of all the others. This, so-called 'streamlining' approach minimizes the distance traveled by work-in-progress (WIP), speeds up the end-to-end process, reduces the amount of WIP — which costs money — and so potentially improves both efficiency and effectiveness.

All of these, and more, might reasonably be considered to be systems engineering. Moreover, the replacement strategy might take place in stages, and so effectively be satisficing, with each bite of the cherry moving further towards whole solution. (That approach evidently runs the risk of provoking counterintuitive responses during the ongoing process of providing each part-solution. . . .)

Self-organizing Systems

Some systems are self-organizing, that is, they develop their own internal organization — processes and structure — without external direction, control or influence. The Sun is self-organized into three concentric spheres, with heat energy being progressively transferred from the hot core by radiation and convection, to the surface.

Self-organization can be observed in much humbler circumstances. If a beaker of hot coffee is allowed to settle for several minutes, before carefully adding creamer and sugar then, if the conditions are right, the creamer, instead of mixing, may form into small cells, about the size and shape of an ear of corn (maize). These small cells form into concentric rings on the surface of the coffee. Usually, the effect is fleeting, but occasionally several rings form, before unexpectedly and suddenly, the cell walls collapse and full mixing occurs within seconds.

Human activity systems (HASs) may be self-organizing in some circumstances. Groups of people, 'thrown' together and motivated by fear, reward, or a desire to achieve, have been observed to form into teams to take on different tasks that the whole group feels need to be undertaken. Moreover, unlike managed organizations, such teams changes their size and composition according to the size and duration of the tasks they undertaken with individuals migrating between teams as the need is perceived.

A second example of people self-organizing is to be seen at cocktail parties, or pre-prandial gatherings, where people socialize before moving on to eat. Initially the assembly room is empty. As people arrive, they form into small groups, talking. As more people arrive, more groups form, but the size of each group tends to reach a steady state, so that newcomers do not join a fully formed group, or if they do, a current member leaves to join another group. Eventually, the room is filled with people in equally sized groups, with individuals flitting from group to group ('circulating'). The reason behind this phenomenon appears to be simple: as more people arrive, the background noise of people talking rises. Anyone within a group trying to hear others in the group talking has to be able to hear over the background hubbub. As the group size increases, there comes a point where the diameter of the group is too great to permit audibility, so people will leave to join another group. In support of this hypothesis, group sizes tend to be smaller if there is an orchestra playing; this increases the background noise, such that only smaller diameter groups permit audibility.

It is a moot point as to whether, or not, bringing a group of people together and encouraging/motivating them to self-organize could reasonably be described as systems engineering. . . however, there remains a suggestion that self-organizing systems may be well ordered, adapted to the tasks, efficient, flexible, and effective, particularly in changing, or high-pressure, situations.

System of Systems

A system is made up from parts, often themselves systems: they are generally referred to as subsystems. To be a subsystem is to satisfy the definition of system, i.e., to be a whole, and to be made up from complementary, interacting parts. In this sense, then, every system is a system of systems, although the term has rarely been used, since it would be a tautology.

It has been the practice for many years, for instance, to form an avionics system for a modern aircraft by synthesizing it from various extant, off-the-shelf systems: navigation; automatic flight control, attitude sensors, pitot-static sensors and instruments, communications, radars, altitude sensors, stores management systems, etc., etc. These might be made to work as one by interconnecting them through special interface units and by introducing a common data highway and some central processing system to collect, correlate and coordinate information sources and sinks. Would an avionics system developed in this way be termed a system of systems? It seems unlikely.

In recent years, the term has arisen to mean something rather different. A number of extant, operational systems, often businesses or enterprises, may be brought together under one umbrella, and referred to as a system of systems. But is the term valid?

For a system of systems to exist, there would have to be a whole. The various system parts, effectively subsystems, would interconnect in some degree to create synergistic interactions, such that the parts worked together, cooperated and coordinated their activities with some degree of unity.

So, a system of systems would, to deserve the epithet, display the basic characteristics of any system. In forming a system of systems, by bringing together extant systems, it is unlikely that full and complete interaction would be attained, such as would be necessary to produce optimum emergent behavior. Instead, it is to be expected that there will be limited interaction. This does not mean that an overall system cannot be formed; rather that it is unlikely to be optimal. However, it may well proffer significant advantage in terms of synergy, efficiency, and effectiveness.

An example of a 'system of systems' might be a city-wide integrated transport system, bringing together under one umbrella the previously independent transport systems: intercity railways, suburban railways, underground railways, buses, ferries, riverboats, taxis, coaches, etc. Simply calling disparate transport elements an integrated system would not make it so (politicians take note). To be a system of systems, the various elements would need to be 'harmonized,' in several respects:

- Timetables would be harmonized so that the flow of people was not interrupted by having to wait when disembarking one mode of transport before embarking on another.
- Interchange points between modes of transport would be constructed/improved to facilitate passenger flow and comfort.
- Ticketing and charging would be made seamless, so that passengers could buy tickets that would carry them across all the transport media.
- Capacities of each of the modes of transport would be adjusted to preclude queues forming at the various entry, boarding and interchange locations.
- Possibly, measures would be required to control or reduce road congestion caused by private cars and other vehicles. For instance, deliveries to shops might be permitted only during the night and at weekends, when passenger demand for the integrated transport system was low.

Evidently the various modes of city transport can, in principle, be welded into a whole, and one with emergent properties, too: in this case, properties of reduced mean journey time over the whole system, increased capacity of the whole system, increased simplicity of perception and of usage by passengers, and hence of an overall improved 'commuting environment.'

There are many other examples that might reasonably be termed 'systems of systems,' although we might not think of them in that way. For instance, the criminal justice system can be viewed as a circular pipeline system of systems, consisting as it does of the following systems:

- policing system for the detection and apprehension of suspected criminals
- custody system to hold suspects during investigation
- crime information system
- trial and judgment system;
  - prosecution system;
  - defense system;

- punishment system;
- rehabilitation system;
- probation system;
- policing system for the detection and apprehension of suspected criminals.

The policing system appears both at the start and the end of the list of systems: many wrongdoers are recidivists, going around the criminal justice cycle not once, but many times–hence, a circular pipeline. Although this overall process of crime, detection, punishment, probation. . . is well understood to be a whole system, it is generally not treated as such, or even viewed as such, by governments. That this is true is evinced by the fact that increased spending in one area, e.g., rehabilitation, would reduce spending in others, e.g., recidivism, and hence the numbers in the whole loop. Which perhaps explains why the various elements of the cycle are paid for generally out of different government purses, and why the whole is treated only as separate parts.

Despite this, the whole does have emergent properties: the number of supposed wrongdoers in the whole system; the sum cost of the whole (which is generally not publicized); the removal from society of 'undesirable elements,' with the consequent feeling of security and justice that society supposedly enjoys. Evidently, the criminal justice system is not so much separate from society, but a controlled and regulated compartment within society.

Other examples of the notion 'system of systems' are to be found in defense, where an army might be made up from infantry, armor, cavalry, engineers (sappers), intelligence, etc., etc. Clearly, the whole is made up from complementary, interacting parts, and the whole exhibits emergent properties, capabilities and behaviors of strength, coordination, synergy, mobility and many more, recalling the principles of war. . . . So, an army is a whole, and is a system within any sensible definition. That this is true would be particularly apparent if one of the principle 'subsystems' were absent, e.g., an army would hardly be an army without infantry.

There are various arrangements where the term 'system' does not appear justified, however. Several like enterprises may be categorized with each other under some umbrella title, or holding company. Is this a system of systems? It cannot be described sensibly as a system if the parts do not interact, do not cooperate, do not coordinate, and do not, together, exhibit properties of the whole. That is not to suggest that there is no potential advantage; there may be some economic or marketing advantage, for instance. However, the arrangement would not sensibly be a system as such.

Similarly, it is to be expected that any system of systems worthy of the tautology would exhibit emergent properties, capabilities and behaviors of the whole, i.e., not exclusively attributable to the rationally separable parts, i.e., the constituent systems.

Why is it important to identify whether or not a grouping of systems forms a system of systems, or not? For the important reasons that, if the grouping is really a system, then it has the potential to be in harmony with its environment, to adapt to that environment, to be more than the sum of its parts in terms of capabilities, performance, efficiency, effectiveness, to exhibit synergy, homeostasis, dynamic stability, to evolve, and so on. To mis-classify, say, a family of systems as a system of systems, is to misunderstand, and to form unreasonable expectations of capability, performance and outcome.

Bottom-up Integration

It is possible, and often practicable, to synthesize technological systems by bringing parts together and making them work together through interfaces. This is sometimes called 'bottom-up integration.' For instance, the average personal computer (PC) can be formed in this way, using various parts that are readily available in the market. By choosing between variants of available parts, PCs with different capabilities — processing speed, capacity, overall size, etc. — can be created.

Bottom-up integration, as the example suggests, is a useful, pragmatic approach to synthesis where the parts are understood, where they do not adapt significantly to each other when interacting, and — in short — where the synthesis process employs the so-called building-block approach.

For instance, a new radar might be synthesized by bringing together a transmitter, a receiver, an antennae, etc., each of known and well-understood characteristics, and so producing a new radar.

Bottom-up integration is not without risks, which can become evident when, for example, software packages are employed such that only part of a whole package is required. A particular application might require only one module of, say, a complex word-processing program. Counterintuitive problems may subsequently arise when that module is used as part of another system; one reasons for such problems might be that the module depends for its good operation on interactions with other parts of the original world-processing program, no longer accessible. In such a case, it would be prudent to include the whole of the program, even though only part of it was apparently needed. . . .

In the general case, interactions between the parts, which are expected to be linear and consistent, may result in counterintuitive results. Emergent properties, including potentially undesirable ones, are due to interactions within and between the parts, including energy, information and timed interactions. Sometimes the building-block approach can overlook such interactions, such that the 'whole wall' may not behave as anticipated. Owners and would-be builders of PCs may be painfully aware of the limits, as well as the advantages, of the building-block approach.

However, bottom-up integration, properly planned and executed, may reasonably be viewed as systems engineering. Some organizations go to considerable lengths to produce discrete building blocks that function linearly and predictably, which have standard interfaces, which do not adapt to inflows or outflows, and which exhibit, for most practical purposes, immutable transfer functions. Their view of systems engineering might be that the system they are creating is not so much the individual parts, but more the standardization and interchangeability of parts so that they can be readily put together in various ways to achieve different objectives, i.e., they are creating a construction system. And it seems to be a valid viewpoint. . . one with which the architects and builders of the Great Pyramid, with their regular, rectangular blocks of limestone, would have sympathized.

Completing the 'Whole' of Systems Engineering . . .

Earlier, it was noted that there was more to Apollo than the mission system and the mission control system. The whole system included the ability to 'self develop,' or evolve over an extended series of missions, from the first orbital steps through to the full-blooded Moon exploration missions. If we look at major defense and other national projects, we may see that the initial system as delivered is sometimes obsolescent at delivery; or, if not, may soon become so. Large projects and complex systems take time, sometimes decades, to conceive, design, develop, make, test, prove, deliver and work up. During this time, the original problems that the project was designed to solve might well have morphed; the operational environment may well have changed, too. So, the delivered solution system has a good chance of being outdated and outmoded at the point of delivery.

It is also the case, as with Apollo, that there is much more to be delivered in conjunction with a typical mission system (an aircraft, a ship, a tank, or some other platform). Additionally, there will be a concurrent need for a maintenance and servicing system; a repair and replacement system; a crew selection, training and continuation training system; and so on.

Not so apparent, however, but just as real, there has to be a system in place for reviewing the original problem, as now morphed, and evolving the design of the solution system such that it is better able to address the problem as it presently is, not as it was. In defense and other government circles, the outcome of such deliberations is generally a midlife update, a refit, or similar. The facility that undertakes this re-examination of the problem, redesign and rebuild is conventionally considered to be separate from the mission system and all the support systems; but; should it be separate, or is it more properly part of the whole?

For the whole system to include within it the ability to adapt and evolve would be more in keeping with the organismic principle, i.e., would result in a more unified whole. This unified whole would be open, it would exhibit homeostasis, it would adapt to its environment (i.e., exhibit intelligent behavior), and it would in principle have increased longevity, increased harmony both internally and externally, and emergent properties, capabilities and behaviors that kept in line with those required to address the evolving problem.

The concept of system viability is illustrated in Figure 4.2, where a system is seen as viable if it is able to sustain itself. An open system is internally dynamic as it receives inflows, distributes the inflowing resources, converts some of them to work, disposes of waste, and creates product, etc., as outflows.

For these archetypal processes to execute, the interacting parts within the system have to work in some degree of harmony, to cooperate and coordinate: this is marked as synergy in the figure: no synergy, no viability.

To remain viable, a system also has to survive, which generally implies that it has to be able to avoid, neutralize, or accommodate external threats.

Similarly, since it has working parts, it has to be able to maintain itself in the face of internal threats, i.e., detect, locate, excise and replace parts and/or interactions that are not acting as they should.

It has, too, to maintain some level of dynamic stability, or steady state, marked as homeostasis in the figure, affording a balance between inflows and outflows. And finally, it has to be able to evolve in line with a changing environment, with changing threats and opportunities. acronym

As the figure shows, these five aspects form the acronym S-MESH (pronounced 'es-mesh;' it is catchier): they form a necessary and sufficient set of features for a system to remain viable. Looking at the whole solution system in this way, it becomes evident that 'maintenance,' does not maintain the mission system alone; additionally, it maintains all the other (sub) systems, including the maintenance and servicing systems (sic), the training and recruiting systems, and (what might

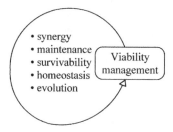

Figure 4.2   Viability management. Initial letters may be formed into a useful viability acronym — S-MESH.

be called) the evolution management system, etc. (Maintaining the maintenance systems raises the specter of Gödel's Incompleteness theorem, implying that there may be limits to how completely this can be achieved; however. . . .) And this wider 'whole' would interact with other systems within the environment which would supply resources, dispose of waste, etc., to maintain dynamic stability. The whole can be seen as organism-like, existing in harmony with its environment. . . .

There is a case, then, for the whole solution to any problem to include within itself, not only the ability to adapt, but to evolve as well, so that the evolving solution system can solve, resolve and dissolve the continually morphing problem in the changing environment. This process of evolution would be further facilitated if system conception, design and implementation had this potential for evolution in mind throughout. Including the ability to adapt and evolve within the organismic whole suggests that the whole may, in the final analysis, be self-organizing, adaptable, optimal, flexible, enduring, in harmony with its environment, and intelligent.

Summary

Systems engineering has been around since before the Pyramids were built. It was not recognized and codified as a discipline until the 20th century, however. Recognizable examples of systems engineering include the air defense system for England during the Battle of Britain, and Apollo, which set the style and substance of practical systems engineering.

World War II provided a major stimulus for the development of systems engineering, with defense and military systems becoming more complex, and with the concomitant need to understand and manage that complexity. A host of well-publicized disasters demonstrated to the world that complex systems could be problematic, and that there was a serious need to understand systems, counterintuitive behavior, and how best to manifest 'new' systems that were able to resist threats both from without and within.

The systems approach was seen as the best way to understand, and systems engineering was seen as the best way to solve, resolve or dissolve complex issues and problems in virtually all walks of life: sociotechnical systems, social systems, healthcare, industry, agriculture, ecology, economics, international peacekeeping, etc., etc.

Arguments have waged back and forth about the substance of systems engineering since the 1950s: is it applied systems science, or is it engineering? Such arguments arise since solutions to many problems include technological parts, within sociotechnical systems such as defense, policing, air traffic management, and the like. Engineering organizations may then conceive and design technological elements of such socio-technical systems, and so believe that they are 'systems engineering.'

On the other hand, the systems approach, of regarding a system only and always in the context of being open to, and adapting to, its environment, is at the heart of the conception and design process, as is the concept of emergence. It seems that there may be, in essence, two different ways of conceiving, designing and developing technological solutions:

- the classic engineering approach, decomposing the whole and synthesizing the whole from its parts such that the whole precisely equals the sum of the parts;
- and the systems approach, regarding the solution as open, adapting to environment, dynamic and potentially exhibiting requisite emergent properties, capabilities and behaviors — where the whole can be greater than the sum of its parts.

The wide variety of definitions and descriptions of systems engineering conceals a singular philosophy behind them all. Despite differences in language and application, they concern themselves with wholes, finding answers to whole problems, analysis and synthesis, innovative design, management of complexity, integrity, quality, risk management, planning and decision-making. These considerations suggest a catch-all definition of systems engineering:

Systems engineering is the art and science of creating whole solutions to complex problems.

The objectives of systems engineering are seen as being to solve, resolve or dissolve problems (suggesting that systems engineering may have more in common with doctoring, and detective work, than with many other disciplines). In so doing, systems engineering tasks itself with conceiving and manifesting the optimum solution, i.e., the best solution in the circumstances, where best implies most able to undertake the mission of the system, making best use of available resources. The optimum solution will be one where the parts combine to create requisite emergent properties, capabilities and behaviors, i.e., where the whole is greater than the sum of the parts; emergence offers more from given resources.

Self-organizing systems occur in the natural world — the Sun is self-organizing; so, too, is the solar system. Human activity systems may similarly be self-organizing; that is people may form themselves into groups, teams, etc., adopting different roles and tasks, even reconfiguring themselves as the whole group seek to achieve some goal, defend against some threat, etc.

System of systems is a relatively recent concept, referring to the bringing together, under one umbrella, of a number of companies or organizations to form a larger whole. If the term is to have any meaning, then this larger whole should, presumably, exhibit the characteristics of a whole, i.e., the open parts should interact and complement each other such that the whole is greater than the sum of its parts.

Instances of groupings referred to as 'system of systems' do indeed conform with these aspects of a system — a city's integrated public transport system for instance — while others seem to be more groups, classes, or families of systems, rather than systems of systems: the difference is significant. A system of systems may reasonably be expected to exhibit requisite emergent properties, capabilities and behaviors, while associations and families may not; moreover, the 'design' of a system of systems is practicable using the same notions, methods and practices as for any other complex open system.

Bottom up integration refers to the practice of synthesizing technological solution systems from engineered parts, which are generally designed not to adapt to each other, although power, information and signal may pass between them. The parts do not interact, as such, but transform, and the whole is intended to be precisely the sum of the parts. This is a pragmatic approach to the construction of a variety of different technological systems from a basic set of building blocks, and is employed both in brown goods and white goods manufacture, amongst many others. This, too, may be regarded as systems engineering, in that the system that is being created is not so much the product or artifact, but is a construction system, not unlike Lego™ or Meccano™.

Finally, a look is taken at what should constitute a whole in the context of delivered systems which are expected to have unlimited existence, such as some defense systems, companies and corporations, etc. The delivered set of systems for, say, some defense project, may include the mission system (e.g., a new fighter aircraft) plus the ground support facilities (maintenance, servicing, repair, etc.) plus recruiting and training of both mission personnel and support personnel, not forgetting trainers of the trainers, and so on.

Importantly one other element may be added to create a viable whole — the system for evolving the whole, i.e., for examining the original problem and the way in which it is changing, for conceiving revised mission system and support system designs and for implementing those designs. If this element, let us call it the evolution system, is included in the whole, delivered solution system, then that whole need have no limit to its existence, since it will be able to adapt, and accommodate change, as it arises. Similarly, it will be able to 'redesign' itself, 'reconfigure' itself, and so can evolve its emergent properties, capabilities and behaviors. In principle, by adding the evolution system to complete the viable whole, that whole becomes organic, self-organizing, an intelligent system of systems, enduring, continually optimal, and in balance with its environment: worthy goals, indeed.

Assignments

1. Develop coupled definitions of a system, and of systems engineering, without including any mention of method, process, sphere of application, etc. Define engineering, using the same rules. Identify areas of potential overlap between systems engineering and engineering as you have defined them. Discuss.
2. Systems engineering supposes there to be a best solution to any problem. Would a best solution consider the potential cost of that solution, or is the concept of 'best solution' confined to that which solves, resolves or dissolves the original problem most precisely and completely. Justify your view.
3. a) Systems engineering is applied systems science.
   b) Systems engineering is simply engineering done properly.
   c) Systems engineering adopts the systems approach.
   d) Bottom-up systems integration is systems engineering.
   e) Emergent properties are to be avoided — they indicate poor engineering. . . .

   Discuss these statements, identifying an appropriate Weltanschauung and context for each, together with likely motives behind, and validity of, each statement.
4. Explain what you understand by the term 'viable system?' What properties, capabilities and behaviors would a viable system necessarily possess/exhibit? Would you consider a viable system to be — necessarily — successful, effective, reliable, durable? Explain, with examples.
5. Define and illustrate the term 'system of systems,' with examples from your experience both of groupings of systems that warrant the term, and others that do not. Explain the differences. Do you consider the differences to be important, or not? Justify
6. Using the catch-all definition of systems engineering given in this chapter, identify the generic activities/steps that you would consider were needed to manifest an optimum solution, starting with a given problem space. You should identify no more than twelve steps, no less than six. Explain your rationale; justify your result in terms of completeness. Is the process you have used Cartesian Reduction? Explain and justify your response.

5

System Models

He his fabric of the heavens
Hath left to their disputes, perhaps to move
His laughter at their quaint opinions wide
Hereafter, when they come to model Heaven
And calculate the stars, how they will wield
The might frame, how build, unbuild, contrive
To save appearances, how gird the sphere
With centric and eccentric scribbled o'er
Cycle in epicycle, orb in orb.

John Milton, 1608–1674, *Paradise Lost*

An understanding of systems, concepts, systems thinking and systems engineering is facilitated by the use of simple models. This chapter contains a variety of models, starting with the simplest and most obvious and working up to the more complex.

The Open System

All systems are open in some degree; that is, they are open to their environment and interact with that environment. Examples of open systems abound: a human cell, a person, a factory, a radar, a transport aircraft, an economy, etc., etc.

Figure 5.1 shows a notional diagram of an open system. It receives energy and resources, and it 'emits' dissipation (of energy as heat, light, sound, etc.) and residue, which might be waste or product, or both. All of these are conserved quantities. Any open system may also receive information (e.g., from sensors), which it can store and utilize, and it may also emit information; these are not conserved quantities.

The figure shows an open system boundary, with an environment outside, and an environment within the boundary; these may be different environments, and the boundary may be tangible, fuzzy, permeable, or notional.

The existence of features within the open system may be deduced, and are shown within the notional boundary. An open system is internally dynamic, as inflowing resources are processed,

---

Systems Engineering: A 21st Century Systems Methodology   Derek K. Hitchins
© 2007 John Wiley & Sons, Ltd

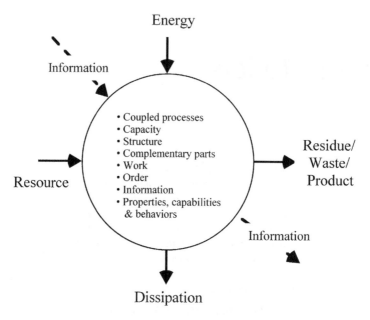

Figure 5.1   General view of any open system, showing a notional boundary.

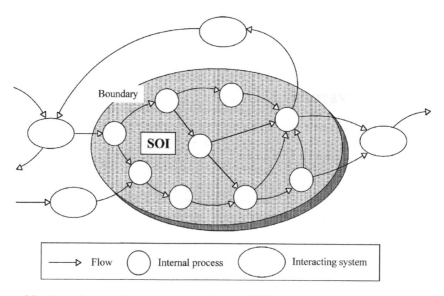

Figure 5.2   Alternative view of any open system-of-interest (SOI) in context, showing internal processes and flows. For homeostasis, the sum of all outflows must equal the sum of all inflows; this may be achieved by inflow demand, suggested by the 'feedback' system at the top of the figure, which signals demand for inflow. In fauna, this might be hunger, promoting demand for food. In a factory, this may be output promoting a demand for the supply of more 'goods inward' to be assembled, machined, etc.

stored, transported, converted and emitted: work is done; heat is generated. There are coupled processes connecting inflows to outflows. The processing parts cooperate. Functions are performed. Outflows occur, as waste, or as product.

All open systems possess these features. In addition, there may be features relating to viability — see page 98 — which may be deduced. For instance, survivability may be enhanced by a containing boundary, as in a human cell or a tank; and, synergy, maintenance and homeostasis will be evinced by the continuing, steady state existence of the open system. Evolution, the remaining aspect of system viability, may be deduced only over an extended period of time as the open system adapts to changes in its environment.

Figure 5.2 shows an alternative view of open systems, in which the internal processes are evident. In animals, for instance, such flows would include food ingestion, digestion, energy and nutrient extraction, and waste excretion, i.e., through the alimentary canal. There will be many other routes too, including those that use the extracted energy to do work and dissipate heat, and those that use nutrients to maintain, reconstruct and expand the internal parts of the SOI. Following the concepts expressed in Figure 5.2, an open system may be visualized as a flux concentration, with continuous lines of flux passing through it, not unlike the magnetic flux lines of a bar magnet.

This view of systems as open and internally dynamic is significantly different from the view usually presented of technological systems; such views tend to emphasize structure, interface and function, with power supplies, and firm boundaries.

Simple Nesting and Recursive Models

Open systems do not exist in isolation, of course. Figure 5.3 shows a number of open systems mutually interacting, such that the residues of some are the resources of others, and the dissipations of some are the energy of others. Information is shown passing through the lattice, 'updating' systems as it passes through. Only three open systems are shown — there might be many more. In the simple example, however, the three systems are mutually complementary, such that a boundary can be drawn around them as shown. Looking at the boundary as an open system, it can be seen that it receives energy, resource and information, and it emits residue, dissipation and information — as does any open system.

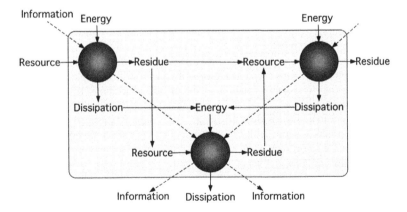

Figure 5.3 Interacting open systems in a recursive lattice (Hitchins, 1992).

This, then, is a recursive model, in which open systems are 'contained' within open systems, which will be further 'contained' within higher-level systems not shown, *ad infinitum*. Note that the shaded containing system has all the characteristics of an open system as shown in Figure 5.1. It does not follow, however, that this containing system is necessarily viable, even though the system parts within it are viable. For the containing system to be viable as a system, an appropriate selection of open systems would have to come together in the right way, such that the whole could exhibit homeostasis, for instance. So, a collection of interacting cells need not constitute a viable organism: a selection of interacting commercial units need not constitute a viable business.

Systems comprised of open, interacting systems need not behave linearly; indeed, the more parts and interactions, the more potential there appears to be for nonlinear behavior. Complex systems may behave linearly, but may accommodate nonlinearity, too. For instance, many open systems receive inflows and emit outflows intermittently. Internal interactions and processes may be conducted linearly, however, using various methods of storing energy and resources within the system, effectively smoothing out peaks and troughs in supplies and services. As the human body stores energy as glycogen or body fat, so a factory has a goods inward and a goods-outward store, and may have smaller sets of parts-to-be-processed stored by processing machines, so that the machines can operate continuously.

Social Genotype — a Notional Model

Figure 5.4 shows a notional model of the social genotype: see page 50. In the model, the familiar DNA helix is recast, with roles and relationships replacing the familiar nucleotides and hydrogen bonds. As different DNA can result in widely different flora and fauna, so the social genotype can manifest itself in a wide variety of social systems. According to the hypothesis, shared culture and beliefs bind the various roles together, such that culture and beliefs can survive the passing of individuals; those that replace them are obliged to 'fit in' to their new role by the established roles and relationships with and through which they find themselves interacting. The social genotype becomes set and virtually immutable with time. Since such social groups exhibit characteristic group behavior, there is a suggestion that the analogy with DNA can be extended to include a social phenotype. It has been previously noted by Jung (see page 12) that large organizations can behave in ways that would be alien to the individuals in the organization, for example. It is not unreasonable to view this emergent behavior as phenotypic. Indeed, from a starting position of viewing the social genotype/phenotype as simply analogous to the biological genotype/phenotype, it is no great leap to consider that the social concept may be more in the nature of an extension to the biological concept, as the organization might be seen as an extended super-organism. . . .

We can see the durability of the social genotype in the way in which people cling to their culture and their beliefs, passing stories and legends down through the generations. It seems also to happen over time within companies and organizations: initially fluid and malleable, start-up companies soon start to form into more-or-less rigid groupings — divisions, departments, sections, etc., and soon after that they produce company literature, handbooks, etc., which describe and instruct how the company is intended to operate, work, conduct various processes and procedures, etc. By this stage, the organization and its social genotype have gelled, making subsequent change difficult to impossible: at least, not without seriously disrupting the whole role–relationship structure.

The social genotype also offers potential explanations for the ability of social systems, communities, etc., both to resist change, and to survive trauma. Further, particular role relationship structures may form the basis for assessing and evaluating social capital — see page 50.

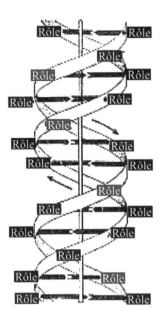

Figure 5.4 Conceptual social genotype, with roles and relationships replacing the nucleotides and hydrogen bonds of DNA.

Cybernetic Models

Cybernetics is the science, or study of communication in organisms, organic processes and technological systems, where it is often associated with methods of control through feedback.

Figure 5.5 shows a classic feedback model, in which an output is matched to a desired input by differencing them and using the difference as an error signal, which is then driven to zero. This may be described as closed-loop control, since the information pathway forms a closed loop, as shown. Control may be less than precise, especially if the desired output is changing constantly, if the environmental disturbance is changing, and if the cycle time around the loop is significant. Control theory has developed to understand and address such issues; with the result, that fine, precise control of technological devices can be highly effective. Using such concepts for control of people is, however, less likely to be effective, if only because people do not present conveniently stable transfer/transmission functions and behavior.

Despite this, there is a widespread management practice of control by setting goals, measuring perceived shortfalls against such goals and using the gap between desired and achieved results to drive further change. It seems reasonable: so much so, that managers (and governments) seem not to notice that the practice creates its own counter-reaction, with people feeling undervalued, being considered as 'just another brick in the wall,' and generally feeling alienated. So engrained is the practice, that many managers see no alternative. . . .

Figure 5.6 shows the futility of the approach. We suppose there to be puppet master above, out of view, who is controlling the top puppet. This, in turn, controls the second tier puppets, which then control the lowest tier of puppets. (This structure represents, or perhaps parodies, a tiered management system, in which the boss wishes to micro-manage, or control, the workers.)

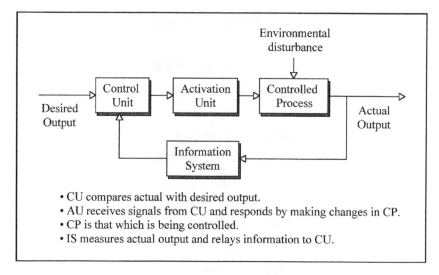

Figure 5.5   Simple cybernetic model (Hitchins, 1992).

The puppet master works through the tiers, so that his instructions are sent downwards to make the lowest tier of puppets do his bidding. If he wishes to control them, problems arise. He cannot see what they are doing in real time, and even if he could, he could not issue corrective instructions in time because of the delays in information going up and down the cords — which are, in any event, spongy and slow. Control, in the cybernetic sense, would be quite impractical.

Models of Systems Architecture

Open systems that mutually interact may form networks of systems. Open systems that are comprised of open interacting systems may form patterns. These patterns, or paradigms, may be referred to as (systems) architecture. Architecture paradigms are evident in Nature — see Figure 5.7. The figure shows sections through four simple creatures. For the simplest unicellular creatures, there is a semi-permeable outer cell wall, through which energy and resources are drawn in, and waste matter is discharged. As body plans become more complex, the direct interchange of substances across a single membrane is inadequate, and waste matter — for example — has to be carried from internal parts towards the outer membrane. With more complication comes the need to circulate (e.g.) blood, to carry energy and waste, and for homeostasis. The model with three cell layers and a body cavity is topologically quite different, as the cavity can provides another surface through which to ingest food and to pass out waste.

The human figure, with its alimentary canal, is topologically the same as the three-cell layer creature with the body cavity — both are topological 'donuts.' The human has a jointed skeleton, which is part endoskeleton and part exoskeleton. The skull, for example is effectively an exoskeleton, providing protection for the largely fluid brain. The spine is jointed and provides protection for the bundle of nerves, the spinal column, which passes through it. The spine also gives shape to the soft organs and outer skin. The ribs are dual purpose, protecting the soft inner organs while at the same time operating as the 'bellows' to expand and contract the lungs.

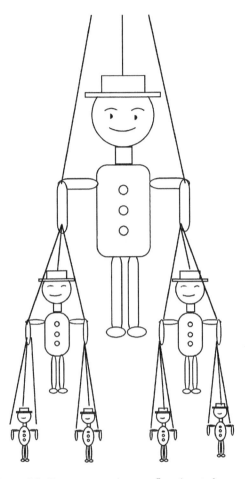

Figure 5.6 Puppets on a string — a flawed control concept.

In architectural terms, the skeleton may be considered as characterizing the physical architecture: the arterial and venous systems form around the bony skeleton, carrying energy to, and waste from, the extremities. Similarly, the central nervous system is 'superimposed' on the skeletal structure, with both sensor and motor nervous reaching from the extremities to the cerebral cortex.

So, there seem to be at least three physical 'architectures' in animals: skeletal, nervous and arterial. Recent research is suggesting that 'functional architecture' is evident in the brain itself. (Ramachandran and Oberman, 2006.) When an individual decides to do something involving a set of coordinated activities, such as walk, or swing a gold club, a set of neurons fires in parts of the brain. For different actions, a different set of neurons fires. It seems likely that these neurons in turn initiate and coordinate the set of activities, so that the set is conducted 'subliminally.' Primates seem to develop a series of neuron templates corresponding to certain actions, and these templates can be activated at will by the 'conscious mind' simply by choice. We decide to start walking

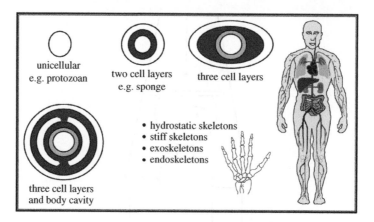

Figure 5.7   Natural structures and architectures.

and it happens without our having to work out which muscles to activate in which order, how to maintain balance, etc.

The neurons in question not only fire when we initiate an action: they also fire when we see others taking that same action; for this reason, they are called 'mirror neurons,' and they may signify empathy. Research has shown, too, that different sets of mirror neurons fire according to the intent of the actions that we take and observe. So, anticipating familiar behavior on the part of another person may not be, as has been supposed, primarily a logical process of deductive reasoning, but instead may be built into our 'neural templates' and neural pathways. In other words, it is built into our functional architecture.

Human functional/physical architecture is, apparently, straightforward. Our remote sensors, eyes and ears, are mounted on the head to facilitate sensing at a distance, and to encourage correlation between the two sensors. The nose is above the mouth to smell food before ingesting, and to permit correlation with the eyes when looking for food. The body as a whole is not the fastest, strongest, most agile in the animal kingdom, but it does seem to be the most adaptable: the human is essentially an open system, ensuring that we humans can interact, adapt to each other, cooperate and coordinate our actions to a degree unequalled by other social animals on the planet. And our body architecture provides the foundation for all of this. The human body architecture, both functional and physical, provides a valuable guide to the architecture of systems that we may wish to create. . . .

Beers' Viable Systems Model

Stafford Beers developed a considerably more sophisticated cybernetic model: 'If cybernetics is the science of control, management is the profession of control — in a certain type of system.' (Beers, 1972)

Beers developed a model of management control based on his understanding of the human nervous system. Evolved over a number of years, this is the Viable Systems Model (VSM); one version of which is shown in Figure 5.8. He deduced that viability was maintained by:

1. 'engaging in different activities;
2. keeping them from interfering with each other;
3. managing them together,
4. focusing on the future;
5. and, doing so in the context of an identity within which the interests of the whole over time could be considered.'

This, he believed, is how the human nervous system works, and also how successful collective enterprises work. Figure 5.8 has become something of a VSM icon, with the various levels in the figure corresponding with the list of five aspects above. The model is recursive: within the circle can be seen three small replicas of the right-hand part of the diagram; within the circle part of each small replication, can be seen two even smaller replicas. Each replication contains the same conceptual structure and process as the whole.

This may become clearer with example. Suppose that the icon represented some large corporation. Then the circle might contain three divisions of the company, with the *inset* circles being the divisions *per se*, while the remainder — the three boxes above each circle — represented the management. Each division, with its management, interacts with its environment — Division A with environment A, Division B with environment B, and Division C with environment C. The Divisions operate in different, but related, business environments, so there is some overlap between

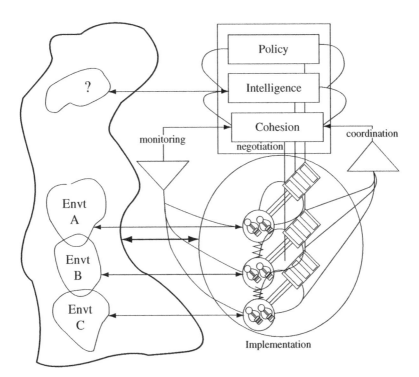

Figure 5.8   Beers' Viable Systems Model in recursive organizational form.

environments; the difference represents essential variety. Similarly, the divisions cooperate with each other, shown by the 'springs' connecting the inset circles. Returning to the whole icon:

- The large circle represents System 1 of the five systems.
- System 2 coordinates the actions of, and prevents oscillations within, the System 1 units, and is represented by the upward-pointing triangle marked 'coordination.'
- System 3 makes decisions about the day-to-day running of affairs, about resource allocation between System 1 units, about synergy between those units, etc., and is about overall cohesion — as shown in the figure.
- System 3* has a special auditing function, shown as monitoring and represented by the downward-pointing triangle.
- System 4 is concerned with looking forward into some future environment (question mark in the figure) so that opportunities may be seized, pitfalls avoided, etc. There is also a System 3/4 homeostat, indicated by the connecting curves, charged with keeping a balance between present activities (3) and future activities (4).
- System 5's functions include setting context, establishing corporate identity and 'providing closure to internal dialogues.' System 5 also monitors the homeostatic balance between System 3 and System 4; this can be important, as the optimum balance point tends to fluctuate over time.

Each of the Systems 1–5 has its own internal structure, too, following the general lines of the whole. So, System 4 might concern itself with identifying future opportunities, e.g., marketing intelligence; it would then follow that the people concerned with this function would comprise a system of interacting people and that system would, again, repeat the iconic structure — all within System 4.

Presenting the VSM as being concerned with business may make it easier to comprehend, but risks undervaluing and underestimating the VSM, which can be applied to almost any complex management structure. It has, for instance, been applied to the organization of a honeybee hive (Espejo and Harnden 1989). Beers himself famously applied it at national level to the nation of Chile, in support of Chile's President Allende, although that enterprise was cut short.

VSM is interesting and insightful, but has not proved universally popular within management circles. Perhaps this is because, although based on the human nervous system, the VSM appears to be somewhat cold, calculating and inhuman; or, put another way, it make little allowance for human initiatives and frailties. Instead, VSM has become associated, perhaps unfairly, with authoritative, even totalitarian, management styles and regimes. Perhaps, too, the VSM is rather opaque in some respects, and to understand the model and its implications takes time and effort. Nonetheless, Beers' VSM stands out as and beacon for faithful followers of the cybernetic control style of management.

Open-loop Control Models

Even the most cursory examination of managers at work within commercial organizations shows that control is not applied in the cybernetic, real-time feedback sense. Figure 5.9 shows a typical scenario, in which a 'boss,' or director gives bare instructions to a worker, or 'actor,' who then acts as instructed by the director. As the figure shows, the director is unlikely to instruct the actor in any detail, for two reasons: first, he may not know as much as the actor about the task to be done; second, it would take far too long, and would appear to the actor as micro-control, against which the actor would almost inevitably rebel. (I say 'almost inevitably,' as the actor could take

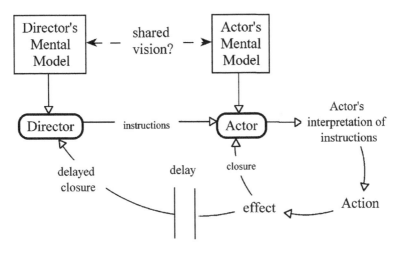

Figure 5.9   Open-loop control and direction.

the alternative course of action and simply wait for the full set of instructions and obey them to the letter, taking no initiative of his own. This is a not-untypical reaction when being directed by a 'control freak,' for instance, and eventually results in problems as the director fails, or is unable, to give complete instructions to his many actors/subordinates.)

Given bare instructions, the actor then acts in accordance with the director's wishes, expanding on the instructions to perform the task using his own knowledge and experience. There is no immediate feedback; indeed, the director may trust the actor sufficiently not to check that the actions have been taken and the work done.

As the figure shows, this type of control, dubbed open-loop control, is not really compatible with cybernetic concepts: it depends for its effectiveness on the director and the actor sharing approximately the same mental model of the actions required and the work in hand. So, the director might be a foreman, asking a plumber — the actor — to install a new sink unit in an apartment. The foreman may not be a plumber, and may not know as much about installing sinks as the plumber; but, the foreman knows that the plumber understands what is meant by 'install a new sink,' which might include acquiring the sink plus taps, hot and cold isolator valves, feed-pipes, traps and wastes from store, fitting the unit, connecting it up to the services in the apartment, earthing the metal pipe work according to regulations, etc. And the foreman might expect to hear back from the plumber only if there had been any problems. . . . This is a long way from cybernetic management control; yet this is much more the style of management control in general use. . . although, this type of management control would not be incompatible with System 1 activities in the VSM model of Figure 5.8.

The 5-layer Systems Model

A different kind of model, the 5-layer system model looks at the different layers for the application of systems engineering.

Layer	Generic title	Sphere
5	Socio-economic systems engineering	Legal and political influences. Government regulation and control
4	Industry systems engineering	National wealth creation, the Nation's 'engine'. (Japan operates at this layer)
3	Business/enterprise systems engineering	Industrial wealth creation. Many Businesses make an industry
2	Project systems engineering	Corporate wealth creation. West operates at this layer.
1	Product/subsystem engineering	Artifacts. To some the only 'real' systems engineering. Many Products (can) make a system

The 5-layer systems engineering model is of the 'nesting' variety, i.e., each layer sits 'inside' the one above. So, many products (may) make a system, many projects may make a business, many businesses may make an industry and many industries may make a socio-economic system. A top-down view reveals that the nesting is incomplete: there is more to a socio-economic system than just industries; there is more to an industry than a group of businesses; and so on. However, the nesting concept serves its purpose of showing different aspects of systems engineering.

Layer 1: product/subsystem engineering

Layer 1 of the 5-layer model of systems engineering concerns itself with the systems engineering of artifacts which, generally, form a subsystem, or part of a subsystem, of some greater whole; alternatively, such artifacts may be seen as complete products.

The process starts with identifying the symptoms of some problem, which leads to an exploration of the problem space and an elucidation of the whole problem, and its roots or causes. Once the problem is understood then ways of tackling the problem may be addressed, including resolving, dissolving and — in this case — solving the problem. A variety of conceptual solutions is then generated: each is considered (value judgments) and one or more preferred solution concepts selected; the selection is made by assessing the relative merits of each conceptual solution in terms of likely effectiveness, cost, ease of construction, availability of parts, likely timescales, etc.

As the figure shows, the purpose, role and intended behavior (requirement) of the preferred conceptual solution can then be identified. Achievement of the purpose may be described in functional terms — if the solution system performs a function or functions, it will achieve its purpose. In addition to primary functions, the solution system may also perform other functions designed to neutralize threats and anticipate risks in its future environment.

The functions-to-be-performed may then be elaborated into interacting subfunctions, and clustered to form a functional architecture, elements of which can either be undertaken by some technology, or performed by a human operator, user, etc. The elaborated functional architecture may then be tested to see that it does, indeed, perform the requisite overall functions — this is necessary to ensure that the interactions between functional elements are appropriate, of sufficient capacity, etc.

A physical architecture is then generated which sees the functional architecture partitioned and apportioned into physical compartments which may be designed to fit into some physical space, and operate in some future and environment. In creating the physical architecture, choices will be made about the technology to be used to achieve the functions, about the physical containers, about the interconnections between the containers, and about the interactions between the physical structures and other systems in the future environment. As before, value judgments will select for required fit, form and function according to effectiveness, ease of construction, availability of technology, timescales, cost, etc. There may, for instance, be many ways to instantiate a functional architecture using different technologies, and these various ways might incur widely differing costs, while affording differing effectiveness, reliability, etc. Finally, the assembled whole, including the human operator/user, will be tested and trialed to show that it does, indeed, neutralize all the original symptoms and address all the perceived threats.

Note that the process, loosely described in the figure, elaborates and maintains the interactions between functions, both in the functional architecture and, importantly, in the physical realization. Note, too, that the process is one of synthesis without reduction — unless elaborating a function into a number of interconnected subfunctions might be considered reduction. (It would be reduction only if the subfunctions were considered independently of each other, which should not be the case when elaborating.)

A simple example of the process outlined in Figure 5.10 might concern a problem, the symptoms of which emerged from a residential area where there were annual increases of blocked rain-gutters, and of vehicle accidents. Upon investigation, the problem turned out to be caused by the annual shedding of leaves by the many trees around the area. Several ways were suggested to alleviate the problem:

- Cut down the deciduous trees, thereby removing the source of the problem, i.e., dissolving the problem, making it 'go away.' This was unacceptable to the residents
- Put netting over the gutters, and place warning signs by wet leaves on roads, i.e., resolving the problem (netting gutters is known to have only limited effect). This was deemed ineffective.
- Design a cleaner of some kind to both clean gutters and roads, to remove the leaves, which might then be used for compost.

Following the outline of Figure 5.10, several conceptual options would include a blower for blowing the leaves out of gutters, and some kind of vacuum cleaner that could suck up water as well as leaves. This latter was the preferred concept, since it both captured the leaves and removed water from blocked drains and gulleys.

Residents in the area valued their quiet neighborhoods, so it was felt that a noisy cleaner would not be acceptable. Moreover, some of the blockages only presented themselves late in the year, when the leaves could be held in plugs of ice and snow — the vacuum pump would have to be able to deal with such environmental conditions. Further, it was estimated that the area council would need at least a dozen of these items in the first instance, but there would be a potential for sales of several hundreds of the devices if they proved successful. There was, then, the prospect of a business venture, suggesting that the vacuum pump should be robust, reliable, powerful, able to operate under a wide variety of conditions and power sources, inexpensive to manufacture, aesthetically pleasing both visually and audibly — and capable of being transported and operated by one suitably trained person.

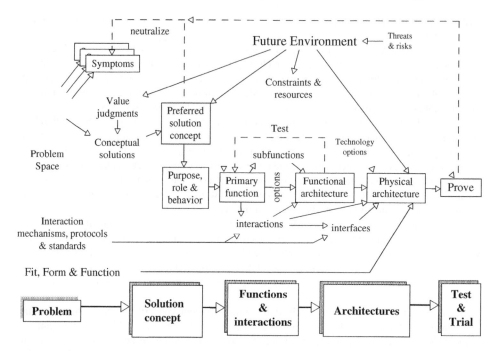

Figure 5.10 Typical Layer 1 systems engineering paradigm: creating an innovative artifact/product The figure is a conceptual process model, starting with some problem, shown on the left in a problem space, and ending with a proven solution to the problem on the right that, as the dotted line shows, is proven by its ability to neutralize the symptoms of the problem.

So, role, purpose, behavior and primary function were established. Elaborating subfunctions saw the conception of various nozzles and tubes for sucking, vacuum chambers, filters for catching material, chambers for compressing and storing extracted material, transformers for supplying power to the electric pump (chosen for its low noise output) and so on. In addition, the design incorporated a heated blower for melting ice on roads and in gutters, gulleys and drains, prior to sucking out water and fluid: this was, effectively, the vacuum pump in reverse; the same subfunctions, but rearranged and with the addition of a heating element in the flow. The subfunctions were clustered to form physical chunks, the physical architecture being fitted on to a small, standard vehicle trailer, which could also carry an off-the-shelf petrol–electric set for those occasions when a direct electrical supply was either not available or was inadequate.

The whole was assembled from bought-in parts except for some pipework, a chassis and a smart plastic cover and control panel, which gave the appearance of the vacuum-blower–pump being a unit — which in essence it was, since the whole was made from 'complementary interacting parts, with properties, capabilities and behaviors emerging both from the parts and from their mutual interactions to synthesize a unified whole.' Which is the definition of a system.

The process outlined in Figure 5.10 goes from problem space to solution space and conceives and synthesizes an innovative solution system that is effective, compatible with, and adapted to its environment.

Layer 2: project systems engineering

Table 5.1 presents a typical systems engineering process outline at Layer 2, in tabular form. Note that, although using different words, the overall pattern of activities is very similar to the Layer 1 version of Figure 5.10; systems engineering is, after all, systems engineering, at whatever layer — it is the domain, nature of the problem and hence the substance of the solution system that change; not the methodology.

Table 5.1 again starts with the problem, and conceives solution options which are traded against the criteria for 'good' solution, where good means satisfying values of effectiveness, reliability, affordability, and so on. The model presumes that overall design may be partitioned into manageable parts, which may be separately developed before being brought together, and integrated into a unified whole. This involves a degree of reduction in the overall process, which is ameliorated by ensuring that the compatibility between the parts is maintained throughout development. This in turn implies that functions and functional interactions implicit in the overall design are preserved; there is a risk of this not happening, or at least of being less than perfect. To ensure that nothing has gone wrong during this procedure, the integrated whole is proved in a dynamic simulation which is designed to show that the solution system can indeed solve the problem, i.e., neutralize all of the symptoms which characterized the original problem.

Figure 5.11 presents Layer 2 systems engineering more from a process management viewpoint, showing more of the organization that surrounds and supports the systems engineering process. In the model, the objective is to design the whole system (product and process) for the whole life cycle. The associated Weltanschauung, then, is that the delivered system solution will have an operational

Table 5.1 Tabulated conceptual systems engineering process at Layer 2: project systems engineering

	Identify the problem Understand the need Develop solution concepts	
Determine criteria for a 'good' solution		Conceive a range of solution options
	Trade options against criteria	Design, develop, create and prove simulated test environment
	Select preferred solution	
Maintain compatibility between developing parts	Partition conceptual design into manageable parts	
	Develop the separate parts	
	Bring the parts together	
	Prove the integrated system in a simulated environment	
	Commission the system	
	Support/upgrade in operation	

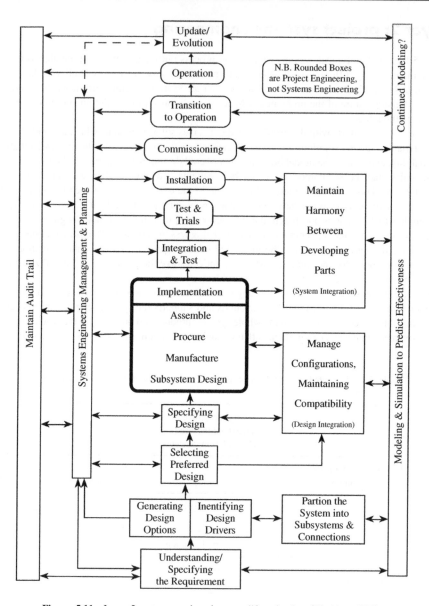

Figure 5.11 Layer 2 systems engineering — a lifecycle view (Hitchins, 1992).

life, during which it will develop and evolve. This is 'designing the whole system' including documentation, maintenance and servicing facilities, maintainer training facilities, operator training and crew training facilities, etc: the design of the whole would include, for instance, self-test, built-in test (BITE) and test access for maintenance and fault diagnosis, training modes within operational facilities, and the ability to upgrade the system during its operational life, such that both performance might be enhanced and operational life extended.

Note that the model contains a highlighted square in the centre, marked Implementation, which includes subsystem design, indicating that this is a Layer 2 systems engineering model; conventionally, subsystem design would be a Layer 1 activity.

The systems engineering model of Figure 5.12 suggests an ability to manage change. As situations unfold, and the problem to be solved becomes clearer, it may be that the way to best solve the problem may be perceived differently. In the model, the process maintains an audit trail; the customers and designers often have to make decisions early in the project cycle, when insufficient information is available. Later on, perhaps several years later, customers have been known to query the reason for certain design features, and even to suggest that the contractor has made a mistake when, in fact, the customer was party to the original decision.

Such selective memory sometimes arises as the customer's representatives may change several times over the developmental timescale of a major project. There is a consequent need for change control: because of change in the system, there may be several build standards of the system-to be-delivered, which would be a recipe for disaster if not properly managed during development, since interconnecting elements from different build standards might well result in non-operation at least, malfunction, or damage. Part of the practical systems engineering process is configuration management, which attends to build standards, compatibility between elements, interfaces, infrastructure, etc.

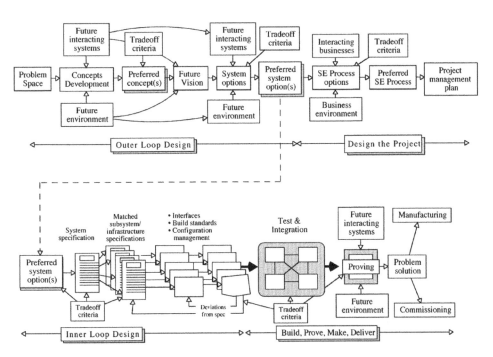

Figure 5.12 Layer 2 systems engineering procedure model, incorporating business and project management aspects.

Finally, note that in the model of Figure 5.11, (customers' and users') requirements are established as part of the overall systems engineering process. This can be achieved effectively only by understanding and addressing the customers' and users' root problems. There may, of course, be a difference between what people require and what they need. It is not unknown for customers and/or users to want something that is less than ideal for their purpose and which may not be what they really need to address their problem. Indeed, it is not unknown for the customer to presume his problem to be something other than the real problem... hence, it is important for systems engineering to explore the problem space and to examine the prospects for dissolving, resolving and solving the problem. And, since integrity is a watchword in systems engineering, it may prove necessary to explain to the customer what the real problem is and what the best solution might be....

Figure 5.12 shows a third view of systems engineering at Layer 2; this might be called a systems engineering procedure view, and it incorporates systems engineering at two distinct layers, in that systems engineering 'designs the process' that is to be used to design, develop, create, test, integrate and prove the whole solution system. This 'designing the project' is done in conjunction with project management and takes into account business factors and the business environment in which the work is to be done.

Figure 5.12 starts, top left, with the problem space and finishes, bottom right, with the problem solution. Proving the problem solution is performed using the same, or similar, representations of future environment and future interacting systems that were employed at the start to develop concepts.

The procedure model is arranged in two rows: the top row shows the development of future vision (requirements) and the preferred system solution — referred to here as the Outer Loop Design (see page 293). This is used to choose between various systems engineering process options (waterfall, concurrent, spiral, etc.) and thence to develop a project management plan.

The bottom row sees the design and specification of the whole solution system, together with the elaboration and partitioning of the overall design into interacting functional and interacting physical parts and their specifications — including the specification of the interactions and interfaces, both within the whole solution system, and to the future environment with its future interacting systems. This is referred to as Inner Loop Design (see page 293). When the parts are developed, they will then be progressively integrated according to a plan, before testing the whole in a representative environment, within which the problem solution should be seen to solve the original problem.

The essence of systems engineering

Figures 5.11 and 5.12 encapsulate many ideas about the 'how' of systems engineering, in terms of how it might be organized and managed, but neither should be mistaken for the essence of systems engineering. The essence of systems engineering is in: selecting the right parts, bringing them together, orchestrating them to interact in the right way and so creating requisite emergent properties, capabilities and behaviors of the whole. Essential systems engineering is executed such that the parts and the whole are operating and interoperating dynamically in their environment, to which they are open and adaptive, while interacting with other systems in that environment. See 'In pursuit of Emergence — the Generic Reference Model of any system' on page 124 *et seq.*

Layer 4: industrial systems engineering

Figure 5.13, ostensibly industrial systems engineering, shows three layers at once, indicating in practical terms how the nesting of the various layers of systems engineering presents itself. The figure shows two links (businesses, or enterprises), which form the final two tiers in a supply chain, an instance of industrial systems engineering. The final tier is the 'lead business,' so called because it interacts with the end customer or the market, or both. The lead business might be an assembly plant, assembling vehicle parts into complete, working vehicles for sale, or assemblies into aircraft, or boards into computers, or parts into washing machines, or modules into hi-fi equipments. . . . The first tier supplier might be one of several operating in parallel, feeding the lead business with assemblies of varying kinds that go together to make a vehicle — there might be an engine and gear train supplier; a seating and internal furnishings supplier; an instrument panel and lighting supplier; a bumper supplier; a vehicle security system supplier, etc., etc. Each of these second tier suppliers will be fed, in turn, by third tier suppliers, and so on, until at the start of the supply chain there may be raw resources, or perhaps recycled materials.

Goods flowing in at bottom left from third tier suppliers, not shown, are assembled or produced, or both, and then move on to Parts Supply, i.e., to the next tier; this is termed Manufacturing Systems Engineering in the diagram. This may be a continuous flow, as in mass production, or a continual flow, as in lean volume supply. Completed products go into the market to be sold, or not. If they are not selling too well, then there may be a need for variants, or perhaps for a new product. The purpose of Project Systems Engineering is to conceive, design and test a derivative, variant

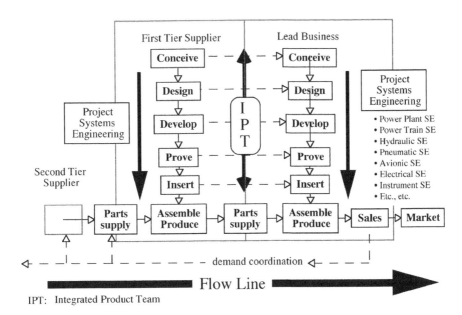

IPT: Integrated Product Team

Figure 5.13 Two links in industry supply chain systems engineering, showing two tiered businesses/enterprises, each with its own project systems engineering capability for feeding variants or new products into the continual product flow line which runs from left to right through the figure to meet demand, which 'flows' from right to left. . . .

or replacement of some product that is not selling too well, and to insert it smoothly and without interruption into the continual/continuous flow of goods passing through each business. Typically, this would include not only the new parts to be assembled, but tools, jigs, fixtures, bandoliers, etc., needed for assembling them, too, where these differed from what went before. Project systems engineering is essentially Layer 2 systems engineering; in an aerospace company, for instance, specialist disciplines might include aeronautical systems engineering, avionic systems engineering, hydraulics and powered flying control systems engineering, electrical power distribution systems engineering, etc., etc.

The flow through of products can be thought of, like a river, as having both breadth and depth. Depth might refer to as the number of a particular product passing through per period. Breadth might refer to the number of different product types passing through. Financial return relates to the breadth times depth per period, i.e., to the flow volume.

Supply chain and other industrial systems are not seen as having a life cycle, *per se*. That is not to say that they can never fail; rather, that there is no particular reason for them to fail or wear out, since they can not only continually change what they manufacture and supply, but they can also recreate themselves using the financial return on sales to undertake research, identify, design and make innovative new products, and continually replace and update their equipment and facilities. In this respect, they are similar to organisms such as the human body, which extracts nutrients from the materials flowing through it, with which to continually rebuild and replace the parts of the body at cellular layer. Of course, a supply chain that fell out of synch with its market would not sell, would run out of money and would cease to trade; but then, that would be due either to carelessness, or some major trauma, rather than any incipient aging factors in the design of such systems, many of which operate on global scales. In practice, lean volume supply chains create their own market demand by innovating at such a rate that they tend to operate at, or near maximum capacity, with the aim of manufacturing only what has already been sold.

Layer 5: Socio-economic systems engineering

Socio-economic systems engineering works, as may be imagined, on a rather larger scale even than industry systems engineering. The former USSR developed 5-year economic plans that, according to communist principles, strove to gain from each according to his ability and to give to each according to his need. So, some nations within the regime were required to produce goods and services against quota limits, and to supply them to other nations; each nation had a different plan, with the various plans being complementary such that, in principle, everyone should be satisfied.

Clever people, using smart methods, worked out the 5-year plans; but they did not work. Nations did not produce what they were tasked to produce: they might produce more, often less. Food and other essentials were often in short supply.

In contrast, free market economies do not plan ahead in this manner, neither do they set quotas, nor predetermine prices. Instead, production and interchanges are generally unplanned and chaotic. It turns out that this is much more robust than a planned economy, since it is soundly based on financial motivation, i.e., capitalism. It may not be without its faults, however. There have been 'wine lakes' and 'butter mountains,' excesses of food production, yet people around the world are still starving. It seems that, while free market economy is the best we have so far, it may not be the ultimate approach.

It is interesting in this context to note that the ancient Egyptians had a robust economy with virtually no manufacturing, as we would recognize it, and with no imports or exports to speak of. During the Pyramid Age, for instance, they operated a widget economy, in which the principal manufactured objects were monuments and statues. They did, of course, make ropes, and they knew how to irrigate their farmland, so that food supplies were ample.

Figure 5.14 shows the issues. Although money had not been invented, taxation was in evidence in the form of goods and services. The goods were food, clothing and particularly donations to the temples and to craftsmen to build temples and statues, to make fine murals and other artwork, etc. So long as the Nile inundation enabled the production of food, the population both grew and developed socially, in part because of the high degree of cooperation needed to trap the Nile flood water in the various irrigation systems; it needed only one error for the water, with its precious load of nutrient-rich silt, to leak away.

The economy was kept alive, too, by faith and fear. It was vital to maintain the level of the annual Nile inundation, on which food and farming depended; offerings were made to the many gods to ensure this. Disease, being eaten by wild animals, childbirth and many more were sources of danger and death which might only be averted by appeasing various gods, too. Every time something went wrong, it might be considered that a new temple, a new statue, a new way of worship to the gods in question might prevent recurrence. So, classes of bureaucrats emerged, together with the priesthood, and the businessmen who plied their trades up and down the Nile — which acted as the main artery of the fledgling nation, binding the various parts together into a unified whole under the king, or pharaoh as he would eventually be called. It was, after all, only the king who could commune directly with the gods, and who could — in the afterlife — ensure the annual flood of the Nile; hence, the pyramid — a way of ensuring immortality for the king, and hence for the land and people of Egypt. Building pyramids was an act of motivated self-interest on the part of the people.

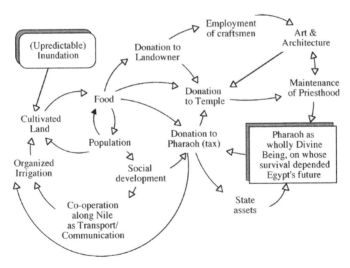

Figure 5.14 Ancient Egypt: socio-economics in the Old Kingdom; a highly successful economy without money, and without manufacture of utilitarian goods.

In Pursuit of Emergence — the Generic Reference Model of any System

Whence emergence? Can we purposefully 'design-in' emergence?

Emergence stems from the interactions between parts, which together (parts and interactions) create properties of the whole. Might it be possible to purposefully create emergent properties, capabilities and behaviors for some system, such that the whole was not only greater than the sum of the parts, but also 'greater' in a way, and to a degree, that was determined by the design and creative processes for some solution system? In other words, is it possible to design a solution system such that it exhibits requisite emergent properties, capabilities and behaviors?

To manifest solution systems that exhibit requisite emergent properties, capabilities and behaviors has been, is, and should be, the goal of systems engineering: it is the *raison d'être* of systems engineering, whether systems engineering at Layers 1–5, systems engineering of technological solutions, sociotechnical solutions, human activity systems, processes-as-systems, indeed virtually any system that we might wish to develop, synthesize, modify, repair. . . . Despite that avowed intent, it has not always been evident to practitioners just how they might go about 'designing-in' requisite emergent properties, and conversely how to avoid inadvertently designing-in unwanted emergent properties and experiencing counter-intuitive behavior.

Previous models in this chapter have not really contributed to any clearer understanding of how to manifest requisite emergent properties, capabilities and behaviors. Cybernetic models predict behavior, but to a very limited degree, and largely with the intent of minimizing unwanted behavior. Beers' VSM has some potential in respect of management of large organizations, although experience shows that attempts to control groups of people tends to have limited, short-term success at best, and moreover, by encouraging conformity, stifle initiative and creativity. Models of systems engineering at Figures 5.11–5.13 showed conceptual process, or management of process, or a combination of the two, without revealing just what the particular processes might be that would result in requisite emergent properties, capabilities and behaviors. Are such processes, for instance, different in every system design case, or is there a straightforward route to creating particular emergent properties, capabilities or behaviors?

One way to find out, to explore the potential for 'designing-in' emergent properties, capabilities and behaviors, is to elaborate the inner workings of a system, so that the discrete parts and interactions can be highlighted, and their effects traced as they make emergence manifest. For example, when one atom of nitrogen combines with three atoms of hydrogen, electrons from the four separate atoms associate to form a set of outer electrons encompassing the whole trigonal, pyramid-shaped molecule, NH_3. The outer electrons of the molecule form a pattern characteristic of ammonia, and contribute, with the other electrons and nuclei, to ammonia's emergent properties, including its (lack of) color, pungent odor, thermal capacity, refractive index, etc., etc.

So, we know how the emergent features of ammonia arise by bringing together the right atoms in the right way. Can we, perhaps, develop a generic approach, i.e., one that works for any system, such that by bringing (generic) parts together and enabling (generic) interactions, we may create (generic) emergent properties, capabilities and behaviors? To find out, we need first to create a generic model of any system: one that shows the inner workings, the parts and their interactions . . . .

The generic reference model (GRM) concept

That there could be a single, generic model to represent any system does not, initially, seem likely. After all, there are so many different kinds of systems, in so many different forms, guises, etc., and there must be many throughout the universe, and within the innermost workings of biological and ecological systems, of which we know nothing. Surely, it is arrogant to even consider that there could be a generic model?

It may not be possible to create a universal, scientifically provable, systems model of any system; however, it may be possible, using deductive reasoning (with its attendant properties of being falsifiable), to establish a model that is widely applicable and useful. The value in creating such a model is that it may be used as a reference model, with a variety of useful characteristics:

- The entities and interactions that are represented in a generic reference model should also appear in any real-world system, and in the complete design for a real world system. If they do not, then the real-world system, the design, or both, is probably deficient. A generic reference model, then, can be used as a valuable means of checking for completeness, both in terms of entities within an overall system, and of interactions within that system
- A generic reference model can also be used in a more active and creative manner, by using it as a template for the conception and design of some real world system. So, each part of the generic reference model, and each interaction, is instantiated and realized in the design and creation of the real world system
- If the generic reference model can be represented in a dynamic simulation, then instances of the model, optional designs, can be considered as hypotheses, where different hypotheses are optional solution-system designs to solve some problem. By dynamically simulating these various hypotheses in a representative simulation of their future environment, it may be possible to test each hypothesis, to evaluate which of them gives the best result, and to 'tune' the design as to give the 'best' overall design. ('Best' implies that the design solves the original problem as well as it can within constraints of capabilities, time, resources, etc., and 'best' is often referred to as 'optimum.')
  - This is the scientific method at work. As usual with the scientific method, it is not possible to prove that a hypothesis is correct: only to prove that it is incorrect. When the generic reference model is used as a dynamic design template to generate various hypotheses (designs), these cannot be proved 'correct,' but they can be shown to be incorrect. That ability would be invaluable for systems design, allowing weak and ineffectual designs to be 'weeded out,' leaving only the optimally effective to be considered.
- A generic reference model should reveal the generic bases of emergence, allowing requisite emergent properties, capabilities and behaviors to be both understood and, potentially 'designed-in' by instantiating the generic elements

All of which depends on the ability to create a generic reference model. The objective is to deduce what exists/happens 'inside' any open system. The method used is to consider widely disparate system types, and to see what common aspects can be deduced as necessarily existing within them.

Figure 5.15 shows a start point in the deductive process. We may think of any system as having existence, or being. This idea seems to work for an earthworm, or for NASA's Apollo (the two archetypal extremes of open systems that I, personally, use to check out such deductive reasoning; let us call them the reference archetypes).

Figure 5.15   The notion of a generic reference model.

Being, or existence, implies negative entropy — negentropy in the figure — so that some sense of order is inevitably associated with being, or existence. From negentropy to form is a seemingly reasonable step; but does every system have form? What about a system of ideas? What about a process-as-a-system? These have form only in the loosest sense, in that the various parts are necessarily interconnected in some way, suggesting that the whole has shape, even though that might be ephemeral, or constantly shifting, amoeba-like, as parts associate and re-associate.

All systems, then, have being, and that being may be recognized by the existence of order, at least in some degree. So, the solar system has being; it has the Sun at its focus, with planets orbiting the Sun, and with satellites orbiting the planets in familiar self-similarity. But the solar system does not appear to have any particular function, nor does it exhibit much in the way of behavior. Like many natural systems, the solar system just 'is.'

So, while most systems have form, not all of them exhibit either function or behavior. Evidently, the generic reference model has to be applied sensibly: it has to be understood and applied in context.

Some systems do appear to perform functions. The earthworm eats soil, extracts nutrients, and excretes waste; it also moves, and can undertake mini-missions, such as crossing a patch of open ground in search of shelter or food. So, its actions appear purposeful. NASA's Apollo likewise performed functions, appeared purposeful and undertook missions. For both archetypal references, the performance of functions implies tangible structures (muscles, sinews, fluids, organs, machines, sensors, effectors...) to prosecute the function — these tangible aspects will appear under the Form heading.

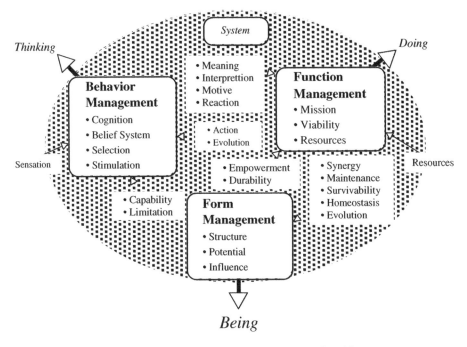

Figure 5.16 The generic reference model at Level 0.

Figure 5.15 also shows Thinking. Not all systems think. It is unlikely that the Earth thinks, although you can get arguments on that score from proponents of the Gaia concept, which sees Earth as a sentient being. A system that reacts to stimulus is not necessarily thinking; this might be a straightforward reflex. So, the earthworm is not generally thought of as thinking when it avoids moving over dry ground, but instead skirts around dry areas and stays, and moves instead on wet leaves.

Instead, evidence of thinking is to be seen where repetition of the same stimulus does not always result in the same reaction. NASA Apollo missions changed in response to what was being achieved, and what was being discovered, on successive missions.

So, all systems exist, some are active and perform functions, while some think, or respond to stimulus. Figure 5.16 shows the next stage in the development of the generic reference model, or GRM. The figure shows Being, Doing and Thinking, as before, but with these various aspects being interrelated and elaborated. This is the GRM at Level Zero. It purports to show various aspects of the 'internals' of any system, with three aspects being introduced as the first level of elaboration: these three aspects will be further elaborated in following paragraphs. At all levels and for each-and-every aspect, the elaboration is intended to ensure minimal completeness, i.e., to be both necessary and sufficient.

■ *Function management.* The functions-to-be-managed are made manifest in the tangible elements of form. The GRM presumes that, for those open systems that perform functions, there must be within the system some means for 'managing' the various activities. Human locomotion employs physical structures, but the process is managed cerebrally. Activities expend energy, and use resources (energy, substance). Open systems may be thought of as undertaking missions — looking for food, putting a man on the Moon, choosing a breakfast cereal, hitting a tennis ball, etc. Open systems must remain viable — a concept we have encountered before as fundamental to open systems. Completeness is implicit in that functions can be performed, actions taken, missions pursued if, and only if, there are resources within the system to energize it, and if the open system remains viable throughout.

■ *Behavior management.* For those open systems that 'think,' the GRM presumes that there is some capability within the system to sense and recognize stimuli, to 'consider responses to stimuli,' etc. See below.

■ *Form management.* Where there is tangible form, where the various parts of a system are cohesive, there may be structure, contained energy/power, and a balance between various influences that contributes to homeostasis.

Function management

Function Management may be elaborated as shown in Figure 5.17, which shows threats and change as endemic, and which is best understood in parts, as follows:

■ *Mission management.* At the top of the figure can be seen an operational environment; this is presumed to be that environment in which missions are undertaken. So, for the earthworm, it is the space where it moves, finds food, hides, etc. Function management, in its minimalist aspect, extracts information from the operational environment, uses that information to set/reset objectives, formulate/revise strategies and plans, and execute plans, deploying functions and resources in cooperation with others if necessary. This simplified mission management process applies, for instance, to command and control systems, individuals walking around on their daily business, armies, and of course, the archetypal references: the earthworm and NASA's Apollo.

■ *Resource management.* Conceptually very similar to Mission Management, Resource Management operates within a conceptual resource environment. Fundamentally, all that Resource Management can do is acquire, store, distribute, convert and dispose of resources. Conversion refers to doing work of some kind, which generally involves converting energy, with consequent generation of product and or waste.

■ *Viability management.* Viability Management is maintaining the dynamic steady state of the open system. Completeness in viability management is achieved by considering all threats to viability: internal threats, external threats, and the balance between internal and external factors, both short term and longer term. It further elaborates as:

   ● *Synergy.* Cooperation and coordination ('orchestration') of parts complementary within the open system to produce some desired external effect: implies interaction, direct or indirect, between all parts of the system: relates directly to emergence.

   ● *Maintenance.* Keeping everything in working order. Detecting, locating, identifying, excising, replacing and disposing of parts that have failed; or, continual replacement of each and every part such that no part fails prior to replacement. This second method is (part of) the biological approach to maintenance by continual cell replacement, and requires the ability to synthesize precise replacement parts using DNA as a template. It is also employed, for instance, in aircraft maintenance, where critical components are 'lifed,' and replaced before their life expectancy has run out, and before any defects or failures have occurred; instead of DNA, lifed components are manufactured to design templates, but the notion is the same. Also the identification and neutralizing of pathogens, or entities that are potentially dangerous implies sensing correct operation and 'authorized presence' throughout the system, which in turn implies some means of sensing incorrect operation and unauthorized presence.

- *Evolution.* Accommodation and adaptation, generally over periods long compared with rates of fluctuation of dynamic stability.
- *Survival.* Seen as: avoidance of detection, self-defense, damage tolerance, and damage repair. These form a set: if detection is not avoided, then there is self-defense; if this is inadequate, then there is a fallback to damage tolerance; and, in many systems, full survivability can be restored after damage through damage repair
- *Homeostasis.* Maintaining the internal environment of the open system such that all the various parts may continue to act and interact effectively. Generally implies balancing

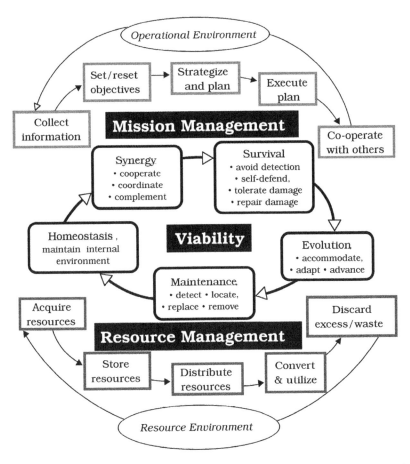

Figure 5.17 Function management elaborated to show, from top to bottom, Mission Management, Viability Management and Resource Management. The Operational and Resource Environments are 'outside' of the open system: all other aspects are 'inside' the open system. Boundary, if there is any, is not shown.

inflows and outflows to maintain the level of some internal quantities, states and conditions. Does not imply cybernetic feedback, but does not preclude it either

- *Together*. As the figure shows, these five aspects are mutually supportive: maintenance, for instance, must maintain the bases for synergy, homeostasis and survival. Survival in the short term will depend upon homeostasis, and in the longer term will depend upon evolution; and so on. Hence the useful acronym already introduced: S-MESH

Each of the aspects above may be further elaborated using the same approach of ensuring minimalist completeness in the elaboration.

Behavior management

As Figure 5.16 showed, Behavior Management elaborates into cognition, belief system, selection and stimulation: these are somewhat arbitrary aspects; Figure 5.18 shows the Behavior Management aspect in full.

Modeling behavior is not straightforward. Psychiatrists and psychologists seek to understand behavior, and few would claim their understanding as a precise science. The Behavior Management model is, at best, a greatly simplified view of behavior, and the model should be seen as a 'work in progress.' Despite these reservations, it may nonetheless be sufficient for purpose as a reference

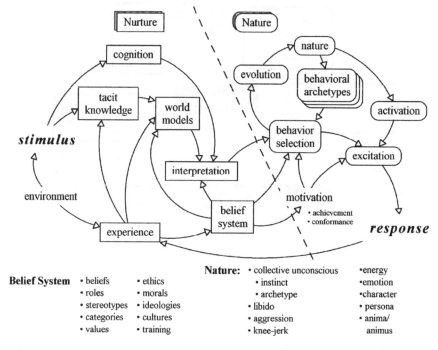

Figure 5.18   Behavior Management — an aspect of the Generic Reference Model at Level 1.

model, when employed as an integral part of a whole GRM representing an open system interacting with other open systems.

As the figure shows, the behavior management model is founded in the Nature vs Nurture paradigm. Nature is shown as an evolved/evolving state, with some of its many psychological aspects including libido, aggression, energy, emotion and character. Also shown is Jung's collective unconscious, with its instinct and archetypes: see 'Understanding open system behavior' on page 12 *et seq.* In the diagram, Nature also incorporates 'behavioral archetypes,' indicating that it may be in the nature of a system (e.g., a person, a regiment, a corporation) to behave in characteristic ways in response to stimuli, e.g., to fight or to flee, to be angered and agitated, to confront, to ignore, to celebrate, to be excited, and so on.

Nurture is shown as a continually changing state, open to the environment from which comes both experience and stimulus. Central to Nurture is the Belief System, a complex of categories, beliefs, roles, stereotypes, values, ethics, morals, ideologies and training. These many aspects are conditioned by culture, and may be looked at as the expression of culture in the individual or group.

Training is important. Behavior is 'colored' by Nature: training can in effect become second nature. So, a platoon of soldiers may be made up of soldiers, all of whom separately would find killing abhorrent, yet through their training they have come to believe as a group that killing is necessary and that their way of killing is effective, justified and incurs minimal risk to themselves and their platoon. In the right circumstances, then, justified killing has become second nature, and has overridden the abhorrence that seems natural in many humans.

To understand the model, consider a stimulus arriving at the point marked Cognition. Cognition recognizes the stimulus, if at all, from memory: tacit knowledge and world models (see page 44) enable the stimulus to be sensed, categorized and interpreted. Belief systems may also be involved in this process, so that a stimulus might be interpreted, not so much for what it is, but more for what an observer believed it ought to be, or was expected to be.

The interpretation of stimulus is presented to behavior selection in the diagram, where a choice is to be made of reaction to the stimulus. Choice of behavior is influenced by experience, belief, motivation, constraint and nature. Fastest of these appears to be nature, as observed in the so-called knee-jerk reaction, where an individual reacts almost as an instantaneous reflex; this may be evolution's contribution to survivability, where to survive necessitated instant response.

Knee-jerk reaction may happen, but often a more considered approach suppresses the knee-jerk, as the individual or group considers the ramifications, the constraints, and the likely impact of typical behaviors. So, considered reaction to stimulus may be colored by belief, by culture, by ethical and moral considerations, by considerations such as 'not my place to respond,' or 'my responsibility — I must do something,' and so on.

These many factors permit the individual or group to behave in considered fashion, choosing one of the behavioral archetypes available, and exciting that selected behavior to a greater or lesser degree according to nature, motivation and constraint. So, an individual may feel that the stimulus just received (a serious personal insult, perhaps) necessitates an angry response, but is aware that the person delivering the insult is senior and influential. Knee-jerk reaction may be to strike out, but the considered approach, having suppressed the desire to hit, and weighing the merits of appearing very angry, may on balance decide instead to smile and enquire after the insulting senior's health. This will, in turn, evoke a response from the 'insulter,' who may 'lose face:' now it is the insulter who is faced with choosing between archetypal behaviors: he could back down and apologize; 'raise the stakes' and deliver another insult in the hope of provoking violent reaction; withdraw, realizing that he had been outwitted; and so on. Behavior evokes behaviors, sometimes in an interacting, escalating chain.

Excitation, in the figure, is 'powered' by Nature and by Motivation. Psychologists suggest that motivation comes in two forms: achievement motivation and conformance motivation. These two need not be in accord. A bright schoolchild may be motivated to achieve high scores in tests. However, not wishing to be seen as 'different,' the same child may be motivated to conform to the class norms, and to score much the same as the others in his peer group. Bright children have been known to deliberately 'flunk' tests just to be part of the 'in' crowd.

Examples used to illustrate the workings of the model have been of individuals. However, the Behavior Management model is for systems as wholes. The model may be applied to any whole that exhibits behavior, provided it meets the definition of 'unified whole.' For example, it would be reasonable to apply the Behavior Management Model to a corporation comprised of several thousand people provided that the corporation behaved as a unified whole. It could act so either because, through training and shared culture, the people all behaved as one. More likely, however, the behavior of the whole would be governed by the behavior of a few, perhaps a board of directors and senior managers, who would be receiving stimuli from outside of the corporation and who would also be responsible for interpreting and responding to those stimuli. So, if the central management group were a 'unified whole,' then the behavior of that group may typically be the behavior of the corporation; junior personnel within the corporation might be (and generally are) blissfully unaware.

The generic reference form model

The Form Management Model, or Form Model, is illustrated in Figure 5.19 as a Venn diagram. It consists of three aspects: structure, influence and potential.

Structure is what might be expected of a physical model, and consists of:

- *Boundary*. The limit within which the whole can be said to exist. Boundaries may be evident and clearly defined, as for a human, or a tank. On the other hand, they may be fuzzy, fluid, or virtually nonexistent in any tangible sense, as for a socioeconomic system, a political system, a military discipline system, a process-as-a-system, etc.
- *Subsystems*. Viable, open, interacting subsystems exist within the boundary, if there is one, or a 'sphere of influence,' if there is not; this sphere would be identified by the existence, or not of relative order, or reduced entropy. Each subsystem will contain intra-connected sub-subsystems. The boundary of any subsystem may be as fuzzy as that of its containing system. Subsystems perform functions that the whole system has to perform to pursue its many missions/to achieve its purpose (if any)/to solve its problem/to realize its concept of operations (CONOPS) — and to remain viable. Multi-functions may be performed as cooperative, series/parallel sets to achieve emergent effects, activated, coordinated and stopped by Mission Management, i.e., synergistically. Generically, subsystems will comprise sensors, effectors, processors, transformers, infrastructure, etc., arranged into a generic structure of clusters and linkages called architecture.
- *Connections*. Media for interactions between subsystems, including communications technology, CCTV, conversation, resource interchange, etc. Includes connections with external environment. (Interactions, *per se*, are activities, part of functions)
- *Relationships*. Spatial, hierarchical, moments, centers of inertia, lift, mass, gravity, etc. Also relationships in terms of personal interactions, formal and informal, as reinforced by Influence (see below)

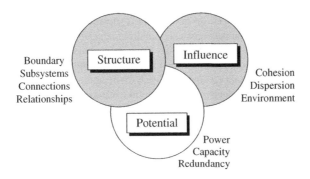

Figure 5.19   The Generic Reference Model Form (Hitchins, 1992).

The *complexity* of structure can be measured to compare architectures, and hence to identify potential options. A useful, insightful measure is configuration entropy, the degree of disorder in pattern. (Hitchins, 2003)

Potential. A system represents some degree of order (local reduction in entropy), which equates to (potential) energy. The GR (Form) Model presents this generically as:

■   *Power.* The rate at which internal energy can be converted to do internal work.
■   *Capacity.* The ability and amount of storage of both energy and resources. Together with Power, Capacity indicates endurance and the capability to do work on stored resources.
■   *Redundancy.* Replicated facilities providing a cushion against defects, failures and damage. Also enables reconfiguration.

These three present a necessary and sufficient set, encompassing all the energy, in its various forms, within the system.

Influence. There have to be various influences at work within the whole system:

■   *Cohesion.* The 'forces' binding the parts (subsystems) of the whole together.
  •   include physical and non-physical influences such as struts, mutual protection, discipline, loyalty, shared beliefs/culture/ideologies, etc.
■   *Dispersion* The 'forces' tending to push/pull the parts (subsystems) apart:
  •   include physical and non-physical dispersives such as rust, sabotage, computer viruses, psi-ops, divided loyalties, factions, spies, panic . . . .
■   *Environment.* The medium within and through which the Cohesive and Dispersive Influences act.

These three also form a necessary and sufficient set. Within a 'containing system,' stability/homeostasis can exist only based on some parity between dispersive and cohesive influences, the effects of which are mediated by environment, which may also be physical, and/or psychological.

The Generic Reference Model in List Form

The GRM may be used, as described above, as a checklist to ensure that a system design, or system solution, is complete, at least in terms of entities.

A typical example of the GRM in use in this way is shown in Table 5.2. There would be three such tables: one each for function management, viability management and form management. Only the function management table is shown, and has been instantiated with entries from the design of a notional command and control system, as might be employed by some mobile force.

The table is set out in three main columns, one each for the three aspects of function management: mission management, viability management, and resource management. Each of these three columns is then further subdivided into two columns, headed GRM and SOI respectively, for generic reference model and system of interest. Each GRM column is pre-filled in with the aspects for mission, viability and resource management respectively.

To check the completeness of a design, or of an extant system, the process is one of instantiation: for each GRM entry, enter the corresponding feature in the real-world system, or design as appropriate. This has been done perfunctorily in the table. The result in the example would be less than satisfactory. For instance, no counter-entry exists against 'management of conversion,' which refers to resources being converted within the system as the system does work, resulting in product or waste or both.

Similarly, responses against other entries would be inadequate. Homeostasis, for example, refers to more than the air temperature and humidity. It refers to the numbers of people of the right skills working within the C^2 system; to ability of the mobile formation to retain its operating environment when packed up and on the move; i.e., to maintain communications, to continue to receive sensor and intelligence information, to continue to process the information and produce strategies, decisions, orders and instructions on the move. Homeostasis is about maintaining a satisfactory internal steady state, albeit perhaps a dynamically varying one.

A command and control expert going through the table, not forgetting the behavior and form management tables accompanying it, would find much food for thought in the paucity of the answers, which would provoke questions and investigations, as a result of which hopefully the

Table 5.2 GRM architecture generation.

Internal architecture generation table					
Mission management Management of...		**Viability management** Management of...		**Resource management** Management of...	
GRM	SOI	GRM	SOI	GRM	SOI
...*information*	Communications. and imaging centre	...*synergy*	• Formation management C^2 Headquarters communications infrastructure	...*acquisition*	• CPRM Base supply Training
...*objectives*	CPRM	...*evolution*	Performance recording and analysis systems Post-operations evaluation	...*storage*	Logistics support vehicles Ready-use stores
...*strategy & plans*	C^2 Intelligence C^2 Operations	...*maintenance*	Mobile maintenance teams	...*distribution*	Mobile distribution fleet
...*execution*	C^2 Operations	...*survival*	Formation management Self-defense system	...*conversion*	?
...*cooperation*	C^2 Operations	...*homeostasis*	Climate control	...*disposal*	CPRM

CPRM: contingency planning and resource management. C^2: command and control.

design would be fully fleshed out. That, after all, would be the aim of employing the tabular GRM in this way.

The GRM and the Systems Approach

The generic reference model purports to show the many 'internal' aspects of a system. To use it in any dynamic sense, such as a simulation, it would be appropriate to set the GRM in the context of the systems approach, i.e., to view the GRM as an open system, receiving inflows and providing outflows, to its environment and interacting with, adapting to, other systems in that environment.

Figure 5.20 shows the basic notion. One GRM, at left, is interconnected with a second, at the right. This second might represent the environment in which the first is to operate, and from which it may draw its resources. This, then, is consistent with the systems approach. The left GRM might be a cardiovascular system, the right-hand GRM could be the whole body within which that cardiovascular system functions. Changes in the body affect the cardiovascular system, invoking changes in the cardiovascular system that change the body, in a continuous, orchestrated symphony. The work being done by heart and body expends energy that is continually re-supplied through the circulating blood.

The left GRM could be a naval destroyer engaging an opposing GRM/destroyer at right. As the left GRM fires upon the right GRM, the latter is damaged, so reducing its firepower, such that it becomes less effective. However, it still inflicts damage on the left GRM, which is also rendered less effective, and so on. Each affects the other, affects themselves, in a continuous interchange that, once started, may continue until resources run out, until one or both is rendered useless, etc. Unless of course, the ships are re-supplied and are able to repair themselves....

Figure 5.21 shows such a situation, in which two protagonists interact in some operational environment, while at the same time — or between rounds — each is also provided with logistics support and may be upgraded at the same time. Note that the GRM has been opened out compared with the previous graphic: function management has been elaborated into mission, resource and viability management, so that the external logistic systems can be seen as interacting with resource management. Such interactions could replenish consumables, such as power, energy, materials and throughput, and could also supply spare parts to replace those that had failed.

Figures 5.20 and 5.21 are helpful in presenting and understanding how the various interactions might take place, and in proper observation of the systems approach, but they are not best suited to dynamic simulation. For that, the GRM is best presented in layered form.

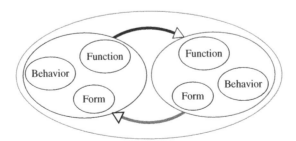

Figure 5.20   Interacting GRMs.

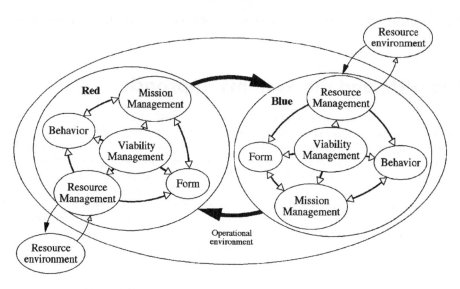

Figure 5.21   Interacting protagonists supported by non-combatants.

Figure 5.22 presents the GRM in layered form, with a total of five panels. The three central panels are, from the bottom, Form, Behavior and Mission Management at the top, part of Function Management. The other two parts of Function Management, Resource Management and Viability Management are at left and right respectively, and are shown as interacting with all of the central layers.

To understand the layered GRM, examine first Mission Management, top center. Within the layer, going from left to right, are: Collect information, Set/reset objectives, Strategize and plan, Execute (plan) and Cooperate with others. This layer contains the same basic five elements from the initial GRM, shown here interconnected into an information processing system — the interconnections signify both the flow of information and synergy. Note that Collect information is linked to Systems in the Form layer, indicating that information is collected through some system — a communication system, a sensor, perhaps, or a database, an intelligence source, or all of these. Note, too, that Execution is linked to Systems in the Form layer, indicating that information from Execute causes some system in the Form layer to perform some function — to move, to switch on or off, to hit a ball with a club, or whatever.

The Form layer, then, contains various tangible elements of a system; sensors, processors, effectors, structures, compartments, shields, power sources, connectors, teams of people, instruction manuals, etc: these tangible elements are able to perform functions. Since the GRM could represent a human as a system, the Form layer will contain all of the tangible parts that a human body contains: skeleton, organs, arteries, nerves, etc. The GRM could also represent an earthworm, or NASA's Apollo, the archetypal references, so what appears in the Form layer, by deduction, is the generic features common to both of these extreme archetypes, since both of them exhibit Form.

That would not necessarily be true for the center panel, annotated Behavior: we may not consider that the earthworm exhibits behavior, at least not in the classic sense of providing different responses to the same repeated stimulus. Many of the systems we may wish to represent might not exhibit behavior, in which case some, or all, of the layer may not apply.

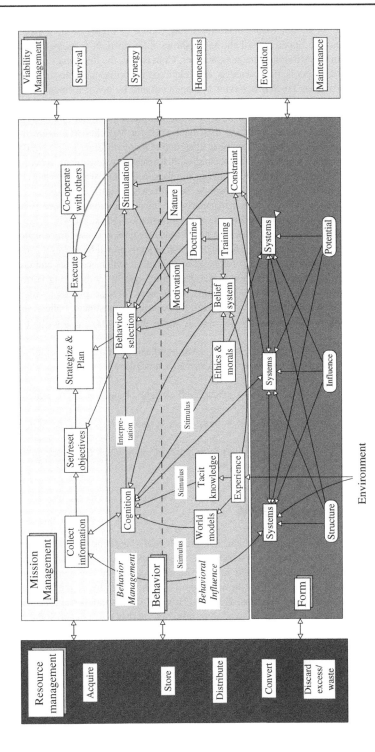

Figure 5.22 The Generic Reference Model in layered form.

The Behavior layer is similar to the basic model as shown in Figure 5.18: it has, however, been rearranged so that the upper part of the layer shows a left-to-right process in parallel with the information flow in the Mission Management layer. Conversely, the lower part of the Behavior layer has been arranged as 'pools of knowledge' that relate to the various tangible subsystems in the Form layer. The two parts are labeled Behavior Management and Behavior Influence respectively.

Some of the interconnections that must exist between aspects and entities in the layered GRM are not shown: some interconnections will be specific to particular instantiations of systems. Not shown, for example, are all the connections that must exist for Viability Management, i.e., for maintenance and for synergy, since these can be instantiated in many different ways. Similarly, there will, of necessity, be routes and storage associated with Resource management, where resources might include spare part replacements, food, energy supplies, personnel, waste products, inflowing parts and outflowing products, deliverable weapons, new process designs, etc., etc.

Some interconnections are shown. Note particularly the interconnections between Behavior and Mission Management. Typically, sensors (Form) collect information (Mission), which then has to be recognized and categorized (Behavior). Based upon this *interpretation* of sensed information, appropriate behavior will be selected (Behavior), based on which objectives will be set/reset and strategies and plans reviewed and updated (Mission), followed by execution which activates and orchestrates the various functions which can be executed in the interacting systems constituting Form.

Mission management, then, is not some cold, mechanical, precise process; rather, intelligence, objectives, strategies and plans are influenced by perceptions, instincts, beliefs, motivation, training, doctrine, ethics, morals, ideology, etc., etc. It seems that we may be describing emergent behavior. . . .

The GRM in layered form is insufficient on its own as a basis for simulating complex systems; instead, it has to form part of at least three components in simulation.

1. The GRM-based simulation of some instantiated system of interest, SOI, which is to be understood in context.
2. The GRM-based simulation(s) of one or more other systems with which the SOI is, or will be, interacting
3. The environment(s) in which items 1 and 2 exist and through which they interact. This could be a single environment, or many, according to context.

Instantiated layered GRM

Given these three, it is possible to simulate some open SOI, operating in and adapting to its environment, and interacting with other systems — as shown conceptually in Figures 5.20 and 5.21. A graphic of such an arrangement would contain elements too small to read, but the partly instantiated GRM model of Figure 5.23 shows items 2 and 3 vestigially at the bottom, marked as Operational Environment, with attendant Neutrals, Hostiles, Going and Terrain. As these terms suggest, this partial instantiation is of a naval ship, prepared, perhaps, to engage some enemy — or not, as the case unfolds.

The Form layer shows a variety of technological facilities, including: radar, navigation and communications, weapons management, internal systems sensors (to sense damage, fire, breakdown and failure), ship control, weapons control and weapons, together with various displays and controls to be used by personnel, whose actions and behaviors are represented in the other layers. These technological facilities are some of the complementary parts forming the whole system. Choice of parts is determined by examining the problem space, conceiving remedial solutions, establishing

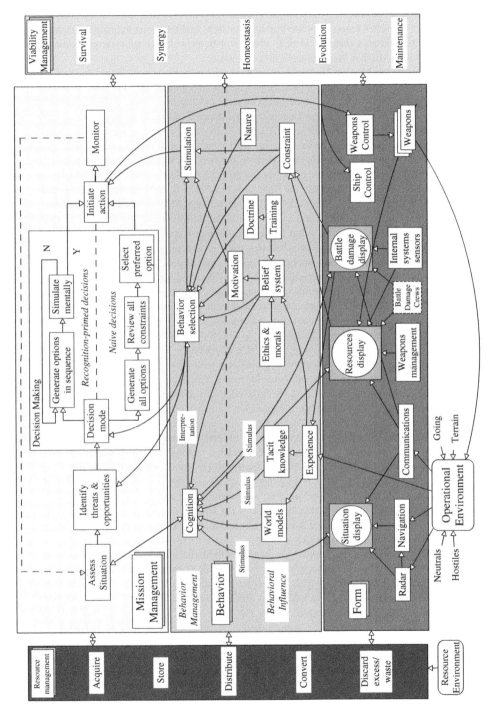

Figure 5.23 Partly instantiated layered GRM of a naval ship engaging an enemy at sea. Interacting systems, and particularly the ship(s) to be engaged, may be represented by one or more other, interacting GRMs — not shown.

solution system purpose, developing solution system concepts to achieve that purpose, together with a CONOPS, which identifies requisite prime mission functions and threat neutralization functions. These, then, are the functions that are needed, which will be provided as subsystems within the Form layer, and which Mission Management will orchestrate.

'Initiate Action,' in the Mission Management layer is shown as connected directly to Weapons Control in the Form Layer indicating that the action to be initiated, in this instance, is the launching (or not) of some weapon; there will be many such connections to manage many different activities and functions. Whether a weapon will be launched would depend, of course on circumstances and on the contemporary Rules of Engagement (ROE), part of Doctrine, shown in the Behavior layer. This interchange indicates how the various aspects in the three central layers operate with each other, both serially and in parallel, while the aspects in the two lateral panels ensure the ability of the whole system to continue to pursue its mission and purpose. Evidently, here, we are identifying the basis for emergence; within the model, the right parts have been brought together and caused to interact such as to create properties of the whole . . . . Note, too, that Mission Management conducts many actions and missions in parallel and in series: a naval destroyer may be fighting on three fronts, air, surface and subsurface, and it has to continue to float and move, as well as fight.

The Behavior layer may appear as standard. However, in instantiating it for the purposes of simulation, it would be necessary to establish the nature and substance of the various elements in the Behavior Management model, including tacit knowledge and world models appropriate to the particular instantiation. Similarly, it may prove important to know about the moral and ethical attitude, which would color behavior and might, for instance, either inhibit the selection of some archetypal behaviors, or perhaps favor others. For individuals or groups who are trained and operate to some standards, there may be some formal doctrine which dictates behavior expected from the whole system under different circumstances. The UK's Royal Navy, for example, is not alone in posting formal statements of doctrine to which captains may refer when necessary; the existence of formal doctrine suggests some conformity of behavior between different ships and crews, such that the behavior of the whole may be predicted by knowledge of the appropriate doctrine — always supposing the captain and crew adhere to it. . . captains in history have sometimes made their names by not so doing!

Mission Management in the figure shows an elaborated version of the basic model, with different approaches to making decisions: recognition-primed decision-making (RPD), in which an expert decision-maker uses experience to quickly find an answer that is good enough (Klein, 1989); and, naïve decision-making where the idea is to generate all the options, to separately generate all the criteria which would classify a good answer, and to trade one against the other to find the best solution. Naïve decision-making takes time and, if the would-be decision-maker(s) is/are not experienced, may either fail to generate sensible options, or may not properly associate options with their attendant and consequential risks.

Learning from experience-adaptive behavior. Note that the Behavior layer includes an element, Experience, which is informed (in this example) by both a Battle Damage Display and by information from the operational environment. This allows the simulated version of the ship-GRM to be aware both of how many simulated hits have been inflicted on the enemy, how many have been received, and what damage they have done. Using this comparative knowledge, of hits inflicted vs hits received, allows the (simulated) ship to modify it behavior, according to the Rules of Engagement.

For example, if the ship were taking a pounding, with much of its firepower lost, then the ship may withdraw from the engagement, undertake repairs, refuel and rearm and re-enter the fray — all as part of the simulation. Perhaps surprisingly, this makes the simulated GRM 'intelligent,' in the classic sense that it can change its behavior with experience of repeated stimulus. A variety of potential behaviors can be built in to the simulation, to enable it to behave intelligently: similarly, the enemy can be afforded intelligent behavior. The result of such inclusions is that the behavior of the whole (all participants in their contexts) is emergent, and may prove counterintuitive. This is evidently true for naval ships engaged in combat: it is equally true for other GRMs representing, say, two or more people cooperating, an enterprise competing with other enterprises in a market sector, a new transport system operating with other transport systems in a city, an artificial organ introduced into a body, etc., etc.

Does the GRM capture emergence?

We have seen that the instantiated GRM, with its specific parts arranged in a specific way so that specific interactions occur does exhibit emergence; in the example, a model of a naval destroyer exhibited emerging self-preserving behavior. By making the appropriate connections in Figure 5.23, it would be possible to synthesize capabilities of self-defense, offensive action, navigation, etc., all of which would be emergent in that they would be 'capabilities and behaviors of the whole not exclusively attributable to any of the rationally separable parts.' Not only would such connections coordinate the functions of various pieces of technology to cooperate, but the human operators would be taking a central part, too, making judgments and decisions, developing and implementing strategies that could be realized only if all of the parts of the whole — technology, people and process — worked in unison, i.e., as a unified whole. Importantly, this synthesis of emergent properties, capabilities and behaviors is sensibly undertaken only when the whole system is open to, and dynamically interacting with, other systems in an environment. As the example shows, too, the whole will adapt as a result of this continual interaction, and the emergent properties, capabilities and behaviors may change as the whole system adapts. It seems inescapable that such synthesis should be conducted in simulation, at least in the first instance, unless those conducting the synthesis are able to conduct such simulation mentally: for simple systems, this may be feasible; for complex systems, not.

   We may conclude, then, that while the GRM offers the basis for designing-in emergent properties, capabilities and behaviors of the whole, the realization of *particular* emergent properties, capabilities and behaviors requires some architect or designer to ensure that the necessary parts are present within the whole, and that realization also requires that he/she/they identify and make the appropriate connections so that the various functions can interact. In effect, the designer is identifying and implementing the pathways necessary for synergy, where synergy is defined as the coordination and cooperation of parts within the system to produce some desired external effect.

Comparing the GRM

The GRM has 'been around' for over 15 years (Hitchins, 1992), although the Behavior Management aspect was developed only recently (Hitchins, 2003), being rather more complex that either Form or Function Management aspects. The GRM bears comparison with Beers' Viable Systems Model — see page 110; as the respective model icons suggest, they are quite different.

   VSM is based on Beers' understanding of the human nervous system controlling the various parts and functions of the human body. It is, as Beers' described it, a cybernetic model of management. It accepts that the environment exists and may influence the direction an organization follows, under management guidance, but it is not open to environment, it does not adapt as it interacts with other systems, and so on. In the short term, and from a corporate management perspective, this might seem to be good. However, it may be some distance from defining a system being in harmony with its environment.

The GRM, on the other hand, may be thought of as modeled more on the whole body with its many physical, functional, organic, behavioral/psychological aspects. There may be a nervous system, but it is not overt, and control in the cybernetic sense is less in evidence. Instead, synergy and homeostasis are to the fore, with the open system managing inflows and outflows of substance, energy, resources, information, etc., such that a high-energy internal balance and steady state are maintained. Homeostasis ensures a stable foundation for systems to exist, function, behave and endure.

By choosing particular subsystems and interactions, an instantiated GRM could be induced to behave in much the same way as a cybernetic management 'machine,' but that would trivialize the GRM's potential. The GRM is very much about systems that are open to their environment, interactive with other systems, mutually adaptive with other systems and potentially both complex and intelligent: and, it is about systems that are viewed, analyzed, understood and synthesized in their dynamic, interactive context. Above all, it is about synthesizing requisite emergent properties, capabilities and behaviors; in that respect, the GRM is unique. The synthesis of requisite emergent properties, capabilities and behaviors should be the goal for systems engineering of every system, of every kind, in every sphere of knowledge, understanding or endeavor.

Summary

A series of models has been presented, all relevant in some way to systems engineering. They include a 5-layer model showing systems engineering conceptually applied at different levels from product through project, enterprise, industry and finally socioeconomic system; process models, management models, cybernetic models, Beers' Viable Systems Model (VSM), and the Generic Reference Model (GRM). Each of the models has its purpose, application and merits.

The essence of systems engineering is posited as being the synthesis of requisite emergent properties, capabilities and behaviors in context, as the solution to some problem (see The essence of systems engineering on page 120). The Generic Reference Model is unique in offering the potential for such synthesis. As such, it offers a new and better method for synthesizing solutions to complex problems and issues, i.e., for the systems engineering of all kinds of systems, and will act as a linchpin in the Systems Methodology to be presented in Part II.

Assignment

1. You are tasked with examining the management organization and communications for a major international corporation of your choosing. Develop a model of the whole organization using Beers' VSM as a template. Compare the model with the way in which the whole corporation is actually organized and structured. Identify significant differences and apparent omissions, either in the model, or in the corporation, and suggest reasons for the differences.
2. Figure 5.10 showed a diagrammatic representation of systems engineering at Level 1 of the 5-layer model of systems engineering. Compare the approach and method indicated in the figure with classic engineering approaches. Identify major differences and explain why they might exist, and what might be their significance.

3. Develop a layered Generic Reference Model for the human body, assuming the individual to be occupied with normal daily activities of working with other people in a team, etc., excluding association with any technology such as tools or transport.

4. Using the GRM from 3 as a guide, develop a second layered GRM for a group of people, working in some shop of factory, all trained to perform different aspects of the same task.

5. For the human body, develop a dynamic model of resource management, considering only the essential resources of food, water and oxygen.

6. Repeat 3, but for a Formula One racing car, identifying what resources you consider essential . . . .

Case A: Japanese Lean Volume Supply Systems

Introduction

The Japanese industrial phenomenon started with World War II. At the end of the war, Japan was a defeated nation, with the remnants of a high-capacity vehicle manufacturing industry, but with limited markets for any manufactured vehicles, since postwar arrangements dictated by the US did not allow export. Meanwhile, the US dominated global markets with mass production, after the style of Henry Ford. Faced with only domestic markets for their vehicles, Japanese manufacturers faced a bleak choice: amalgamate; go out of business; or, find some way of keeping going with a greatly diminished market.

The Toyota Motor Company was one such organization. After the war, Taiichi Ohno and Eiji Toyoda pioneered the concept of lean production (Womack, *et al.*, 1990). Ohno obtained some second-hand body presses from the US and experimented with them. The body presses were fitted with male and female dies, the shape of the required body part: these had to be accurately set up, else both the sheet metal and the die could be damaged in the process. In the US, changing dies to make a different body part would take 1–2 days, using highly skilled labor. Ohno's experiments cut die-change times down to three minutes (!), and moreover deskilled the task, using the same staff that worked the press to change the dies.

This one change afforded great opportunity: whereas with mass production it was essential to have long runs of identical items to amortize the cost of the machinery, with Ohno's approach it was feasible to have short runs, and to change dies frequently and easily. This, in turn, opened more doors. Instead of lengthy and expensive marketing assessments to assess likely sales potential for some proposed product, it now became simpler, quicker and much cheaper to make a spread of new items, present them to the market, and see which sold best. Toyota could now make a range of vehicles, cars, lorries, pickups, etc., for the domestic market, with economical short production runs on each.

Such a small market was soon satisfied, however. Repeat order sales would come only if the owners of vehicles had to replace them for some reason: if the vehicles wore out, or were unreliable, or rusted. . . . A regime was introduced where vehicles were allotted a life of, typically, six years, after which they had to be scrapped; this gave the motor industry a guaranteed turnover, but at a cost to the motorist. Initially, Japanese goods, including cars, were viewed as of poor quality: to use the contemporary expression, they had 'built-in obsolescence;' simply, they were designed to wear out quickly. This did, indeed, mean that the purchaser of a Japanese car would soon need a new one, but the buyer was very likely to choose a vehicle from a different manufacturer in the hope of getting a better deal.

Quality was an issue. Poor quality deterred buyers. Good quality meant the vehicles did not wear out, leading to vehicle retention and delayed repeat business. How to improve the quality and increase the business turnover at the same time . . . .

The answer was innovation: if people who had already bought a good-quality, reliable vehicle/computer/hi-fi/washing machine/etc., could be persuaded that there was a newer model or version now available that did more things, had more capability, went faster, or whatever, then they would buy the new one, regardless of the fact that their existing version was still perfectly good. So, the solution was a combination of high quality and continual innovation: together, these powered market demand ('market pull'), which lean volume supply systems depend upon for continued existence.

None of the above explains the leanness aspect of lean volume supply, however. Leanness derives from many different, interacting factors, including:

- Market pull (see above), as opposed to the production push of mass production: leads to . . .
- . . . making and assembling only that which had effectively been sold already,
- thereby eradicating stocks of unsold produce.
- Minimizing work in progress (WIP) within factories, in goods outwards, goods inwards, traveling between factories, between processes within factories, etc., etc. WIP costs money to buy, and makes no profit — in essence, it becomes a recurring cost, or an overhead.
- Just-in-time (JIT), part of minimizing WIP; parts to be machined, assembled, etc., are not stocked, are not held in batches awaiting processing, etc., but arrive just as they are to be processed.
- JIT and minimizing WIP dictate that parts being moved between factories are moved individually, rather than in lorries which wait to accumulate a full load before traveling. Motorcycle couriers may transport parts instead, as this may prove more economical overall, when the cost of the WIP-parts is considered.
- Small batches in innovative manufacture show up defects quickly, allowing errors to be corrected: also eliminate need for large inventories.
- Highly skilled and motivated workforce, multi-skilled, able and willing to tackle any job, resulting in high productivity per worker, and consequent low labor costs.
- Value engineering to establish the essential function of various parts/subsystems and their costs. Operates hand in glove with . . .
- . . . 'market price minus:' establishing an attractive market price early in the design process and working back from the overall price to the value, function and price of the various parts.
- Outsourcing complete subassemblies such as bumpers, dashboards and instruments, interior furnishings and furniture, etc. This is as opposed to the mass production practice where a bumper might be sourced as, say fifteen different parts, each of which could be separately designed by the mass producer, and separately sourced, with the attendant risk to the assembler that the various parts were, in some way, incompatible.
- Multi-sourcing, to guarantee supplies, and to shorten supplier reaction times

Culture played a large part in the Japanese success story, too; culture is often underrated in Western evaluations. The effects of culture are not always evident. For instance, it has been estimated that there may be as many as thirty times the number of personnel employed in counting stock of various kinds in a western factory as in an equivalent Japanese factory. Why? In the West, anything that is not secured is likely to 'walk;' i.e., be pilfered (stolen) by employees. Tires, engines, wipers, windows, seats, etc: every part, and even the machinery, will be stolen if not safeguarded. In a traditional Japanese company, all of the employees are 'family:' one does not steal from family, so, no pilfering. Hence, costs in a Japanese company will be significantly less, both because there is no need to replace pilfered stock and there is no need to employ small armies of stocktakers and security guards.

The culture allows different practices. In a western car manufacturer, tires might be supplied in bulk to some goods inwards store, and paid for as a load. In a Japanese company, the tires are supplied just in time (to be fitted to wheels and assembled), but the supplier is not paid until a vehicle rolls out of the door — at which point he is paid for five tires. This has several effects: first, the supplier is motivated to supply only what is needed, when it is needed, else he is funding WIP without return; second, the supplier becomes part of the team since he becomes actively interested if there are holdups on the assembly line, and will contribute to resolving holdup problems, which would delay his receipt of payment.

All of which presupposes a highly motivated Japanese workforce. Workers are given 'jobs for life:' they are paid by seniority for doing any given job, rather than by the nature of the job; were they to move to another company, they would then start at the bottom of the seniority ladder, and would take time, perhaps many years, to match their previous earnings. Workers become short-term fixed cost, like machinery, but cannot be depreciated. Instead, it was seen as good to continuously enhance workers' skills, and to *use* their knowledge and experience: in top Japanese companies, every worker actively contributes ways of improving processes and products, and is rewarded for so doing. The rate at which each worker contributes 'ideas and inventions' may be an order higher per worker than in western companies.

Symptomatic of the East–West dichotomy is the Japanese approach where workers report on the performance of their team leaders, while in the West, managers report on the performance of their workers. Japanese workers tend to work in teams, with a leader. Successful leaders attract the best workers to their teams, enabling such teams to compete successfully with other teams. Successful teams may be paid bonuses, which are for the whole team, not for the individual; generally, each individual in a team might receive identical shares of the team bonus. This has a cohesive effect within the team.

Suppose, for example, that one of the team were to be temporarily ill, off color, or perhaps suffering from personal problems. The prospect of a team bonus, rather than an individual bonus, motivates the other members of the team to pitch in and help the lagging individual, so that the whole team continues to perform. This is in stark contrast to 'performance-related pay (PRP),' popular in some control-centric western management circles, where individuals are earmarked for higher pay based on their manager's assessment of their performance. (PRP can be divisive, unfair, and is wide open to abuse, favoritism and bias. While the intent of PRP may seem obvious and rational, it is essentially an individual control mechanism that may demotivate, inhibit and fragment the team, rather than achieve its intended goal.)

Cultural differences are to be seen in contracting arrangements, too. Mass production sees the end supplier, the lead company, designing all the parts of some future product, and then putting the parts out to competitive tender so that each part can be procured at the lowest price. In this way, it is hoped to manufacture the end product for the lowest overall cost. It does not work well. Bidders, eager to get the business, offer to make parts at a price that is less than they can really afford, i.e., one that affords insufficient profit margin; if they do not, they may be unlikely to get the business. Once they have got the business, and the customer is supposedly committed, then the supplier will attempt either to make the part for less, perhaps by reducing quality, using different materials, finish, etc., or will try to raise the price per part, or both. This results in aggravation, confrontation, strikes, etc., and a general tendency for quality to go down and price to rise throughout a long production run.

Japanese lean production arrangements can be quite different. The design of some new product, or a new version of an existing product, may be a joint affair, with suppliers of various subsystems/subassemblies taking part. Once an overall design is generated, together with a selling price that will guarantee sales, then reverse engineering sees the design and specification of the various subsystems. The price of each subsystem is also agreed with the main contractor, and is set by agreement so that the respective supplier/subcontractor may anticipate a reasonable profit. This may involve a revised design of the subsystem, perhaps using different materials, with the active assistance of the main contractor, so that it may be manufactured to the right quality for the previously agreed price.

During a production 'run,' should a supplier find a way of making the part for less, then that supplier stands to make a greater profit. Alternatively, this could mean that the selling price for the whole product could be reduced, thereby enhancing sales prospects. One approach, then, is for the supplier to reduce his price to the main contractor, for the main contractor to reduce the overall price, for increased profit to accrue from increased sales, and for the supplier who lowered his price to benefit from the increased profit on sales. With this approach, it is in the suppliers' best interest to drive manufacturing costs down, and quality up, so that the supplier as well as the end contractor may increase profit. This arrangement may be included in the contract arrangements, with the result that the price of products may reduce throughout a production 'run.'

So, contracting arrangements for lean supply encourage cooperation and coordination between suppliers and end contractors — quite different from mass production, with its history of confrontation, discontent, strikes and rising prices. Culture in the West seems to be set into a pattern of confrontation and aggravation: lack of trust is endemic in both directions between governments and defense contractors, subcontractors and main contractors, and so on. Such a cultures will not promote and sustain lean production, at least not Japanese style.

Attempts in the US and in UK during the 1990s to introduce lean acquisition practices into national defense procurement were scuppered by the extant bureaucracies in both countries, each of which seem much more comfortable with the atmosphere of mutual antagonism and distrust that exists between government and the defense industry. Besides, the two bureaucracies would have been seriously undermined, to the extent that neither would have really been required at all — which would have saved each nation untold billions of dollars/pounds, and made thousands upon thousands of civil servants unnecessary and unemployed. Just as turkeys do not vote for Thanksgiving/Christmas, of course, bureaucracies exist chiefly to defend themselves . . . which, in this case, they did *par excellence.*

Investigation

It is not the intent of this case study to explore the whole of the Japanese lean volume supply system; others have done that very well, and have written excellent volumes on the subject. Notable among these is 'the machine that changed the world' (Womack *et al.,* 1990), which is based in a 5-year, 5-million dollar MIT study on the future of the automobile. Nor is it the intent to suggest that all Japanese industrial companies are as good as the best; they are not.

Instead, it is the intent to look into some of the phenomena that are at the heart of lean production, to see if the detail matches the hype, and to find rational — perhaps even scientific — support for the Japanese approach to systems engineering at enterprise, industry and socio-economic levels.

The open system viewpoint

Figure A.1 shows an individual business entity, part of some notional lean volume supply system, presented in the form of a GRM. The enterprise is shown shaded, in the center, with the standard GRM aspects, which include culture within Behavior management. Parts from various suppliers, shown as Resource environment are passed through the enterprise, leaving at right as integrated products into an Operational environment, generally a market of some kind. A link from the Operational environment back to the parts supply at left suggests that flow of resources through the system is in response to demand for parts, and that such demand is accompanied by revenue from sales; in simple English, sales pay for more parts to make more products. For each enterprise in a chain, the next enterprise may be seen as its market.

The figure shows that there are two quite distinct categories of resource. Center bottom of the figure is a second resource pool, coupled to Viability management; these are resources that establish and maintain enterprise viability: synergy, maintenance, evolution, survivability, and homeostasis (S-MESH). Of these five, synergy and homeostasis are, perhaps, the most significant. Inflows and outflows will include men, machines, materials and money; with recruiting, retirement, training, research, development, planning, administration, security, internal communications, supply chain communications, outsourcing, goods inwards, goods outwards, sales, etc., etc. Maintaining

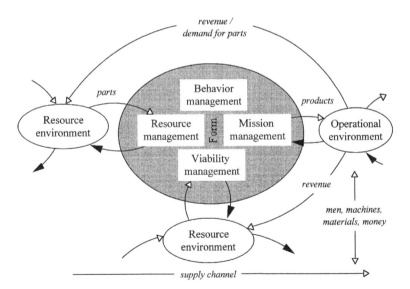

Figure A.1   An enterprise, represented by a GRM, in context as one link in a lean volume supply system.

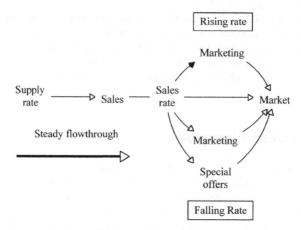

Figure A.2 Heijunka. Flow smoothing by varying marketing and sales efforts to manipulate demand, leading to a steady flowthrough and consequent minimized unit production costs . . . ; solid arrowhead indicates *reduced* marketing effort.

homeostasis implies inflow/outflow balance between a wide range of many different features, some tangible (e.g., numbers of employees) and some less tangible (e.g., skill levels).

Figure A.2 shows a more pragmatic view of a notional enterprise, this time at the head of a supply chain, i.e., interfacing with the market and the consumer. The figure shows the bare bones of an end supplier or main contractor, with the manufacturer making a range of products, and being fed with parts by a range of suppliers for these various products. Each of the various products will experience peaks and troughs in demand; part of the manufacturer's plan will be to try to maintain the sales rate of each product in the range at a steady level: it can be shown that increased WIP, and hence potentially increasing costs, accompanies both a rise in manufacturing rate, and a fall in manufacturing rate; the lowest WIP is associated with steady manufacturing rates. Smoothing the flow rate into the sales channel is *Heijunka*.

Advertising tends to increase sales, demand and flow rate. Conversely, reducing advertising reduces demand, sales and flow rate. It is possible, therefore to use advertising as a 'control valve,' turning it on to anticipate dips in production flow, and similarly turning it off to anticipate rising demand. If these rises and dips can be effectively counteracted, then the revenue stream from the combined flows of all the products in the supply channel can, in principle, be maximized.

Of course, a time will come when a product within the product range is losing its consumer appeal to such an extent that neither increased advertising, nor 'special deals' can maintain sales flow: this product will have to be replaced in the product range, either by a totally new product or, perhaps, by introducing a new version of the previous product, one with enhanced 'bells and whistles,' perhaps even one that was pre-planned to deal with just this eventuality.

When the sales rate for a current product falls below a threshold, despite best efforts at advertising, etc., then the top loop in the figure will kick in, to conceive, create, test, prove and insert a new product or new version of a product into the production flow stream. In an ideal world, the transfer from the old to the new would be seamless; there would be no cessation of manufacture, no dip in sales, etc. For this to happen, the new product/version will have been advertised in advance to build up demand, so that it immediately has a sales flow upon introduction.

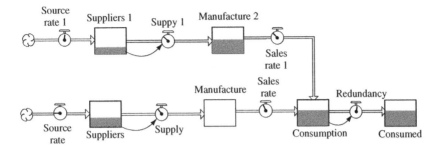

Figure A.3 STELLA™ simulation model of Heijunka, or production flow smoothing.

The switchover situation is simulated, as in Figure A.3, which shows the original product in the supply chain at the bottom, and its replacement product at the top. As demand for the first product wanes despite sales promotions and marketing, a point will be reached at which manufacture of the first product falls away and manufacture of replacement product takes over.

Simulation results are shown in Graph A.1, where line 3 — the 'sum production rate' — is evidently not perfectly flat, indicating a dip in the sum of the production rates for the two products. It is difficult to bring about the ideal flat curve, even in a simulation: in the real world, there will be many imponderables. How much advertising, and of what kind, is needed to create demand for the new product, such that the rising curve of demand matches the falling curve of demand for

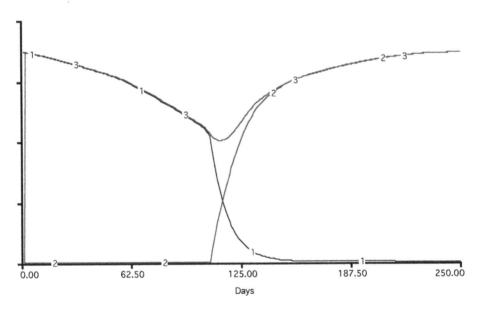

Graph A.1 Heijunka — product switchover: STELLA™ simulation. (X-axis is elapsed days). Line 1 is the production rate for the product being replaced. Line 2 is the production rate for the replacement product. Line 3 shows the sum of their production rates. In an ideal world, line 3 would be flat and horizontal, showing a constant production rate with no 'hiccups.'

the old product? How much anticipation is needed, such that the replacement product is ready for manufacture in time, without being either too early or too late.

Accurately achieving the ideal switchover would require great skill, and not a little luck. On the other hand, the general approach could (and does) offer a reasonably smooth transition, and may be acceptable in the context of there being many other products in the product range that were continuing to be sold through the supply channel, unaffected by the switchover.

There are many mechanisms for maintaining homeostasis. Heijunka is one: supply assurance is another, in which forward planning determines the needs for future parts, including those associated with a new product or variant, and the new process that its manufacture will necessitate. New parts will have been designed jointly with the supplier, who may also provide assembly jigs, bandoliers for automated assembly of electronic devices, test harnesses, etc., etc., as well as the new parts to be assembled. Inserting the new product or variant seamlessly into the supply channel requires careful planning, testing and coordination. Given that, the overall result is another aspect of homeostasis, in that the overall inflow outflow balance is maintained — an essential for any open system to endure.

Market pull vs production push

The lean volume supply chain operates in a 'market pull' regime as opposed to a 'production push' regime usually associated with mass production. In production push, production rates are maximized and held constant in order to amortize the cost of machinery and manpower across a large number of products, so keeping the unit production cost (UPC) to a minimum. This works well, provided there is a ready market for the product. When there is not, finished products accumulate in storage, at great expense to the manufacturer who has paid for the materials, the production and now the storage, all without any return. We are all familiar with the sight of large storage areas filled with brand-new, parked cars awaiting shipment and sale.

Market pull operates quite differently, by first creating demand for a product, and then meeting that demand. Demand is created by continual innovation, with new products, or new variants of existing products, being advertised and brought to the market with sufficient frequency to fill the production capacity. It has its problems: creating too many variants too quickly — 'product churning' — can cause customer disenchantment, as their bright, shiny new 'toy' is superseded by an even newer, even shinier model before they have had time to show theirs off! There is also the potential issue of delay: if the manufacturer is genuinely responding to demand, and only starts to make a product after it has been sold, then the customer will have to wait for the production time to elapse before receiving the product. For car production on mainland Japan, this delay may be a few weeks; for global enterprises, the delay would be biased by the time taken for parts to travel between various factories in the supply chain. Production push starts with a production target, which sees parts made in, say, the third tier being sold to second tier enterprises, which process and sell their products to first tier, the lead company, which then sells into the market.

In contrast, market pull starts market demand, creating a market discrepancy, which demands products to sell from the first tier. This creates a first tier discrepancy, which demands product parts from the second tier suppliers, and so on, from lead company back up the supply line.

Graph A.2 shows results from running a supply chain simulation in the production push mode. Line 3 shows market demand for the product going up and down vigorously, before leveling out. What happens within the supply chain? Lines 1 and 2 show typical first and second tier accumulation of parts and products, works in progress, which are 'caught' in the system by the downturn in demand. The second tier in particular sees a dramatic rise in 'unsold' product

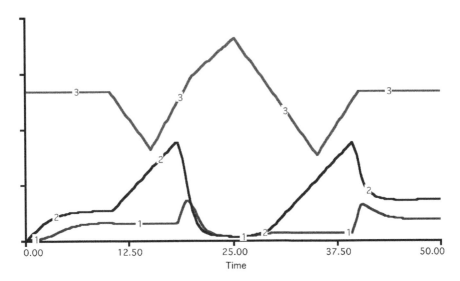

Graph A.2 Production push. Line 1: first tier WIP. Line 2: second tier WIP. Line 3: variable market demand for product. Variation in demand sees significant increase in WIP, increasing overheads and product costs.

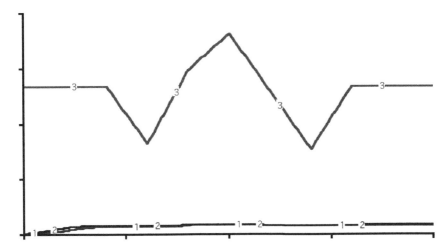

Graph A.3 Market pull. Line 1: first tier WIP. Line 2: second tier WIP. Line 3: variable market demand for product. Variation in demand has little effect on WIP within the supply chain.

parts, since (in the simple model) the rates of production have continued unabated. Even when market demand returns to its previous level, lines 2 and 3 do not return to their original, lower, levels.

Graph A.3 shows the same supply chain simulation, but this time run in market pull mode: the results are significantly different, with negligible amounts of WIP being captured, despite the same major changes in demand occurring.

The simulation is very simple, of course, but it does illustrate one aspect of lean production. While in mass production, dips in demand can leave the manufacturer with very large inventories of unsold stock — all of which has to be paid for — there is no such problem with market pull.

That is not the whole story, however. By changing the nature of demand such that it always outstrips production capacity (i.e., no dip), the simulation suggests that production push is at least as effective as market pull, and perhaps even more so, since there can be less delay in the production push system. It is reasonable to conclude, then, that market pull affords leaner production in variable markets, but not necessarily in markets where demand is continuously rising to outstrip supply.

Kaizen and the assembly line

Taiichi Ohno's approach to assembling vehicles on an assembly line was unlike that of his mass production contemporaries. It was their general practice to keep the production/assembly line moving, that is to avoid any holdups. Their objective was to make as many of the product in the shortest time and to amortize the cost of the expensive manufacturing facilities by keeping them in continuous operation. So, the intent was to get the assembly line working at full speed from the beginning, and to keep it going.

Potential problems would arise when mass production assemblers tried to fit a part that either would not fit properly, or would not work correctly when fitted. To correct this *in situ* would hold up the line at that point, creating a queue behind, and rendering the assemblers ahead of the problem idle. This was unacceptable, so the offending part was 'tagged,' i.e., marked with a label to say what was wrong, and the assembly process continued unabated. Subsequent to completing the assembly process, tagged assemblies were taken to an area where work was done to repair the defect or deficiency. The repaired assembly could then pass quality control/inspection.

This did not work well. In practice, some tagged parts were buried under other assembled parts, and could not be accessed without significant disassembly; in consequence, defects that did not materially affect operation might be left *in situ* and 'forgotten.' As a result, mass produced cars, for example, could be less reliable than expected and hoped for. Moreover, post-assembly repair was not a rare event: on some assembly lines, the floor area set aside for post-assembly repair could be as much as 40% of the overall floor space.

Taiichi Ohno installed an alarm over every assembly station in the line. When any of the parts to be assembled/installed did not fit, or was in any way unacceptable, the assembler at that point raised the alarm, and the complete line stopped, while the defect was investigated. With all the other assemblers now unoccupied, there were plenty of people available to undertake the investigation. If the problem lay with the supplier of the part, then a team would visit the supplier to investigate the problem, help the supplier put things right, and then return to the assembly plant, to restart the assembly line.

With a significant number of parts to be assembled, the prospects for delay were manifold. If the line were not stopped at the first assembly station, then it would be at the second, or the third, and so on. Taiichi Ohno's contemporaries suspected that he would never assemble a single car in this way, since none would ever make it to the end of the line. However, he persisted, and each time that problems were sorted out, the assembly process went further along the line until, finally, it reached the end without any holdups. In the process, most, if not all, of the issues of supplied part quality had been resolved with the various suppliers, who were now welded into a composite team.

So successful was Ohno's approach that post-assembly quality control proved unnecessary. There were no tagged parts, no 'buried problems,' so the end vehicle was of good quality and reliable throughout. It also became characteristic of lean production assembly plants that they take time to reach their maximum assembly rate. Whereas mass production sought to sustain maximum assembly rate from the start – and failed — lean production assembly followed a rising curve as problems were eliminated, leveling at a maximum rate that could be sustained. And there was no need for a large, post-assembly repair area.

Ohno's approach is consistent with *Kaizen* — the philosophy of continuous improvement. Kaizen is to be observed in many areas of lean production, and may be seen as recognition that it is not possible to get everything absolutely right first time. Instead, it may be better, while trying to get things right, to continuously improve aspects of process, of design, of practices. If Kaizen is adopted in every aspect, then the end product must, eventually, be superlative. Moreover, if the nature of the demand changes, then Kaizen will oblige the product to follow the change. Kaizen is to be seen in successive models of Toyota cars, for instance, where each successive model overcomes deficiencies that may have emerged during the operation of the previous model. Kaizen has been criticized by those who say that Japanese car designs lack the flair of, say, a Ferrari, with more attention to detail than to the whole. On the other hand, Japanese car designs are continually improving and moreover Toyota's cars are likely to spend relatively little time in repair and servicing. The proof of that pudding is in the sales figures. . . .

The dynamics of the Toyota-style assembly process, its behavior if you will, are not difficult to simulate; creating such simulations helps to understand how queues and delays can occur, how they can be ameliorated, and so on. Figure A.4 shows a STELLA™ simulation of two stages in a notional multi-stage assembly process.

The partly assembled composite is seen entering Sector 2 at top left, marked exit GP server (where GP is Good Parts, as opposed to BP: Bad Parts.) These Good Parts pass through the Coupler into Queue 2, as directed by Operations Control, which coordinates activities; injected into the flow from Sector 1 are Good Parts 2, from the supplier of parts to be assembled in Sector 2. Also entering Queue 2 are Bad Parts, again from the Stage2 supplier, although these are not yet known to be bad.

On attempting to fit the flow of mixed good and bad parts, the good parts enter GP server 2 into Good Parts Assembly 2, while the bad parts are detected as unfit and are diverted into Bad Parts Diverter 2. As soon as this diversion occurs, Operations Control 2 stops the flow of both good and bad parts into Sector 2, and — not shown on this small part of the overall simulation — also stops all the other stages. Mean Bad Parts 2, bottom, is a reservoir of bad parts that is then reduced, simulating the process of eradicating the defect at source, through Rate Decrease 2. When the defect has been eradicated, the line starts up again — all stages in synchronism. Note that Mean Bad Parts 2, bottom, is fed with Degrading 2, which reintroduces defects — Bad Parts — over time, to represent lapses in quality against which no suppliers may completely protect themselves.

Figure A.4 represents only two stages in an assembly line: there could be many more, and the whole may be simulated by stringing together a sequence of such stages in series and, in some case, in parallel too; this would be facilitated by introducing more Couplers.

Graph A.4 shows the result from simulating just three stages. The y-axis shows the number of products emerging from the end of a three-stage assembly process over time. To start with, nothing emerges, as the initial defects are sorted out, stage by stage. Then, when early production starts, there are continual interruptions, as less obvious defects are discovered and, possibly, some new ones have crept in. The interruptions become less frequent, finally petering out to leave the assembly line operating at a set assembly/production rate. With more than three stages, there is

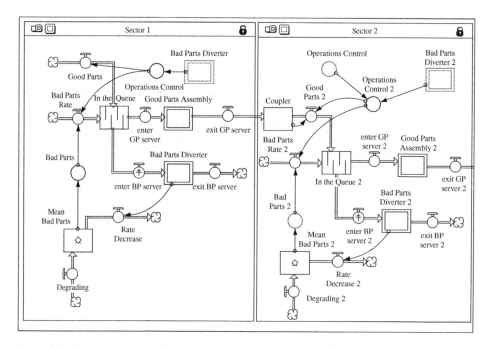

Figure A.4   Two stages in a multi-stage lean production assembly line. Both stages are identical, as are subsequent stages in the simulation. A mixture of 'good' and 'bad' parts is received from suppliers. Bad parts, ones that will not fit or work correctly, are detected and diverted. The line is stopped (Operations Control), while diverted bad parts are investigated at source, where the deficiency is corrected, so reducing the incidence of further bad parts. Part of a STELLA™ simulation

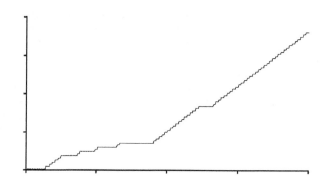

Graph A.4   Increasing production rate profile, lean production simulation.

potential for more initial delays; these lead to the characteristic slow start-up dynamic of many lean production exercises.

Processes are continually made leaner though kaizen: finding ways to do a particular job more quickly, or with less effort; identifying more suitable, less expensive materials, perhaps; simplifying

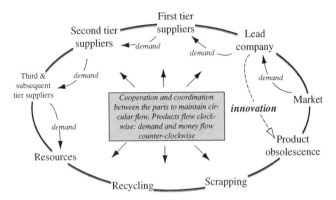

Metrics: 1. *Flow rate around the system*
2. *Proportion of circulation time resources stay in Market/use*
3. *Recycling proportion*
4. *Product value to environment ÷ product cost to environment*

Figure A.5 A lean volume supply system formed into a circle, in which obsolescent products are recycled to minimize the need for raw materials. Flow rate of materials around the system corresponds to flow rate of money in the opposite direction around the system. Coordination around the circle may be augmented by a dedicated IT system, triggering synchronized parts manufacture in response to demand. Cooperation is endemic in the culture, the supplier contracting arrangements, and in the Keiretsu — see text. Note the idealized metrics.

the assembly process into straighter lines, perhaps; and so on. As a result, there is a noticeable 'learning curve' effect, in which suppliers' costs fall through production — in stark contrast to mass production experiences.

Keiretsu

Japanese lean volume supply organizations (LVSOs) operate within a financial and commercial environment that is significantly different from that of the West. Figure A.8 shows a notional LVSO, which has been formed into a loop by recycling obsolescent products, to reduce product costs and to reduce dependence on raw supplies. The lead company supplies to the market in response to market demand: products and materials flow clockwise around the circle. Revenue from sales goes with demand, anticlockwise around the circle. Products are made obsolescent by the advertising of innovative new products currently, or imminently, available from the company/manufacturer. It is the company/manufacturer that creates the demand to which it must then respond.

A LVSO may consist of a number of companies linked into a fan-in, or loop structure. A *keiretsu*, an association of commercial entities operating a system of circular equity holding, may support the industry. The entities would include suppliers, the lead company, banks, trading companies, insurance companies, even competitors: some twenty in all. The entities are arranged in a conceptual circle, in which each entity holds equity in the next around the circle. So, each member of the keiretsu is financially dependent on the others in the circle, and it is of course in everyone's interests to support all of the others. Western organizations, hoping to mount a stock exchange 'dawn raid' on a Japanese company might find, to their disappointment, very few shares available to buy, since other Japanese companies, in whose view the circle is considered to be defensively strong, already own them. Should one of the companies in a keiretsu feel inclined to dispose of their shares, then they would be likely to find shares in their own company being disposed of by others in the same keiretsu. Hence, shares move very little in and out of the keiretsu, to and from non-keiretsu organizations.

Unlike their predecessors, the zaibatsu, which were essentially holding companies, keiretsu are not legal entities. The member companies are not legally united, but are held together by a sense of reciprocal obligation. Zaibatsu were dismantled by the US after World War II. Keiretsu have vast financial resources, which underpin the integrity of member companies. This arrangement might well be deemed socio-economic systems engineering, Japanese style.

Summary

The contrast between the traditional approach to mass production, as exemplified by the US manufacturing industry, and the lean volume supply, as exemplified by Toyota, Sony, etc., is best seen in tabular form:

Traditional USA	Comparison	Japan
Profit	Objective	Survival
Free	Competition	Between chains/circles
Free market	Regulation	Indigenization
Production push	Assembly	Market pull
Cost plus	Pricing	Market minus
Adversarial	Contract	Synergistic
Specialist	Defense	Homogeneous
Hire and fire	Labor	Jobs for life
Specializations	Skills	Multi-skilled
Lowest bid wins	Suppliers	Vital source — protect
Supplier stocks	Inventory	Nobody stocks

This stark comparison has been somewhat blurred in recent years, as some Japanese companies have been influenced by US practices, and as some US companies have adopted traditional Japanese practices. In any event, not all Japanese companies are as good as their leading companies. Today there are car manufacturers in the West who can match the performance of the best Japanese suppliers; to do so, however, they have had to instill a different culture throughout their organizations, one much more akin to that of the best Japanese industrial giants.

Attempts in the West to transfer the benefits of Japanese-style lean volume supply to acquisition and procurement in the defense sector have not been successful, however: in both the US and the UK, attempts have been frustrated by government bureaucracies. The culture of secrecy, mutual distrust and antagonism that pervades government defense acquisition and defense suppliers is alien to, and seems unable to accept, any sensible degree of cooperation and trust. Governments take their responsibility for safeguarding public monies to include tight control over defense research and development, and even tighter control over the costs of manufacture, test, integration, etc. The result is the opposite of what is required — rising costs, extending timescales, and concealed performance shortfalls, reminiscent of the problems experienced with classic mass production. Governments seem unable, or unwilling, to recognize that tight control of human activity invariably results in counterintuitive behavior, as control creates its own reaction.

The Japanese approach to lean volume supply throws a different light on systems engineering. The leading Japanese industries, rather than conduct systems engineering at artifact and project levels as in the West, operate instead at business, industry and socio-economic levels. Of course, artifact and project systems engineering are to be found, but the real focus is on the lean volume supply system as a system, i.e., as a whole. See Layer 4: Industrial Systems Engineering on page 12.

■ The lean volume supply system is designed to be homeostatic; to maintain a steady flow, to balance inflows and outflows, to self-maintain, and to auto-reconfigure as needed to maintain homeostasis.

■ Synergy is promoted throughout the chain or circle by creating an information system to coordinate activities, such that market demand for a product is signaled back through the tiers/anticlockwise around the loop, triggering the appropriate suppliers to make parts and assemblies for onward delivery at the right times such that JIT is achieved without queues, holdups or any unnecessary WIP.

■ Lean volume suppliers, operating globally, accumulate wealth sufficient to undertake their own R& D, and enabling them to conceive, design and create innovative new products, and to spread the applications of new technology — in this manner, the lean volume supply system is able to evolve, to switch between markets with agility, even to create new markets and to satisfy them

■ The ultimate business objective of lean volume supply systems is survival. To survive includes making a profit, of course, but short-term profit is not the goal — instead, it is survival in the long term. The leading organizations accumulate wealth, researching, innovating, adapting and evolving their product ranges, driving costs and process of products ever downwards, and responding with agility to changes in market tastes and climates.

The leading lean volume supply organizations 'press all the viability buttons.' If there is an Achilles' heel, it has to be in their ability to promote and maintain the culture on which their performance and success fundamentally rests. For more traditional western industrial organizations, the good news is that the culture can be created, and can take hold, within their organizations. Conversely, attempts to import many of the Japanese ideas, without first establishing the supporting and enabling culture, are unlikely to succeed.

Part II
Systems Methodology

6

Overview of the Systems Methodology

Though this be madness, yet there is method in't.

William Shakespeare, 1564–1616, *Hamlet*

What is the Systems Methodology?

A methodology is a system of methods, or a body of methods, rules and postulates used by a discipline. So, a systems methodology is a body of methods, rules and postulates used by systems practitioners to investigate, understand and address systems, their issues, problems, behaviors and contexts, and — where appropriate — to moderate, modify, or otherwise address and solve, resolve or dissolve issues and problems.

Essentially, a systems methodology is the 'how' of systems engineering; but that explains very little. The systems methodology incorporates methods, tools, procedures, processes, practices, and the skills and experience of practitioners who may be formed into teams, with leaders and managers; the whole then becomes the systems methodology.

Arthur D Hall III saw it as follows: 'what is envisioned is a new synthesis, a unified, efficient systems methodology: a multi-phase, multi-level, multi-paradigmatic, creative problem-solving process for use by individuals, by small groups, by large multi-disciplinary teams, or by teams of teams.' (Hall, 1989)

Hall went on:

> applied systems methodologists have been around for 40 years; they are called systems engineers, operations researchers, management scientists, systems analysts, policy researchers, value engineers, ecologists, and cyberneticians. They include . . . those from all the applied sciences, even if not usually associated with the systems movement. . . .
>
> A methodological umbrella big enough to cover systems sciences and the applied sciences practically mandates the development of a process that is context free. This makes a high level of awareness of process necessary for good systems work. . . .

Systems Engineering: A 21st Century Systems Methodology   Derek K. Hitchins
© 2007 John Wiley & Sons, Ltd

The Social and Economic Potential of the Systems Methodology

Our world is a complex, dynamic place in which the tempo of life and conflict seem to increase inexorably. Our social systems become progressively more complex as we diversify our activities: create new organizations, businesses and industries; bring new technologies on line, enjoy our freedom to innovate, become ever-more materialistic; as our populations continue to increase, as our hunger for power and energy continues to grow; and as the waste products of our lifestyles accumulate to pollute our biosphere and prejudice the biodiversity of the planet. Small wonder, then, that some see this profligate lifestyle as counterproductive, even dangerous, and hanker after a simpler, more spiritual way of life: disagreement between holders of such divergent viewpoints is inevitable.

Our world has been made smaller by enhanced communications, by ease and speed of travel, by improved infrastructures, etc. This has had the effect of coupling many open systems that were previously only in the loosest of indirect contact. Close coupling increases the rate of interchange between systems, causing their behavior to become more dynamic, even chaotic. Technology has not only enhanced communications and travel, but has materially affected our ability to make war, so that the 'footprint' for our weapons is ever increasing, delivery vehicles go further and faster, and so on. After some 100 000 years of *Homo sapiens*, we are still not wise enough to manage human affairs without resorting to conflict and war.

Not surprisingly, as our many complex systems dynamically interact, change and evolve, issues and problems arise. Many of these are familiar to us, even if there seems to be no way of addressing them: meeting the global demand for energy; preventing the biosphere becoming even more polluted; stemming the observed diminution of species diversity; accommodating the burgeoning human population of the planet; and so on.

Many problems and issues are familiar to us, too, at a more local level, in the organizations where we work, perhaps: concern about an organization's morale; mixed sets of management objectives, disagreement and a need for a unified plan; a high-level briefing required urgently to rebut parliamentary criticism of a project; a partnership at risk of breaking up owing to lack of shared vision; differing views of causes for lack of performance, effectiveness and/or efficiency; a complex technological equipment repeatedly presenting inconsistent fault symptoms in high-altitude operations; concern over both research and teaching quality and the need to improve quality management in a university; how to resolve inventory problems in a large, diverse, defense engineering company, leading to inefficient logistics, poor response times, and customer dissatisfaction.

Then, of course, there are the major national and international projects, notably in defense, civil engineering, nuclear energy, aerospace, infrastructure, etc. These are seemingly always late and over budget, with continual wrangling between those who buy and those who build. Not to mention (but of course, I shall) software intensive projects, national information technology projects for national insurance, immigration, health, law, police and judiciary, etc., which in many countries have risen to the level of national jokes. . . . Large, software intensive projects appear to be either beyond the wit of man to control, or are they, perhaps, beyond the wit of man to realize that they cannot be controlled — ever.

What if . . . there was a way of addressing all of these problems, indeed of addressing virtually any complex problem? What if there was a way of solving, resolving or dissolving any complex problem, so that even problems that have no solution might still be 'dealt with,' or at least understood . . . ?

Systems Methodology – a Paradigm

The systems methodology, then, is postulated as a generic, or universal, problem solver. Can there be such a thing? The short answer is probably 'no;' there must be issues and problems that are so intractable that solution is not possible.

However, it might be possible to create a systems methodology that addresses a wide variety of problem types, magnitudes, contexts/environments, complexities, etc. For the systems methodology to have such wide applicability necessitates that the systems methodology, *per se*, must be independent of problem type, context, etc. Instead, it has to fall into the category of 'a way of doing things' — a paradigm, or pattern that forms the basis of the systems methodology.

Aspects of the Systems Methodology

The scientific dimension

'Systems' can be just about anything that fits the definition: i.e., 'an opens set of complementary, interacting parts, with properties, capabilities and behaviors emerging both from the parts and from their interactions to synthesize a unified whole.' The definition is system-type, -scale and -context independent, and it evinces scientific roots in the terms 'open,' 'emerging,' 'synthesize,' and 'unified whole.' These are terms from the systems approach, of open systems, and of systems science, the science of wholes.

The systems methodology is necessarily founded in systems science, in that its operation recognizes systems, works with systems, configures and reconfigures systems, recognizes dysfunction in systems, and finds provable ways to 'repair' those dysfunctions. Provability is key: unless there is an established scientific underpinning to the systems methodology, any results or prognostications emanating from the systems methodology will lack credibility.

To be credible, the systems methodology 'adopts' the scientific method, i.e., the scientific method is 'built in.' The systems methodology incorporates the four steps of the scientific method, which are generally presented as:

1. Observe and describe a phenomenon or group of phenomena.
2. Formulate one or more hypotheses to explain the phenomena.
3. Use of the hypotheses to predict the existence of other phenomena, or to predict quantitatively the results of new observations
4. Perform experimental tests of the predictions by several independent experimenters and properly performed experiments

For the systems methodology, modifications may be necessary to the text, but not the principles, of the scientific method. So, the 'systems scientific method' becomes:

1. Explore the problem space: observe and understand the causes of dysfunction and disorder, i.e., the nature and behavior of the problem within its dynamic context. (This is the systems approach: understand and explore the system as an open, interactive part of some whole, where the part adapts to the whole, and vice versa)

2. Conceive one or more potential solutions, resolutions and/or dissolutions (remedies) to the problem that, if introduced into the context, would remedy the whole problem. (These are the hypotheses.)

3. Represent (visualize, model, prototype. . .) the potential remedies interacting with other systems in the solution space, such that both the potential remedy and the other interacting systems may mutually adapt. Modify/enhance the representation(s) to render the best, whole 'answer' to the whole problem in the circumstances — as constrained by the solution space. (This is, or can be, quantitative prediction, together with optimization.)

4. Test/prove the modified representation(s) in a variety of relevant contexts. (This is testing the robustness of the remedy.)

Systems science pervades the systems methodology: it is built-in to the methods and tools such that practitioners using such methods and tools observe the principles of systems science/applied science as a matter of course. Practitioners may be expert in the context of the problem space and/or the solution space so will bring their own scientific knowledge to bear, together with their experience and practices.

The logic and epistemological dimensions

The systems methodology is necessarily logical and rational where both terms indicate, *inter alia*, an absence of cultural bias. The knowledge upon which the systems methodology is founded must be sound; epistemology is important; else, the results from the systems methodology may be invalid, yet the practitioners applying the methodology would believe them valid: as Will Rogers, the lariat-twirling vaudeville raconteur, famously said: 'It's not what you don't know that's the problem — its what you know that ain't so!' Part I of this book hopefully enables readers to appreciate the systems theoretic bases of the systems methodology.

In particular, the generic reference model (GRM) has been presented earlier (and will appear again in the systems methodology): it represents a system as having three aspects; being, doing and thinking, relating to form, function and behavior respectively. The thinking/behavior aspect is of concern, epistemologically — the 'do you know what you think you know,' and 'how do you know what you know' considerations Many engineers are uncomfortable with 'softer' issues such as personality, psychology, social anthropology, etc., none of which is addressed in many engineers' education and training. Moreover, psychologists, anthropologists and others are not in total agreement themselves about human behavior, mores, psyche, etc.

That which applies to the remedial system in the way of psychology, anthropology, and other soft factors, applies equally to the systems methodology in operation, where it is an undoubted complex system in its own right. The systems methodology is comprised of methods, tools and processes being used and conducted by individuals, teams and teams of teams, all with behavioral characteristics, all operating in a dynamic, interactive, creative cauldron. A vitally important contribution of science, the scientific method and epistemology within the systems methodology, is to ensure that the cultural, political and other biases, which might be present within individuals and team members, do not find expression in the application of methods, in the selection of information, in the choice and acceptance of strategies and options, in the identification and neutralizing of threats, etc.

Within such a minefield, to talk of logic is, perhaps, tantamount to treading on a mine. The approach taken in the systems methodology, however, has been to limit the extent of the methodology such that its content is logical. For instance, we may not fully understand how different behaviors arise in different people in response to different stimuli; but it is logical, rational and reasonable to deduce and state that behavior is response to stimulus, without necessarily knowing which behavior responds to which stimulus. So, it is logical to perceive and anticipate causality.

In many situations, it is logical, rational and reasonable to limit the range of possible behaviors that might result from a particular cause. When an army commander finds that his army is surrounded, that he is running out of provisions, water and ammunition, there are no reinforcements, etc., many emotions may be running through his mind, but only three stark options facing him: fight to the death; surrender; try to break out. Of course, he could always attempt a bluff....

By operating in this manner, the systems methodology may prove logical, rational and reasonable even when dealing with soft and wooly issues.

The time dimension

The systems methodology contains processes and tasks at its heart; some processes necessarily involve sequence. Without knowing the problem first, for instance, it is not sensible to predicate a remedy. So, for some aspects there is a natural, inescapable, logical sequence, which will take time to perform.

Complexity appears to emanate from at least three aspects: connectivity, variety and 'tangling,' or the degree of interweaving of strands and parts of any system. (Hitchins, 2000) The greater the complexity of an issue, the more time may be needed to unravel the knot, to identify and characterize the various open systems, and to understand how their various ramifications interact and interdepend.

The time dimension, then, relates to the time taken to 'apply the systems methodology;' in the affairs of man, the kinds of processes, tasks and methods that are at the heart of the systems methodology will generally be subject to 'management.' Timescales for completion of some project may be set and predetermined, perhaps with less than due regard to the nature of the problem to be solved, the degree of complexity to be unraveled, etc., and also — importantly — without due regard to the dynamics of the problem space. So, setting too short a period may allow insufficient time to apply the systems methodology. Set too long a time, on the other hand, and the original problem may have morphed by the time the systems methodology has been applied, such that the final remedy is applicable to a patient who has recently died, or at least to one with a new, and different, problem.

Having said that, the very existence of a systems methodology, and of a group of practitioners practiced in its application, offers the best hope for a sound remedy to any undefined problem in the shortest practicable time. Moreover, knowing that the problem may morph during the conduct of some process places upon the systems methodology the responsibility for remedying, not the original problem, but the continually morphing problem.

The time dimension relates also to the remedial system (i.e., the solution to, the resolution of, or the dissolution of, the problem): its dynamics, its 'metabolic rate,' if you will. Many problems are cyclic in nature — for example, many businesses have an annual cycle, while some social systems have a 'generation' cycle, affecting the cyclic reappearance of fashions in music, clothing and in crime, for instance, and the periodic swings between cultures of 'freedom of expression' and 'repression of expression,' which can be seen in education. Remedial solutions have to be applied at the appropriate stage in a cycle, else they may make matters worse rather than better, and have to be able to adapt dynamically at least as fast as the problem they are addressing.

The cultural/political/behavioral dimensions

Culture has no place in hard science. However, culture does have a place within systems science and hence within the systems methodology. This arises for a variety of reasons, perhaps best demonstrated by example: solutions to problems have to be culturally acceptable to those who need the problem solved. Similarly, predictions of (particularly) human systems behavior are necessarily culturally colored. This is not to permit cultural bias within the systems methodology per se, but to recognize cultural aspects in both the problem and the solution spaces.

Politics is generally an issue with complex systems, too; those holding political views cannot be appealed to, in general, by logic, for instance. Political bias causes an observer to emphasize and accept information that jibes with his or her own affiliation, while at the same time de-emphasizing and rubbishing information that is discordant with his or her own beliefs: for many people, it seems, political beliefs color their viewpoints, their Weltanschauungen; epistemology is not a consideration. For such people, the logical, rational, reasonable remedy to a problem may be quite unacceptable; hence, the problem cannot be solved, at least, not in their eyes. In such cases, dissolving the problem may be the only way forward, i.e., moving the goalposts in such a manner that the politically biased, of whatever persuasion, have no grounds for complaint. Sometimes, however, moving the goalposts may herald a whole new raft of problems. . . .

The moral and ethical dimensions

The systems methodology must exhibit integrity, in both senses of the word, i.e., must adhere to high moral principles and professional standards, and must remain complete, undivided, sound and undamaged.

Can a systems methodology be moral? It can be moral in the sense that the methodology has a duty to be open, honest, unbiased and complete. It can also exhibit a moral responsibility in proving that solutions to problems are valid, optimum and complete, as well as culturally acceptable.

A key attribute of the systems methodology, then, is integrity. Traditionally, systems architects and systems engineers have jealously guarded their integrity, choosing always to tell the truth even when it may be unpalatable. In industry, telling the truth may, at times, be prejudicial to continued employment; sometimes the systems engineer may have to be sparing with the truth, but a true systems engineer never lies.

Is there an ethical dimension to the systems methodology? Yes, there has to be, in the sense that the systems methodology can be employed to address problems and issues of all kinds, including those that are politically, economically, environmentally, socially and even security sensitive. The systems methodology is ethically bound to be objective, impartial, unbiased and culturally aware, yet culturally unbiased, too.

The systems methodology operates within a global environment where waste and pollution are of growing concern. As in the medical profession, there is an ethic for the systems methodology that solutions to problems 'should do no harm' to the biosphere and to the environment.

The social dimension

The systems methodology incorporates individuals, teams and perhaps teams of teams, all with their various skills, capabilities, cultures, tastes, etc. These practitioners form into social groups both informally, and formally, perhaps as part of some organizational structure. The applied systems methodology operates within a societal environment where there may be problems, those who want the problems addressed, and those who have an interest in the social impact of solutions to problems. The systems methodology therefore has a substantial social dimension, with various social systems acting, interacting and counteracting.

Despite this, tensions between the various social groups, and the various cultures that might influence activities, judgments and choices within and without the systems methodology should have no effect on the integrity of the product from the systems methodology. This necessitates that absolute, objective measures of the conceptual solution system be applied at key points in any procedure; relative measures, those that might compare one conceptual solution with another, may be useful, but will prove insufficient to ensure integrity and freedom from social influence and cultural bias.

The organizational dimension

To perceive and comprehend the systems methodology in operation, then, envisage it as part of a wider whole, i.e., adopt the systems approach. The systems methodology is not a fixed entity, but — as a system open to its environment — adapts to its environment, yet remains viable, with all that implies: synergy, maintenance, evolution, survivability and homeostasis (S-MESH).

Nonetheless, there has to be within the systems methodology an identifiable, characteristic 'way of doing things,' a paradigm that transcends any adaptation, in that its identity, principles, integrity and values remain unchanged. The methods, tools and practices used for a particular application may be selected from a range of suitable candidates, as a surgeon selects the right scalpel, or a craftsman selects the tool appropriate to the job in hand: so long as the methods, tools and practices are context free.

There is, in all of this, evidence of the need for organization within the systems methodology: organizing data into information; practitioners into teams; teams into roles, such as problem exploration, conceptualizing, architecting, designing, assessing threats and risks, proving, etc. It is in the nature of people that some are more adept at starting and innovating, while others are more at ease with managing and coordinating, while yet others are more at home with finishing. It is pragmatic to recognize these predilections and to organize the systems methodology into corresponding structures and phases.

The economic dimension

It has become the practice in systems engineering circles to be concerned about the business aspects of potential solutions to problems, about the cost to customers, and about the importance of stakeholders. This might seem to be both natural and reasonable, but it may also inhibit the discovery and creation of the optimum solution. If, for instance, there is some premature decision about how much a solution is to cost, then options and alternatives will be reduced, the resultant solution is unlikely to be optimal, and may be less than effective. (Conversely, less expensive ways of achieving the solution system may be overlooked, or the simpler solution 'over-egged,' or 'gold-plated.')

In such cases, the systems methodology may be viewed as operating in the 'satisficing context,' i.e., resolving the problem by producing a resolution that is 'good enough.' Although this is not 'solving the problem,' it may prove both pragmatic and useful in the short term. Satisficing may result in a succession of resolutions, producing first one resolution to the problem, then a second, and perhaps a third; each resolution learns from the operation of its predecessor, such that the whole problem may eventually be solved, and the goal eventually achieved (*cf* Apollo). While clearly slow and probably expensive, such approaches have the merit of generally being quicker and more affordable at the start, even if prolonged and expensive overall. By biting off just a piece of the cherry each time, the overall process may also prove less risky, as each step forward is founded on prior experience, and is made with better knowledge of the risks. . . .

Undue emphasis on the cost of things early in the problem solving process may also be quite inappropriate for reasons that are more fundamental. Even the most cursory look at the history of man's achievements will show that the greatest and most memorable are rarely the result of financial caution. Apollo would not have happened if the budget had been constrained from the start. The Channel Tunnel joining England and France would not have been started had the financiers been aware of the overall cost at the start. The Scottish Parliament building in Edinburgh, a masterpiece of civil engineering, was many times over budget. Look around the world at the great achievements, and this same picture emerges on all fronts — the greater the achievement, the more likely it is to have been financially unconstrained.

This observation can, of course, be seen from different viewpoints. One view might be that the goal, the problem to be solved, is so important that cost is not a sensible consideration: that was the view at the time of Apollo, and it is often the view during periods of intense warfare, where survival is paramount. Another view is that people are incredibly poor at estimating the cost of solving complex problems, and underestimate through lack of understanding of the complexity and complication in creating a solution.

A third view might be held by those who have a good idea of the overall cost of solving some problem, but are also aware that customers, buyers and backers would be deterred if they knew the real cost; instead, the end cost is concealed, and those concerned with cost are subject to 'mushroom management:' kept in the dark and fed horse manure.

There is, too, an important economic dimension to the systems methodology in the sense that the applying or conducting the systems methodology requires time, skill, energy and money. To judge whether or not the systems methodology offers 'value for money' in any situation, it would be prudent to consider the cost of not using the systems methodology. If the systems methodology can solve complex problems that cannot be solved in other ways, then the alternatives to the systems methodology would seem to be guesswork coupled with trial and error; how costly might that prove, and how long might it take?

In reality, the cost of applying the systems methodology is directly related to the nature of the problem or issue, the threats contained within the environment, and the constraints imposed by the solution space. In some instances, an individual will have the necessary skills to apply the systems methodology on his or her own. Such an individual would find that observing the paradigm, applying the methods, tools and practices, encouraged objectivity, creativity, innovation and risk management. The individual would create an audit trail as proof of the bases for design, of solution optimality, etc., and would if required be able to prove that the solution system solved, resolved or dissolved the original problem as appropriate.

It is also true, naturally enough, that the individual would take time to become used to the methods and tools, and proficient in their application and interpretation — there is a learning curve. Once this curve has been climbed, the individual would not only create comprehensive solutions to complex problems with the highest integrity, but would also do so in short order. It is in the nature of a methodology that it contains the underlying process as an integral part of the paradigm, and that this process constitutes the critical path from whole problem to whole solution.

Similarly, teams and teams of teams operating as an integral part of the systems methodology benefit from its inherent order, structure and strategy. It is also true that teams of people will go through a learning curve in applying the various methods, tools and practices; additionally, they will take time to develop person-to-person relationships and understanding, so that a team will take longer to become practiced than an individual. On the other hand, of course, the team will be comprised of many complementary skills and capabilities, such that the team would be expected to prove more capable and more proficient than the individual.

The technological dimension

The systems methodology employs/applies context-free methods and tools; none of these need be technology-based or supported. However, vast amounts of information may be generated in many applications, and information technology may prove invaluable in handling this information.

The essence of 'system' being order, there will be occasion to assess or evaluate entropy, and to reconfigure systems to minimize configuration entropy, e.g., to reveal intrinsic architectures. Again, this may be accomplished without technology, but suitable tools may both ease and speed up the processes, to the point that some complex issues — although they could be addressed by hand — would be best addressed using some kind of information manipulation tool.

It is, too, in the nature of open system problem solving, particularly on the grand scale, that the dynamics of the problem, the situation, the environment and the solution become of paramount importance. While it may be possible for those of superior brainpower to conduct the necessary mental gymnastics, for most of us it is likely that we will resort to dynamic open systems modeling. Such modeling has the benefits that it allows many different aspects of the open-system problem, its potential solution and its environment, to be simulated, to interact, and to find their resultant balance points in ways that even the brightest person might find difficult to match. Moreover, dynamic simulations provide experimental learning laboratories, permit the application of the 'scientific method,' demonstrate solution validity — or otherwise — to practitioners and customers, and serve as a cornerstone of proof.

The technological dimension may arise as a part, or occasionally the whole, of the solution to some problem. In the general case, the solution system is likely to be socio-economic, sociopolitical, sociotechnical, or any combination of the three. Some systems engineering practitioners recognize systems engineering as concerned only with technology; they emphasize 'engineering,' to the virtual exclusion of 'system.' On the other hand, some systems engineering practitioners concern themselves with social systems, particularly businesses and enterprises, and they emphasize (soft) 'systems,' to the virtual exclusion of 'engineering' — except in the sense of 'to create, to cause to happen.' As so often in such cases, the truth lies somewhere in the middle: systems engineering is best viewed, perhaps, as a portmanteau word, i.e., a singular expression with singular meaning, rather than two words with separate (divergent?) meanings.

Systems engineering, then, is creating optimum solutions to complex problems: it is neither systems, nor engineering alone; it is both systems and engineering at the same time. In the technology context, solutions to complex problems may be sociotechnical. For instance, long-range ground radar is (part of) a sociotechnical system: the whole radar-as-a-system includes operators, supervisors, controllers, maintainers, etc. It is also an open system, so interacts with the corporate system (air traffic management, perhaps), which adapts to the radar, just as the radar adapts to the air traffic system/whole.

This is generally the case: purposeful solutions to problems are highly unlikely to be exclusively technological, although some may contain a technological element which others may chose to see in strictly reductionist terms as a discrete, closed, artifact-to-be-engineered. Such situations may occur, for instance in defense procurement where, for contracting simplicity and tight managerial control, a technological artifact may be considered, designed and procured as an isolated entity: this is intended to concentrate development effort, and to prevent post design change, which can bedevil development and manufacture. Such practices do not bode well, however, for the operational performance of such entities when subsequently introduced to a complex sociotechnical system of which they will form an inflexible (and possible obsolescent) part.

Systems Methodology: Conceptual Model

The systems methodology is, essentially, a problem-solving methodology, and as such is formed around a problem-solving paradigm. One such is shown in Figure 6.1, which shows a problem-solving process in diagrammatic form. It starts at the top with 'Issue' — a topic of concern. The issue is presumed to stem from intrinsic problems, which can be used to model an Ideal World, i.e., a world in which the problems would not exist, and therefore the Issue would not arise.

Having generated an Ideal World model, it is compared with the Real World, so that the difference between the two can highlighted and used as an agenda for change, to move from the present Real World towards a future Ideal World, in some undefined way. However, the figure shows that any future improvements can be judged by their ability, or otherwise, to address the original problems.

This is one version of the General Problem Solving Paradigm (GPSP); it is simple in concept, yet can be powerful in application, especially when incorporated into the systems approach. The systems approach adds strength to the GPSP by allowing that changes may cause adaptations, both

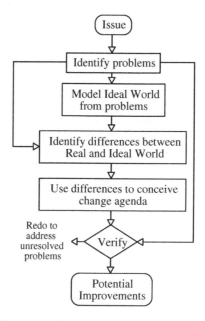

Figure 6.1   General problem-solving paradigm.

of the 'issue systems,' and of the corporate whole of which they form part. In this way, the 'agenda for change' takes account of the dynamics of open systems interactions, and the 'acid test' — does the changed world solve the original problems — can act as a backstop, or safeguard against any errors, misapprehensions or misconceptions along the way.

Using the GPSP in association with the systems approach addresses the dynamics that are so often overlooked in trying to address problems, where the solution causes further, often counterintuitive problems, which may on occasion prove more severe than the original problem that was to be solved.

The GPSP is not the only problem-solving paradigm. There is another in wide use — the systems engineering paradigm. One version of this approach is shown diagrammatically in Figure 6.2. It starts by defining, or circumscribing, some problem space. There follow two independent activities: conceive solution options; and, independently, identify ideal solution criteria. Then, tradeoff between options and criteria to find the optimum solution and, finally, formulate strategies and plans to implement. Although this forms the nucleus of many systems engineering processes, it is much more widely applicable and is used by engineers, managers, financial advisers, and police. . . in fact, most people seem to have used this approach to finding the best answer.

Note, too, that it proposes finding the optimum solution — this is solving the problem, Ackoff-style, and is suggestive of some evaluative process. Interestingly, there are arguments even amongst seasoned systems engineers about optimization, some declaring that it is not a necessary part of systems engineering yet, at the same time, insisting that trading between options is a central tenet of the discipline and practice. As Figure 6.2 shows, tradeoffs are about finding the optimum. . . although often by non-mathematical, or quasi-mathematical means.

Comparing the two paradigms, it is evident that they are significantly different. Figure 6.1 emphasizes the problem, but has little to say about the solution, while Figure 6.2 emphasizes the solution, and how to get it, but with little to offer with respect to the problem, per se. It is possible, then, to combine the two paradigms into one.

Figure 6.3 shows the two problem-solving paradigms combined into one. Together, they form the basis for one strand within the systems methodology — the strand concerned with solving complex problems. There are two other strands that must also be addressed: one concerned with resolving

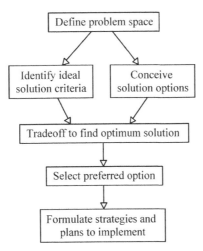

Figure 6.2 The systems engineering problem-solving paradigm.

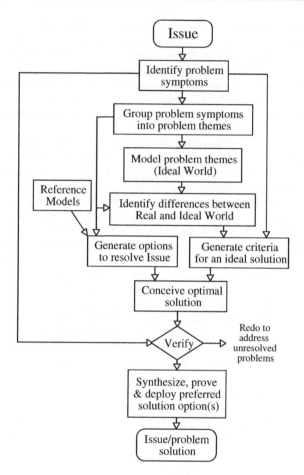

Figure 6.3   Combined general problem-solving paradigm and systems engineering problem-solving paradigm. The combined paradigm aims to solve problems, rather than resolve, or dissolve them — see text.

problems, i.e., satisficing, or finding an answer that is 'good enough'; and another concerned with dissolving problems, or 'moving the goal posts' such that the issue or problem effectively disappears. Having said that, the figure does represent the strand in the systems methodology that many engineers would regard as systems engineering. . . .

Notice in Figure 6.3 the appearance of Reference Models. These are models such as the Generic Reference Model – see page 124 *et seq*. Reference Models are invaluable when generating options, to promote completeness of each option, such that like is compared with like, including emergent properties, capabilities and behaviors.

A conceptual model of the systems methodology is shown in Figure 6.4; it starts with an Issue or Problem (the terms are interchangeable in this context). The problem is evident owing to problem symptoms, perceptible in the problem space. These symptoms may be instrumental in directing exploration of the problem space, in the same way that a patient's symptoms may direct a doctor towards a diagnosis. The 'diagnosis' promotes a conceptual remedial solution, or remedy, one that

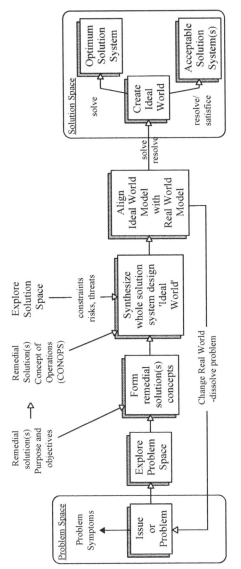

Figure 6.4 A conceptual model of the systems methodology.

if it were applied might reasonably be expected to neutralize all of the symptoms – this is the acid, or decisive, test of any posited solution at any stage in the overall process. A conceptual remedial solution is defined by its purpose, and by its emergent properties, capabilities and behaviors: that is, what it is intended to do, how it will work, and what effects it is meant to have; there may be little about structure, content, or viability at this early stage.

The conceptual systems methodology shows next the synthesis of a system design, such that the design, if implemented would neutralize all symptoms, creating in the process the so-called ideal world; i.e., the best world in the circumstances. Note that the ideal world design is not necessarily some completely new system — at this stage, it is simply the design of an ideal system that, if realized, would in operation, while interacting with and adapting to, other open systems:

- generate none of the symptoms of the problem;
- exhibit requisite purpose, emergent properties, capabilities and behaviors;
- be viable in the solution space.

The next stage is to align the ideal world model with the real world model, i.e., to see where they match, and where they do not match, and to understand the relevance of the differences. This is a decision-point. Potentially, the issue can be addressed, i.e., the symptoms eradicated, by solving, resolving or dissolving:

- 'solving' implies creating the ideal world:
  - by altering (modifying, reconfiguring, connecting/disconnecting, enhancing/diminishing/balancing, etc., etc.) that which exists already, or. . .
  - by creating and introducing a new system to interact with those already in existence;
- 'resolving' also implies creating the ideal world indirectly, by satisficing, i.e., finding an answer that is 'good enough' or, a series of answers that home-in on the ideal world solution, perhaps, over time.
- Finally, dissolving the problem is seen, in the figure, as effectively addressing the problem space in such a way that the problem effectively disappears (see The Systems Approach on page 16.).

A choice will be made, generally by exploring each avenue and its implications, and comparing their relative merits in terms of feasibility, practicability, timescale, effectiveness, reactions and knock-on effects, cultural acceptability, likely costs, etc., etc. and, it may be that the ideal solution is to be found in some solving, some resolving and some dissolving all at the same time. . . .

Note in Figure 16.4 that:

- the remedial solution has purpose and objectives, which will include neutralizing the problem symptoms;
- a concept of operations (CONOPS) is developed for the remedial solution system, to show how it is intended to operate; there may be several competing CONOPS associated with any remedial solution, and indeed several competing remedial solution systems;
- exploration of the solution space is necessary to identify constraints, limitations, risks and threats that might prejudice the synthesis and manifestation of an ideal world design and solution system(s);
- according to situation, it may include functional-to-physical mapping, i.e. partitioning the overall design into physical compartments, with interconnections and interfaces.

An applied version of the systems methodology is shown in Figure 6.5 as a behavior diagram. The diagram shows functions or processes in the center panel, with essential inputs to each function or process at left, and outputs from each function or process at right. Inputs and outputs form logical patterns, too, such that the whole behavior diagram is self-checking, both vertically and horizontally. Note that the behavior diagram presumes that the objective is to create an optimized solution to the problem, rather than to resolve or dissolve the problem; moreover, some particular

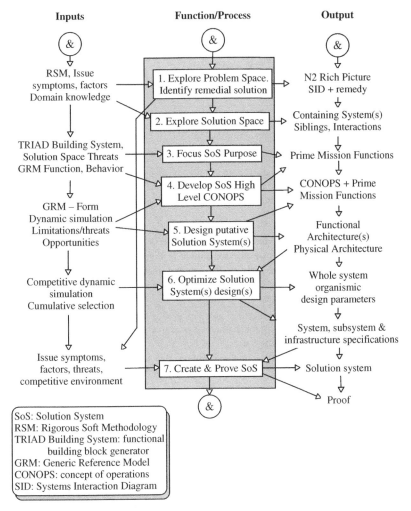

Figure 6.5 The systems methodology as a high-level behavior diagram. In this example, particular tools and methods have been shown. In the general case, tools and methods may be selected from a suitable range, according to the nature, magnitude and extent of the issue or problem. The behavior diagram shows both inputs to, at left, and outputs from, at right, any central function/process. Using the GRM underpins solution system emergent properties, capabilities and behaviors. Cumulative selection enables optimization. Note the built-in acid test, that the solution must solve the problem.

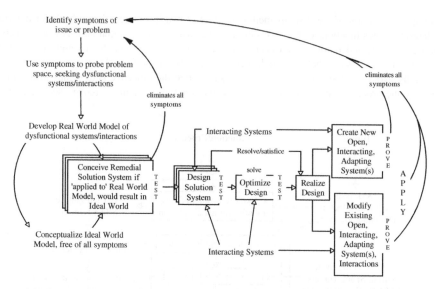

Figure 6.6 Organismic view of the systems methodology, showing solving and resolving, and alternatives in design realization. Note the continual test/prove theme, requiring the developing designs to demonstrate their ability to eliminate all of the original symptoms.

tools have been nominated. Typical tools and methods will be presented later: subsequent chapters will expand on the seven steps/stages/processes shown in Figure 6.5.

Figure 6.6 shows an organismic view of the systems methodology, as a closed-loop system, one seeking closure. Note that there are two distinct systems approaches to addressing the issue or problem:

- create some new, additional, or replacement system or systems that has to be realized and introduced as one or more entities; or. . .
- modify that which already exists in the problem and solution spaces, by reconnecting, by reconfiguring, by (re)balancing, but essentially not by introducing new systems, entities and/or artifacts.

Engineers may recognize the first of these as systems engineering. Systems practitioners may recognize the second of these as reorganization, proactive management, or as systems engineering, too. Both groups are correct, although neither may be too happy to concede the others' territorial claims to 'ownership' of systems engineering.

Create a (Better) Systems Methodology?

The systems methodology is a paradigm: an archetype, pattern or model, forming the basis of a disciplined basis to solving complex problems and issues. There may be other paradigms, other ways of creating solution systems. If there is such a thing as a 'best solution,' then competing paradigms would necessarily have in common that, in starting from the same problem space, they would end up with the same 'best solution:' they need not have followed the same route, however, suggesting that competing paradigms would differ principally in the demands they imposed on skills and resources, and the time and cost that they incurred in the process.

On the other hand, the systems methodology presented above, this particular paradigm, has a degree of flexibility built in, since the various activities shown, for instance, in Figure 6.5 can be undertaken in different ways, using either manual techniques or with the support and aid of tools-supported methods, according to the nature of the problem, the volumes of information being handled, the experience and proclivities of the practitioner, and so on. The key to such a methodology is to so design, organize and arrange each process/function/method that the output from the previous activity is exactly the input required by the current activity, and that the output from the current activity is exactly the input required by the following activity. Each process, then, may be seen as a transformation, and set of processes as an 'information transmission matrix,' in the sense that disorganized information about the problem space, the situation and the solution space enters at the start, undergoes a series of transformations, and emerges as highly organized, specific information about an optimum solution system. The end-to-end process is one of progressively reducing entropy, to such a degree that a specific solution can be realized, all within the context of the systems approach.

The systems methodology presented at high level above, and to be presented in more detail later, is, then, capable of improvement, in that different and better methods and tools can undoubtedly be conceived than those currently available. If newer, better methods are substituted individually, then the paradigm remains constant. If, on the other hand, all of the processes, methods and tools are changed together, then that might constitute a paradigm shift: the paradigm is in the pattern of skills, processes, practices, tools and methods.

Create an Intelligent, Auto-adaptive, Evolving Solution System?

If it is possible to conceive, design and realize an optimum solution to a complex problem, and if that problem is continually changing, then it should also be possible to continually conceive, design and realize; that is, to continually reconfigure the system so as to maintain its ability to solve the continually morphing problem.

Figure 6.7 shows a notional approach to designing such an auto-adaptive system. The design envisages that the system contains a model, a dynamic simulation of itself, operating in its simulated environment and interacting with other simulated systems in that environment. So, the system is presumed to exist in some hostile environment, which it is able to sense, and to which it presents its emergent properties, capabilities and behaviors, indicating that it is open and interactive. The system is presumed to be 'designed' using so-called genetic methods, i.e., there are 'genes' coding for different elements and configurations of elements that go to make up the system.

The problem-solving process operates by simulating the environment, and by injecting into that simulation a representation of the system as open and interacting with others, in which state it is supposedly achieving its goals optimally. The problem to be solved is one of determining whether or not the current configuration is, indeed, optimal, and if not, what should be done about it. . . . Offspring from the current configuration are generated for the simulation, such that each offspring differs from the current configuration in some respects and in some small degree. Should a simulation using one of the offspring show a significantly better performance than the run using the (model of the) current configuration, then the current configuration will be rearranged to match

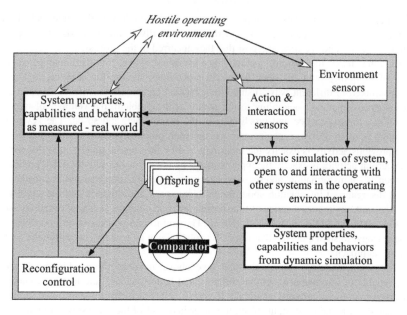

Figure 6.7 An auto-adaptive system reconfiguring to maintain its balance with other systems in the environment. See text.

that of the most suitable offspring. At that point, the selector will activate the reconfiguration control, reconfiguring the (real) whole system to match that of the preferred offspring model. In reconfiguring, the system's open, interactive and adaptive emergent properties, capabilities and behaviors are restored to optimum for the contemporary situation.

Limitations are evident in the conceptual design. The environment simulation is not the environment; unless the simulation is sufficiently representative, decisions based upon it may be invalid. Even if the simulation were adequate, there is a risk of continually reconfiguring in response to environmental perturbations; some threshold would be necessary, below which reconfiguration did not occur, implying that optimization, as such, may be impractical in a dynamic environment, since the current configuration is likely to trail behind the true optimum as the environment changes. Furthermore, highly dynamic environments may change at rates which the auto-adaptation cannot sustain — perhaps as the dinosaurs were unable to adapt to rapid climate change.

Nonetheless, the basic notion of including within the system the ability to continually solve the problem of its own re-optimization is conceptually interesting.

Auto-adaptation and the intelligent enterprise

If a system such as an enterprise, a business, or industry contained within it the ability to address its own problems and create solution system designs to optimally solve those problems, then the whole system could be auto-adaptive in the manner of Figure 6.7. When problems are substantial, enterprises tend to employ external consultants using such methodologies, but it does not have to be that way. Using the systems methodology, an intelligent enterprise, for example, would continually scan the environment, assess the situation both externally and internally, and would continually rebalance, reconfigure, adapt itself so as to remain viable, and to survive in the longer term. (I choose 'survive' rather than 'make a profit' since an intelligent enterprise would realize that the enterprise that survives is the winner in socio-economic terms of wealth creation and wealth accumulation. Of course, it has to make a profit along the way, but it also has to be prepared to invest some of its revenues in remaining robust and viable, and in continually enhancing performance.)

Summary

A systems methodology is feasible and practicable, and would afford a paradigmatic, archetypal, generic approach to realizing complete solutions to complex problems and issues. Such a systems methodology would, in effect, incorporate the 'how' of systems engineering, taking it out of the realm of sometimes-dubious practice, and into that of provable solutions of high integrity.

A wide range of issues, incorporating a wide variety of problems, environments, situations, technologies, cultures, societies, etc., can be addressed thoroughly and objectively using just the one paradigm, such that the solution systems are provably sound as well as culturally acceptable. Using the systems methodology, societies would have at their disposable an invaluable tool, not only for solving problems, but also for planning, estimating, managing, organizing, strategizing, optimizing, etc., etc. The systems methodology would allow the rational and logical analysis and solution of problematic issues such as national and global energy shortages, national and international disasters, disaster relief, diminishing species diversity, atmospheric pollution, and many, many more.

At the same time, the systems methodology would support the rational and logical analysis and solution of problematic local issues: policing and the rule of law in democratic societies; affordable national defense procurement; enterprise and industrial performance; national resource management; waste disposal; etc; etc. It is not possible to solve every problem; some are simply too intractable. However, a common systems methodology would allow diverse groups and cultures to work together within a rational, science-based framework, pooling knowledge and experience so as to develop provably sound solutions to a wide range of important problems.

The systems methodology is a composite of skills, domain knowledge and experience, tools, methods, practices and processes, together enabling an individual, a team, or a team of teams to explore a problem space and to realize a whole solution system to the whole problem, in some solution space. There are many aspects, or dimensions, of the systems methodology: scientific, logical, epistemological, temporal, cultural, political, behavioral, moral, ethical, social, organizational, economic, and technological have been mentioned — there are many more, building to an ontology.

The systems methodology progressively reduces entropy observable in the problem space to the point at which a real-world solution can be specified, in detail, and realized. Since the solution system need not be technological, realizing the solution system may involve (e.g.) recruiting and training teams of people, reconfiguring an organization or process, writing a new instruction manual, or even changing a culture, as well as creating a new or replacement sociotechnical system, or a technological artifact.

So, systems engineering is more, much more, than creating new technological systems. It is more, too than solving problems: additionally, it can include resolving and dissolving problems.

The systems methodology incorporates a range of tools and methods that may be selected, as a surgeon selects a scalpel, according to the job in hand. The various tools and methods can be viewed as transforms, in that they handle and organize data and information about the problem space, the solution space and the environment. Each transformation can be thought of as taking data as an input and providing information as an output, where 'information is to data' as 'more ordered is to less ordered.' Successive transformations treat the previous transform's output as their input, and so on, creating an 'information transmission matrix,' which takes the raw, disordered data from the problem space and provides well structure, ordered, specific information about the solution system as an output at the other.

The systems methodology is not some linear-predictive information machine, however: choices and decisions have to be made continually throughout the process, since much of the capability is derived from the skill and knowledge of the systems practitioners, as well as from the problem and solution domains. Different individuals, teams, or teams of teams would produce different solutions to the same problem: some may be better at exploring the problem space and extracting domain data, some may be more familiar with the solution space and the threats contained therein, some might generate — and design into the solution system — unique strategies for addressing such threats, and so on.

So, while different teams with different capabilities might all produce provable solution systems, nonetheless some solution designs may be richer than others, address a wider range of threats, employ more cogent strategies, contain greater diversity, employ different kinds of solution system (e.g., a human activity system solution as opposed, say, to a largely technological solution), and so on. It is important, too, to remember that solution systems have to be culturally acceptable; different cultures, in different situations, with different values, may see different solutions systems as the optimum solution to their problem. This is not to say that the systems methodology will operate in a culturally biased fashion: it should, however, recognize that solutions have to be culturally acceptable; else, those being handed a solution to their problems, may neither recognize it nor accept it. An optimum solution is, after all, one that is best in the circumstances of the problem space and the constraints imposed by the solution space.

The systems methodology may be viewed as a means of solving problems to be applied whenever a major issue arises; in such circumstances, it might 'be applied,' if that is an appropriate term, by some consultant external to the organization with the problem. However, the systems methodology can also be used 'in-house' as a way of ensuring that an organization (enterprise, business, industry, socioeconomy) is operating to its best advantage, i.e., optimally. In this guise, the systems methodology is set to continually sense symptoms of dysfunction, indicating an incipient problem either within the organization, or with its interactions with other systems. Once detected, these symptoms will indicate the source of dysfunction and identify the root problem, which can then be addressed and solved. In this manner, including the systems methodology within an organization renders it auto-adaptive, so that it may continually adapt, reconfigure, even redesign and reinvent itself as the problem morphs. Such auto-adaptive systems hold the promise of operational success and survival.

Assignment

1. Justify the view that the systems methodology is a system in its own right. Identify a set of emergent properties, capabilities and behaviors that the systems methodology would evince in operation.
2. If the systems methodology is a system, as suggested in 1, then can it be conceived and created using the systems methodology? Would this flout Gödel's Incompleteness Theorem? Discuss.

3. Consider the systems methodology as an operational system in an environment, interacting with other systems. Specifying an environment of your choice, identify the threats you envisage to the systems methodology and to its continued operation.

4. The systems methodology embodies the systems approach. How, then, can the systems methodology, as an open system, interact with, and adapt to, other systems in its environment? Explain, with examples from Layer 3 of the five layer model – see The 5-layer systems model on page 113.

7
SM1: Addressing Complex Issues and Problems

Mankind always sets itself only such problems as it can solve; since, looking at the matter more closely, it will always be found that the task itself arises only when the material conditions for its solution already exist or are at least in the process of formation.

Karl Marx, *A Critique of Political Economy*, 1859

Problem-solving Paradigms

We have already encountered two problem-solving paradigms on page 172 *et seq*. There are many ways to address problems, depending on the nature of the problem. Doctors, for example, have an effective way of finding out what is wrong with patients by looking for symptoms of variation from the norm, such as temperature, blood pressure, pallor, and so on.

Complex systems, such as the human body, major enterprises and socio-economies, exhibit problems generally as dysfunctional behavior, that is by something not behaving as it should, or as it has previously (and satisfactorily) behaved. Dysfunctional behavior may emerge because a system, or systems, are not functioning as they were, or as they should, or because of some interaction irregularity.

Confusingly, and especially with complex systems, the symptoms may emerge where the problem isn't. We are familiar with this in our personal health. A headache need not indicate anything wrong with the head — it could be something we ate. A pain in the upper left arm need not indicate anything wrong with the upper left arm — it could be the onset of a heart attack: then again, it may not be . . . .

Such confusion arises within our bodies, and within complex systems in general, because of the coupling and interactions that occur between subsystems/parts of complex systems, which result in the behavior of each subsystem both depending upon, and impacting, the behavior of other subsystems, both directly and indirectly through intervening systems — so-called transitive effects.

With complex systems, trying to solve problems by treating the symptoms would be tantamount to treating smallpox by putting calamine lotion on the spots — not a lot of use. However, this is what we tend to do in everyday life. Politicians in particular, and managers too, may see (or choose to see?) a symptom of some issue or problem, and address that symptom in isolation. Not only is this ineffective, but it can mean that precious resources are wasted in salving rather than solving.

Linear, Complex, Nonlinear and Intelligent System Behavior

It is often much easier (cheaper, safer, expedient) to recognize the symptoms of some issue than to dig deeper and find the underlying or root cause. And with complex systems in particular, it may be difficult to find the root cause, even if the investigator were willing to dig and delve.

Systems can behave in unexpected ways. We expect linear, cause and effect type of behavior: pay more money, get better results; depress the pedal further, the car goes proportionally faster; shorter exposure freezes motion more precisely; and so on. Life, as we know, does not turn out like that: pay more money, results stay the same, but staff wastage reduces; depress the pedal further, wheels skid, lose traction; shorter exposure admits less light, picture underexposed, motion-freeze indiscernible.

Behavior, then, is rarely linear — even within our digital computer based systems, linearity is something of an illusion as we sample, round-off, work to N digits, etc; generally, behavior is nonlinear in the real world. (See Nonlinear Systems Thinking on page 78.) In addition to the kind of nonlinear, but continuous, behavior exhibited when foxes, rabbits and grass interact, breed, eat, predate, etc., there are also chaotic systems, where the next instance of behavior may be related to the previous instance, but in an unpredictable way (like the weather) or may even be random (like a roulette wheel or a lottery).

There is also a tendency, again notably amongst politicians, although we are all guilty, of presuming that there is a singular cause to any problem. We look for something, or someone, to blame: we look for the one silver bullet that will put everything right. Looking back on past problems may convince us that there are no silver bullets: complex issues generally involve a number and variety of source problems, each of which has to be addressed if we are to have any hope of resolving the whole issue.

As though to confuse a confused issue even further, there are intelligent systems, where the next behavior may be based on the previous behavior, but is also considered in the light of context, situation, strategic alternatives, tactics; see Behavior Management on page 130. An eye for an eye may not always be the best response: sometimes, turning the other cheek works miracles.... On the other hand....

System Dysfunctions: the POETIC Acronym

The kinds of situations and systems where we might reasonably want to address issues and solve, resolve or dissolve problems are generally complex, invariably nonlinear, sometimes obscure, generally sensitive, and involved. In such circumstances, we may need to be objective, impartial, insightful, sensitive, culturally aware, etc. Indeed, we may not only need to *be* these things, but be *seen* to be these things — else we may be deemed unacceptable as problem-solvers.

In such circumstances, method is important. Method not only helps us to organize our approach to problem-solving, it also inspires confidence in others, particularly where the method can be presented to interested parties, so that they can see for themselves that it is rational, sensible, reasonable, objective, etc.

Nonetheless, method on its own is of limited value — practitioners also need domain knowledge and experience, neither of which is embedded in a method or methodology. Experience of addressing complex problems in businesses, enterprises and industries suggests that many have their dysfunctions rooted in areas indicated by a useful acronym — POETIC.

- *Politics*, sometimes with an initial capital letter, where politics refers to interrelationships, tactics and strategies involving power, authority, influence and manipulation. Many problems, or so it seems, arise from entrenched positions arising within corporate boardrooms, leading to standoffs, failure to take action, and stagnation.
- *Organization*. The manner in which an organization is set up, the way it is configured, a surfeit or shortage of internal communication, coordination and synergy — all of these and many more can contribute to systematic issues and problems
- *Economics*. The amount, availability, distribution, withholding, investment, movement, etc. of finances, together with a desire for short-term profitability at the expense of growth and stability
- *Technology*. Access to, availability of, incorrect choice of, cost of, reliability of, capability of, unreasonable expectations of, and blind faith in, etc., etc.
- *Inertia, Inactivity, Indolence*. Open, complex systems are active and interactive, or they are inconsequential. Issues and problems arise not so much from doing the wrong thing as from not doing anything . . . .
- *Culture*, or 'the way we do things here.' Cultures are palpable within organizations with major divisions: different cultures will pervade each division. Culture need not be bad, in that it encourages conformity — which can be valuable in traditional engineering companies, for example, where innovation in the wrong place, or at the wrong time, can cause havoc. However, it may effectively stifle innovation, inhibit intelligent behavior, and cause an organization to become so 'set in its ways' that, like the dinosaur, it becomes outmoded. Other companies, not so shackled, may continue to develop and improve, leaving the isolated culture to its fate. It is possible, too, to encourage an open, enquiring, thinking culture, to encourage rather than to blame, to be open to new ideas rather than rely on experience . . . .

Soft Systems Approaches

Methods of addressing complex problems and issues, particularly those involving people, are sometimes called 'soft' methods, where soft refers to the lack of hard, concrete material, evidence, etc. A soft systems method, then, would be one that considered the people working, perhaps, in some organization, their individual interests, objectives, attitudes and mutual interactions, perhaps. A soft system is one that, although comprised of people, and therefore technically 'manmade,' nonetheless does not have a clear, singular purpose: instead, it may have many, conflicting purposes, lack synergy, etc. A soft situation might be described as 'messy.'

A hard system, by contrast, would have a clear, singular purpose, and would have all the parts within that system contributing towards that singular purpose. One might say, then, that the objective of a soft system method is to convert a messy, soft system into a coherent, hard(er) system.

Some people use the terms 'soft' and 'hard' to refer not to the coherence of the system in question, but to the predominance or otherwise of technology in the system. So, they might describe a Boeing passenger aircraft as a hard system, or an avionics system within the aircraft, perhaps, on the basis that the system in both instances is very largely technological in nature. While understandable, this would be a misuse of the term, and might obscure the fact that some technological systems can be 'soft,' while some human activity systems can be 'hard.'

It is also the case that some see 'soft' and 'hard' ideas, systems, methods, etc., as quite different, and having little or nothing in common. That, too, would be inappropriate: 'soft' and 'hard' might better be viewed as the extremes of a spectrum, not of color, but of coherence, clarity and purposeful behavior — in other words, an entropic spectrum, where soft corresponds to greater disorder, or higher entropy, and hard corresponds to greater order, or lower entropy. It is the objective of systems engineering, in this context, to move from disorder to order, from higher to lower entropy. This might be achieved, methodologically, by first using soft systems methods to bring some degree of order to soft situations, issues and problems, and then by using appropriate methods to progressively increase order to the point that specific (i.e., hard-ish) designs and solutions can be manifested.

Degrees of intervention

It is not the case, however, that a specific hard solution is appropriate to every problem, or issue. Many commercial organizations, enterprises and industries would be reluctant to entertain the idea of reconfiguration, or indeed any significant change, simply to resolve some internal issue. Instead, they might seek the services of some consultant to help them understand the issues, and to suggest some sensible way of moving forward. (Notice how the words and expressions used are quite different from the ideas of firm specifications of whole solution to complete problems — this is moving from disordered to slightly less disordered — maybe!)

Bringing in some external 'help' is sometimes called 'intervention:' such events come in various guises. A consultant may visit, work with some managers, suggest some changes to organization, process or procedure, some retraining, or whatever, and leave: this might be a relatively minor intervention.

A complete redesign of the organization and the implementation of that design might be called a major intervention. Such events are less common, but do occur from time to time, under the banner: 'revolution, not evolution.'

A halfway house may exist, in which a complete redesign is done from first principles, and then compared with that which currently exists; this is Ideal World against Real World. Comparing the two to see the differences can result in minor changes in the Real World to move it more in the direction of the Ideal World: alternatively, the two worlds could be so far apart that a decision is taken to replace the Real World lock, stock and barrel with a new, and hopefully Ideal, World — the major intervention. Or, there may be a middle path, where elements of the Ideal World design are implemented and added to, or inserted into, the Real World.

As an example of this 'middle-of-the-road' approach might be the introduction of a quality assurance system to an existing assembly line, a new division to an existing, multidivisional organization, a new anti-rain mode to a ground control radar, a new, high-precision, anti-armor weapon to the range of weapons carried by a ground attack aircraft, a decision support system to an existing command and control system, or a neonatal unit to an existing hospital. In each case, the pre-existing Real World is left relatively unscathed, except for the need to adapt to and accommodate the new addition envisaged as part of the Ideal World. The resulting whole is a changed Real World, and may — or may not — substantially equate to an Ideal World, according to whether the intent is to resolve, or to solve the original problem that prompted the redesign. The term 'intervention,' however, is associated with human activity systems, their organization and management, rather than with technological systems.

Consultants, or consultancies, conducting these so-called interventions bear a heavy responsibility, since their recommendations, if acted upon, could make or break an organization. They need method, and such methods seem to come in two forms.

Consultants may have a preconceived idea of the ideal functions, form and behavior that a successful organization should possess. Such preconceptions may have been developed by examining organizations deemed to be successful, usually in financial terms, and observing the value of various key parameters. These might include turnover per employee, profit-to-turnover ratio, percentage of profit dedicated to research and development, percentage of employees in various age brackets, innovative ideas generated per employee, average span of employment, and many, many more. The consultant will use 'approved' values for many of these parameters, determine the equivalent values for the organization under investigation, and recommend changes to bring these in line. This can be a somewhat procrustean approach, since in reality 'one size does not fit all,' and it is more applicable to commercial organizations than to those, say, in the public sector.

Consultants of a different breed concern themselves with finding out what makes the organization 'tick,' where any dysfunctions might be and how best to restore function, behavior and form to its (presumed) former good state. Consultants will be aware that they know little about the organization, while the people with whom they will interact within the organization may have decades of experience, knowledge and understanding of the organization, its characteristics, markets, competition, limitations, etc. The consultant, then, draws this information from the organizational members and effectively enables them to perceive their own problem; he may then advise them as to how best to ameliorate the situation.

Because this second approach concerns itself with the people in the organization, it may be deemed 'softer' than the first approach. However, it seems to be the case that many organizations are made up of groups with different purposes, such that the whole has no clear purpose, while the groups pull in different often conflicting directions. Sorting this out, so that all parties 'face in the same direction,' may be a 'people problem.' So, while the first approach carries with it a preconceived Ideal World, it may be less than ideal for a particular organization and moving toward that supposed ideal may require surgery. The second approach, on the other hand, is more likely to invoke a lifestyle change, one that is culturally acceptable to the organization as a whole. Both will involve change and change management.

Consensual Methods

The second of the two approaches to intervention requires that the consultant conducting the intervention should draw upon the knowledge and experience of the directors, managers, employees, etc., of the organization, helping them in the process to recognize their own problems and how to go about resolving them. Consultants may draw upon a variety of methods to extract such information, the idea being to develop a consensus from a representative group of individuals, and possibly then to use that group as future 'agents for change.' The following sections introduce some of the methods in use.

Brainstorming

This is a well-known approach in which a selected group of people is encouraged by a moderator to come up with ideas in response to a topic or a trigger question. Ideas may emerge at random around the room, and may be recorded on a flip chart.

The method is so well known as to require little description: it may not work too well, however. Problems can arise in the choice of people; junior members of the group are likely to defer to senior members, or may be unwilling to come forward for fear of ridicule. Brainstorming sessions can be slow to start, and difficult to moderate, particularly if some senior member decides to flex his authority. Although popular, brainstorming sessions can be limited in practice.

Nominal group technique (NGT)

Nominal group technique is also well established as a method of developing consensus, and is used to develop plans for future organizational activities and developments. It can be used to great effect in conjunction with Interpretive Structural Modeling — see below.

A topic is presented to a representative group of people; the topic may be concerned with, for example, some problematic situation facing the organization either internally or externally. A moderator, often the consultant, then conducts a discussion of the topic, before asking participants to write down their ideas on sheets of paper provided for the purpose.

After a suitable delay for people to generate their ideas, the moderator will invite participants to read out their first idea in turn, while the moderator, or assistant, copies them on to a flip chart or board. Each participant will then proffer their second idea in turn, and so on until all the ideas have been collected on to flip chart sheets, which may be posted around the room. The intention is to distance each idea from its originator, so that it may be considered objectively.

The group will next discuss the various ideas, combining those that are essentially duplicates, and reaching a group understanding of what each idea really means — not always obvious. Each member of the group is now afforded a rank scoring opportunity, often from ten to one. Participants are invited to rank order the ideas they preferred by allocating a 'score' of ten to the most favored, nine to the second most, and so on, down to one. The moderator then aggregates the 'scores' for the group. Some ideas may receive no scores, and will be dropped. The remainder will be reproduced as a rank-ordered list, from which will be selected perhaps the top twenty or so. These constitute the ideas — responses to the topic — that have been produced by the group as whole. Hence the term, 'nominal group.'

These various favored ideas may then inspire strategies to achieve them, and the formulation of plans to implement them. If the consultant/moderator has done his job well, he will have contributed little in the way of intellectual content to the resulting plans, which will have been conceived and drawn up by the participants, and which, therefore, are most likely to endorsed by them. The participants may then become the agents for change, as well as the proselytizers of the plan to the other members of the organization.

Idea writing

Idea writing takes NGT a little further. After discussion of a topic, participants in the group are invited to write their ideas, suggestions, etc., on a sheet of provided paper, using the provided pencil. After only two or three minutes, the moderator will request that each participant hand their sheet to the participant, say, two to their left. The receiving participant can then see the ideas already written by the providing participant, which should trigger a new set of ideas in his or her mind. After a further short period, the papers are handed round once more, this time by a different number of spaces. The process repeats for maybe thirty minutes, or until the moderator sees that most participants have run out of ideas.

The purpose of this convoluted procedure is twofold: to stimulate different strands of ideas within the group; and to conceal the source of any particular idea so that each may be treated objectively during subsequent discussion, without *ad verecundiam*, or *ipse dixit* (appeal to revered authority), and without disdain; such attitudes may be brought about by knowing that an idea was generated by someone senior/experienced, or junior/inexperienced respectively. Using standard paper and pencils, requiring responses to be printed in upper case, and complicating the path that sheets have followed from participant to participant will hopefully muddy the waters sufficiently.

Thereafter, the list of generated ideas may be handled as with NGT, and developed into a strategized plan of action to which all should agree, since all have participated and are then most likely to 'buy in' to the process and the product.

Warfield's interpretive structural modeling (ISM)

John Warfield proposed interpretive structural modeling (ISM) in 1973, and it has proved a most powerful approach to understanding complex relationships and situations, for formulating complex strategies and plans, and hence for making sound decisions. (Warfield, 1973, 1989, and Janes, 1988).

ISM has been described as a computer-assisted learning process that enables individuals or groups to map complex relationships between many elements, so providing a fundamental understanding, and providing the route to developing courses of action and solving problems.

ISM does not essentially require computer support — it can be executed by hand, using pencil and paper, but the process can be somewhat laborious, so computer support is often used to reduce the labor. There are few calculations, per se, within the processor — it is used largely to store, handle and reconfigure the information that participants in an ISM 'session' generate. ISM is, essentially, context free, as should be any sound systems method/tool.

ISM starts with a set of entities, between which relationships are yet to be established; in this respect, it may fit well with NGT and Idea Writing, both of which can be used to produce the list of entities. ISM typically produces a network, or tree of the entities as nodes, with the relationships forming the branches between the nodes. These trees, or networks, may be typically of four types:

- Intent structures, where the entities are objectives, and the links are 'helps to achieve.' The tree can then be read from root to tip as 'objective A helps to achieve objective B, which helps to achieve objective C,' and so on. The root objective is the one at the bottom of the tree — if that cannot be achieved, then none of the others is likely to be achievable. The objective at the top of the tree is the one that all the other objectives help to achieve — it is generally the mission. The development of intent structures is a powerful means of coordinating the various objectives and purposes of seemingly-disparate interests within an organization
- Attribute enhancement structures, where the entities 'contribute strongly to each other.' This is an invaluable way to establish the rationale for project or system designs, for example.
- Precedence networks, where activities precede each other. Precedence networks can be used to formulate rational, logical project plans early on in a project, at a time when the durations of

various activities may not be known. (Conventional project management tools require activity durations as an input.)

■ Priority structures, where projects are 'more important than each other.' This can prove a valuable way of establishing priorities where there are differences of interests and opinions within a group. Every member of the group contributes to the end result, and is morally obliged to buy into the result, even if it is not what he or she expected or wished for at the start.

ISM is an invaluable tool for individuals, working out their own priorities and plans. In group practice, a moderator or facilitator may employ ISM in a session, where information is projected on to a screen in front of the 'nominal group.' For an intent structure, the information would appear as a pair of objectives, with the question pose 'does objective A help to achieve objective B?' The group has to decide if the answer is yes, no, or not related, which may take some discussion. The screen will then present the question, 'does objective B help to achieve objective A?' i.e., the first question reversed. So, to each pair of entities there may be four responses — A helps to achieve B 'Y/N,' and B helps to achieve A, 'Y/N.' The results are stored, and the process is repeated, but with entities B and C as a pair, C and D as a pair; and so on. There could be scores of objectives.

The resultant reachability matrix is comprised of ones and zeros, and can be mapped directly into the corresponding structure. The structure, or network, may be drawn up on a board using sticky notes, and it may surprise the participants, even although they contributed all the information from which the network was constructed. Such surprise is curious, but not uncommon — participants are unable to forecast the outcome, even although they have participated fully in the process. (Using ISM, which I do almost daily, I have found — to my surprise — that I am greener than I would have supposed, i.e., I have a far greater concern for the environment than I had suspected. A tool/method that can teach you things about yourself clearly has hidden depths! (Hitchins, 1992, 2003.)) The participants will then be invited to change anything in the structure with which they disagree by rearranging the sticky notes — providing other participants agree to the change.

The whole process can be time consuming, particularly where there are divergent view about what is causally related to what, and in which way . . . . However, the strength of the method lies partly in this time consumption, since it allows participants time to understand, to recognize the merits of others' arguments and to move towards a consensus. On the other hand, it seems likely from experience that the time taken for a session rises approximately with the square of the number of people participating, so smaller groups may be preferable to larger groups, especially where addressing a large number of entities.

It is also noticeable that some people can become impatient with the whole process, especially where they feel that it is likely to interfere with their decision-making opportunities, or to provide a considered response where they prefer to manifest a visceral response. Such people are, of course, not particularly interested in consensus . . . .

Checkland's Soft Systems Methodology (SSM) in Intervention

The problems that surround issues of purpose are well appreciated. Stafford Beers, creator of the Viable System Model (see page 110), once remarked, 'the biggest confusion in which I was ever professionally involved concerned the purpose of a health system to which there are as many answers as interests involved'. The ease with which an organization can examine and deal with issues related to purpose depends on the view it holds about the nature of organizations. That view will fall between the two extremes recognized in academic research: positivist and interpretivist. According to the positivist view, organizations are rational and goal seeking. This makes it straightforward to identify the purpose of an existing system within an organization or to design a new system. The positivist view also simplifies the task of change management. The disruptive influences that might form the basis of opposition to the new system are considered to be aberrant, but amenable to management action: opposition can be overcome by changing some elements of the system or by replacing the people.

The interpretivist view sees an organization as being built up from social relationships between the individuals and groups within it; as the individuals and groups interact, they construct and reconstruct the network of relationships that defines the organization. Purpose is harder to expose in this environment, not least because the various groups are likely to hold differing views as to the purpose of the organization or a component system.

Problems in exposing the purpose of a system make it harder to establish what any supporting system is actually supposed to do and any project carried forward in these circumstances risks failure. Change management also becomes problematic. If some system is implemented based on a flawed understanding of the purpose of the wider system it is supposed to support then that relationship will be damaged; however, in this case redesigning the system or even changing the people is unlikely to improve the situation.

SSM was designed to help managers address unstructured problems that relate to the purpose and actions of an organization or subcomponent. (Checkland, 1972, 1981; Wilson, 1984). It arose out of work done by staff at the Systems Engineering Group at Lancaster University, England. They embarked on a program of action research — a combination of organization interventions and academic reflection — that started out by trying to apply traditional systems engineering methods to address managerial problems. Members of the Group found that their efforts were regularly frustrated because they could not always determine the purpose of the system or organization they were trying to assist. SSM represents their development of systems thinking which they believed would address the problems they had encountered.

SSM is built around the seven-stage model shown in Figure 7.1. The key point to note is that the analyst is required to address the problem situation from two perspectives: parts of the method require the analyst to relate to what is actually happening in the situation being analyzed (the Real World), whilst other parts should be driven only by logic and systems thinking (toward an Ideal World). The aim of the first two stages is to try and build a picture of the problem situation as a precursor to identifying a number of possible purposes for a system. The system can either be a new system designed to alleviate the problem or a redefinition of an existing system and the statements of purpose might take account of a number of different viewpoints. In the third stage, a root definition is developed for each system that describes six key aspects of that system:

- ■ 'Customers' of the system — the victims or beneficiaries of the transformation that the system carries out.
- ■ 'Actors' within the system — those who carry out the transformation.
- ■ 'Transformation process' carried out by the system — what the system does in converting the input to the output.
- ■ Weltanschauung — the worldview that makes the transformation meaningful in the context of the system.
- ■ 'Owners' of the system — those with the authority to stop the transformation process.
- ■ 'Environmental constraints' — elements outside the system that it takes as given.

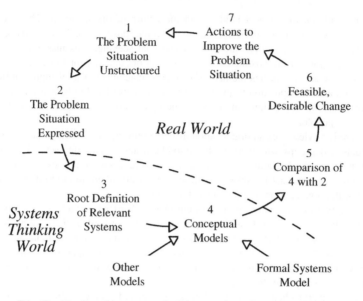

Figure 7.1   Checkland's soft systems methodology — the outline process (Hitchins, 1992).

In the fourth stage, each root definition is elaborated to produce a conceptual activity model that includes the core activities required to service the needs of the root definition. The elaboration can be pursued to any level of detail — each activity simply acquires its own root definition — and can be done in a single step or through a number of iterations so that the analyst ends up with a set of hierarchical models. In either case, the elaboration should be the result of systems thinking rather than of explicit reference to existing organizations and processes and should expose only those activities that are logically necessary.

Stages five, six and seven are primarily concerned either with change management or with situations where the system design effort forms part of a wider (e.g.) business reengineering program. The aim of stages five and six is to develop courses of action that are both feasible — i.e., can be started and hopefully carried through given the existing culture in the target organization — and desirable — i.e., they will bring about beneficial change. The courses of action are derived by comparing the models developed in stages three and four with existing systems and processes in the organization. In stage seven, the developed courses of action are put into practice. A single iteration of SSM is unlikely to solve a problem: it will alter the situation that caused the problem to surface, i.e., action will simply create a new situation that may benefit from further analysis and intervention using SSM. In this respect, one iteration of SSM may resolve the problem, while several may be needed to solve the problem — which may morph in the meantime, requiring continual intervention.

Given the description of SSM it is clear that its contribution to the implementation of a system is during the problem solving and design effort, which it can support in a number of ways. At the most fundamental level, it can be used to develop and clarify the purpose of a proposed system by allowing analysts and stakeholders to examine the implications of any number of root definitions. Having developed an agreed purpose, activity modeling — stages three and four from the model in

Figure 7.1 — can be pursued to the point at which the information required to enable the system to function and the necessary transformations can be meaningfully described. This information can be used to drive later stages of the design process. Alternatively, given an assumption of agreed purpose, the activity modeling from stages three and four can be pursued in the manner just described to generate an abstraction of existing processes and activities. The risk in this approach is that the assumed purpose is not the real purpose and that any implemented change may encounter significant resistance.

There are two major points that differentiate SSM from other methodologies or methods that might be used at the start of a design exercise: the ability to abstract and the ability to surface issues related to the purpose to be served by the wider system. Models in SSM are made up of sets of dependent activities; they make no assumptions about the structure of the organization that will carry out those activities or about the precise nature of the information artifacts that will be produced or required as inputs. Analysts can instantiate the models by assigning activities to existing or planned organizational elements and by naming the artifacts required to help define the way a system should be implemented, how it will fit in with other, existing systems and what functionality it should provide.

SSM offers the analyst an opportunity to examine the purpose of the wider system within which some new system will be implemented. Whilst there is nothing to stop an analyst doing the same thing using another method, it is not required: indeed, it requires a conscious decision on the part of the analyst. SSM, then, pursues the Systems Approach and considers a system as open, adaptive and an interactive part of some whole.

There is one other, perhaps more minor point that sets SSM aside from other means of depicting systems and that is its accessibility to the layman. Nonspecialists are more comfortable with the 'softer' approaches to depicting systems, rather than 'harder' approaches: whilst the harder methods may produce the detail required for detailed system design, they are not as effective tools for communicating with the untrained.

The value to any organization of such an intervention/investigation depends strongly on the view it holds about the nature of organizations generally. If it believes that it is a rational, goal-seeking body then the investigation will be appear to be of limited value — system purpose will be obvious and unambiguous. Organizations that recognize the more social view will also recognize the value of the intervention/investigation.

Hitchins' Rigorous Soft Method (RSM) in Intervention

Like SSM, the rigorous soft method (RSM) is based around the General-Purpose Problem-solving Paradigm (page 172). RSM is intended for addressing complex problems and issues, and for supporting the conception of potential remedial solutions. It may therefore serve as the first 'stage' in the systems methodology. RSM is context free — analysts and investigators working in the problem domain, and having intimate domain knowledge, bring information about the issue or problem into the method. Investigating the sources of dysfunction within some complex system can generate large amounts of data and information; unlike SSM, RSM employs tools and processing methods to handle, organize and process the information, where 'process' implies the progressive reduction of entropy as disordered source data is transformed into specific solution information.

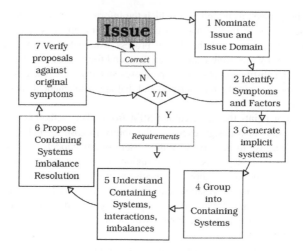

Figure 7.2   Rigorous soft method (RSM) — the process icon.

As Figure 7.2 may suggest, RSM is a self-checking system for conceiving remedial solutions to complex problems. So, how does it work? It uses processes that are analogous to the way in which a doctor may diagnose a patient's illness.

A doctor is trained and experienced in diagnosing dysfunctions in the human body — one of the most complex systems on the planet. To do this, doctors have established norms for various aspects of the healthy human body, according to age, sex, size, occupation, working environment, etc. Symptoms of dysfunction for an individual, then, are seen as divergences from the norms for such an individual.

Suppose that a patient came into a doctor's surgery complaining of feeling 'out of sorts,' but having no idea of what was wrong. The doctor might follow the following procedure:

■ Ask the patient about their background and recent activities, visits abroad, etc., to identify recent environment and exposure to risks

■ Check for symptoms of variation from the norm: temperature, respiration, pulse, blood pressure, pallor, skin lesions, perspiration, agitation, etc. If none shows unusual variation, then check deeper, testing urine and blood for variations, inclusions, infections, imbalances, etc.

■ Wherever a symptom is identified, relate it to the potential dysfunctions within the body that could cause that symptom. Skin rash, for instance, could be caused by some infection, by food poisoning, by organic disorders, etc. High pulse rate could stem, similarly, from a variety of organic causes, as could raised temperature, and so on.

■ The objective, then, is to identify as many symptoms as possible, and to find the organic dysfunction or external factor that is common to them all. The more symptoms, the more likely that, although each symptom may point to a variety of causes, the whole set will point to only one or two causes, as being common to all symptoms.

■ Where more than one possible cause is perceived, the doctor will then consider how one of these may generate the other, i.e., he/she would mentally model the behavior of patient's body, seeking causal relationships.

■ Note that doctors rarely have to go through such an elaborate procedure; experience of having seen conditions before generally leads them to a quick diagnosis. Rarely, however, problems arise with obscure disorders, and where a patient is suffering from more than one complaint, each of which is generating symptoms.

■ Having determined what the source dysfunction(s) may be, the doctor can then propose a remedy, which may vary from doing nothing, through medication to surgery. The doctor will consider at all stages the risks involved from any course of action in relation to the patient's condition, and will endeavor to 'do no harm,' where unsure of the efficacy of the remedy.

Figure 7.3 shows the diagnostic process generalized, with the proposed remedies being tested for minimal side effects. The diagnostic process is embedded in the RSM process, shown graphically at Figure 7.4.

1. RSM starts with the identification of an issue of problem, or perhaps...
2. ... with the identification of symptoms that point to some issue of problem. It would not be unusual to redefine the nature of the supposed issue once a number of symptoms has been identified and explored.
3. Each symptom 'implies' the existence of a set of so-called implicit systems contained within the issue/problem space; these are open, adaptive, interactive, functional systems that must exist for the symptom to emerge — and that emergence signals that something is dysfunctional within the set. Structure within the issue domain is generally irrelevant.
4. The sets of implicit systems (one set per symptom) are unified, clustered/aggregated to identify implicit containing systems. A hierarchy shift results in the highlighting of problem themes within the issue; this stage presents a rich picture of the whole issue.
5. The clustering process reveals potential interaction imbalances...
6. ... resolution of which would be required as a minimum to neutralize the original symptoms
7. Conceive and test candidate remedies to see if they would, if implemented, eliminate/eradicate all of the symptoms identified in 2, and all the imbalances of 6. Candidate remedies would also be tested for cultural acceptability.
8. Nominate one or more sensible and acceptable remedies – provided there are any, which is not guaranteed

RSM can be used by individuals and by teams. The quality of the output, is, of course, limited by the quality of the input, in particular by the knowledge skill and insight of those exploring the problem space and exposing the various symptoms and factors. Factors are those aspects of an issue that make it 'special,' unusual, out of the ordinary, perhaps even unique, but which are not of themselves symptoms of dysfunction.

Working with symptoms of dysfunction makes it is possible to use them to formulate the requirements for addressing and remedying the issue — in effect, the problem (Real World) suggests its own remedial solution (Ideal World).

Figure 7.5 presents the rigorous soft method in the form of a behavior diagram, which some may find more familiar, and which lends itself to the establishment of a resourced process model for formulating a remedial systems solution, or solutions, to a particular problem. These are conceptual only, at this early stage in the systems methodology, but interestingly and importantly conceptual remedies to the whole issue or problem are suggested intrinsically by the application of RSM — effectively, they 'emerge' during the process of creating Ideal World views.

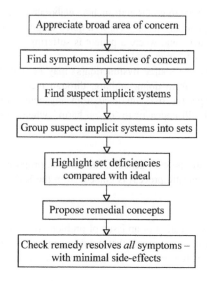

Figure 7.3 Outline diagnostic procedure.

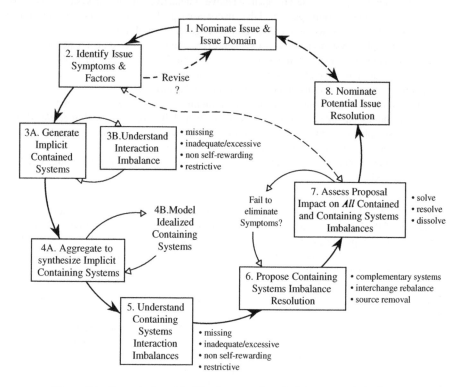

Figure 7.4 Conceptual model of the rigorous soft method — process view. See text.

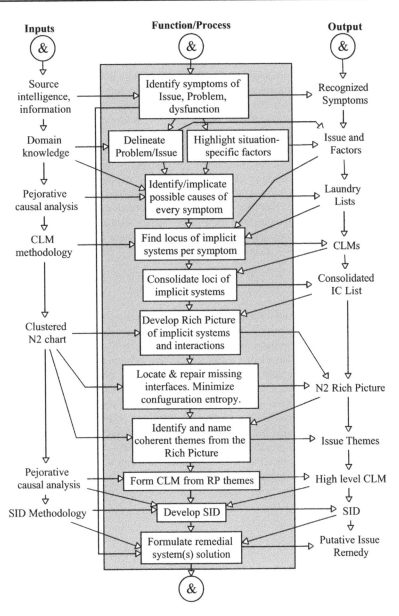

Figure 7.5 The rigorous soft method as a behavior diagram, showing sequential/parallel functions in the center panel: these are enabled by inputs at left, and by the outflows from previous functions/activities, resulting to outputs in the right hand column. SID is systems interaction diagram. IC is implicit system. Note that remedial system solution concepts 'emerge' as a natural byproduct of the RSM process.

Summary

There is a variety of ways of that complex issues and problems may be addressed. Some of those associated with interventions are addressed, including Warfield's interpretive structural modeling, Checkland's soft systems methodology, and Hitchins' rigorous soft method, with the latter being designed specifically as the front end of the systems methodology.

Interventions are often associated with issues that arise in organizations from disagreements about the purpose of the organization, with different people holding different ideas as to purpose. Two views are cited: the positivist view, which sees organizations as purposeful and goal seeking, so that differences may be ironed out, if necessary by removing people; and the interpretivist view, which sees an organization as made up of people and groups of people interacting purposefully, but not necessarily with the same purpose. An interpretivist view would see the social systems within the organization realigned, such that they worked together harmoniously. Both views are expressive of systems engineering, with the positivist view being 'harder' than the interpretivist view; neither is right or wrong, *per se*, but one may be more appropriate than the other according to situation.

Assignment

A 10-storey building is proposed for a beach 'beauty spot' where no such buildings presently exist. The building is to provide apartments on the upper stories, while the lower stories are to contain offices, gymnasiums and a large indoor swimming pool for access by the public, with water chutes, artificial tides, etc. The plan has alerted interested parties, including the tourist office, the beach preservation lobby, the local council — which is divided over the plans — local estate managers, prospective owners of apartments, several local sports clubs who would like access to the facilities, and even the police, who are keen to encourage youth facilities to reduce crime in the area.

Using role play to present yourself as each of the interested parties in turn, define the purpose of the project as your role might see it, identify a set of objectives that your role would like to see achieved in relation to the project, which — you are advised — is likely to go ahead in some form or other. Aggregate the objectives to form an intent structure in which 'lower' objectives are linked to, and help to achieve, higher objectives.

Comment on your results; hence determine whether you consider yourself positivist or interpretivist, and why.

Case B: The Practice Intervention

Situation

A consultant was invited to 'intervene' in the operations of a partnership in the building services professions. The partnership was comprised of engineers with a variety of specialisms: heating and ventilation; elevators and escalators; power distribution systems; and so on. The business was surviving, but not flourishing; a recently joined partner encouraged the idea of bringing in a consultant to help identify and sort out the problems.

The consultant had had no previous dealings with the partnership, and knew nothing of them. He decided that, before visiting the organization and conducting any sort of intervention, he needed to know more. He elected to explore the issue facing the partnership initially off-line, using the rigorous soft method, which he would apply on his own, to probe the psyche of the group and to clarify his thoughts. He asked for, and received, the organization's marketing materials, and he also asked for each of the partners to write down a single response to a question. Each partner was asked to complete the question: 'How can we. . . ?'

Using this information, the consultant sought to establish what was going on in the organization, at least in structural terms, before he came face to face with them all during a more formal intervention.

There are, then, three parts to this case:

- The first part shows what the consultant received by way of answers to the 'How can we. . . ?' questions, and how he used the rigorous soft method to investigate further, but in a 'hands-off,' off-line manner.
- The second part shows the intervention in action, with nominal group technique, idea writing and interpretive structural modeling in action.
- The third part compares the first two, to see what they separately revealed, together with weakness and strengths.

Systems Engineering: A 21st Century Systems Methodology   Derek K. Hitchins
© 2007 John Wiley & Sons, Ltd

Off-line Informal Investigation using the Rigorous Soft Method (RSM)

After a short delay, the consultant duly received the marketing and publicity materials and the 'How can we. . . ' questions; these are tabulated at Table B.1, together with the consultant's observations about each questioner, as revealed by the nature of the question.

The issue facing the partnership was fairly evident, or so it seemed, from the questions — each of which came from a different partner. The general tenor of the questions was one of concern, not about the performance of individuals, but about the manner of their working as a cohesive, coordinated, integrated group. So, the issue might be stated as:

> *Concern over the Practice's ability to remain viable and survive the current economic recession*

The next stage in the off-line analysis — the whole of which took about half a day — was to turn the responses to the questions into symptoms; this is an almost trivial process as shown in Table B.2

The next stage in the RSM process (see page 195) is to use each symptom as a 'probe' to investigate the probable cause(s) of dysfunctional behavior within the system (in this case within the partnership). Within the RSM method this is done in a formalized manner using several techniques which are best explained by example.

Figure B.1 shows a crude example — the consultant's first attempt whilst at home, before breakfast (!) The procedure employs a standard proforma. At top right is entered the symptom to be explored — in this case, low efficiency: wherever practicable, the symptom is described using pejorative terms. A so-called laundry list of possible causes for low efficiency is then drawn up, also using pejorative terms — we humans are much more skilled at criticizing using pejorative

Table B.1   Exploring the problem space — looking for symptoms

The 'How can we. . . ? questions	Consultant's observation
How do specialist activities on the periphery of general building services design integrate within an organization to form an efficient functioning unit. . .	Despite requests for consultant partners to ask only one question each, this partner sought to ask two questions in one. Both indicate disquiet about whether, or not, the enterprise is one system or a collection of separate, nonintegrated parts.
. . . and how can effective management communications be achieved?	Concern over management, or lack of. . .
How can the diverse talents and personalities within the practice best be brought together to maximize performance in the present economic climate?	Again. Concern that lack of cohesion and coordination between individuals is prejudicing business performance of the whole
How can we market our experience?	Limited marketing know-how
How can we achieve common aims?	Lack of integration
How can we keep afloat in these trying times?	Tear-jerking — is someone messing about?
How can we obtain maximum benefit from our assets to sustain growth, success and profitability?	Good question — the real issue?

Table B.2 Turning questions into symptoms.

Core question	Symptom
How can we take advantage of our different specializations within the practice?	*Poor specialist variety cohesion*
How can we become more efficient?	*Low efficiency*
How can we improve management communications and coordination?	*Ineffective management communications*
How can we optimize our business performance?	*Non-optimal business performance*
How can we best market our experience?	*Ineffective marketing of experience*
How can we achieve a common aim?	*Lack of common aim*
How can we use our capabilities to maximize performance?	*Ineffective application of assets to maximize performance*
How can we present out varied specializations to customers?	*Poor unfocused self-image*

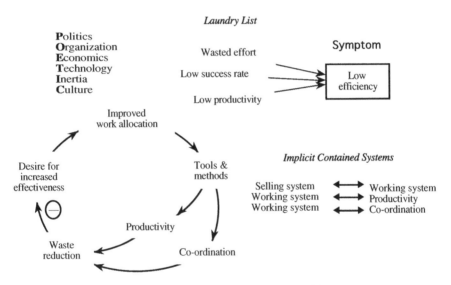

Figure B.1 Locus of probable causes — low efficiency.

terms than praising using positive terms — this propensity to criticize is turned to advantage in this technique, called 'negative assertion.'

Note that the causes of low efficiency could only be guessed at in this instance. The acronym POETIC is presented on the proforma to remind investigators of the likely causal factors.

Given a list of possible causes, it is reasonable to assume that they are related, since they all refer to the same system. Next, they are formed into a causal loop model, bottom left, dropping the pejorative terms in the process, but adding additional elements to support the essential logic in the loop. In this particular case, the consultant also interpreted the list of possible causes, and used his experience to form a causal loop model that made sense in the particular context of low efficiency.

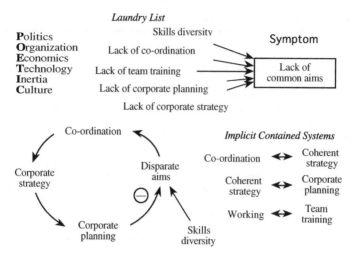

Figure B.2   Locus of probable causes—'lack of common aims'.

In so doing, he created something of an Ideal World model, showing what might be expected in an efficient organization. This method, even although not strictly executed in this case, has seemingly magical abilities to form Ideal Worlds from pejorative, real world symptoms.

Lastly, at bottom right, the consultant wrote up the implicit contained systems that must exist, and that had imbalances resulting in Low Efficiency

The next symptom to be tackled was 'Lack of Common Aims'– see Figure B.2. It follows the same plan: pejorative symptom, top right; pejorative laundry list, top center; CLM bottom left; Implicit Contained Systems, bottom right.

The consultant's coffee was having some effect by this stage; he identified possible causes of the symptom, and recognized that they were all related. The CLM, bottom left, dropped the pejoratives and left a CLM that indicated an Ideal World in which disparate aims were brought together under the mantle of coordinated corporate strategy and planning. This led to the Implicit Contained Systems at right, which evidently must be dysfunctional; else, the symptom at top right would not have emerged. . . .

The next symptom looked more to the heart of the issue facing the practice: poor, unfocused self-image. The laundry list of possible causes required some thought, and the CLM was rather more complex than usual. This arose because the practice was comprised of a number and variety of specialists, each considered expert in his own domain. It was not surprising, perhaps, that there were problems with the image of the whole. The CLM, at this point, switched from being simply analytical, and moved toward suggesting potential solutions — shown in the dashed, arrowhead lines, which suggested themselves while examining the CLMs for the first two symptoms.

By this stage, the consultant was beginning to see how the various symptoms, although they might appear quite different, were tending to point in broadly the same direction, and indeed were identifying the same implicit systems as being imbalanced (or nonexistent as it turned out — see later.)

The next symptom confirmed his view — see Figure B.4, nonoptimal business performance. Forming the pejorative laundry list was not difficult, but the nonpejorative, Ideal World CLM proved a little less tractable. Eventually, he created a double loop CLM, with the left-hand loop

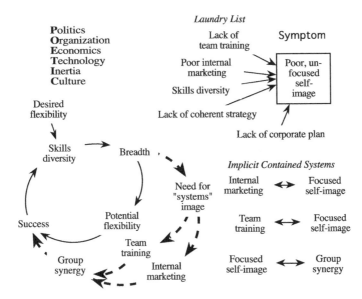

Figure B.3   Locus of probable cause — unfocused self-image.

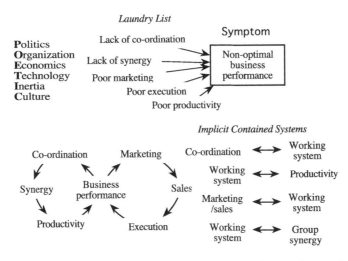

Figure B.4   Locus of probable causes — 'non-optimal business performance.'

concerned with what went on 'inside' the practice, while the right-hand loop was focused on relationship with 'outside' the practice. The consultant finished off the remaining symptoms in the same manner, and then drew up a list of all the implicit contained systems: see Table B.3.

The table was then used to form an N^2 chart, Chart B.1. This a straightforward process, in which all the entities — the implicit contained systems — are selected as the leading diagonal elements in the chart, and the interactions between them are recorded in the chart at the appropriate

Table B.3 Symptoms and implicit systems.

Implicit contained system A for...	Symptom	Implicit contained system B for...
Focused self-image	Poor specialist variety cohesion	Business performance
Self-interest	Poor specialist variety cohesion	Mutual self-reward
Team training	Poor specialist variety cohesion	Self interest
Selling	Low efficiency	Working
Working	Low efficiency	Productivity
Working	Low efficiency	Coordination
Office management	Ineffective management communications	Procedures
Procedures	Ineffective management communications	Discipline
Office management	Ineffective management communications	Coordination
Managing skills diversity	Ineffective management communications	Procedures
Coordination	Nonoptimal business performance	Working
Working system	Nonoptimal business performance	Productivity
Marketing/sales	Nonoptimal business performance	Working
Working	Nonoptimal business performance	Group synergy
Focused self-image	Ineffective marketing of experience	Marketing strategy
Marketing strategy	Ineffective marketing of experience	Marketing investment
Marketing investment	Ineffective marketing of experience	Promotion methods
Promotion methods	Ineffective marketing of experience	Self-image
Skills diversity	Ineffective marketing of experience	Focused self-image
Coordination	Lack of common aims	Coherent strategy
Coherent strategy	Lack of common aims	Corporate planning
Working	Lack of common aims	Team training
Coherent strategy	Ineffective application of assets to maximize performance	Corporate planning
Corporate planning	Ineffective application of assets to maximize performance	Coordination
Skills diversity	Ineffective application of assets to maximize performance	Focused self-image
Internal marketing	Poor, unfocused self-image	Focused self-image
Team training	Poor, unfocused self-image	Focused self-image
Focused self-image	Poor, unfocused self-image	Group synergy

locations. If the relationship between a pair of implicit contained systems comes up twice, then it is recorded as '2,' otherwise as '1.'

The N^2 chart contains no new information — it is simply a different way of representing the same information as that in the table. However, the chart can be reconfigured to reveal structure within the group of entities; this can be done by hand or, as in this case, using a simple computer tool — see Chart B.2. The chart shows three so-called Implicit Containing Systems, which have been identified in two ways:

■ First, the various implicit contained systems have associated with those others to which they are functionally connected, so automatically forming functionally interconnected groups

■ Second, the interfaces between the blocks have been selected so as to 'cut' very few inter-group links, i.e., to recognize loose coupling

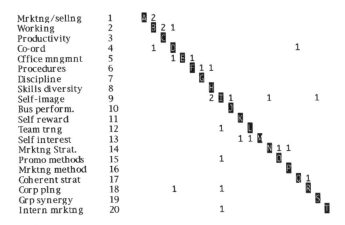

Mrktng/sellng	1	
Working	2	
Productivity	3	
Co-ord	4	
Office mngmnt	5	
Procedures	6	
Discipline	7	
Skills diversity	8	
Self-image	9	
Bus perform.	10	
Self reward	11	
Team trng	12	
Self interest	13	
Mrktng Strat.	14	
Promo methods	15	
Mrktng method	16	
Coherent strat	17	
Corp plng	18	
Grp synergy	19	
Intern mrktng	20	

Chart B.1   Unclustered N-squared (N^2) chart printout, showing relationships between implicit contained systems. N.B. The tool used to accumulate the N^2 chart has truncated the names in the left-hand column: the full titles can be seen by reference to Table B.3.

These three groupings are indicative of three 'problem themes:' 'group organization and method,' the analysis suggests, is dysfunctional; so, too, is the 'group business development system;' and, less obviously perhaps, there appears to be a dysfunctional 'motivation system,' which concerns itself with self interest (motivation), team training and reward. Note that all three of these containing systems/problem themes refer, not to individual partners, but to the system/practice as a whole. Note, too, from the chart that there are clear nodes (signified by cross patterns formed from the interfaces) indicating the sensitivity of (Method of) Working and Coordination in the upper block, Group Organization and Method, and (Lack of) Focused Self-image in the Group Business Development System.

The N^2 Chart B.2 may be presented as a Causal Loop Model, making it easier to understand the interactions dynamics between the problem themes — see Figure B.5. The three causal loops

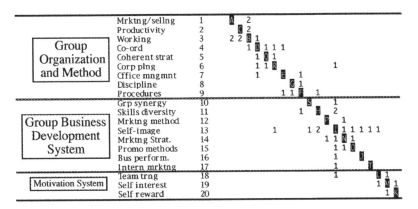

Chart B.2   Clustered, N^2 chart, showing implicit containing systems in the left-hand column. Making the chart symmetrical about the leading diagonal also makes the interface patterns more evident.

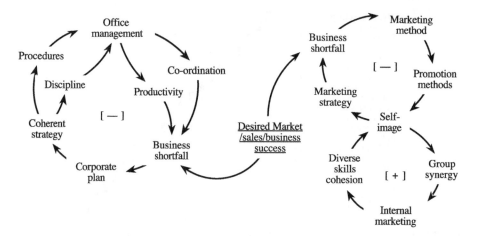

Figure B.5   CLM developed from Chart B2.

represent the three groups from Chart B.2, which in turn represent three so-called problem themes. The left-hand loop is about straightforward management of business operations — organization and method. The top right loop is about marketing, including the promotion and projection of self-image. The bottom right loop is also about self-image, but this time it is about internal marketing, i.e., about convincing the staff that their very diversity is their strength. . . .

The first two loops present as control loops, with control directed towards reducing business shortfall — clearly a concern to the practice. The third loop is a self-reinforcing loop (positive feedback) and this has the potential to either 'spin up,' in which case self-image will rise and the top right loop will operate freely, or 'spin down.' In this event, the current poor self-image will deteriorate further, preventing the proper operation of the marketing loop and effectively scuppering the practice.

Using the 'How can we. . . ' questions and the marketing material as inputs, and applying RSM albeit somewhat crudely, the consultant now had a reasonable idea of what might be going on — sufficient, at least, to know what to look out for.

Note. It might be thought that the same conclusions might have been reached simply by examining the original 'How can we. . . ?' questions. Using RSM has enabled the consultant to lay out his thoughts, his rationale, and his ideas clearly on paper, as an audit trail from which he has learned, and with which others may agree, or disagree — but at least they ideas are out in the open for discussion.

In the event, the consultant did not show his simple, 'rough and ready' RSM analysis to the partners, preferring instead to see if the results of the ensuing intervention confirmed his suppositions, or not. . . .

Hands-on Intervention — using NGT and ISM

The consultant later met up with the members of the partnership on 'neutral ground,' in the nearby city center. The atmosphere was somewhat frosty, and it became evident that there was some hostility within the group to the whole 'intervention exercise.' Surprised, but unperturbed, the consultant introduced himself by stating that he knew little about them as an organization, and certainly did not know sufficient to advise them about their business. Instead, what he proposed to do was to introduce them to Auto-Intervention: together they would undertake a Voyage of Discovery into Self-Consultancy. At the end of the session, which might stretch over a day or two, they should have learned enough never to need him, or any other external consultant again. The consultant observed that he now had their attention.

He then outlined his plan: to so conduct an intervention that the group effectively conducts its own analysis, and develops its own strategy and plan. His approach was to adopt the traditional facilitator rôle, but at the same time to present several interactive methods to the group so that the group chose both the method and the direction of the sessions.

The consultant presented three optional 'trigger questions' which, he observed, seemed to address the issues of concern:

1. What objectives would you like the practice to achieve over the next 5 years?
2. What do you consider to be the attributes of a successful partnership?
3. What do you consider are the most important tasks/projects that you should undertake within the practice?

As he hoped and expected, the group unanimously chose the first trigger question: this jibed with his RSM analysis.

Next he offered them a choice of addressing the trigger question using either brainstorming or idea writing, which he described as follows:

- *Brainstorming*. The group is given some creative tasks, e.g. 'conceive as many ways as possible to move water uphill'; once the creative juices are flowing, the subject of interest is introduced. *Pros*: simple, quick. *Cons*: easily dominated by individuals, good ideas can be ridiculed, difficult to maintain positive-only attitudes, creates internal group dissension
- *Idea writing*. Group members write their answers to a trigger question on a sheet of paper. After a short time, group members are asked to pass papers to the person next to them. Recipients see providers ideas and add more of their own. Procedure repeated several times. Resulting ideas then presented one at a time, in turn from each member. *Pros*: suppresses dominance, generates wide range of ideas, hides 'owner' of ideas. *Cons*: large number of ideas requires handling by facilitator and group.

There was some discussion, after which the group chose idea writing. This led into a session using nominal group technique, with the generation of a host of objectives, followed by an allocation of scores to end up with a ranked list of group objectives — see Table B.4.

It became clear during the discussion of the table that one of the senior partners had contributed objective 27 with a view to disrupting and discrediting the process, with which he did not agree. On observing that objective 27 had received only one vote (not from him!) he had reconsidered his view, and was now less inclined to dismiss the process.

The group then agreed that they need consider only those objectives that had received a score, and they elected to drop the rest.

Next, the consultant introduced them to interpretive structural modeling, and the group entered into an ISM 'session,' with a view to developing an intent structure (see Warfield's interpretive structural modeling (ISM) on page 191).

Table B.4 NGT: table of group objectives with scores.

Serial	Group objective	Score
1	Achieve standing within the industry and with clients	29
2	Identify what we are selling	20
3	Develop an organization with recognized acquisition value to others	11
4	Financially secure practice	27
5	Short-term survival	27
6	Establish wide client base to weather difficult times	30
7	Growth of wealth	35
8	Establish how we are going to sell	30
9	Provide a basis for developing a second career	
10	Continual improvement in technical competence	11
11	Remove the work pressure on individuals which adversely affects their private lives	11
12	Concentrate expertise into new client base	1
13	Establish a niche market	12
14	A means of widening expertise	
15	Develop structured approach to projects	8
16	Increase the number of disciplines in the practice	
17	Quick access to general information	2
18	Improved financial control	17
19	Establish how partners and associates team	
20	Better quality accommodation	6
21	Identify focused self-image	22
22	Join A.C.E.	
23	Coordinated, effective communications	6
24	Retirement in 5 years	
25	Improved standard operating procedures (SOPs)	3
26	Establish R&D capability for solving design problems	
27	Take over another practice	1
28	Develop export markets	
29	Introduce office automation	
30	Improved staff training	
31	Understand our market	21

For each of the objectives taken in pairs, they were asked the question 'does objective A help to achieve objective B, or is it the other way round, or do they both help each other, or is there no relation?' There were four possible answers to each pair: responses were accumulated in a matrix, as shown in Chart B.3; and an initial practice intent structure was drawn from the chart, Figure B.6. The whole process took several hours, and was accompanied by much discussion between the participants (protagonists?), during which there was a discernible development of understanding and consensus within the group.

The initial intent structure was developed on a whiteboard, using sticky notes for objectives and colored pencils to draw the lines: this allowed the group to consider changes to the results. The group did, in the event, change a number of entities, before creating the final result at Figure B.7.

```
Info access         1    1 1 1 1 1 1 1 1 1 1 1 1 1 1 1 1 1 1 1 1 1
Structured projec   2    0 1 1 0 1 1 1 0 1 1 1 1 1 1 1 1 1 1 1 1 1
Finance control     3    0 0 1 0 1 1 1 0 1 1 1 1 1 1 1 1 1 1 1 1 1
Know market         4    0 0 0 1 0 0 0 1 0 1 1 1 1 1 1 1 1 1 1 1 1
Accommodation       5    0 0 0 0 1 1 1 0 1 1 1 1 1 1 1 1 1 1 1 1 1
SCPs                6    0 0 0 0 0 1 1 0 1 1 1 1 1 1 1 1 1 1 1 1 1
Communication       7    0 0 0 0 0 1 1 0 1 1 1 1 1 1 1 1 1 1 1 1 1
Identify product    8    0 0 0 0 0 0 0 1 0 0 1 1 1 1 1 1 1 1 1 1 1
Work pressure       9    0 0 0 0 0 0 0 0 1 0 1 1 1 1 1 1 1 1 1 1 1
Technical compete  10    0 0 0 0 0 0 0 0 1 0 1 1 1 1 1 1 1 1 1 1 1
How to sell        11    0 0 0 0 0 0 0 0 0 0 1 1 1 1 1 1 1 1 1 1 1
New clients        12    0 0 0 0 0 0 0 0 0 0 0 1 1 1 1 1 1 1 1 1 1
Wide client base   13    0 0 0 0 0 0 0 0 0 0 0 0 1 1 1 1 1 1 1 1 1
Niche market       14    0 0 0 0 0 0 0 0 0 0 0 0 0 1 1 1 1 1 1 1 1
Self image         15    0 0 0 0 0 0 0 0 0 0 0 0 0 0 1 0 1 1 1 1 1
Finance security   16    0 0 0 0 0 0 0 0 0 0 0 0 0 0 0 1 1 1 1 1 1
Take over          17    0 0 0 0 0 0 0 0 0 0 0 0 0 0 0 0 1 1 1 1 1
Wealth growth      18    0 0 0 0 0 0 0 0 0 0 0 0 0 0 0 0 0 1 1 1 1
Achieve standing   19    0 0 0 0 0 0 0 0 0 0 0 0 0 0 0 0 0 0 1 1 1
Acquisition value  20    0 0 0 0 0 0 0 0 0 0 0 0 0 0 0 0 0 0 0 1 1
Survival           21    0 0 0 0 0 0 0 0 0 0 0 0 0 0 0 0 0 0 0 0 1
```

Chart B.3 ISM reachability matrix developed from NGT table of objectives. Names have been truncated: see Figure B.6 for full titles.

The group identified features from the intent structure, with some prompting:

- Pervasive or systemic objectives were at the bottom, and were of most immediate concern, since unless these objectives were achieved it would be impractical to move further up the tree.
- 'Financially secure practice' and above were seen as a 'wish list' of highly desirable outcomes, provided lower-half objectives could be achieved. The contentious objective 27 was lumped-in with other 'wish list' items
- The intent structure revealed (the need for) four implicit systems:

  - a general information handling system;
  - a corporate management system;
  - a marketing/selling/image system;
  - and a system for addressing the client base directly.

The group elected to develop strategies to achieve lower-half objectives. Next, the consultant introduced them to two ways of developing those strategies. The first method used the idea of overcoming threats.

The method is shown in Table B.5, and is part of the TRIAD Building System (see page 225 *et seq.* and Figure 9.1). It involves identifying potential threats to achieving an objective, and then developing strategies to overcome the threats. However, the younger partners seemed less comfortable with this method, suggesting that they lacked the experience of older partners at identifying threats. The consultant introduced them to causal loop modeling as an alternative approach and this was more successful. (The first method had effectively allowed the older partners to dominate the group; younger partners were more comfortable using CLM, which prevented further domination.)

Four of the CLMs developed by the group with facilitation are shown in Figure B.8. The technique presents each respective objective in negative terms, as a concern. Next a rich laundry list is developed, also in negative, pejorative terms, so that the final causal loop model — which, of course, drops the negative, pejorative terms — then presents a positive strategy to achieve the

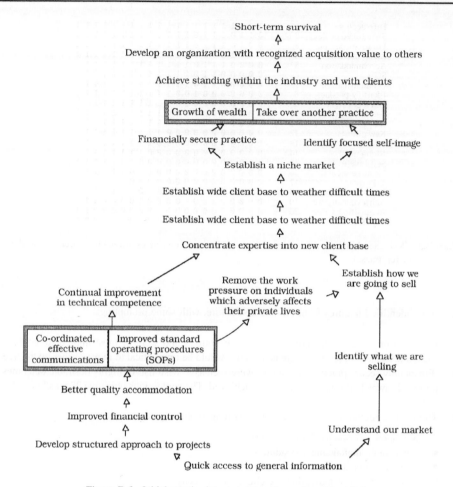

Figure B.6 Initial practice intent structure drawn from Chart B.3.

objective. This simple process of switching between negative pejoratives and positive assertions is surprisingly powerful.

A further advantage of developing the CLM is that the process promotes completeness and closure: it will often be found that the items in a laundry list are insufficient to develop complete, logical, closed loops; additional elements have to be added to complete the logic, and these turn out to be items that should have been in the laundry list had it been complete. In this way, the laundry list can be enriched and completed.

The group then went on to develop a range of strategies by working from objective to laundry list and omitting the causal loop model, seemingly unaware that they had, in effect, reverted to the first method. . . . These included:

- CQI — continuous (business) quality improvement;
- Corporate plans — annual, budgeting, forward investment, setting market targets, etc., which had not been previously done;

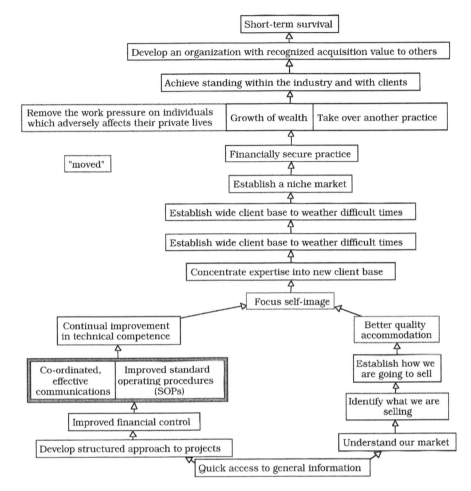

Figure B.7   Modified final practice intent structure.

- Partners' meetings — every week without fail! It seemed that there had been no regular partners meetings;
- Monthly project business reviews. Each member, in turn, each month, presents a project for peer business review.

A table of strategies was drawn up, bringing together all of the ideas; see Table B.6.

The group now had a number of self-generated strategies to achieve the pervasive or systemic objectives from the lower half of the intent structure. Some members of the group expressed concern that strategies were all very well, but. . . would the group implement them, or would the whole exercise be forgotten in a few weeks time, leaving them as they were at present?

The group asked for a further exercise so that the strategies could be developed into a plan, which could be held by everyone and checked for progress at the partners' meetings that had

Table B.5 Developing strategies to overcome threats to achieving objectives.

Objective	Threat	Strategies
Quick access to general information	• Poor information sources • Poor retrieval systems • No dedicated resources • Plenty of data, less information	Beef up office management internal technical marketing
Understanding our market — parallel working and assessment of satisfaction	Engineers like solutions rather than finding and meeting needs	Positively research clients needs — respond! Develop 'needs' questionnaire — use with clients to establish their perception of practice Follow-up post-job to determine client satisfaction/shortfall

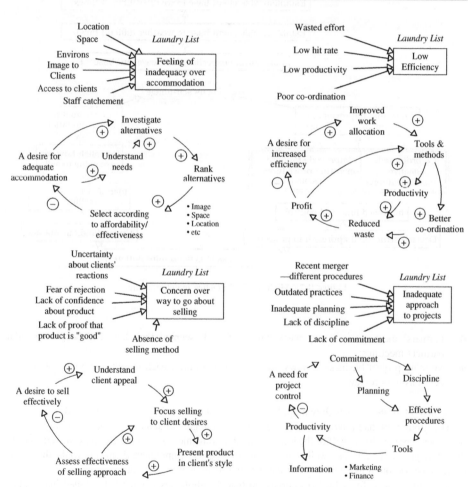

Figure B.8 Four CLMs developed by practice partners to identify strategies to achieve objectives.

Table B.6 Proposed practice strategies to be adopted to achieve objectives from intent structure.

Serial	Strategy
1	Beef up office management
2	Positive research of clients' needs
3	Seek client feedback
4	Improve work allocation
5	Improve office productivity
6	Focus selling on to client appeal
7	Assess selling effectiveness
8	Explore cost-effectiveness of new accommodation
9	Improve project procedures
10	Generate focused self-image
11	Improve financial control
12	Instill commitment and discipline
13	Introduce corporate planning
14	Introduce partners meetings
15	Introduce project reviews

been proposed. The consultant, who had thought his work done, fired up the ISM program and facilitated the development of a precedence network, which used as its trigger question: 'does strategy A precede strategy B, or is it the other way round, or must they coincide, or is there no relation?'

The resulting precedence network was drawn out into an outline strategic plan at Figure B.9. Several junior partners then developed the precedence network into a time-based strategic plan, which they drew up and printed off using their in-house facilities. They purchased some boards and

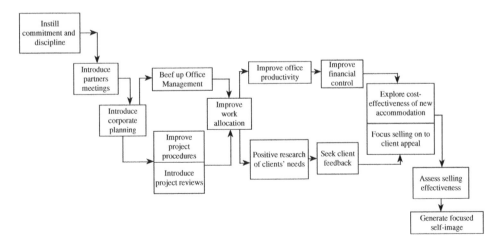

Figure B.9 Precedence network for practice strategies to achieve objectives. The network served as a basis from which to develop a time-based strategic plan/GANTT Chart to implement the various strategies.

pinned the process plan on the boards, adding a 'time now' marker, which was initially set at zero. The plans were then mounted on the walls in the senior partners offices, a further indication — if one were needed — that the junior partners expected the senior partner to 'drive' the plans forward.

Group discussion followed. The original intent structure (before adjustment) and the precedence network both showed that Image, which they felt ought to be a systemic objective, consistently emerged as an end product. Was it symptomatic of their heritage as engineers? Was it a problem? Should they recognize the networks as correct, indicating group 'emergent properties'? If the group was exhibiting 'undesirable' emergent properties, should they/could they do something about it?

(At this point, the consultant considered the 'auto-intervention' to have been effective, and took no further active part. The group was now actively involved in system thinking.)

Comparing the Hands-off RSM Investigation and the Hands-on Intervention

The RSM enquiry was conducted using only responses to questions and practice marketing literature provided, without visiting the organization. It gave remarkably good insight into structural and functional group deficiencies, as revealed later during the Intervention. However, it gave little insight into ways to improve performance that would be acceptable to the practice — particularly with respect to group motivation

The intervention was quite different; for a start, it operated with the hidden agenda of developing a group dynamic towards consensus. The process revealed structural and functional deficiencies that the RSM investigation had suspected, but added the dimensions of team building and gave the group a plan for improving their situation in which all had *participated* and to which all were *committed*. The intervention thus affected what RSM could only identify: group motivation.

- *RSM* helped to understand the problem in surprising depth, and particularly how problem elements interact;
- *RSM* provided a sound basis for modeling present situation and proposed changes;
- *Intervention*—as practiced in this instance — developed a potential solution to the problem, perhaps with less understanding of the internal structure, but with commitment to the plan through participation in its formulation;
- The two methods are complementary. The plan developed during the intervention could usefully be modeled using the Ideal World structure developed under RSM.

Note that at no time did the consultant inject his own knowledge or experience into the proceedings. He did, however, use insights gained from the RSM exercise to formulate appropriate questions to focus the group's attention on issues of relevance during the intervention. In this manner, he used only the knowledge and experience derived from the group to determine and introduce the solution to their own problems. His role as consultant, facilitator and moderator rolled into one, was to help them recognize and organize their own knowledge and experience, and particularly to overcome the interpersonal boundaries inside which each of them, as specialist consultants, operated.

He also helped them to develop a practice self-image, not as a collection of specialists, but as a whole, integrated system in which complementary variety within the parts was an essential ingredient. So, they moved intellectually from a viewpoint where their different specializations were a problem, to a viewpoint where their specializations fitted together to provide a complete set/system. From this new perspective, it became clear to them that they needed a few more specializations, rather than less, and they proceeded to develop expansion plans.

Summary

The case study concerned itself with a minor intervention in a consultancy practice. The members of the practice, the partners, were each expert in their own specialized fields, yet the enterprise as a whole was not performing well in business terms.

The consultant conducting the intervention prepared beforehand by asking for, and receiving, answers to a carefully selected question: each partner had to separately complete the question 'How can we. . . ?' The purpose of the question was to reveal, either each partner's aspirations, or each partner's concerns, or both. The use of the pronoun 'we' encouraged the partners to respond in the context of 'we, the whole practice,' as opposed to 'I, the individual.'

In preparation, the consultant applied the rigorous soft method, using the responses to the questions as 'symptoms' of the issue facing the practice. This informal, hands-off exercise provided a clear view of the problem themes facing the practice, together with an Ideal World model of how it could and should be working. Although based on a seeming paucity of corporate information, the output from applying the RSM was sufficient to provide the consultant with some confidence that he understood where the dysfunctions within the practice might lie.

The hands-on intervention was unusual, in that the consultant elected to show the partners how to conduct their own 'intervention,' such that they would need no further outside assistance. He introduced them to several methods at each stage of the proceedings, letting them choose which method to pursue. In the event, the group started with a choice of trigger questions, explored the chosen trigger using idea writing, moved on to nominal group technique and then to interpretive structural modeling, creating an intent structure for the practice. The group then proceeded to develop a strategic plan for implementing the intent structure, without any further assistance from the consultant. He had achieved his objective: they were thinking for themselves as a group.

Reviewing the whole exercise, is it not unreasonable to label it as 'systems engineering?' An optimum solution to the issue facing the practice had been conceived, designed and was in course of being implemented. True, there was little in the way of technology: there were no shiny new products or artifacts, but the structure of the practice/system as a whole had been revised, new functions and process improvements introduced, with synergy evident as the partners started to pull toward common goals. Essentially, a group of virtually separate parts had become an optimum, integrated system operating as a unified whole — achieving that is surely systems engineering.

8

SM2: Exploring the Solution Space

Nothing puzzles me more than time and space; and yet, nothing troubles me less, as I never think about them

Charles Lamb, 1775–1834

Introduction

At the end of Chapter 7, we saw that the application of methods such as Checkland's SSM and Hitchins' RSM could help to organize efforts to explore issues and problems. RSM, in particular, presented one or more so-called remedial systems solutions at the end of the process, echoing medical diagnosis of organic dysfunctions. In keeping with the medical analogy, it should be remembered that there may be no solution to some problems, and that it is entirely possible to misdiagnose, especially where there is insufficient hard data and information to go on. Like any other method, SSM and RSM obey the 'garbage in, garbage out' dictum — except, of course, that RSM will reject as inappropriate any remedial solution system concept that fails to neutralize all of the original problem symptoms. This 'decisive test' is designed to prevent either misdiagnoses, or diagnoses that address only part of the whole problem, from being further implemented.

Generating a remedial system solution concept may be necessary, but it is far from sufficient. If the potential remedy is to become a real-world solution, then the remedy will have to work within real-world constraints, which may be physical, temporal, cultural, financial, etc. Bringing a conceptual system solution together with a representation of the solution space in which the solution must be viable and operational is the first step towards molding the remedial system solution concept into a real solution, resolution or dissolution to or of the original problem.

Approach

The approach is outlined in Figure 8.1, a behavior diagram that follows on in sequence from Figure 7.5 on page 199; note the input, top left, of 'remedial solution system(s)' was also the

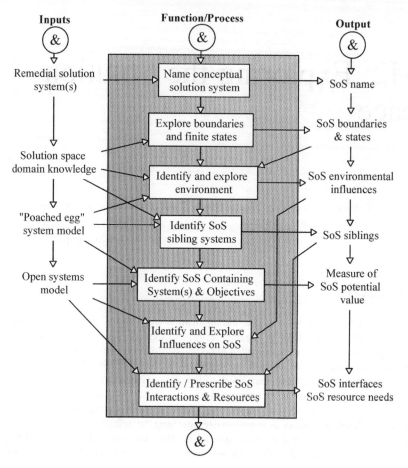

Figure 8.1 Behavior diagram — exploring solution space constraints. (SoS is 'solution system,' which is conceptual only at this early stage.)

output shown on the previous figure. The first step is to name the conceptual solution system: this is more important than it may appear; having the right name for a potential solution system gives it identity and substance in the minds of those looking for a solution. Conversely, giving it a poor, or inappropriate, name can damage the fledgling's prospects of being appreciated and realized.

Boundaries and finite states

The process of exploring the solution space is conducted methodically, first by looking for boundaries and any finite states that the conceptual solution system is likely to exist within, or may be required to observe: real-world systems are generally perceived as existing within clear boundaries, and they may also operate within a variety of finite states.

For instance, a nuclear power station will be well bounded, and may have strict limits place upon where it may be sited: it may also function in a number of different states, or operational modes, including an offline test mode, a standby mode, a backup mode, a work-up mode, a full power generation mode, a routine servicing mode, an emergency close-down mode, and possible several more. These boundaries and states are not inherent to nuclear power generation; rather, various administrative and regulatory features impose them in the solution space.

Similarly, a proposed new radar system may be required to have a number of different operating modes, including maintenance, standby, test, low transmission power (to minimize interference), high power (to overcome jamming interference), etc., all of which may be imposed by the solution space, and none of which is apparent from the conceptual solution of 'new radar system.'

Sometimes, of course, boundaries are simple, physical boundaries, as when three crewmen and their equipment have to fit within an oddly shaped capsule; when an avionics system has to fit within a limited aircraft hatch; or, a manufacturing process has to fit within a five cubic meter space.... Then again, a range of mountains, or a river, may bound a floral and faunal site. Motorways and similar highways may create unique ecosystems, in which rare species may thrive without interference from man, who rarely if ever visits the areas bounded by the dual highways. Or, a boundary could be associated with a culture, a moment of inertia, or whatever. The objective of the activity is to identify limits to the solution system that will be imposed by the solution space.

Environments, influences and interactions

Next in the sequence from Figure 8.1 comes 'identify and explore environment.' Simplistically, environment may be thought of as that which pervades the boundary in which the solution system will exist. The environment may affect the manner in which the various parts of the solution system interact to cooperate and coordinate their activities, and in so doing may affect synergy — in either a positive, or a negative sense. Of course, for large, diverse or complex solution systems, a particular environment in which they will eventually exist and operate need not be uniform, or even consistent. This may impose requirements on the solution system, i.e., that it be able to operate in a variety of local environments and remain viable (S-MESH applies: synergy, maintainability, evolution, survivability, homeostasis).

For instance, a manned trip to Mars will see the various space vehicles comprising the mission package traversing between the planets, passing through radiation belts, perhaps experiencing the outflow from solar flares, entering the atmosphere of Mars at high speed, and finally settling on the surface to experience high winds and dust storms. The astronauts will, hopefully, exist and operate within a relatively controlled environment inside their vehicles and inside their space suits. But, all environments must be accounted for, and all may impact various features of the conceptual, remedial solution system as it moves from pure concept towards practical realization.

Environments can also be cultural, and even legal, as for example where solution system will have to operate within, say, the legal framework of US, English, Swiss, or Shariah law. Were the conceptual remedial solution, for instance, a new penal institution, a new branch of an international bank, a division of an enterprise, or a factory to manufacture products under license, then the relevant national legal and cultural environment within the country may prove relevant, and may shape the development of the remedial solution system's future design.

Structure and dynamics

Next in sequence comes 'identify SoS sibling systems.' The eventual solution system will interact with other systems in its future environment(s). There will be interaction routes, media, etc., and there may be need for communications protocols, interfaces, etc. The sibling systems referred to are those other systems that will exist in the future environment such that they all, together, constitute a wider, or containing, whole. In this context, siblings may already exist, may be expected to exist, or may even be necessary to create, to realize a containing whole.

This activity is, essentially, recreating/identifying the so-called poached egg — see page 77. It is important for the eventual solution system to be set into its containing system and to interact with its siblings, else it will not be viable, will not operate correctly, and will not have the effect which its realization intended. Undertaking this activity will enable the designer of the solution system to envisage that system in operation, interacting and adapting, acquiring resources, exchanging material and information, disposing of waste, consuming and dissipating energy, etc., etc.

Identifying the containing system, of which the remedial solution system will become part, is important for another reason. The containing system will have objectives, and the value of the remedial solution system will be measured by the degree to which it contributes to those objectives, in concert with its siblings. This degree is likely to be maximized by optimizing the developing remedial solution system's design. (Hitchins, 1992)

There is yet a third reason for this activity: it provides the foundation for the eventual simulation of the solution system in its environment, adapting to those other systems with which it will interact; it is then, an essential aspect of the systems approach, without which the performance and effectiveness of the eventual solution system could not be sensibly determined.

Resource needs

An important output from these activities is an assessment of the resources that the remedial solution system will draw upon in the solution space. This becomes apparent when considering how the remedial solution could be viable in its future environment, and is most easily envisaged when considering a brand new, or 'green-field' system.

For instance, suppose the conceptual remedial solution system were a new sensor array to be located on the far side of the Moon, away from man-made radio interference. Once constructed and set to work, such a sensor array would need to remain viable, perhaps over extended periods. Invoking the S-MESH acronym again reminds us that there are many factors to consider, each with resource implications.

■ Synergy between the array elements and the sensors will have been built into the design and construction, but must be maintained in a hostile environment of wide and sudden temperature shifts, cosmic ray impacts, etc.

■ Maintenance will require the provisioning of spares, and perhaps automatic repairs, or self-healing — which raises the issue of maintaining the self-healing systems. . . .

■ In the light of operational experience, the array and sensors may prove to be capable of improvement or expansion, so that modifications and adaptations can be foreseen.

■ Then again, the whole array must be able to survive damage from its harsh environment, perhaps through damage tolerance, or perhaps through some self-defense mechanism, which may itself be subject to maintenance.

Evidently, without knowing much about the remedial solution at this early stage, prior to design, it is still possible, and important, to consider what influences will affect the future system, and what resources will be needed both to operate the future system, and to maintain its ongoing viability.

Perhaps the most significant of the components within the S-MESH acronym — homeostasis — is worthy of special attention. Evidently, a solution system will remain viable only so long as it can maintain homeostasis; which can imply many different factors all maintaining balance at the same time. Consider, for instance, a remedial solution system as a new manufacturing plant. Once set up, it will be vital to maintain not only staff numbers, but also the necessary skills for staff; that, in turn, may necessitate either continual recruitment of people with the requisite skills, or a training system, or both. Homeostasis is required also in financial dealings for the new plant. It requires payment for its output such that it can require, and pay for, materials and parts from suppliers that can be manufactured and assembled to form new output for sale. Revenue from sales must also pay for employees, recruiting, redundancies, training etc., and also for maintenance of continual upgrade of plant and machinery: there is a need for homeostasis in manufacturing capability, too, which implies that there should exist machine vendors which can support continual upgrade.

Homeostasis is important for technological systems, too. As a simple example, consider the case where the remedial solution system concept involves a more powerful computer within the confined space of some vehicle equipment bay or hold. Homeostasis might be concerned, *inter alia*, with maintaining the temperature within the bay, such that the environment is kept within sensible limits for all the equipments in the bay. If the new computer is going to dissipate more heat than its predecessor, then there may be a need to improve the heat extraction facilities, which may in turn affect the thermal signature of the whole vehicle: in the wrong circumstances, this might prove problematic. Such considerations could materially affect the remedial solution system concept, to the point of rendering it unacceptable.

Pursuing the future need for viability in general, and homeostasis in particular, identifies not only future resource needs, *per se*, but also future 'complementary systems:' these are systems, other than the conceptual remedial solution system, which will either already exist, but may need modification, or that do not yet exist, but will be required to maintain solution system operations and viability.

The need to maintain internal balance, or homeostasis, pervades most aspects of any system. In an enterprise-as-a-system, homeostasis affects materiel, personnel, security, finance, training, production, administration, buildings maintenance, facilities maintenance, communications, information services, business intelligence, etc., etc. External resources may be subject to competition. Maintaining the steady state in the system of interest may require the cooperation of other systems, complementary systems for repairing shortage or for shedding excess. In some instances, complementary systems may be new; on the other hand, it may be that the appropriate systems exist already, but may need to change to accommodate a modified remedial solution system.

Summary

The solution space provides opportunities and imposes potential constraints on the remedial solution system, even in its conceptual state. It is important to explore the solution space early on in the systems methodology, so that the credibility of competing remedial solution systems can be assessed. It may be the case that some solution concepts would be seen as impractical, while others would invoke the need for new, or adapted complementary systems, significantly expanding the sphere of influence, complexity and potential cost of some conceptual solutions. On the other hand, it might also be the case that the remedial solution may be able to draw upon, and make good use of, resources that would be available in the solution space, so simplifying the solution system design and creation task.

In any event, the solution system will have to be 'tailored' into its future environment and will interact with other systems in that environment. Exploring the solution space enables the identification of the solution system's siblings, and their mutual containing system. This will provide the basis for assessing the absolute and relative values of competing remedial solution system concepts.

In the process, both a static ('poached egg') model, and a dynamic functional model, of the solution system in context are necessary to understand the constraints (interactions, interfaces, structures, boundaries, environments, etc.) that the eventual solution system will operate and be viable within. Understanding these at the conceptual stage allows successive stages in the systems methodology to mold the developing concept and design so that the eventual solution will operate in harmony with other within its environment.

Assignment

In addressing a complex issue, a conceptual remedial solution system has been identified as including a long-term, manned installation on the far side of the Moon. The role and purpose of this facility will be twofold: a) to act as an outbound staging post for manned missions to Mars; b) to act as an inbound staging post from deep space, zero gravity missions, where astronauts may acclimatize themselves over a period of time before returning to the full gravitational impact of Earth.

Using Figure 8.1 as a guide, explore the solution space within which this facility will exist and operate, remembering that, although it will be on the far side of the Moon for technical reasons, full communication and coordination will still be necessary, both with mission crews and with Earth Mission Control. You should include in your response, those items and influences that you consider will impose constraints or, afford opportunities, or both, for the conceptual remedial solution system.

9

SM3 and 4: Focusing Solution System Purpose

The Englishman never enjoys himself except for a noble purpose

A.P. Herbert 1890–1971

SM3: Solution System Purpose

The previous chapter explored the solution space, seeking constraints and opportunities that might influence the concept, design and, indeed, the validity of a conceptual remedial solution system. Using that knowledge, it is feasible to explore the nature of the conceptual solution system. Since, in general, we will be addressing man-made systems (in the broadest sense), then it is appropriate to focus attention on the purpose of the conceptual solution system: exploring and expanding on a system's purpose will enable that system's effect and effectiveness to be envisaged, and will also concentrate design effort towards a specific end.

A useful approach to the consideration of a system's purpose is to appreciate its prime directive — a term incorporated from the life sciences. The prime directive of a system is a statement of its ultimate purpose. Such a statement gives no indication of how that ultimate purpose is to be achieved. For Man, it may be described as propagation of the species, where propagation requires not only procreation, but also nurturing and protection of offspring until they, too, are of sufficient maturity to sustain the propagation process. That example shows the value of the prime directive: it rises above the details and gets to the nub of purpose.

However, that example is less than ideal in another respect: all animals on the planet have the same prime directive; they all seek to propagate their species, making that particular prime directive rather poor as a differentiator. We must be more specific to identify the purpose of a particular system solution.

For instance, the prime directive (PD) of a government might be: 'to provide a secure, prosperous environment within which the population may develop and flourish.' The PD is deliberately bare, and contains only one verb, to ensure only a singular purpose. Similarly, the PD of a marketing department might be: 'to position the company so as to flourish in a competitive market;' indicating that the role of the marketing department is one of positioning, not one of selling.

Systems Engineering: A 21st Century Systems Methodology   Derek K. Hitchins
© 2007 John Wiley & Sons, Ltd

Developing a PD for a new system is a significant aid to objectivity, but by itself, it is not enough. To be useful, it has to be associated with a strategy, identifying how the ultimate purpose expressed in the PD is to be achieved. There may be alternative strategies, or ways to achieve the prime directive. Moreover, each of the strategies might incur different risks, might experience threats. So, there are three parts to a 'useful' PD: the statement of ultimate purpose; the statement of strategy for achieving that purpose; and, the threats envisaged that prejudice the strategy.

Looking back at the PD for Man, we can see that the strategy adopted by early man was to develop family units, and for groups of families to form communities for their mutual protection and for cooperative hunting, gathering, etc. Threats included natural disasters, predation by wild animals, shortage of food, and possible raids by other social groups, so that early man chose to operate by day, and shelter by night. Evidently, the strategy generally overcame the threats, so that the PD was observed — else we would not be here.

Threats and strategies

There may be one PD, but many threats and many strategies. It can be seen that strategies come in two 'types:' those to achieve the PD 'unopposed;' and those to overcome the threats to achieving the PD.

It is often useful in practice to analyze the PD, using a technique called semantic analysis: this enables the PD to be envisaged as a sequence of objectives. Take, for instance, the PD for a national air defense system, which might be: 'to neutralize enemy air incursions into national air space.' This might be semantically analyzed as follows:

To neutralize. . .	To eliminate the threat from. . .
. . . enemy. . .	. . . those declared by government to be hostile who. . .
. . . air incursions. . .	. . . enter by air without permission. . .
. . . into national airspace	. . . into the airspace legally defined and internationally promulgated as sovereign, national airspace

Implied means: national air defense assets — surface-to-air missiles, airborne interceptor aircraft, beamed energy weapons. . .

Semantic analysis expands the PD, term by term, into an almost legalistic expression. Using the semantic analysis allows a sequence of objectives to be identified: delineate airspace boundaries; detect air incursion across boundaries; identify intruder; classify intruder as friendly, neutral or hostile; neutralize hostile intruder. Neutralize is a military euphemism, which could mean any of an escalating range of actions, according to the 'temperature' of the situation and the nervousness of the defenders: warn off, escort out of airspace, force to land; render intruder's weapons ineffective, shoot down. . . .

Figure 9.1 illustrates the developing process in the form of a behavior diagram. The diagram center panel starts with developing a prime directive for the conceptual solution system. This PD is then semantically analyzed and a sequence of objectives is derived, the final one of the sequence being the 'mission' objective. Threats to achieving each objective are considered.

Strategies are then conceived for achieving each objective. (The conception of effective strategies requires an understanding of the solution space/the operational environment in which the solution system will seek to achieve its purpose.) There are two kinds of strategy: those for achieving the

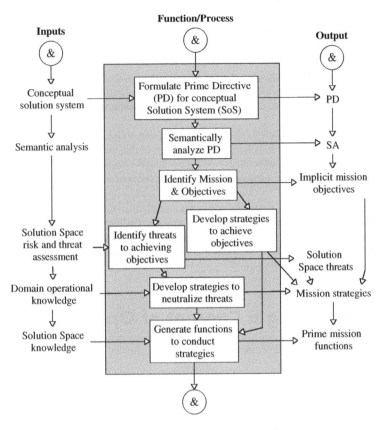

Figure 9.1 The TRIAD building system: purpose and threat management.

objective unopposed; and, those for overcoming threats to achieving the objective. All of these strategies become the mission strategies because they all have to be effective in order for the mission to be accomplished.

Finally, the various strategies are addressed as prime mission functions that the solution system has to be able to perform to achieve its mission in the threat operational environment. The overall process is referred to as the TRIAD building system (Hitchins 1992, 2003) because of the repeated occurrence of the three elements: objective; strategy to achieve objective; and, strategy to overcome threat to achieving objective. The technique is powerful, creative and insightful; moreover, it generates a considerable amount of information pertinent to the design of the overall solution system.

(The TRIAD Building System can be used as a method or technique for addressing, not only the developing system solution, but also for addressing the systems methodology itself. If the application of the systems methodology is considered as a system project, then that project may be envisaged as existing/surviving in an environment — often that of the organization or company that is 'hosting' the project. It is feasible, then, to identify threats to the continuation of the project, strategies to achieve the projects objectives, and strategies to neutralize the threats at the same time.)

The range of threats that is identified and neutralized will vary for any given solution system concept according to the capability, experience and insight of the team members concerned in applying the systems methodology. In practice, there may be a delicate balancing act to perform. On the one hand, the experienced team may perceive many more threats and generate prime mission functions to address them, resulting in a more complex and potentially more costly solution. On the other hand, the inexperienced team may perceive fewer threats, may generate less credible strategies to address those threats, and may consequently generate a simpler, and potentially cheaper solution.

Which is better? All things being equal, the more capable solution might be preferred, but may be less affordable. However, it is not necessary to judge at this early stage. Later in the systems methodology opportunities will arise to 'try out' the solution system design, operating in a typical environment containing a range of threats, and it will then be possible to assess the value of features of the design that have been incorporated specifically to address particular threats. Moreover, the cost of the end solution system is less related to the complexity of the design, more related to the manner of its creation. For technological elements, in particular, the same prime mission functions may be performed by different technologies of widely different costs: it is too soon in the systems methodology process to be unduly concerned with cost, and besides, integrity requires that, initially at least, we conceive the best solution.

SM4: Developing a Concept of Operations (CONOPS)

Focusing on Purpose would be incomplete without identifying how Purpose was to be achieved: the way a solution system is expected to work, operate, create the desired effect, etc., may be referred to as its concept of operations, or CONOPS. It is quite possible, even usual, for a remedial solution system to be associated with more than one CONOPS. Competing CONOPS then vie to be the best, or preferred CONOPS, where best is judged according to derived measure of effectiveness — value judgments.

As a concrete example of CONOPS, see NASA's Apollo on page 84: there were many competing CONOPS, each being tested step-by-step throughout a simulated or imagined mission, until the best, most likely to succeed, emerged as the preferred CONOPS.

At a more prosaic level, consider a remedial solution to a particular national energy problem, where the solution system has been conceived as an extensive wind turbine system. Simply stating 'wind turbine system' is clearly insufficient. How it is it expected to work, as a generator of electrical power in the context of national power generation, with other sources of power also available? Where is it to be sited? ? Wind power, of course, works best when there is a wind, and wind, as we are all aware, is a highly variable commodity. What happens when there is no wind: is the locale deprived of power? What happens when there are gales and hurricanes? Must the wind vanes be feathered, to avoid damage from over-rotation in strong winds? Would feathering result in no power generation, normal power levels, or what? What if peaks in power generation coincide with troughs in demand? Can excess generated power be stored? How?

By stepping through processes of generating wind power under a range of environmental conditions, possibly storing excess power, coupling to the national grid, and drawing upon previously stored energy reserves, it becomes clear that there are many options along the pathway, and that some are preferable to others in terms of risk, response times, expense, performance, effectiveness, efficiency, etc. For example, excess energy might be temporarily stored in batteries, by raising water into a reservoir, in large flywheels, etc., etc. It also becomes clear that, in many cases, generating wind power may turn out to be much more about regulation, control, switching and storage, than about windmills — they are the relatively simple, if highly visible, part.

So, the development of a CONOPS may be seen as creative, with competing CONOPS being compared using value judgments to select that which is best of those on offer. Both aspects, the selection of CONOPS and the process for judging between competing options, require that the analysts/architects have an intimate knowledge and understanding of the solution space, the operational environment and the constraints, opportunities and likely threats within it.

Figure 9.2 shows the process for developing a high-level CONOPS, in the form of a behavior diagram; this diagram follows from that of Figure 9.1, above: note, in particular, that the input, top left to the first activity, top center, includes solution domain knowledge, mission objectives and prime mission functions. With these inputs, indicating what the purposeful solution system is expected to achieve, it is possible to generate one or more concepts of operations, initially perhaps in the form of a storyboard: sketches, rich pictures, sequence diagrams, even cartoon-like strips, etc., as needed to illustrate 'how things are meant/expected to work.'

With any complex solution system, containing system and environment, it is likely that there will be several CONOPS, competing for the top spot. Choosing between them requires, next, the development of a robust set of so-called measures of effectiveness (MOEs). These are values for the solution system, and particularly for its performance, that will be prized by the owner/user of

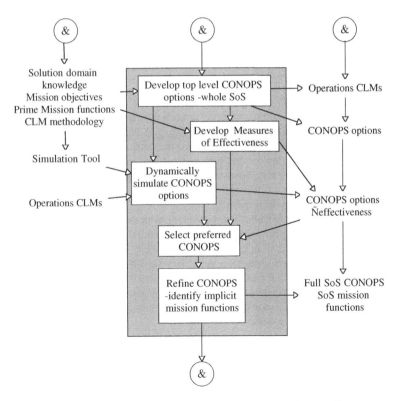

Figure 9.2 Developing a sound concept of operations (CONOPS).

the solution system. According to the mission, they may include probability of mission success, timeliness, viability, degree of effect, reliability, survivability in the face of threat, affordability, integrity, security, resources and logistics, etc.

The high-level CONOPS options may next be simulated dynamically, to assess the impact of interactions within and between systems, timeliness, resource implications, etc: this is an important aspect, since it allows the various system parts to interact and to adapt, to draw upon resources, to demonstrate potential performance and effectiveness. Simulations of competing CONOPS provide a cogent basis for choosing between them, setting simulated performance against the MOEs as basis for rational choice. Not only does the simulation process provide the basis for comparative judgment, it should also provide an absolute measure, since some (or all) options may be seen not to achieve the mission satisfactorily. If none is satisfactory, then either a refinement process is needed to improve the preferred option(s), or new and different CONOPS may be generated and simulated.

It is by no means certain that an acceptable CONOPS can be found. Where that is the case, then it may prove necessary to revisit the process of identifying the solution system: in such cases, it is possible that the answer is neither so solve the problem, nor to resolve it, but rather to dissolve it, i.e., to move the goal posts.

Behavior diagrams, such as Figures 9.1 and 9.2, may give the impression that the systems methodology is linear, but the reality is that the processes tend to be repeated as their outputs are seen to be incomplete, inadequate or even incorrect. Happily, the facility for seeing that the outputs are incorrect is built in to the processes, in this case into the dynamic simulation of the competing CONOPS, since the acid test for the simulation must always be whether it demonstrates that the original problem symptoms will be neutralized, or not.

Assignments

1. Formulate a prime directive for a university, semantically analyze the PD, identify a sequence of objectives and a mission for the university, suggest threats to the achievement of those objectives, and strategies that will both overcome the threats and achieve the objectives. How might you establish credibility for the strategies that you have proposed?
2. Formulate a prime directive for a horse-race gambling system, and repeat the activities above, in 1.
3. As part of the solution to an impending national energy shortage, a robust and extensive advertising and publicity campaign is proposed by government to persuade the people and the anti-nuclear lobby that nuclear energy will be an essential part of the future national energy strategy. Establish a concept of operations for the publicity campaign-as-a-system, presuming the campaign to extend over at least three years. Develop MOEs by which to judge the efficacy of the campaign. Consider how the campaign might sensibly be simulated dynamically, and discuss the value of such a simulation were it to be undertaken. Consider, also, whether or not your personal attitudes to nuclear power impose themselves during this exercise, and how you deal with any such imposition.

Case C: The Total Weapon System Concept

C(1): The Battle of Britain Command and Control System

Introduction

At the outset of the Battle of Britain in 1940, the German plan was to employ their superior air power in what has since become classic fashion: they would obtain air superiority over southeast England, as a prelude to land invasion — their Operation Sea Lion that, in the event, never took place. Gaining air superiority meant eradicating the RAF's fighter aircraft; this was to be achieved by a campaign of bombing airfields and factories, to destroy fighter aircraft on the ground, and to halt production. So, southeast England faced a bombing campaign aimed, not at centers of population, but principally at fighter bases and factories.

The UK was ill prepared and lacked experience with which to take on the might of the German war machine. Germany was particularly proud of its air force, the Luftwaffe, which had gained invaluable experience and become battle hardened during the Spanish Civil War. Germany had built up its airforce strength, so that at the start of the Battle of Britain the Luftwaffe could muster over 1200 aircraft on the French side of the English Channel, compared with only some 300 fighters, mostly Hurricanes with a sprinkling of Spitfires, facing them in southeast England.

The Royal Air Force (RAF) did have one ace up its sleeve: Chain Home radar. The UK had not invented radar — that seems to have been a German invention, but they used it at sea, where it seemed to work best. The German command apparently had no idea that the UK had a number of Chain Home radar stations dotted along the southeast coast, facing out to sea. The UK also found that their radar worked best over the sea, with land creating so many echoes that it was difficult to pick out and track an aircraft flying over land: however, detection and ranging over water were possible. . . .

Air Marshall Hugh ('Stuffy') Dowding had introduced the RAF's radar system in 1936 as part of his efforts to improve the UK's air defense system. Dowding also encouraged the formation of an extensive network of visual sighting posts of the Royal Observer Corps (ROC). Civilian

Systems Engineering: A 21st Century Systems Methodology   Derek K. Hitchins
© 2007 John Wiley & Sons, Ltd

volunteers manned the posts, which are still to be seen at high points all around southern England to this day. Each post was equipped with an optical sighting device mounted on a plinth, using which an operator could estimate the direction (track), range, height and speed of an aircraft, and could also identify aircraft type, how many aircraft in a raid, etc.

The manned ROC posts connected by a network of telephone lines (an intranet?) to a filter station, which received air intruder reports, often from several ROC sites at the same time or in quick succession. From these reports, filter stations were able to determine raid details, if there was more than one raid, and — if there were — where the raids were headed. The filter station then passed this raid information on to sector operation centers, which had large plotting maps of southeast England, on which operators placed and moved markers to indicate the progress of raids.

Air defense fighters were located at a number of airfields: aircraft could be 'scrambled' in response to an incoming raid, and given a vector to fly once airborne, so that they would intercept the raid. There were some 20 squadrons all told, and these could be rotated so that some squadrons were on duty, while others took time off for both pilots and ground crew to recover and repair. So, another factor in the equation describing the hoped-for balance between the opposing forces, was the tactics to be employed by overall commander Dowding, and by Air Vice Marshall Keith Parks, Air Officer Commanding 11 Group, covering southeast England — a New Zealander by birth, and, as it turned out, key to victory in the UK.

Figure C.1 shows the set up as it was in the summer of 1940, omitting the numerous ROC posts. The area was divided into sectors, each with its sector operation centers located at an RAF

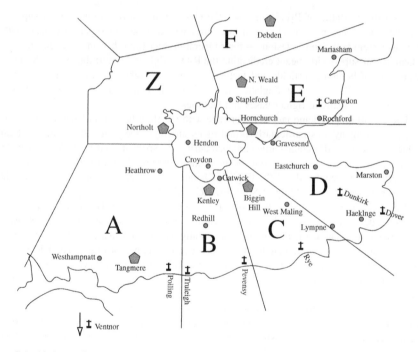

Figure C.1 Notional plotting map of southeast England, No 11 Group, RAF, 1940. Sectors are labeled A to Z. Pentagons are sector operation centers, one per sector. Other airfields are shown as small circles. Chain Home radars are shown as towers distributed along the coastline, looking over the sea towards France (bottom right, not shown).

station. Sectors also had airfields with fighters based at them, although the same fighters could be dispersed, usually forward towards the coast; this had the dual effect of making them harder to locate, so reducing the risk of damage due to enemy strafing while on the ground, and also making them closer to the coast; so, quicker to intercept incoming raids.

Interacting Systems

There were many interacting systems on both sides during the conflict. Figure C.2 shows a simplified view, with RAF No. 11 Group Fighters centre left, and the Luftwaffe with its fighters and bombers to the right. Clearly, they were the protagonists, but there was more to the system than that, else the Luftwaffe, with its 4:1 aircraft supremacy would have simply overrun the RAF.

Both sides expected to lose aircraft and crews during the campaign. For Germany, the way to replace losses was either to get new aircraft and crews direct from Germany, which was relatively slow, or to redeploy aircraft based in Norway, Denmark and the Low Countries, where their role was largely to harry allied shipping.

For the RAF, there were three possible resupply routes: direct from factories, from stocks of fighter aircraft that had been constructed and were held in reserve, and from other groups which were not so hard pressed by Luftwaffe attacks at this time. Obtaining replacement crews could be more problematic, however — replacement fighters from reserves or straight from factories did not come with pilots — these had to be recruited and trained, both of which were time consuming.

One important factor came to the aid of the RAF. The Luftwaffe adopted a tactic of assembling their bombers and their fighter escorts into airborne formations over France before crossing the Channel. While Chain Home radar was primitive by today's standards, it could pick out the large echoes created by these formations, which gave No. 11 Group early warning of an impending raid — also shown in Figure C.2.

The possession of radar turned out to be a 'force multiplier:' without the radar, and therefore without early warning, No. 11 Group would have been obliged to mount continuous daylight air patrols along the southern coast, ready to engage incoming targets; this would have reduced significantly the operational availability of RAF aircraft to meet unexpected Luftwaffe raids. . . .

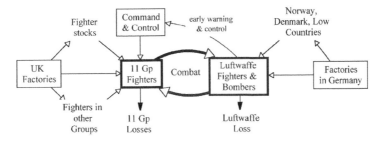

Figure C.2 Battle of Britain, 1940 — the systems interaction diagram (SID). Not shown, but a significant factor, was the weather — a highly variable, typically English, summer.

Working up the System — Operational Systems Engineering

The elements of the UK air defense system were in place before war broke out; the system worked, but it did not work well. In particular, the system took too long to respond to detected raids, so that enemy aircraft would have been able to deliver their bombs on airfield targets with relative impunity. In current parlance, the 'time around the loop' was too long, where the loop was the time to detect, report, decide on response, scramble defensive fighters, vector them on to target, engage and destroy/deter the enemy before bomb-drop. So, while the technological elements of the system were in place and performing, the overall system was nonetheless inadequate.

The Chain Home radar could detect Luftwaffe formations assembling over France some 20 minutes before the consequent raid arrived over the southern coast of England. The requirement for No. 11 Group was to meet those incoming German bombers, with their fighter escorts, as they crossed the English coastline. The key time-response factor, then, was 20 minutes.

Dowding set about improving air defense; his objective was to reduce the time taken by the people in the loop to perform their various tasks. ROC visual sightings, essential for tracking raids over land, were correlated in the filter station: one enemy raid might be reported by a number of ROC sighting posts, with reports being unsynchronized and offering estimates only of numbers, types, bearing, altitude, etc. Consequently, the filter station could be faced with an overload of information, from which to sort out the most likely situation. And suppose there were two raids, or one raid splitting into two sections. . . .

Dowding introduced reporting procedures, to reduce the time taken to describe a raid or enemy position. It is from this period that expressions such as 'Angels 15' come, meaning 'at an altitude of 15 000 feet. By introducing simple verbal codes for reporting various situations, Dowding not only accelerated procedures, but also reduced the propensity for misunderstanding. With practice, the mean time taken to filter and report a raid was reduced to 4 minutes — still a relatively long time, but workable.

The sector operations centers were similarly scrutinized and procedures improved. One of the problems with the map, on which counters were positioned to represent incoming raids and outgoing fighters, was that raid reports might be few and far between, so that the position of a counter may not be updated for several minutes. This could lead to misunderstandings about the 'state of play.'

Dowding and his team improved the Operations Room Plotting Clock, Figure C.3, adding different colors, red, blue and yellow, around the rim of the face, at five-minute intervals. So, the first five minutes after the hour were in red sector, the next five minutes were in yellow sector, and the third five minutes were in blue sector. The colors then repeated in the same order around the clock. When a new plot was received in the sector operations center, its marker was put on to the plotting table (the map) with a color label on it corresponding to the color that the clock minute hand was pointing to when the plot was received. So, a plot received at 3 minutes past the hour would be marked red on the table, another received at 8 minutes past the hour would be yellow, and so on. In this way, operations staff could tell at a glance whether a plot was current or stale. (This was an early example of the management of latency.)

At airfields, aircraft were positioned on dispersals, with their crews sitting beside them, listening to landline messages broadcast over loudspeakers direct from the corresponding sector operations center. Aircraft could be plugged into ground power trolleys, which were anchored so that aircraft could start up and taxi, automatically disconnecting from the trolley cable as the aircraft moved forward.

Figure C.3 Operations Room Plotting Clock Face, showing the minute hand in a 'yellow' sector

Once airborne, sector operations staff could vector fighters towards an incoming raid using the simple radios fitted into RAF fighter aircraft: (German attacking aircraft had no such communications with their bases). RAF fighters had to climb to altitude as quickly as possible, so that they could engage enemy aircraft with at least height parity, if not height advantage. On average, climbing to operational height took some 13 minutes. Together with the 4 minutes for target filtering time, this left some 3 minutes out of the 20-minute early warning period to transit to the coast, see, and engage the enemy. It was still tight for time. . . .

So, Dowding and his team gradually worked up the overall command and control system, with its sensors, processors and effectors (ROC, Chain Home radar, filter stations, sector operations centers and fighters) by taking advantage of the adaptability of the human elements within the system. Over a period of months, they gradually 'raised their game' until the air defense system as a whole could match the Luftwaffe raiders — just. Although the term was not in use at the time, they had developed a 'total weapon system concept,' in which all the parts of the of the air defense system operated organismically, as a unified whole, to achieve a singular, specific purpose — the neutralization of Luftwaffe bomber aircraft attempting to bomb fighter airfields and factories.

Let Battle Commence

The Battle of Britain was brief, but bloody. Graph C.1 shows the relative aircraft losses for Luftwaffe and the RAF on a day-to-day basis.

While it can be seen that the daily pattern of losses is not dissimilar, at least in terms of peaks and troughs, it is also evident that the Luftwaffe were losing many more aircraft than the RAF. And, since many of the Luftwaffe aircraft were bombers with several crewmembers, the loss of

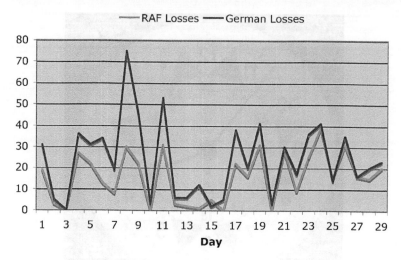

Graph C.1   Battle of Britain: relative aircraft losses, 8 August – 5 September 1940. Losses on each side were broadly in proportion over the 29-day period, with German losses consistently higher than RAF losses. . . .

men was even greater on the German side; this was a cause for great concern to the Luftwaffe, particularly since their aircrews were drawn from the cream of German society.

Moreover, when a Luftwaffe aircraft came down over English soil, any crew who survived would be interned; if the aircraft came down over the channel, the chances were that the aircrew would not survive at all. For the RAF, the situation was different: if a fighter went down, the pilot might bale out, and could be back at his base, ready for operations, in a matter of hours. Not everyone managed to bale out of course. . . , but the advantage, at least in numerical terms was to the RAF. And RAF fighters were not permitted to chase fleeing German aircraft over the Channel, to minimize the risk of losing both aircraft and pilot over the sea.

Graph C.1 is also remarkable for the degree of variability in day-to-day losses. While some of this might have been the result of weather, it is unlikely to have been sufficient to cause such variability. The English summer weather was, and is, notorious in its unpredictability, and the summer of 1940 lived up to its reputation. Moreover, German bombers operated effectively only in clear weather, and could be prevented from operating by mist, low cloud, etc. Poor weather could be a two-edged sword, however, not only preventing raids, but also allowing both sides to repair damage, train up replacement crews, and generally prepare to operate in larger numbers when the weather cleared.

The curious degree of variability in day-to-day downed aircraft statistics is highlighted in Graph C.2. The phase-plane chart is produced from the same figures that gave Chart C.1, but in this case plots the changes in combined aircraft losses from day to day, so that the line on the chart follows that change from day 1 to day 31, in order. Had the statistics varied, say, in a simple wave fashion, then the phase-plane chart would have looked like a circle, or oval. Had the statistics varied chaotically, then the phase plane chart might have looked like the butterfly — see Lepidoptera Lorenzii? on page 35. As it is, Chart C.2 looks nothing like an oval, and can be regarded as vaguely butterfly-shaped only by a major stretch of the imagination.

Did the statistics indicate some fractal characteristic, perhaps? (See Fractals on page 39.) Graph C.3 was formed by sampling the combined loss statistics at different intervals: every day,

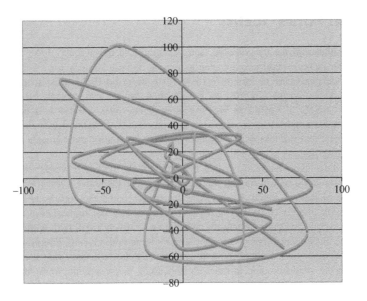

Graph C.2 Luftwaffe vs RAF losses — phase-plane chart, taken from the same historical statistics that generated Graph C.1.

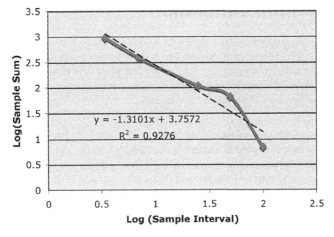

Graph C.3 Analysis of the combined RAF and Luftwaffe loss statistics of Graph C.1, showing linear trend line. See text.

every two days, every four days, and so on, and summing the each set of samples to produce the log–log graph. (This approach highlights the 'bumpiness' of a graph.) Although not conclusive, the results suggest that the pattern of losses may, indeed have been 'weakly chaotic,' or fractal, conforming reasonably well ($R^2 = 0.93$, where 1.0 would be precise conformance) to a power law distribution with a 'fractal' index of 1.3.

AVM Keith Parks' tactics

One possible explanation for the variable statistics might be the actions of AVM Keith Parks. He adopted what, in retrospect, may be seen as classic defense strategy for a besieged force. In particular, he took great care to minimize losses. Not only did Parks insist that No. 11 Group fighters did not pursue enemy aircraft over the Channel, but he also prevented them from rising to the bait when flights of Luftwaffe Me109 fighters came over without their bombers in tow, evidently spoiling for a fight. Parks saw his job as protecting RAF fighter bases and factories from Luftwaffe bombers, and not taking on marauders who, by virtue of their greater resources could better afford to lose fighters in air-to-air combat.

Parks also rotated the squadrons so that the brunt of the defensive work was shared out; a particular squadron might be working all out for several days, then have a rest for a week or more, while crews and planes recovered. In this way, he always had fresh squadrons to hand, should an unexpected peak in activity arise. The number of squadrons on 'immediate operations' might vary, too, with several squadrons sometimes taking on a relatively small raid, while at other times, perhaps only one or two squadrons would face up to a large raid. Overall, Parks' tactics, although questioned at the time, turned out to be brilliant, making the best economic use of his meager resources so that he would be able to meet an impending major assault, while at the same time presenting an entirely unpredictable face to the enemy, who, as it later turned out, were confused about how large, or small, No. 11 Group's resources were.

Battle of Britain Simulation

The simulation described below was developed for a Granada TV program for the History Channel as part of their 'Battlefield Detectives' Series. The Battle of Britain was particularly brief because the Germans, having failed to suppress the RAF in the few short days they had expected, switched instead to a quite different campaign of bombing centers of population such as London.

The TV program set out to investigate three issues:

1. How long could Fighter Command have lasted if the Luftwaffe had continued to attack the RAF bases and radars, instead of switching to bombing London and other cities?
2. Can the Battle of Britain be reasonably classified as a 'win' for the UK, or not?
3. Was the strategy employed by Dowding and Parks, of conserving RAF fighter aircraft resources in anticipation of prolonged activities better, or worse, than the Big Wing concept — which would have contributed most in the long run. . . ?

This last question was occasioned by critics of Dowding and Parks, notably AVM Trafford Leigh-Mallory of No. 12 Group, to the north of London, supported by Douglas Bader, the famous fighter ace who flew throughout the war with 'tin legs,' from losing both legs in a prewar flying accident. Both critics would have preferred to amass their fighter resources into a so-called Big Wing, and to take on the enemy in a decisive major air battle.

The phase-plane chart of performance, Graph C.2 is, in effect, a behavioral signature of a system: a characteristic indicator of emergent behavior. So, what was going on during the intense Battle of Britain to cause this unusual signature? One way to explore the problem space is to simulate the Battle of Britain, and to vary parameters within the simulation until its signature matches that of the real world. Observing the parameter changes that led to this match may shed light on the then situation. . . .

The simulation is presented as a so-called learning laboratory: the user can experiment with many different facets of the Battle to see what the likely outcome might have been. . . . The simulation has three principal parts: representations of:

- the Luftwaffe, a mixed force of fighters and bombers, based in Northen France;
- Dowding's Command and Control System, including Chain Home radars for early warning, ROC filter stations, Sector Operations Centers, and the communications infrastructure that coupled them into a single system
- Number 11 Group, under AVM Parks, with its sectors in southeast England, and its sector operations stations at Tangmere, Kenley, Biggin Hill, Hornchurch, North Weald, Debden and Northolt.

The various parts (modules) of the simulation were built and tested separately, then brought together after the fashion of Figure C.2, and the whole was tested before being used as a learning laboratory.

No.11 Group had some 20 fighter squadrons, mostly Hurricanes with some Spitfires, divided between the sectors and under the control of the sector stations. The 'engine room' of the simulation represents the squadrons of No. 11 Group, accounting for the numbers of aircraft and pilots, the casualty rates, hospitalization and recovery rates, aircraft damage and repair times, aircraft damaged on the ground by Luftwaffe action, etc. Luftwaffe operations are represented in similar fashion, but without any attacks on Luftwaffe bases.

The top level of the multi-layer simulation presents an Operations Map Room in the form of a Group Operations Plotting Table — a map of No. 11 Group. and southeast England. Surrounding the table are controls to alert the fighter squadrons, plus estimates of raid size — From this position, users can play 'Park's Game,' choosing — as AVM Keith Parks may have done — which squadrons, and how many squadrons, to send up in response to a raid. There is also an intermediate layer Control Panel, where 'players' can change the rules — starting conditions, kill probabilities, etc. They may also change the simulated weather, too, making flying more or less likely on a four-hourly basis.

The simulation runs for 1440 simulated minutes, or one day. In real time, simulating one day takes about 3 minutes. Using a sensitivity mode, it is possible to set the simulation on 'auto,' at which time it will run uninterrupted for some 31 equivalent days. The results of each run are accumulated on tables in the Control Center, and then compared with the 'real world' charts of Graphs C.1 and C.2.

Running the BoB simulation

The BoB simulation turned out to be very sensitive: successive runs over a simulated 31-day period could give quite different results, even when variables, such as weather, probability of kill (Pk), Luftwaffe raid size, etc., had not been changed, run to run. This was occasioned, at least in part, by the inclusion of random elements within the simulation. For example, setting a Pk value did not determine that a particular number of aircraft would be downed every time, since the simulation allowed results of engagements to be distributed about mean levels set, in part, by the Pk. Similarly, although the weather could be set to different levels of suitability for flying, the

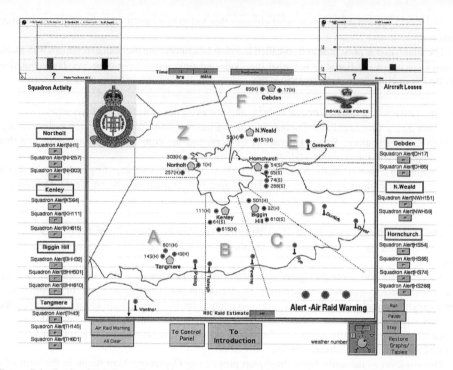

Figure C.4 Battle of Britain — Operations Room simulation. The operations plotting table is shown in the center. Sector Operations Stations are shown left and right, with their respective squadrons. Initials after the squadron number indicate the Sector and the aircraft type: NH is Northolt, Hurricane; KS is Kenley, Spitfire. Illuminated switches show which squadrons are active, and they are also shown as flashing circles on the map. Other squadrons are shown as non-flashing circles. Charts at top left and right are simulated tote boards, and there is a runtime counter at top dead center, showing current time paused at 12 hours 24 minutes. Bottom left are Air Raid Warning switches, to sound the raid alert and the all clear.

periods when a morning or an afternoon were not suitable for flying varied randomly, and so also differed run-to-run.

Sensitivity arose because the variability in results of one engagement determined the number of aircraft returning to their respective bases, to different levels of repair work, to the time available for that repair work before successive missions were due, to the numbers of crew needing medical treatment, and so to the number of aircraft available for the next day's missions... which then affected the outcome of the ensuing engagements, and so on. Such variations accumulated over the 31-day simulated period, so that results were quite variable. Moreover, a large number of 31-day runs, all with the same parameter settings, produced a series of results that did not fall into any obvious pattern. . . .

Running the simulation under virtually any set of preconditions failed to produce results that were comparable with the real-world statistics of Graphs C.1 and C.2. Weather could have a serious effect on the combat outcome: in the extreme, of course, poor weather for the whole 31 days would have virtually stopped all operations, since the Luftwaffe could bomb only in clear weather. Perfectly clear weather throughout, on the other hand resulted in a major flurry of losses on both sides, followed by periods of wound licking and repair, before starting again. The pattern of losses tended, in this situation, to become more cyclic, and again unlike the real-world statistics.

Some success in emulating the real world came by combining the effects of moderately good, though unpredictable, weather with the impact of highly variable tactics on the part of the air defense commander, Keith Parks. The simulation made it possible to try out different tactical schemes to see which might produce real-world-like results.

The first possible tactic was to rotate the various squadrons quite quickly, so that each squadron was on active operations for only one or two days, before being 'rested:' at the same time, only a few squadrons were active — three or four at the most. This enhanced the relative kill rate, i.e., the number of Luftwaffe to the number of RAF aircraft downed. The reason was to do with operational availability: each squadron had about 16 fighters; after several days' rest, all of these, including their pilots, would be available for operations. However, once a squadron had been on continuous operations for several days, some of the aircraft would be damaged awaiting repair, lost, etc., and their pilots likewise. By rotating squadrons rapidly, squadrons starting operations would be to full strength.

The second tactic has been mentioned in the first — to deploy only a very few squadrons at a time. The mathematics is simple: if only, say, 48 aircraft (three squadrons out of 20) are deployed, then — at the absolute worst — only 48 could be lost or damaged. Parks did not want to destroy every attacker in one go; instead, his plan was to break up their formations and pick of stragglers — much as lions do when hunting zebra or wildebeest on the Masai Mara.

The third tactic, and that which brought the simulation nearest to real world statistics, was to be quite unpredictable. To be predictable in the situation facing Parks would be to match the size of the defensive force to the size of the incoming raid. So, if a force of, say, 500 aircraft was approaching — as reported by the ROC — then logic might suggest that Parks should put up twice the number of fighters as when 250 aircraft were approaching. If, in the simulation, that simple rationale is NOT followed, then the simulation results start to look more like real world.

In a way, the sense of this seemingly reverse logic is seen in the previous paragraph — if more aircraft are put up against a bigger force, then more aircraft are put at risk. But, that only makes any sense if Parks believed that the Luftwaffe had little chance of hitting their intended airfield targets, and if he believed that even a small defensive force in the air would put the Luftwaffe off their stride. On the other hand, Parks was determined to preserve as much of his air force as he could in anticipation of the ensuing major German offensive, Operation Sea Lion, evidence of which was to be seen on the French coast with a build–up of barges and other shipping that might be used for invasion purposes. Whether wise, or cautious, or both, Parks' tactics won the day. . . and Operation Sea Lion never happened.

Graph C.4 shows the simulation results for a 31-day run, with moderate weather and the employment of seemingly irrational tactics, which we may notionally attribute to Parks. This graph should be compared with Graph C.1; clearly, they are not the same, but there are degrees of similarity to be seen in the day-to-day variability, in the consistency with which Luftwaffe losses are higher than RAF losses, etc.

Similarity can be seen, too, at the phase-plane graph, C.3, drawn from the same simulation as Graph C.4: the 'signature' in the phase plane is vaguely similar to that of Graph C.2. As with people, no person's signature is the same twice — it would be suspicious if it were: similarly, one should not expect the simulation signature to be identical to the real world.

Graph C.6 provides a closer, more detailed look at the pattern of aircraft losses shown in Graphs C.3 and C.4. Graph C.6 was formed, as was Graph C.3, by taking samples of the combined aircraft losses at different intervals, and summing the samples in each case — this enables the 'bumpiness' of the Graph C.4 to be assessed. The results, shown in C.6, are encouraging. Conformance of

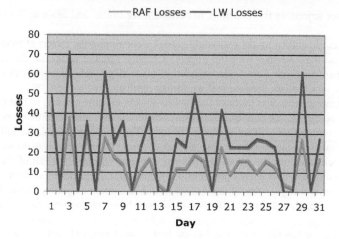

Graph C.4 BoB simulated combat over 31-day period. Weather: moderate. Continual change of defensive tactics.

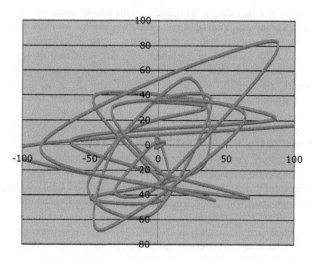

Graph C.5 Phase-plane signature form Chart C.3. Compares with Graph C.2.

the results to a straight line (the R^2 value) is convincing, indicating that the results from the simulation are fractal, while the slope of the line, the fractal index is -1.0312 — different from the real world fractal index of -1.3 — see Graph C.3. Judging by Graph C.6, perhaps Parks really was unpredictable — or perhaps the tactics of the opposing Luftwaffe commander Field Marshall Kesselring, were contributing to the unusual pattern, too. . . .

None of which explains just why the phase-plane diagram — the signature — is the shape that it is — irregular, nonperiodic, nonchaotic. . . . The general form of this signature has occurred before in the book — see Dynamic simulation of phenomena on page 68. In that instance, a phase-plane chart was drawn up to show the variability in population on the River Nile some 5000 years

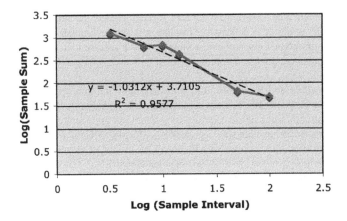

Graph C.6 Analysis of the combined RAF and Luftwaffe loss statistics of Graph C.3, showing good correlation with linear trend line, suggestive of 'weak chaos' and/or self-organized criticality.

ago, caused by variation in the annual Nile inundation triggering feast and famine such that the population rose during feast, but died off during famine.

The variation in Nile population was an instance of self-organized criticality. Graph C.6 is also indicative of self-organized criticality, or 'weak chaos.' And, curious though such a comparison may seem, observing a broadly similar general phase-plane pattern suggests that the interacting systems in the Battle of Britain may also have been in a state of self-organized criticality. In working up ('orchestrating') the various parts of the air defense system into an organismic, cooperative, coordinated unified force, Dowding brought it up to critical performance level — at which level it was, indeed, just able to see off the Luftwaffe.

Which leaves open the other questions posed at the outset, for which the simulation was constructed, and continued running of which suggests the following answers:

I.   How long could Fighter Command have lasted if the Luftwaffe had continued to attack the RAF bases and radars, instead of switching to bombing London and other cities?

   ■   Difficult to be precise, but simulating the next few months, allowing for variations in weather, and assuming that neither side changed its tactics, it seems likely that the Luftwaffe would have effectively run short of aircraft and crews after some six months. . . .

II. Can the Battle of Britain be reasonably classified as a 'win' for the UK, or not?

   ■   In recent years, it has become popular amongst would-be historical analysts to pose such questions. Were a besieged castle to bloody and see-off its attackers, such that they gave up and went to do something different, then there would be little doubt in classifying the attacker as defeated and the defenders as winners. In this case, southeast England was a fortress with a coastline and a channel instead of rampart walls — but the rationale is the same. Of course, it was a victory — certainly, the Germans believed they had been defeated!

III. Was the strategy employed by Dowding and Parks, of conserving RAF fighter aircraft resources in anticipation of prolonged activities better, or worse, than the Big Wing concept — which would have contributed most in the long run. . . ?

- The problem that Dowding and Parks faced included the prospect of a long-drawn-out battle, some of it at least on English soil as the English retreated inland in face of Operation Sea Lion with its landing forces. Even without that prospect to keep in mind, they were grossly outnumbered, and had no real idea of the size of the Luftwaffe reserves.

- Had Dowding and Parks succumbed to the Big Wing concept, then they would have faced two risks: the Big Wing would have been vulnerable on the ground as it mustered fighter aircraft from various stations to assemble in one location; putting the Big Wing up against superior odds was a greater risk, for the obvious reason that the Luftwaffe could afford to trade aircraft with the RAF one-to-one, or even two-to-one, and still turn out the winners, in the style of 'last man standing.'

- As the fight progressed, and as the Luftwaffe lost more aircraft than the RAF, there might have come a time when an all-out attack by the RAF would have made sense. During the time of the Battle of Britain — August–September 1940 — it would have been, and would have been seen as, a naïve mistake.

Despite that last bullet, and despite having won one of the most famous victories in history, within six months both Dowding and Parks were 'moved sideways' (effectively, demoted): winning, it seems, was not enough. . . .

One other point of issue concerns the various systems within the air defense system overall. Would-be experts delight in pointing out the singular 'key to victory.' For some it was the Chain Home ground radar. For others, it was the Royal Observer Corps. For many it was the Spitfire — in spite of many more Hurricanes being involved in the actual battle.

Their enthusiasm for their various causes is admirable, but misplaced. The Battle of Britain was won by the total system, with all the parts operating, cooperating and coordinating as a close-coupled, unified whole. To those who think that is not correct, consider which element of the overall system could be *removed*, leaving the rest to operate. The answer is *none*!

Hence, it is the whole system, or nothing, and it is unreasonable to pick out any one element as the key. That is not to decry the incredible bravery of the fighter pilots, nor the technological innovation of Chain Home, nor the tireless efforts of the ROC volunteers, day after day, night after night, nor the ground crews working all hours, in all weathers and while being attacked by enemy aircraft; and we must never forget the sector operations personnel, the radar personnel, the post office engineers keeping the telephone lines working. . . .

But, it was the system as a whole, and — although the term 'systems engineering' was yet to be coined, it seems entirely reasonable to describe Dowding's development of the air defense systems overall performance, in retrospect, as operational systems engineering of the first order.

As Winston Churchill put it: 'Never in the field of human conflict has so much been owed by so many to so few.'

C(2) The Lightning — Realizing the Total Weapon Systems Concept

Introduction

The end of World War II ushered in the Cold War: the UK became concerned about potential air attacks from the East, and of the potential Soviet use of standoff weapons, which would allow attacking aircraft to launch their attack weapon when still some distance offshore; the weapon would carry on to its destination, while the launch aircraft would return to base unscathed. Soviet aircraft might carry more than one weapon, so might launch at more than one target, and the standoff weapons were smaller and much faster than a manned bomber, so would be much more difficult to intercept. Or so intelligence believed...

Faced with such a daunting threat, the UK developed a postwar air defense system tuned towards the particular threat. New, advanced ground radars were established up and down the east coast of the UK, looking primarily eastwards, out to sea. Operations centers were built underground, in fortified bunkers. The whole of this so-called Air Defence Ground Environment (ADGE) was tied together with a communications infrastructure, including digital data links. Airborne early warning facilities were developed, so that their radars could see beyond the horizon, and they were equipped with data links to report what they had seen to the operations center.

Although using newly developed technology, and facing a different enemy, the architecture of this new ADGE was much the same as Dowding's innovative prewar air defense system — indeed, most of the air defense airbases were the same RAF stations Dowding had used: sector operations centers replaced sector operation stations, data links replaced the hard-wired, voice-operated intranet of the Battle of Britain air defense system, but — yes, the architecture was the same. In place of Hurricanes and Spitfires, there came a succession of jet fighters: Hunters, Javelins and the Lightning.

The Lightning

During the late 1940s and 1950s, the UK was very active in aircraft development, with a host of new aircraft types, engines, wing plan forms, etc. The Fairy Delta Mk2, for instance, as the name implies, was a delta-wing aircraft that led eventually to the development of Concorde. Large bombers were developed, capable of carrying nuclear weapons: the so-called V-bombers, Valliant, Victor and Vulcan, all of which saw service in the RAF.

Among the developments was an experimental aircraft, the English Electric P1. It was a 'notched delta' design, giving it a distinctive shape and an awesome rate of climb to high altitude. With rocket-assisted takeoff (RAT), it is said to have reached 90 000 feet from a standing start in 3 minutes. . . .

The P1's potential made it an ideal candidate for inclusion as the intercept element of the UK air defense system — except, that is, for such trifles as a lack of any sensors and weapons. The aircraft had, in effect, to be redesigned as a practical military interceptor: it needed

- a good radar, to see the threat Soviet aircraft;
- phenomenal speed, to fly out in time to meet the threat aircraft before it could launch its standoff weapons;
- even greater speed to overtake any standoff weapons that had already been launched;
- new kinds of air-to-air weapons to combat the new threat, to reach ahead of the interceptor, to catch the standoff weapon, etc.;
- superior climb and turn capability in case the enemy chose to attack at high altitude

Designers set to create a total weapon system aircraft, initially the P1A, then the English Electric Lightning Mark 1A. This aircraft was to have one purpose — to intercept Soviet aircraft before they could launch their weapons. It was to be part of an air defense system also dedicated to the same purpose, so that the whole air defense system, including the Lighting, was to be a coordinated, cooperative, organismic unified whole.

The objectives were clear: the means of making the P1 into an interceptor much less so. To achieve its outstanding performance, it was, in effect, 'two kerosene burners place one above the other, with wings stuck out either side.' In other words, it was an experimental aircraft, not only without any of the equipments needed to become an interceptor, but also with very few places in which to put them.

The P1, or the Lightning as it would become, needed radar — but where to fit it? The ingenious solution was to create a radar 'bullet' that would fit into the shape of an intake center-body. If this could be sensibly achieved, it would have a hopefully limited effect on engine performance and on aircraft drag. The solution was the Ferranti AI23B, an advanced, 3-cm pulse radar with a sophisticated four-beam antenna system that enabled the radar to lock to a target in azimuth and elevation. The radar was fitted with elegant analog computing facilities that enabled the Lightning to fly intercept paths that were suited particularly to the new missile that was also being developed — the de Havilland Firestreak. See Figure C.5, which shows both radar and missile *in situ*.

Firestreak, based on another research program, Blue Jay, had a solid-rocket motor, infrared direction sensing and fuzing, fragmenting warhead and its own navigation and autopilot. It made use, unusually, of magnetic amplifiers as opposed to the more conventional electronic ones, to overcome problems from heat and vibration. Firestreak had a limitation, however — its sensor was able to pick up only infrared signature emanating from the hot turbine disk of an enemy aircraft, and so was able to attack the enemy aircraft only from behind. This meant that the Lightning would have to fly out towards the target aircraft, and perform a looping maneuver to roll out some five miles, or so, behind the target before firing the missile.

The looping maneuver, which pilots came to call a 'butcher's hook' after its shape, took precious time, and so was a disadvantage. To combat this, the AI23B computer was programmed to fly the ideal 'butcher's hook' relative to the target, and it presented suitable signals both to the autopilot

Figure C.5 A Lighting Mk 1A of No 111(F) Squadron undergoing night OTR (operational turn round.) Note the radome in the nose of the aircraft, the Firestreak Missile with protective covers, center, and the refueling probe at the top of the roundel (photograph by author).

and on the pilot's attack sight (PAS), a smart head-up display, so that the pilot could steer manually if so desired. In either event, the objective was to minimize the time taken to reach a firing solution. . . .

Nonetheless, Firestreak was less than ideal, so an improved version, Red Top, was in development, which would permit head-on engagements. Red Top would not be available with the first Lightnings into RAF service, however.

With no place on the P1 to 'embed' missiles, they had to be externally mounted. This was achieved by fitting interchangeable ventral weapons packs, which could carry Firestreak, Red Top, rockets, or cannon if required.

So, the designers of the soon-to-be Lightning found ways to fit sensors and weapons, but there were still serious problems. The flight duration was brief, to say the least; the aircraft could not carry much fuel in its stainless-steel wings: a detachable ventral tank was fitted, extending flight duration significantly, although durations was still short, especially using full engine power and reheat. Fitting two external missiles and the radar center-body increased drag, which made matters worse, so the hoped for speed of Mach 2.0 was hard to realize. And, having only two missiles did not permit one Lighting to take on many opponents — even supposing it had the time.

Optimizing the Design

The design problem was formidable. The objective was to reach the target when it was as far from the UK coast as possible, and certainly before standoff weapon launch distance. The pilot had to be able to see the target at a distance, day or night. The detection range of the radar was dependent, *inter alia*, on scanner dish diameter: the greater the dish diameter, the greater the radar's range. However, increasing the diameter would increase the cross section of the center body, or bullet, which would increase drag and fuel consumption, and reduce speed.

Similarly, mounting Firestreak missiles externally increased drag and fuel consumption. Increasing the cross-section of the Firestreak might have increased its target detection range, and a larger motor might have increased its firing range, but at a cost to overall aircraft performance.

Some of the tradeoffs are presented in the simplistic model of Figure C.6. The model employs Newton's Law relating force, mass and acceleration. In this case, the force is the difference between (engine) thrust and aircraft drag. The resulting acceleration is integrated to give velocity, and integrated again to give (aircraft) distance traveled. Velocity is used to calculate the drag caused by both radar and missile, such that the larger their cross-sectional areas, the greater the drag and the less the performance.

So, the designers were faced with an optimizing problem, for which the solution lay in balancing the various factors to enable the shortest time to kill. The simple model of Figure C.5, with typical results at Graph C.7, shows that increasing the radar cross-sectional area beyond a necessary minimum gave increasingly poor returns — although a fatter radar might see further, the aircraft would take so much longer to reach the target area, that there would be an overall loss in effective operational range. Besides, there was little point in being able to see much further than the range at which the air-to-air missile could be launched.

The answer had to be to keep the radar and missile cross-sections as small as practicable to minimize drag and maximize aerodynamic performance, while at the same time directing the Lightning on to its target using ground radar, airborne early warning (AEW) or both: these other radars would act as eyes for the Lightning until it was close enough for its interceptor radar to see

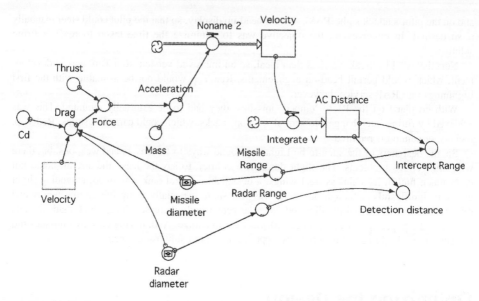

Figure C.6 Simple model to show interaction between radar and missile cross-sectional area and Lightning aircraft operational performance.

for itself. Evidently, it was not sensible to optimize the Lightning on its own — if it was to part of an overall system, then that overall system had to be optimized, and the Lightning would have to be configured as an integral part of the whole weapon system.

Keeping the drag low, and the performance high were essential to deal with another aspect of the supposed threat: Soviet aircraft might try to enter UK airspace at high altitude, above the ceiling of ground-based missile systems. To combat this, the Lightning, with its superb climb rate, would be equipped with a zoom-climb capability. Zoom-climb is a technique for reaching high altitude

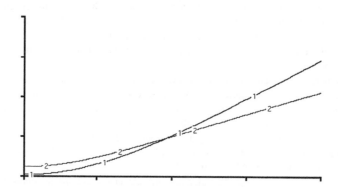

Graph C.7 Lightning potential intercept range (*y*-axis) against time (*x*-axis), from the model of Figure C.5. Line 1: small radar and missile cross-section; Line 2: larger cross-section. Beyond the crossover, smaller radar and missile cross-section would give better overall intercept range. . . .

by accelerating at medium altitudes to very high speeds and then climbing steeply, turning the kinetic energy of forward velocity into potential energy of height. So-called programmed zoom-climb profiles were built into the radar and autopilot, such that the pilot could be guided to zoom climb, topping out in a suitable engagement position. This could be a precise operation, since the Lightning would find it difficult to maneuver at very high altitudes, and would in all probability get only one shot at the target.

The Jamming Problem

If the Lightning interceptor was to depend on ground and AEW radars for directions toward a distant target, then what would happen if the enemy decided to jam the radars, or the radio communications from ground radars to the Lightning, or both? Intelligence indicated that the Soviets were investing in airborne jamming capabilities, so their attack aircraft might carry jammers, they might be accompanied by jamming aircraft, or they might use standoff jammers, i.e., jamming aircraft (or ships) that kept back from the line of interception.

Jamming would have put a major spoke in the defensive wheel — what to do about it? The ground radars (which were to become the Type 84 and Type 85 radars of the Linesman–Mediator system, which combined UK air defense on the one hand — Linesman — with the management of civilian air traffic on the other — Mediator) were fitted with comprehensive anti-jamming facilities and features, including passive detection arrangements that permitted triangulation of multiple jamming aircraft, so they, at least, might be located and, potentially, intercepted. But, that left the problem of directing the Lightning aircraft on to the target aircraft — the enemy might be using communications jamming.

Digital Data Links to the Rescue

Digital data links were in their infancy at the time, and they seemed to hold the potential key to solving the problem of communication jamming. Supposing that target data could be sent automatically from the ground radars, the Types 84 and 85 radars, digitized and transmitted to airborne Lightning interceptors, together with flight control instructions for the interceptor pilots to follow, such that they would roll out just behind the target aircraft in the ideal firing solution.

The system designers went one further: suppose all of this data being sent from the ground radars to the airborne interceptor was not just displayed to the pilot: suppose it was used to automatically steer the interceptor and to operate the engine throttles, such that the pilot — having engaged the system — would have had nothing to do other than monitor the situation. Further, using digital data links, if this arrangement could work for one interceptor aircraft, then it would work for many simultaneously. . . .

A research program was set up to construct and trial a ground to air digital data communication system. To be effective, it would have to be high powered, and the transmitters would have to be on the east coast — preferably near the ground radars, also on the coast.

The research program comprised a number of elements:

■   a processing system at the ground radar to gather data on airborne targets and airborne Lightning interceptors, to calculate intercept paths, speeds, climb/descend profiles, etc.;

■   a processor to fit the data into packets, with headers, addresses and parity, to be sent via data links;

- a multiplexed ground-to-ground data link to pass the data packets over telephone lines to a remote radio transmitter;
- a ground-to-air radio data link transmitter, converting the data stream to modulated radio signals;
- an airborne radio receiver to receive and decode the radio signals, reconvert them into message packets, check parity, and send messages to their respective destinations;
- an airborne interface to use the messages to display target data (range, bearing, altitude, track, velocity, etc.), to control aircraft interceptor direction, altitude, speed, etc., via the aircraft autopilot and autothrottle.

The technology of the day was such that the data link could be unidirectional only: the radio uplink employed the full bandwidth of the airborne UHF transmitter–receiver, a modified version of the conventional set, incorporating a data demodulator. There was insufficient room in the Lightning to fit a second transmitter–receiver: when the pilot selected 'data-link,' the radio received and decoded the data stream only: there could be no voice communication with the ground. To get around this, a special, small voice recorder unit was fitted behind the cockpit with a number of prerecorded voice messages on it; each of these messages could be activated from the ground radar, to say 'Check speed,' 'Check height,' and — most importantly — 'Return to R/T,' indicating that the pilot should revert to voice communications.

Unexpectedly, this recorded voice communication discomforted pilot during early trials. For the sake of clarity, the voice used had been that of a professional woman radio announcer, and pilots complained that the voice messages alarmed them, sounding like there was another person with them in a single-seat cockpit; moreover, the voice was that of a woman! They proposed that the voice messages should be re-recorded using a man, and that artificial 'static' should be included, to render the messages similar to those received over the radio. In retrospect, such trivial complaints were symptomatic of a deeper malaise, and should have forewarned developers of the system. . . .

Developments of prototype equipments took some time, and the Lightning Mk1A had been in service several years before the complete data link system could be set up for trial, in the mid-1960s. A Lightning was fitted with the prototype airborne elements, including a data decoding system that was so large it was referred to as 'the dustbin;' if the trials were successful, a much smaller version would be developed. A ground-to-ground link had been established from the UK midlands, where there was a prototype Type 85 radar, to a more southerly location over several telephone lines. In this more southerly location, a 20 kW UHF transmitter had been set up, with a 20 dB gain antenna, giving an effective radiated power (ERP) of 200 kW; this may be compared with the transmitter power in common use for airborne UHF communications of 25 W.

The trial

The trial was to take place over the English Channel — not far from the coastline that had withstood the worst of the Battle of Britain. The Lightning was to take off and fly over the Channel, be picked up on the prototype Type 85 radar, was then to receive flight control signals from the Type 85, over the ground to ground and ground to air data links, and was to engage the automatic flight controls system (AFCS) for short period of ground-controlled flight. The distances between the ground radar and the ground radio transmitter, and between that transmitter and the aircraft, were representative of the distances supposedly involved in intercepting a Soviet bomber prior to release of its standoff weapon.

Everything was tested on the ground and retested. Came the fateful day, and the Lightning took off as required, transited to the English Channel. . . and later returned. The pilot emerged from the single-seat fighter looking 'unhappy.' It seems that the equipment had worked as it should, but that the pilot was extremely uncomfortable with being flown by some remote agency, and with the manner of the flying.

Unsubstantiated rumor at the time suggested that a scientist or technician at the Type 85 radar site had seen a switch supposedly in the wrong position, and had 'corrected' it, causing the Lighting to perform an instantaneous 180° roll. Be that as it may, and the story may well be apocryphal, the trial proceeded no further, and plans to go to the next stage of technological development were abandoned, apparently without further regard to the Soviet jamming threat.

Conclusions

In spite of the data link fiasco, for that is in effect what it became, the Lightning went on to perform as the principal intercept element in the UK air defense system of the time, but using only voice control — no digital data. Of its type, it was a good example: it had speed, agility and an excellent rate of climb, all features shared by good interceptors. It lacked stamina, and, with only two missiles, it was short of weapon power. A gun had to be fitted retrospectively, when it was realized that it was not a good idea to shoot down every intruder — someone might simply have lost their way: missiles could only kill; a gun was needed, to warn intruders without killing.

In an extension of the iconic BoB image of aircrews relaxing outside their squadron dispersal huts listening to the loudspeaker, Lightnings could be parked on an auxiliary service platform (ASP) at the end of the runway, with their aircraft intercommunications system connected to the sector operations center by 'telebriefing,' through landlines. Pilots could be strapped in, with ground power connected to the aircraft, along with cooled air to feed the pilots' air ventilated suits; one word from the sector operations center was sufficient to scramble instantly, with automatic disconnection of ground power and cooling air and telebrief, followed by takeoff, literally within seconds.

The Lighting had been fitted with a number of automated devices that, in the end, were rarely, if ever, used. Why not? The interceptor radar, Ferranti's AI23B had a computer which would fly the pilot in the ideal 'butcher's hook' to perform a rear hemisphere engagement. Pilots agreed that it worked, but would never use it. When asked why, they had no good reason, other than they wanted to fly the aircraft themselves, in their own way, and they believed that they could do better than any computer — which, incidentally, was not the case. They had a point, though — they were reliable, whereas the AI23B, along with most avionics of the day, was not!

The problem with the data link was, essentially, the same: it was not that the technology did not work, because it did; it was more that the pilots wanted to fly the aircraft themselves, in their own way. They really did not like being sidelined and made to feel useless while some 'faceless boffin' and his equipment flew the aircraft remotely and impersonally from many miles away.

Looking back at Dowding and the way he progressively improved the BoB air defense system, it can be seen that, while he 'tuned' the system by using the adaptability of the human operators, both on the ground and in the air, he did not attempt to change the technology; he trained and organized the people to make better use of their technology. In the Lightning era, it was no longer aviators tuning the adaptive, people-element of the sociotechnical system, but scientists and technicians designing it literally from the ground up; their ambition seems to have been to cut the human out of the loop altogether. It was clear that attempts by scientists and engineers to supplant the human element in sociotechnical systems did not always go down well with human operators. . . .

10

SM5: Architecting/Designing System Solutions

Sir Christopher Wren
Said, 'I am going to dine with some men.
If anyone calls
Say I am designing St Paul's'

Edmund Clerihew Bentley, 1875—1956

Approach

System design is sometimes viewed as an esoteric, even arcane, practice: so much so, that many teachers, references and books no longer refer to system design, choosing instead to talk about 'architecting,' suggesting, perhaps, that design and architecting are substantially the same thing, which may not be entirely correct.

System design imbues a product, the design set, with the essence of systems engineering (see 'The essence of systems engineering on page 120). We have seen, too, how the Generic Reference Model is able to provide a design template, such that requisite emergent properties may be forthcoming (see also 'Does the GRM capture emergence?' on page 141).

To manifest particular emergent properties, capabilities and behaviors in a particular solution system, select and bring together particular functional parts and cause them to act and interact in a cooperative and coordinated manner, invoking requisite synergies. Instantiating the GRM for a particular solution system creates the 'internal' functional architecture, or platform, to select, activate and coordinate the many and various mission functions—see Figure 10.1.

If this seems obscure, then consider the analogy of a man crossing a busy road: his mission is to cross — in one piece. He uses his eyes and ears (sensors) to collect information, chooses (decides) to dodge between traffic (strategizes) rather than use the pedestrian crossing, waits for a gap in traffic (plans) and safely walks across the road (execution) with the tacit cooperation of the motorists — who do not actively try to run him down. The *execution* of the plan requires that

Figure 10.1   Function management (i.e., mission, behavior and resource management) 'orchestrates' prime mission functions, activating them in sequence and in parallel in accordance with the plan and in accordance with the CONOPS.

the brain scan for a gap in traffic, initiate locomotion, control limb movement, maintain balance, and continually scan sensors (primary mission functions); this is orchestration of prime mission functions, which are activated in series/parallel according to the plan (see Models of Systems Architecture on page 108 for functional architecture in the brain).

By analogy, the GRM Function Management corresponds to the cognitive brain and the central nervous system; it is the foundation, the managing platform, enabling the properties, capabilities and behaviors of the whole to emerge. Prime Mission Functions (PMFs) correspond to the sensor sweeps, and to capabilities for movement in an appropriate direction, at the right moment, the ability to dodge, etc., all of which enable the man to cross the road safely, under the direction of his intellect, and in accord with his concept of operations. Function Management activates, correlates and coordinates these PMFs.

A somewhat different analogy might be with a symphony orchestra, with strings, brass, wood-wind, timpani and percussion, etc. An orchestra performs a musical piece by following a score, with each of the players having their respective musical part appropriate to their instrument in front of them. The actions of all the players may be synchronized and coordinated by the conductor, but the orchestration is predetermined, so none of the players is directly controlled in his playing by the conductor. Nonetheless, beautiful music may emerge from the whole.

Comparing the two analogies shows that the term 'orchestration,' as used in Figure 10.1, can be viewed in two ways: for the man crossing the road, the brain planned and orchestrated his sensing and movement in real time, using the central nervous system for motor control, sensory feedback, etc. In the second analogy, the orchestration was preplanned and pre-allocated to the various sections within the orchestra, so that each player was able to play, in principle, on his or her own and in isolation. The whole was then drawn together, synchronized and coordinated by the section leaders and by the conductor.

This difference in approaches to orchestrating prime mission functions is to be found in many systems and situations. In the conduct of warfare, for instance, some European military favor a method called Mission Command, or *Auftragstaktik*, in which an overall commander allocates a set of missions to various subordinate commanders, who subsequently plan and execute their respective missions without supervision or communication with the overall commander. This has the advantage of allowing communications silence, so depriving the enemy of potential intelligence. Each subordinate commander is also aware of the overall commander's intentions, and so is able to modify, or even deviate from, the allotted mission as situations develop, while still trying to achieve the commander's goal. Control is effectively delegated, trust is afforded to subordinate commanders, and they are able to develop their capabilities — and make mistakes.

An alternative form of military command employs continual real-time control of operations, with the overall commander — who is often remote from operations — controlling what is going on; in the US, this has been dubbed 'President to foxhole:' control is centralized. In this scenario, subordinate commanders have less authority and discretion, and there is a heavy dependence on communications, on which real-time control inevitably depends. The potential advantage of this approach is that potentially sensitive decisions are taken by the senior man, rather than by a subordinate commander who may not be *au fait* with political sensitivities. The obvious weaknesses are: (a) that the system breaks down if communications fail; and (b) over time, it undermines the chain of command and the authority of officers within that chain.

Instantiating the GRM requires an understanding of the CONOPS, and the developing nature of the solution system, as evinced by the prime mission functions. For example, a military platform as (part of) a solution system might expect to carry a variety of different weapons, both offensive and defensive, associated with prime mission functions, self-defense, etc. Instantiating the GRM for the platform as a solution system could then throw up a Weapons Management function under the Mission Management heading, and a Rules of Engagement function under the Behavior Selection heading. Why? Because these would be necessary platform functions to 'manage' and coordinate prime mission functions; and, it is on this coordinated interaction and cooperation between functions that emergence is founded.

Similarly, examining a range of prime mission functions for an enterprise might indicate that some are concerned with seeking and anticipating threats, or reducing risks, perhaps by situating the solution system. If so, one would expect to see a marketing, or intelligence function under the Mission Management heading in the instantiated GRM. In general, there are likely to be functions under Mission Management, Resource Management and/or Viability Management that correspond to Prime Mission functions, and which are able to orchestrate these prime mission functions.

The Functional Design Process

So far, in previous steps of the systems methodology, we have identified a range of potential CONOPS and prime mission functions (PMFs) that the solution system will need to perform. The mission features will make clear the nature of the solution system, i.e., what kind of system we are intending to design. The 'external,' or visible functions performed by the solution system find their counterparts among the internal functions of the system. We have yet to identify the many and various functions, which may be thought of as 'internal to the solution system,' that will tie the whole system together, and will cooperate and coordinate, establish and maintain system coherence and viability, etc. We know from the GRM what many of these are, generically speaking; part of design is instantiating, or giving substance to, these generics — see GRM Function Management on page 128, *et seq.*

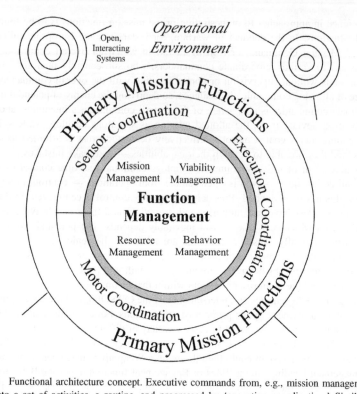

Figure 10.2 Functional architecture concept. Executive commands from, e.g., mission management may be elaborated into a set of activities, a routine, and progressed by 'execution coordination.' Similarly, routine sensor and motor coordination can be delegated, such that primary mission functions can be orchestrated without overloading mission management with detail.

A useful design starting point is to develop a specific version of the GRM, initially in tabular form — see page 133. This process generates all of the elements required for mission management, resource management and viability management. Similarly, it generates elements for cognition management, behavior selection management and behavior stimulation management. The elements are functional subsystems, and will be given names consistent with their function.

So, a functional subsystem that manages internal and external communications might be called a 'communications center' in one context, something else in another; a functional subsystem that senses remote activity might be named 'sensor suite;' a functional subsystem that detects and anticipates competitors activities might be called 'intelligence subsystem;' and so on.

These functional subsystems can be analogous to the 'mirror neuron templates' in the brain, which can initiate, coordinate and interpret series of sensor and motor activities 'subliminally,' so relieving the central, conscious brain from routine processing (see page 110 *et seq.*) This, then, presages the development of functional architecture, with a central functional intent (prosecuting the mission) activating surrounding functional subsystems, which, in their turn, activate and coordinate 'peripheral' Prime Mission Functions. Such a 'tiered' functional architecture minimizes mission management effort, allowing many different missions to be carried out at the same time, if need be, see Figure 10.2.

The overall functional design process is shown in Figure 10.3. In addition to the sequence of functions already discussed above, the figure also shows the tools and the products or 'deliverables'

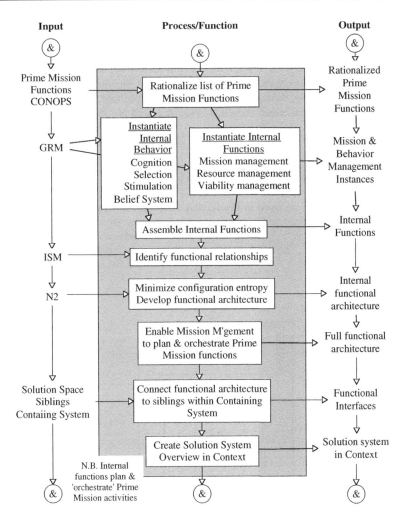

Figure 10.3 Functional design.

that form part of any behavior diagram. As the inputs at left indicate, the various methods and tools employed in design may be 'handraulic,' i.e., developed on paper or on a computer screen, by hand—as, indeed was Figure 10.3.

However, there can be significant advantage, particularly with the design of large or complex solution systems, in using computer-assisted tools. John Warfield's ISM, for instance may be enhanced in use with computer support, and the ubiquitous N-squared (N^2) chart may be automated to significant advantage. In the latter case, the use of clustering algorithms, hill climbing routines and/or cumulative selection methods enable the configuration entropy of a disordered N^2 chart to be minimized: in the process, errors and omissions are highlighted, and clusters are revealed which form the basis of architecture (see Hitchins, 1992, 2003).

Contrary to conventional wisdom, functional architecture and, as we shall see, physical architecture, do not have to be separately designed, or — as some preconceived framework — superimposed on the solution system. Architecture emerges naturally from the problem space, the solution system concept, CONOPS and prime mission functions, and almost as a byproduct of the design process. That is not to say that architectures cannot, or should not, be superimposed, say, for reasons of interoperability, compatibility, damage tolerance, etc., or to reconcile solution space constraints. It is to suggest, however, that superimposed architectures need not be optimal in the context of solution system performance and effectiveness.

The Physical Design Process

The physical design process, outlined in Figure 10.4, involves mapping, or folding the functional architecture, with its subsystems and interactions, on to/into some physical substrate. The solution system is to be manifested as a contained whole, body, organization, enterprise, process, etc: in

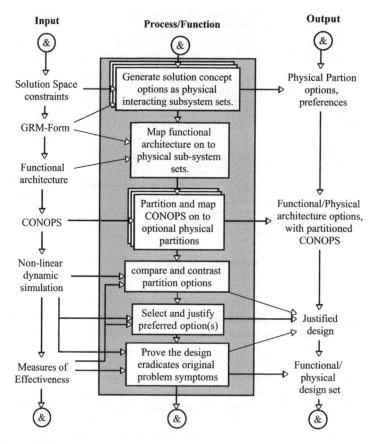

Figure 10.4   Functional/physical design — developing a functional/physical design set.

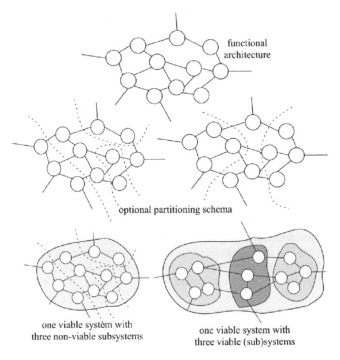

Figure 10.5 Alternative partitioning schema. Complex systems may be partitioned without prejudice to their overall functional behavior, with a view to their future organization or creation. There can be many, different partitioning schema for a complex system. Partitions may contain nonviable subsystems or may, on the other hand, be realized as discrete open, interacting viable systems within a Containing System, which will interact, with other Containing Systems, not shown: this is the development of hierarchy.

the general case, however, a variety of separated parts may be joined to each other by interaction pathways to facilitate behavior as a unified whole.

As Figure 10.5 shows, there may be many different ways to partition a system — in principle the more complex a system, the more ways. To some extent, systems partition themselves, with some functions clustering together owing to more connections between members of the cluster, and fewer with member of other clusters; such functional clusters are candidates for physical association, i.e., for constituting a partition, where closeness facilitates the higher degree of mutual interaction in the cluster. There may be other reasons for partitioning: the overall system may be large and distributed, so that partitions become more manageable; some groups of functions may be quite different in nature from others, suggesting quite different approaches to realization; and so on; and the constraints of the solution space may dictate that the whole system be created as interconnected parts, each of which is able to fit within local constraints.

With so many drivers, it is not surprising that there may be many different partitioning schemes. As Figure 10.5 shows too, there may be two quite different results from partitioning. In one case, the partitions exist within the one system and consist of nonviable subsystems, i.e., subsystems that could not exist, flourish or survive on their own. An example of this might be the human body which has many closely coupled subsystems: cardiovascular, central nervous system, pulmonary system, immune system, etc., etc. Another example in the same vein (!) might be a multi-band analog double super-heterodyne radio receiver, with modular power supplies, radio and intermediate frequency amplifiers and detectors, frequency synthesizers, frequency control loops, etc. In both instances, the many and varied, complementary subsystem parts are closely coupled, mutually interdependent and nonviable, i.e., they cannot function independently.

Alternatively, the result from partitioning might be a number of interconnected, viable subsystems, each capable of existing and operating on its own. An example of this approach was to be seen in Apollo, where the various subsystems were the various craft: rocket, command module, lunar excursion module, etc. While they all fitted together and interacted, each was a viable system, able to exist and function on its own; however, all were needed to fulfill the CONOPS, and, indeed, each of them was able to fulfill part of the overall CONOPS on its own. In effect, the overall CONOPS was partitioned, too, so that each physical partition was allocated a phase of the CONOPS.

A second example of partitioning into interconnected viable subsystems might be a large corporation comprised of a headquarters and a number of divisions, each operating as a separate business with its own marketing, production, sale, etc. Each division may manufacture different parts and products, say, for the aerospace, or the transport industries, such that the various product outputs from the divisions are complementary, with few overlaps. For vehicles, one division might make engines and gearboxes, another transmission and suspension, another bodies and interiors, and so on.

Importantly, the emergent properties, capabilities and behaviors of the whole, are (should be) unaffected by the partitioning into viable subsystems, since both prime mission functions, and the functional architecture mapping, should be unchanged — see Figure 10.4. On the other hand, in creating a number of viable subsystems, i.e., systems in their own right, the whole can now be seen as a Containing System, with its own complementary variety of viable subsystems. Failure of one viable subsystem does not cause immediate failure of another, since they each have their own function, behavior and form management capabilities, and indeed their own CONOPS.

When partitioning into viable subsystem, then, an opportunity presents itself to consider each of these subsystems, with its respective CONOPS (part of the overall CONOPS) as a discrete, open, conceptual remedial solution system, potentially operating in a future environment consisting largely of the other viable subsystems. It may be sensible to iterate the systems methodology, from SM3: Focusing System Purpose; through SM4: Develop CONOPS; and up to SM5: Design; for every viable subsystem, each of which will merit its own instantiated GRM — see Figure 10.6.

Indeed, for a very large and/or very complex system, it may be appropriate to iterate the systems methodology at several levels of this emerging hierarchy of systems within systems within systems . . . and this is what happened, effectively with Apollo and with many other major defense and aerospace projects. This works in practice by creating a 'top-level' systems engineering team, concerned with the whole mission system. Additionally, each of the major subsystems, which may be realized as independent projects running in parallel, has its own systems engineering team, members of which also participate in, or subscribe to, the top-level systems engineering team activities, designs and decisions.

This ensures that any changes, developments, etc. arising at the top level can be cascaded down the hierarchy, and *vice versa* — any unavoidable changes in any one of the major subsystems can be elevated to the top level to see if they would affect interactions and/or the whole. Changes in any subsystem which did not affect its emergent properties, capabilities and behaviors may prove unimportant; changes that did affect subsystem emergent properties, capabilities and behaviors, on the other hand, might lead to major redesign, rebalancing, etc., and could potentially affect the whole. So, each level in the hierarchy had its systems engineering teams; the corresponding systems

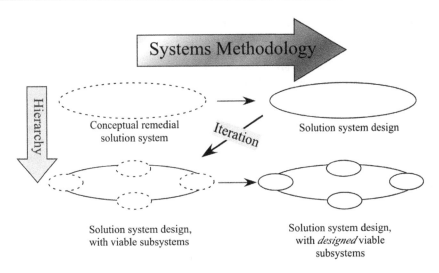

Figure 10.6 Iterating (part of) the systems methodology to accommodate developing hierarchy.

engineering teams were tasked with ensuring that the requisite emergent properties, capabilities and behaviors did, indeed, emerge on schedule, and to the satisfaction of the hierarchy level above.

Output/outcome

Whether or not iteration is appropriate, the end result of SM5: Architecting/Designing the Solution System is an arrangement, or configuration, of interconnected, functional parts. The emergent properties, capabilities and behaviors of the whole SoS, and of each of the subsystems and their interactions, may also be specified. These emerge principally from the nonlinear dynamic simulation of the partitioned SoS as an open, adaptive system interacting with other (simulated) systems in their mutual environments; this simulation is necessary to prove that the functional/physical design resolves (rather than solves) the original problem by eliminating all of the original symptoms emanating from the problem space. It is also used to compare different partitioning schemes. And, when it is proved the design fulfills its purpose, the various parameters, values and structures can be used to specify emergent properties, capabilities and behaviors.

The output from SM5, then, is architecture in the sense of pattern and paradigm, but it is more than that — it also identifies and defines purpose, function, relationship, association and separation, interaction, hierarchy, and emergence. Moreover, by virtue of its derivation, it is largely an Ideal World representation, except in one respect: it is not an optimum solution system design.

To understand why not, consider the orchestra analogy again. Members of the orchestra play their instruments according to the written score, and if their playing is coordinated, the result is music. However, it may not be the best music. To achieve excellence, the conductor has to interpret the score, and this interpretation emerges, partly, as *balance* between the sections of the orchestra. Left to their own devices, the brass may drown out the strings, or the timpani may overshadow the woodwind, and so on. The best, or optimum, performance from the whole emerges when the outputs from the various sections are balanced dynamically as well as coordinated rhythmically. So it is with systems in general — the best performance, the most effective system, and the greatest degree of emergence, arise when the various system parts cooperate, coordinate and balance in operation. So, while the *basis* for emergence has been established by the orchestration of prime mission functions, the optimum *degree* of emergence has yet to be realized.

A second analogy may help to make the point. Consider a Formula One racing car, and suppose we wished to create a new one by bringing together the best parts from the various racing car makers. We might employ the best suspension from Ferrari, the best engine from Renault, and so on. But, would the resulting, new, racing car, automatically be the best? Not necessarily: the various major components must operate in harmony with each other, so that propulsion does not overcome traction, so that weight distribution does not prejudice suspension and cornering, and so on. Again, dynamic balance is the key to getting the best out of the system . . . And we should not forget that the car is 'tuned,' not only to give the best performance on a particular track, but also very much to suit the individual driver. The best car on the day combines the purposeful skills of the driver with the capabilities of the vehicle: a great driver may win with a less-than-perfect car; and *vice versa*.

The last function/process on Figure 10.3 requires justification of the selected partitioning option. This would be achieved ideally by further dynamic simulation modeling, to show that the solution system design, if realized, would neutralize all of the original problem symptoms.

Summary

Chapter 10 has addressed the conceptually difficult issue of systems design, so often overlooked in texts, papers and handbooks as 'too difficult.' Undoubtedly, design of a complex solution system can be a significant task, but it can be undertaken systematically and comprehensively, using the GRM as a design template. In the process, system design does, indeed develop functional and physical architectures, but as an outcome from the processes rather than as an arbitrary input. Architecture is, at its most fundamental, clustering and linking: clusters form during the design process by closer association between some functions (binding) than others; links between clusters (coupling) then appear as a natural outcome. As linkages are instantiated in the design, functional architectures appear without conscious effort. Physical architectures, on the other hand, may be imposed in part by constraints within the solution space, and by other systems with which the solution system will interact. In this text, the term functional/physical architecture is preferred, to plain physical architecture. It is vital, in mapping functional architecture on to physical architecture, to maintain the linkages and interactions between all functions such that full and effective orchestration of the prime mission functions is maintained — else, requisite emergent properties, capabilities and behaviors will not be evinced, or worse, counterintuitive and unwelcome properties, capabilities and behaviors may emerge. Something as simple as inadequate capacity of, or noise on, a seemingly minor communication link, can radically change the behavior of a whole solution system. Similarly, the smallest timing error can prevent coordinated behavior, decoding, synchronized synergistic actions, etc.

Partitioning can lead to the creation of hierarchy, with viable (sub)systems within viable systems. A viable subsystem is one with its own function, behavior and form management. Partitioning an overall system into viable subsystems should not, in principle, affect the functional behavior of the whole. On the other hand, it can mean that failure of one viable subsystem does not materially affect other subsystems, although of course it may impair overall functional behavior and capability. Where partitioning results in a set of open, interacting, viable subsystems within a

Containing System, it may be appropriate to repeat the systems methodology for each and every viable subsystem, considering it to be operating in an environment consisting largely of its sibling, viable subsystems.

N.B. Understanding the practice of system design may prove difficult when it is presented in the abstract form of behavior diagrams, no matter how informative they attempt to be. The best way to understand the processes and methods is to see them at work in examples. Case Study D: Architecting a Defense Capability, is an exemplar of the systems methodology in operation.

Assignment

You undertake the system design for a new, inner city school for children aged 11–18 years, as follows:

- Propose a CONOPS for the school: how is it going to operate, from where will it attract pupils, what will be their age range and ability, what subjects will be taught, what style of teaching will be employed, what will be the students' qualifications on leaving, etc.
- Establish a Prime Directive for the school, and employ the TRIAD Building System, or other means, to propose a set of Prime Mission Functions for the school, including PMFs to address risks and threats to the school, and its achievement of purpose. Limit the number of PMFs to seven.
- Instantiate a Function Management Model (Mission, Viability and Resource) and a Behavior Management Model for the school, using the list form of the models. Include within the lists, instances of functional subsystems to orchestrate the execution of PMFs.
- Using ISM, or similar, identify relationships and interactions between the instantiated Function and Behavior Management subsystems, and develop the resulting subsystem-interface information as an N^2 chart.
- Configure the N^2 chart to minimize entropy, i.e., reveal clusters, links and functional architecture
- Develop the architecture from the N^2 chart as a functional block diagram, connecting relevant subsystems to their corresponding Prime Mission functions.
- Connect the whole to other systems (education administration, social systems, etc.) to form an overview of the school within its future environment.
- Hence, create a design for the new school.
- Penultimately, consider whether, or not, the design that you have created is simply a reflection-in-memory of your own school, or if there are new, and perhaps unexpected, elements in your design.
- Finally, propose measures of effectiveness for the new school, and justify your design.

11
SM6: Optimize Solution System Design

The optimist proclaims that we live in the best of all possible worlds; and the pessimist fears this is true.

James Branch Cabell, 1879—1958

Approach

Optimizing a solution system design means making it as good as it can be in context, particularly in terms of its performance and effectiveness. Optimization, then, enhances and refines, so that optimization may be seen as part of, or perhaps an extension to, solution systems design, making the design even more relevant to the situation in which the SoS will find itself.

As we have already seen, from 'The GRM and the Systems Approach' on page 135, optimizing can seem a complicated business. The system-to-be-optimized interacts with, and adapts to, other systems in their mutual environment, and vice versa. So, change the effect that system A has on System B, and System B is indeed changed. But, that means that the effect that B was having on A may also change, perhaps reinforcing the initial change, perhaps neutralizing it, or perhaps doing something quite unexpected. (This is Newton's Third Law at work, or more generally Le Chatelier's Principle.) And that is just with only two systems interacting; suppose there are three, or four, or a dozen other systems, all mutually interacting dynamically, all changing- and being changed by the others. (For the implications of Le Chatelier's Principle see Understanding open system behavior on page 12, Organismic control concepts on page 20, and the Principle of System Reactions on page 54.)

It would be easy to throw one's hands in the air and declare the problem too complicated to address. Some systems engineers actually disparage optimization, declaring that it leads to 'gold-plating,' i.e., making the solution system expensive and overblown: interestingly, that should be the precise opposite of the truth. Functional, or performance optimization is one of the issues central to the practice of systems engineering; walking away from it is not be a sensible option.

As a less-than-ideal alternative, systems engineers may conduct some tradeoff analysis, considering the solution system as a closed, isolated system. The idea is to generate a range of solution options, and a set of criteria by which to compare the options, and then to conduct a tradeoff to see which of the options best fits the criteria. The problem lies in knowing whether or not the statically selected option will actually deliver the goods when the SoS interacts dynamically with other systems; chances are, it will not. But, suppose it were possible to adjust the system-to-be-optimized while it was operating interactively, and to conduct tradeoffs at the same time, so that the changes in dynamic behavior, performance and effectiveness could be observed, and the 'best' results applied and retained . . . .

Methods, Tools and Techniques

Cost and capability

By this stage of the systems methodology, the 'nature' of many of the design subsystems will be evident. For example, it will be evident that many of the prime mission functions can be conducted by people, or perhaps by some technological system or artifact. While the nature of the subsystem might be evident, the specifics of each subsystem will not be evident: it might be clear that some kind of missile will be needed, but not type or range; radar will be needed, but not which kind, transmitter power, receiver sensitivity, etc; a team of sappers will be needed, but without knowing how many in the team, and their capability; and so on. Part of the optimization process concerns itself with filling in these blanks: at the end of the process, we should know the 'emergent' specifics of the various subsystems.

This presents a problem, in that the process has to be able to 'try out' — in simulation — the effects on the whole of having a bigger, or smaller, missile; a more-or-less powerful transmitter; a more-or-less capable team of sappers, and so on. It may be important, too, to know the relative costs of having more of this, or less of that, since the measures of effectiveness (MOEs) that will be used to evaluate the SoS may well include cost-effectiveness, cost–exchange ratio, profitability, return on capital employed (ROCE), and so on.

One way of addressing this issue is to derive models of such subsystems that relate performance/capability to likely cost. Analysts, who can forecast how much a future aircraft is going to cost by extrapolation from past and present aircraft costs, develop such models routinely. Similarly, there are predictions for growth in computing power, range of weapons, cost of avionics, etc., etc. There is, indeed, an industry dedicated to such speculative parametric analysis, and we can use their output in the optimization process to predict both the capability and cost of, particularly, technological subsystems.

Optimizing the whole

It can be shown that optimization is practicable only when addressing the whole system; trying to optimize part of a system may well result in deoptimizing the whole. Without going into the theory, this is apparent from the orchestra analogy: optimizing, say the brass section of the whole so that it gave the best, and possibly therefore the crispest and loudest, rendition would unbalance it, compared with other sections, and would deoptimize the whole orchestra. It is, indeed, an interesting observation that an optimum solution (orchestra) is generally comprised of suboptimal parts (sections); such an observation is unlikely to impress the musicians! Unless, that is, they realize, that to be suboptimal, in that context, is to listen and adjust your playing so that it is compatible and harmonious with that of other musicians in your own and other sections. In musical terms, an optimal musician might be a soloist, which may render her or him unsuited to play in an orchestra.

Recalling the analogy of the Formula One car made up from the best bits of other cars, it is unlikely that putting the best parts from different cars together will produce an optimum/best result, if only because the parts have not been designed to work together in a balanced, harmonious way. Gear ratios that worked fine with one car might prove less than ideal for the composite car. Traction controls that permitted maximum acceleration at the start of the race in another car, might not give best results in the composite car; and so on. Note, in both analogies, that optimum performance is observed only under testing, dynamic, operational conditions — just where you need those emergent properties, capabilities and behaviors, and just when the whole has to be greater than the sum of the parts.

Seeking an optimum configuration can seem difficult. Consider a system comprised of a number of interacting subsystems: changing the properties of any subsystem is likely to impact on other subsystems, and the change induced in them will both feed back to the initial subsystem and may feed forward, with variations, to other subsystems. Essentially, making a sudden change can result in a series of reverberations, and as the parts reverberate, the emergent properties, capabilities and behaviors of the whole system are changing, too.

Too far fetched? Things don't work like that? Consider the effect of a sudden hike in the price of petrol/gas on a city's commuter transport system. First, there will be protests and outcries, while at the same time long queues will form at the gas stations as people fill up and some attempt to hoard fuel. Next, there will be a marked increase in people commuting into the city by train, and a significant reduction in traffic on the roads. Increased fuel prices will be passed on to retail commodity prices, so people will use their cars to go to the mall, or the supermarket, and stock up on foodstuffs, causing local traffic chaos at the malls, food shortages, panic, and an increase in local traffic accidents.

Gradually, as the raised price persists, and noticing the reduction in commuter road traffic, some people will go back to using their cars for commuting, flirting perhaps for a short period with car-sharing, but eventually going back to solo car travel — after all, they prefer the freedom of being isolated in their own car in a 10-mile traffic jam to sitting or standing on a commuter train crowded with other people. Meanwhile, there is a slow drop in the price of fuel as politicians realize that reelection is looking less and less likely. Next there may be an over swing, as people who usually commute by train decide they have had enough of the exacerbated overcrowding on trains, so they switch to the roads, causing enormous traffic jams. And, finally, things settle back to where they were originally. . . , but with a slightly raised cost of fuel.

Eventually, the commuter transportation system will return to an uneasy dynamic stability, with some overcrowding on the railways and some traffic jams on the roads and highways. It is hardly an optimum, but it is a kind of balance — perhaps 'the best in the circumstances.' And the system illustrates the optimization issue. If you change any one subsystem in the interacting set of subsystems, it will affect the other subsystems, and they in turn will affect it and each other.

One solution to sorting out the tangle is dynamic simulation, not only of the system-to-be-optimized, but also of the other systems with which it interacts at the same time (which is the systems approach, of course). Given such a dynamic systems simulation, it is possible to change any one of the subsystems, to observe its effect on the whole, the interactions with other systems that change everything, and eventually to observe the new point of open system stability — all in simulation, of course; it would be generally be far too expensive, disruptive and time consuming to do anything else. But, how do you find the *optimum* configuration?

One way is to design the solution system very much along the lines of other, existing systems. This has the advantage that it tends to reduce risk, and it employs familiar skills and capabilities: on the other hand, it inhibits innovation.

A second approach is to alter a subsystem or component (easier in simulation), observe the effect on the whole, in terms of, say, cost-effectiveness, and to set the component parameter value to that which gives the greatest increase in effectiveness. Having done that for one subsystem or component, the process may then be repeated for others, gradually improving the chosen measures of effectiveness (MOEs) each time. Although this is a pragmatic approach and can give reasonable results, a little thought may suggest that it need not give the best result, since each change alters the point of balance between the interacting subsystems, so that each change can affect previous changes, rendering them no longer optimal. The situation can be ameliorated by 'going round again,' i.e., by repeating the process of altering each subsystem or component in turn to find the overall best result in context — the optimum.

Another approach is to employ cumulative selection to cut the Gordian knot of complexity. Suppose we have a system, Blue, with N subsystems, operating and interacting with other systems, and the objective is to find the optimum configuration, where optimum is identified as, say, achieving the greatest effect on one or more other systems, Red, in the simulation. We can start by changing each of Blue's subsystem in turn by a small amount, measure the change in effectiveness, and then return each subsystem to its original setting. At the end of N changes (for N subsystems), we should have a record of the changes in effectiveness cause by each subsystem change. One of the changes will be greater (inducing greater effectiveness/performance improvements) than all the others; incorporate that selected change to that subsystem, leaving the others at their original setting. Then repeat the exercise, each time increasing the overall dynamic effectiveness of Blue by a small amount, until it cannot be increased any more.

The result of this cumulative selection process should be the identification of the best configuration of Blue to give it maximum effectiveness while it is operating and interacting with, and adapting to, other systems in their mutually open system environment. There can be problems with such a simple method, however. . . it is possible to find 'local maxima,' i.e., local optima that trap the optimization process and prevent it finding the overall maximum. Furthermore, although the cumulative selection process might be undertaken manually for a very few subsystems, it would become very cumbersome for a large number of subsystem, interconnections, etc.

It is possible to liven up the process by using so-called genetic methods. Using this approach, the design is replicated using 'genes' to 'code for' different aspects of the design; there might be a gene to code for a function, or more likely to code for the 'magnitude' of a function. Another gene might code for the existence, or not, of an interaction. Yet another might code for the capability of some subsystem or part, so that the numerical value of the gene corresponded to the degree of capability (speed, power, energy, reliability, efficiency, etc., etc).

Given a comprehensive set of genes, it is then possible to create an instance of the solution system in a dynamic simulation, to test this solution system instance in operation, and to evaluate its effectiveness when interacting with other systems, also represented in the simulation. The 'genome' can then be 'evolved' to find the optimum design: examples follow.

Disaster relief example

Suppose that the system of interest were an international disaster relief organization, and the objective was to set up a disaster relief project with the purpose of saving as many lives as possible within the disaster area, within a given amount of time. The simulation might then represent the disaster area, Red, with its geography, damaged infrastructure, floods, landslips, dead and dying of thirst, starvation and disease, etc., and it would also represent a notional disaster relief organization with genes coding for agency staff, vehicles, bridge builders, helicopters, field hospitals, secure storage depots, escorting troops, etc., etc; in effect, the full set of genes would code for all of the elements of Blue, the disaster relief organization,.

On first running the simulation, the disaster relief operation would produce initial results, with some lives saved, together with some resources wasted, and others not available in time or location. (A useful way of simulating the in-country activity might be to use intelligent cellular automata, moving, acting and interacting across a terrain map, set to present the contemporary, disastrous, in-country situation.)

The (simulated) performance of the disaster relief organization could then be improved and optimized by providing more bridge-building facilities, perhaps, or by providing less in the way of, say, bedding and shelters, but using the transport to convey more food, potable water and first aid instead. To find the optimum mix, the following procedure might be employed.

Randomize all the 'gene' values, run the simulation, record the results (numbers of lives saved within the set period). Randomize all the genes again, and repeat the exercise. And again; and again. Repeat, perhaps, one hundred times or five hundred times, recording the results each time. Then compare the results and see which combination of genes resulted in most lives saved in the least time. Incorporate this particular gene combination as the new baseline systems design set: repeat the exercise for another one hundred runs, varying the genes randomly about their new baseline values; and so on.

Gradually, and progressively, the simulation accumulates the combination of genes that saves the most lives in the least time. Stepping through the simulation, observing how it works, and understanding why it works better than other configurations can confirm whether, or not, this result is optimum. (Sometimes the results of such simulation runs can be counterintuitive, but are likely to be broadly correct, nonetheless — always provided the simulation is accurate, current and complete.) If the original problem facing the systems methodology was 'how best to organize and equip the disaster relief force,' then the simulation should provide the acid test of the solution system design — does it eradicate the symptoms of the original problem? So, this genetic cumulative selection process would simultaneously justify and optimize/validate the overall solution system design.

The process can be widely applied, and is by no means limited to disaster relief: it is also capable of development. Once an optimum design has been proved, new genes can be introduced to code for facilities that might be considered 'potentially useful;' to see if their inclusion would be justified. Some genes may be driven towards zero values, indicating that they are adding nothing to the design performance of the overall solution system. Affordability can be included, so that optimization might focus on maximizing the ratio of effectiveness to cost, so giving a 'best value for money' solution; or, perhaps, optimization might be aimed towards minimizing casualties in war, enhancing cost–exchange ratios, etc. As many disaster relief organizations are charities, or nongovernmental organizations (NGOs), the way in which they fund projects can form part of the simulation, using publicity from ongoing disaster relief work to raise awareness and encourage donations; this ties together dynamic interactions between the disaster-stricken country and the nations trying to bring relief: and so on.

The naval destroyer example

Figure 11.1 shows the overall procedure within the systems methodology for optimizing whole solution system designs. The process is, in practice, easier to execute than to explain. We have seen how one might optimize the design of a disaster relief operation. Consider, as a different example,

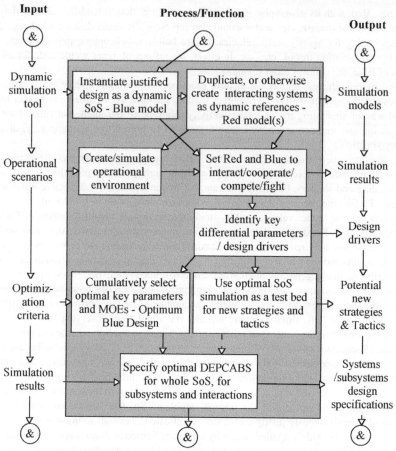

Figure 11.1   Simulation-based dynamic optimization of the solution system design. See text.

how it might be possible to optimize the design of, say, a naval destroyer. Unlike the disaster relief organization, there is no defined Red system, or set of systems, with which to interact. The new destroyer might come up against a wide variety of future opponents, and little is known about future opposition at design time; intelligence on such matters is notoriously uncertain. So, how to proceed. . . ?

One way is to develop a robust dynamic simulation of Blue (see The GRM and the Systems Approach on page 135 *et seq.*), the new destroyer, using as numerical values for the instantiation taken from current destroyers and current equipments and facilities, such that the destroyer performance would be likely to match that of contemporary vessels. Once that simulation model is tested and established as Blue, it can then be duplicated and nominated as Red, an opposing destroyer. Red and Blue can then be cross connected within the simulation such that each can potentially sense the other with radar, communicate with the other via radio and data link, intercept and exploit the other's transmissions, fire at each other when within range, inflict and experience damage, attempt to repair that damage, run out of ammunition, etc. The environment for both ships would include sea states, radio transmission characteristics, and so on. The crews for each ship would be presumed trained to the same degree, operating to the same doctrine, so that command and control for each ship would make the same decisions under the same circumstances.

Once the model had been set up and verified, Red and Blue could then be set to combat, sensing each other at sea, approaching to within engagement range, and then engaging — or not — according to the rules of engagement (ROE). Each ship may engage the other, may inflict damage, and may receive damage. Each ship may try to repair damage, to restore full capability. The instantaneous effectiveness of each ship will, therefore, be constantly changing. Chance and probability dictate that the outcome from any single simulation run is unpredictable; however, since the two destroyers are identical at this point, the results from many engagements should be identical — at least, they should average out to be identical over, say, several hundred simulated engagements.

Such a symmetrical model can be used initially to examine the effects of varying individual parameters. Keeping Red unchanged, individual parameters in Blue may be changed, to see what effect if any the change has on engagement outcome. Not only can material differences, such as reduced number of missiles carried, or increased radar transmitter power, be tested, but also such imponderables as the level of training, and changes in doctrine and ROE can be evaluated, too.

And, the results of such tests may not be as expected. For instance, one might expect increasing radar transmitter power to give Blue an advantage over Red, which remains at nominal power. In a typical simulation, that may be the reverse of the outcome. Why? If Blue's transmitter power increases, Red can detect Blue from a greater distance using passive radar sensors (inverse square law, as opposed to active radar's inverse fourth power law), enabling Red to fire at Blue before Blue is aware of the threat.

Similarly, it is possible to conduct trade studies. It may be possible, at least in simulation, to trade reduced training time, and reduced number of personnel for increased automation and smart technology on the missile system. The trade study would see changes made to Blue only, with an unchanged Red acting as a dynamic interactive reference system.

That is only the beginning, however. Using the same technique of cumulative selection as detailed above, Blue ship's design may be optimized such that it will engage Red with a greater probability of success. Optimization in this context requires an optimization parameter, often a ratio, to maximize or minimize. For two naval destroyers, the optimization parameter might be the ratio of Blue cost to effectiveness, the Blue–Red cost–exchange ratio, the Blue–Red casualty–exchange ratio, or perhaps some combination of all three. To make things more realistic, it may be prudent to expand the number of vessels on each side, introducing support vessels for refueling and rearming at sea, for instance, together with other war machines, surface, subsurface and airborne: see 'Instantiated layered GRM' on page 138. If there is one available, a traditional, trusted, many-on-many battle simulation may be used as a test bed on which to mount the developing, optimizing simulated Blue–Red designs

Having optimized Blue using Red as a dynamic reference, it is possible to turn the tables, freeze Blue's design and then optimize Red to be better than the improved Blue design. This may be referred to as ratcheting the designs, using each in turn to raise the game of the other. (The process has obvious risks that are readily anticipated, including those of violating the laws of physics and creating designs that cannot be built. . . .)

Optimizing supply and logistic systems designs

The notion of developing optimal solutions by setting systems designs into competition is not confined to combat scenarios. See Figure 11.2, which shows two lean volume supply systems competing for market share, perhaps on a global scale, as in motor vehicle manufacture, white goods, or brown goods. If the design and operation of a single lean volume supply system were represented in simulation, including the market with its facility to buy or not, to recycle, or not, then the simulation model of the complete loop could be duplicated and cross-connected, taking account of the availability of skilled labor, material resources, etc. Dynamic simulation of the two, initially identical, competing circles should then result in equal market share. Red may then be employed as an open dynamic, interactive reference, to make changes in the design of Blue, and to test for the likely outcome from those change over a number of simulation runs.

Using the cumulative selection technique allows optimization of the Blue design, still using Red as an open, dynamic interactive reference. Typical Blue optimization parameters, shown in the figure, might include minimum inventory, minimum cycle time, minimum waste and minimum pollution, or a combination of all four . . . .

The result of such a simulation would be the design of the optimum Blue supply circle, assuming that Red did not change — the result would be, in effect, an Ideal World design, and the differences between the present Blue design and the Ideal World Blue design could be used as a driver for change in Blue. However, in the real world, Red (always supposing there is a competing supply chain, which is usually the case) would not stand still while Blue continued to improve its design and operation: on the contrary, Red could also seek to improve its performance, to increase its market share.

It may be possible for Blue to anticipate Red's improvement using the ratcheting technique described above. First, optimize the design of Blue, keeping Red constant. Then freeze the improved

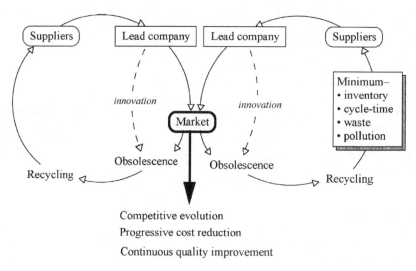

Figure 11.2 Competing supply chains/circles, Red and Blue respectively, mutually enhance each other's performance to the benefit of customers and the environment.

Blue design, and optimize Red. Finally, reoptimize Blue, this time using the enhanced design of Red as the open, interactive, dynamic reference.

In principle, this seems to anticipate changes in Red, enabling Blue to compete effectively over time. However, there may be a better way. If, instead of assuming that Blue and Red are discrete, competing, open, adaptive systems, assume instead that both Red and Blue together constitute a composite supply system, competitively providing goods into a market place or consumer system, which not only consumes, but also gives back obsolescent goods for recycling. The market will have characteristics such as the ability to become saturated, to be energized by innovative products, etc., to be more or less active according to degrees of buyers' disposable income, etc.

With this revised concept, it would be feasible and sensible to evolve the design of the composite supply system, i.e., Red and Blue as one, using the market as the open, dynamic, interactive reference. Unlike the ratcheting method, this approach has the potential to show how the co-evolution of Red and Blue might progress step-by-step, in parallel — which may be more realistic. The optimizing parameter might be different, too. Instead of emphasizing ever-leaner performance, this higher-level approach might consider, for example

$$\text{Product Utility} \div \text{Environmental Impact' } (= \text{ 'Value per Real Cost')}$$

and Energy consumption/dissipation around the make, use, recycle loop, which might have a zero sum energy goal?

Understanding the Design in Context

Once the optimization processes have been completed, the simulation models contain various parameter values which together determine the optimum solution system design, that is the best solution in the context of the environment and the other systems with which the solution system will interact in the future. So, all that has to be done now is to create a real solution system with those parameter values, and job done, right? Not right! Throughout the design process, the ability of the solution system to interact with, and adapt to, other systems has been emphasized. Those other, interacting systems adapt, too, as a concomitant aspect of the interactions. If we had tracked the value of any particular parameter, we would have noticed that it had changed during the optimization process: it may have increased, decreased, varied cyclically, or whatever.

This suggests that, were we to realize a solution system design that had been optimized while interacting with a particular set of systems in the solution space, and were that solution system to then interact either with a different set of systems, or even in the short term not to interact at all, then the optimum parameters we had so carefully identified and manifested would be inappropriate. Reconsider the example of the Blue destroyer with its active radar power being too great, so that an opposing Red ship could detect the Blue ship using passive sensors. Optimum design of Blue would not, as a result, have a reduced transmitter power, since there may be other circumstances where more power was essential (e.g., sensing through heavy precipitation or jamming, operating against ships with no passive sensors, etc.) Instead, the optimized design would enable the radar power to be varied according to situation and circumstance, with the power levels being chosen tactically by operators; perhaps automatically, such that minimum power is transmitted consistent with detecting returns; or, perhaps the degree of spread in the transmitter's frequency spectrum might be varied in concert with the receiver's correlation system; or whatever.

Similarly, optimizing the design of a disaster relief organization such that it was best suited to (e.g.) saving lives in the context of a particular disaster (hurricane on a tropical island) would not imply that the same disaster relief organization would be ideal in other circumstances (tsunami in South East Asia). Instead, what emerges from such considerations is the notion that optimization is context dependant, and since the same solution system may find itself in a variety of contexts, some of them unexpected, it follows that the solution system may have to continually re-optimize, or be re-optimized, according to situation, and at rates corresponding to the rate of change of situation.

To be Linear or Nonlinear: That is the Question

Some systems find their own optimum condition — they are self-optimizing; homeostasis also balances systems, often automatically. The various organs in the human body, for instance, all act and interact, with each 'perceiving' the sum of the others to be its environment. Notably, such organic subsystems are nonlinear, in the sense that reaction to stimulus is not always in proportion to that stimulus. This nonlinearity takes several forms. Often, subsystems for secreting enzymes, hormones, proteins, etc., have a cutoff level: no matter how great the stimulus, the output cannot exceed a certain level. Other bodily systems react, for example, logarithmically, so that they are sensitive to small stimuli, but proportionally less sensitive to larger stimuli. Closing down the iris protects the eyes so that we may see well in strong sunlight, but our eyes can also operate quite well at night with the iris opened wide trying to catch every photon. The ears similarly have a logarithmic sensitivity to sound, such that the dynamic range that the ear can accommodate is indeed wide, from the proverbial pin dropping to nearby thunder. In both instances, the sensory organs adapt to their immediate environment without apparent conscious control. Homeostasis maintains the state of a system without negative feedback: and, there is control through opposing influences or forces, such as agonistic and antagonistic muscles — again, without negative feedback, which tends to linearize systems behavior.

Overall, then, a body develops a point of balance such that all the various organs are active, within their operating range, and the whole might be considered to be in a self-organized, optimal state, with the whole alert and active in its, his or her environment. Situations arise where the sensors and other organs may be called upon to accommodate extremes in the environment, which may tend to induce extremes in some of the organic subsystems. As we have seen, these extreme demands stimulate nonlinear responses, such that no internal damage can be effected by excessive responses. Moreover, the body as a whole moves to a new point of balance, a revised optimal condition, according to each new situation, with little or no conscious involvement.

This is not to suggest that the body's nonlinear, closely couple interactive systems are perfect. Situations can arise where internal sensors can become satiated and insensitive over time, causing problems such as (late onset type 2) diabetes: optimum balance cannot always be maintained, and sufferers may collapse. And, of course, the body can be sensitive to pathogens, which may disrupt operations — hence the body's sophisticated immune system, which absorbs a significant proportion of our internal energy in maintaining our steady state.

Human activity systems are generally nonlinear, and so too, therefore, are sociotechnical systems. Consider, for example, an airliner. Its technological subsystems are engineered, and are generally linear, or at least quasi-linear. Without any crew, the airliner might be seen as an example of linear technology. Add the crew and the whole now becomes a nonlinear sociotechnical system, because its purpose and behavior as a whole are governed by the purpose and behavior of the crew. They may be using linear technology, but the way in which they use it makes the resulting operation of the whole nonlinear.

Looking at Nature's nonlinear systems, and comparing them with our manmade linear technological systems, it is evident that Nature has the edge in many areas. Nature's 'designs' have greater power density (pack more power into a smaller volume): they are often more durable, self-repairing and self-healing, too. It is possible to emulate Nature's successes by conceiving, designing and creating nonlinear technological systems, and some engineers are beginning to do that.

Interestingly, because the systems methodology follows synthetic, organismic and holistic principles throughout, the optimized designs for solution systems that it generates can be realized by manifesting nonlinear systems, perhaps with some nonlinear and some linear interacting subsystems.

Verification and Validation

The use of dynamic simulation models for evaluating and optimizing solution system designs poses the question: 'How good are the models?' It is common practice for both customers and systems engineering practitioners to decry simulation models as inaccurate, and hence invalid and not worth the money that they undoubtedly cost. To some extent, such claims to invalidity can be justified, but they also raise the question 'Is it, then, better not to model, and effectively to guess at the right answer?'

It is important to remember that dynamic simulation models, of the kind referred to throughout this book, are not mathematical models, nor are they engineering models — they are systems models, and moreover they are models of systems that are interactive, mutually adaptable, and open to their environments. If a system were to be instantiated as a team of men, a small company, or a political party, we would not expect the optimal design to be numerically precise and accurate — if it were, we would suspect that the results from the simulation had been misinterpreted. Any such team, company or party is likely to have variable numbers of people, with varying skills and capabilities which will vary over time. A team might have been 'designed' by some systems methodological process, but that would not make it a fixed entity. Teams are not static any more than the individuals comprising the team are static; instead they are developing, evolving, maturing, etc, or 'forming, storming, norming, performing and adjourning' to suggest five stages of team development (see Tuckman, 1965). Moreover, we are all aware that a good team is less about numbers, more about the right mix of personalities, skills and capabilities — as a systems capability simulation would indicate.

Simulating the design of solution systems is, moreover, predicting the future — it is inherently difficult and prone to error, inaccuracy and mistake, if only because all of the various systems of future interest may not turn out as the simulation models fondly imagined they would be.

So, while there is no excuse for inaccuracy, and certainly none for lack of rigor, the goal of validating system design models, in the conventional sense of software validation, may be untenable. It may be possible to verify models in some degree, largely by comparing their behavior with that of extant, real world systems — indeed, the models generally came from the real world in the first instance, courtesy of systems science.

There are some solution design models that cannot even be verified satisfactorily. Consider, for instance, a model that purports to show the restoration of the rule of law in a war-torn country, such as, say, Afghanistan. Suppose that courts, prisons and police stations have been largely destroyed, that there are no longer any judges and lawyers, and that police are few and far between. How might we simulate the progressive restoration of the justice system?

In some countries, it might be reasonable to simulate the dynamics by observing that the provision of sufficient money would result in builders coming forward to build the various premises, recruiters coming forward to produce the required numbers of people, and teachers and educators coming forward to teach new lawyers, judges and police — and all within reasonably predictable timescales. But, the question then becomes: would the provision of money have the same effect in Afghanistan, where the culture is different, where the law is (likely to be) Shariah law — which treats money, and the lending of money, in a way that differs from western practices, etc? Would members of the population who would make good builders, lawyers, judges and police come forward in response to money from, say, the UN and, if they would, how long might they take to do so? Apart from any other consideration, the notion that putting up money and expecting people to volunteer to spend it, is very much one of democratic, free-market economics: it may not apply in a country with many nomads, areas ruled by warlords, others by the Taliban, and so on.

Thinking about such imponderables suggests that system models for such situations must be, at best, uncertain. So, instead of constructing models with a view to validating them, mathematical and engineering style, the concept is to build simulation models as experimental learning laboratories in which, because they are systems models, the solution designs interact with, and adapt to, other systems and their environment. Moreover, such models are used to assess behavior across a wide range of situations, which may not indicate how well they might work in some situations so much as indicate those in which they may not work well, if at all — so-called areas of counterintuitive behavior (see Systems thinking and the scientific method on page 73).

For Afghanistan, the rule-of-law restoration simulation might include the facility to vary the time taken to recruit and train judges and lawyers over considerable limits, for instance, so that different runs of the simulation could assess the different possible outcomes. Running such a systems simulation might indicate the value of importing judges, lawyer and police temporarily from other Moslem nations, for instance — at least as a working hypothesis. Similarly, shortage of prisons, law courts and police stations might be alleviated — at a cost — by introducing temporary (mobile?) accommodation, the value of which could be assessed by running simulations with various options. In this way, the 'design' of the best solution is not just being tested: it is being progressively developed and optimized, too, in simulation, see Figure 11.1.

Summary

Optimizing the design of a solution system is an essential part of 'solving the problem;' optimizing identifies the right parts, brings them together in the right way, to interact in the right degree, to create the requisite emergent properties, such that the whole solution system is greater than the sum of its parts. Optimizing is, then, essential systems engineering.

Optimization can be effected only at the level of the whole system. Seeking to optimize only part of the whole may de-optimize the whole, and is likely to be counterproductive. Difficulties in optimizing the whole may be anticipated, since changing any of the parts not only affects the whole directly, it also affects the other parts with which it interacts, and they in turn may affect other parts (transitivity). Any of these direct and indirect effects may also change the whole.

Because of this potentially complex behavior, optimization often resorts to dynamic simulation of the solution system in context, interacting with other systems so that they are all open and mutually adaptive — as in the real world, of course. Change, action, reaction and transition towards optimal balance are more easily observed in simulation than in the real world, where time, cost, and risk may inhibit, or even prohibit, trial and experiment.

Even in simulation, finding the optimum set of conditions and parameters for a complex solution system design when interacting dynamically with other systems may not be trivial. The potential number of combinations may prove enormous. It may be tempting to reduce risk by designing a new system very much in the image of its predecessors, or other extant, successful systems. An alternative, innovative approach is to use cumulative selection to explore this vast landscape of possibilities, in a greatly simplified version of Nature's genetic approach: in effect, to 'evolve' a solution system design. Examples of how this can be sensibly achieved are presented in outline. (Case studies will be presented later showing this approach in some detail.)

The result of such system optimization approaches is an 'optimum design set' of dynamic emergent properties, capabilities and behaviors (DEPCABs), which are appropriate, in principle, to the whole solution system operating in a particular context. In the event, the solution system may not find itself in that particular context, suggesting that the optimum solution system design should be determined across a range contexts or, alternatively, that the solution system should be designed so that it can optimize itself according to the context in which it finds itself. This is the way in which Nature's systems, such as the human body, operate; in general they rely on closely coupled, nonlinear subsystems which have the ability to find a point of balance such that the whole body is capable over a wide range of contexts. Systems as teams of people, enterprises, etc, and sociotechnical systems, are nonlinear and can effectively self-optimize . . . . Technological systems, on the other hand, tend to be linear and less able to self-optimize in this way, although engineers are getting to grips with the design of nonlinear technological systems, which promise improved power density, flexibility, adaptability and capability.

Design optimization using dynamic simulation models brings into question the validity of such models. Systems models are neither mathematical models nor engineering models, however, and the same rules do not apply. The models of systems are, like the systems themselves, open and mutually adaptive to their environments and to other systems. As a result, the optimized system design has to be interpreted, and should result in a solution system that is also open and adaptable, rather than having one, fixed set of parameters. Given a real world situation like that in the simulated world, the adaptable solution system should be capable of operating with the corresponding optimum parameter set — whether adjusted automatically or by operators. But, for different real world situations, the solution system should be able to find different balance points, different optima, appropriate to those different situations. Solution system design then becomes concerned with how those various optimum configurations can be set up as needed, and what range and variety of situations the design should accommodate.

Assignment

A system is to be set up within a national airline to provide through-life maintenance and servicing support to a family of commercial medium-haul airliners, operating from more than two dozen airfields, large and not so large, spread across the continental US.

1. You are tasked with optimizing the design of this support system; your first challenge is to select three potential optimizing criteria, or parameters, such that the support system affords the best support in the context and circumstances.

- You should keep in mind that the cost of the support system is part of the operating cost of the airline, and as such potentially eats into profits.
- On the other hand, inadequate support could result in major, even phenomenal, expense, in terms of loss of reputation, customer confidence, revenue, and even possibly loss of life — with the consequent potential costs of compensation.

2. Your second challenge is to consider how to employ these three optimizing parameters. Should you use them independently of each other and compare the results: if so, how? Or, should you create a composite optimizing parameter and optimize the support system using this parameter?
3. Explain and justify your preferred approach.

12

SM7: Create and Prove Solution System (SoS)

Prove all things; hold fast that which is good.

Bible, Thessalonians

Introduction

Previous chapters have developed a functional/physical design for a solution system, with purpose, functions, behaviors, architecture, configuration, etc. This chapter sees the systems methodology continuing towards tangible creation and operation — supposing that, for the time being, to be the requisite outcome.

The general approach is outlined in the behavior diagram of Figure 12.1, and commences by confirming the system and subsystem design specifications, which were an output from prior activities. The systems methodology develops a test environment, perhaps an open, dynamic, interactive simulation model, which represents other systems with which the SoS will be interacting, plus context, environment, scenarios, etc. While in some cases this simulation might be a complex, real-time computer-based virtual reality, in others it could be a number of people, perhaps in teams, adopting the role of other (people) systems: in yet other cases, it may be possible to interact with real world systems. Importantly, the test environment, simulated or real, must be representative, open, adaptive and reactive in real time; else, it will not serve to test the design effectively.

The simulated SoS, with its subsystems and interactions, is brought together with the dynamic test environment. As the various subsystems interact, they should exhibit the full range of dynamic emergent properties, capabilities and behaviors (DEPCABs) anticipated in the design, as the whole SoS interacts dynamically with the test environment. If they behave as expected, then the design specifications are confirmed: otherwise, investigation should follow, and changes may be needed to the subsystems design specifications. The decisive test, as always, concerns itself with the ability of the SoS to neutralize all of the symptoms of the original problem. If a significant time has elapsed, allowing the original problem to morph/evolve, then the decisive test may be that of solving the evolved problem: SM practitioners will have been monitoring the changing problem situation and

Systems Engineering: A 21st Century Systems Methodology   Derek K. Hitchins
© 2007 John Wiley & Sons, Ltd

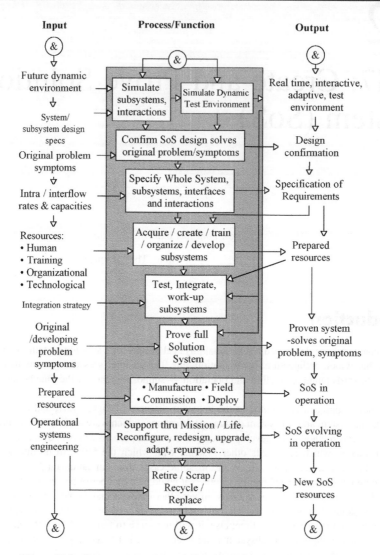

Figure 12.1 Create, prove and set solution system (SoS) to fulfill its purpose.

may/should have allowed for such changes in the developing designs. For instance, the SoS design may incorporate the ability to redesign itself. . . .

Requirement Specifications

Once confirmed, with any changes incorporated, the design specifications may be transformed into a complete 'matched set' of requirement specifications, which should specify, in detail, and with appropriate precision, all aspects of the whole solution system, of its subsystems, of the interaction infrastructure that binds and couples them, of any interfaces within the solution system, or between the solution system and the external world. And, since we are specifying the requirement for an open, adaptive, interactive solution system, it is important also to state the solution system's purpose, the problem it is seeking to solve, the CONOPS, containing system, interacting systems, contexts and scenarios; else, the designs and the specifications will be incoherent.

Manifesting Different 'Kinds' of System

The next stage in the procedure is to realize the various subsystem specifications. How this is achieved will depend on the nature of the subsystem. Some subsystems may be comprised of people operating as a team, a Human Activity System (HAS), such as a management team. Others may be people operating as a team with technological support, such as an Air Traffic Management (ATM) aircraft flight control team, where controllers operate largely via their radar consoles, control panels and radios, to identify, track, direct and ensure the safe passage of transport aircraft: this would be considered a sociotechnical subsystem. Some solution systems emerge as processes, with groups of subprocesses/activity groups, to be organized and implemented. Yet again, there might be subsystems that are purely technological, such as an automated chemical process plant, an intelligent robotic device for cleaning radioactive contamination, or a radar receiver, part of a complete radar.

The manner of realization is dependent on the nature of the system or subsystem. For HASs, it may be appropriate to recruit people with appropriate skills and experience, bring them together in the working environment and train them to work together as a team: this does not always work; sometimes an individual is not a team player, and will not fit in. Being part of any team involves communication, coordination and cooperation with other team players to promote synergy: team players are open, interactive and adaptive subsystems. . . as a result, the 'performance' of a team can develop and improve over time, and the team can adapt, develop and improve as it undertakes different tasks and interacts with other, different HASs.

For HASs, subsystem synthesis involves configuring, organizing, managing, educating, training, encouraging and leading individuals to manifest functional capabilities as required by the design requirement specification. For a maintenance organization, this might mean establishing teams of trained mechanics, technicians and engineers to detect and locate defects and failures in some operational system, to replace defective parts, to confirm the repair, to 'mend' the defective or faulty part, and perhaps to service the operational system according to a prepared schedule. There is, then, the concept of a process, or procedure, that the people follow so that the team as whole performs the necessary functions and exhibits the requisite emergent properties, capabilities and behaviors. (This is analogous to the implementation of, say, an electronic subsystem design where signals/data flow through a series of transfer functions, organized in such a way as to perform some function, part of the whole's emergent properties, capabilities and behaviors.)

For a military command and control system, synthesis would involve the formation of teams of, hopefully experienced people to undertake intelligence gathering, intelligence analysis, operations analysis, tactical and strategic planning, logistics management, decision support, decision-making, execution, liaison, etc., etc. Within each of the teams, a process would be developed and evolved for undertaking the various tasks, and for achieving the objectives of the team — which contribute, of course, to the goal of the whole command and control system. Synthesis would include developing the processes, and practicing their employment to work-up team capability; it would also include coordinating and communicating between the various teams, which would also be practiced and worked up.

The same approach to subsystem synthesis can be true for sociotechnical systems, such as the ATM flight control team. The behavior of each team member may be circumscribed by interactions with the technical elements — the integrated control console, with radar and data displays, controls, communication systems, identification systems, track labels, warning systems, procedural software, etc. So, although the human members of the team are potentially flexible and adaptable, the degree of such flexibility is intentionally limited by the technology: lives may depend on coherent, predictable and repeatable behavior on the part of the controllers.

So, the realization of a 'new' ATM flight control system might not rely entirely on the human ability of the team members to develop their team performance over time; on the contrary, team performance is required to be at top level from the word go, and the technology is employed, partly at least, to ensure that required consistency of performance. The procedures and processes built into the ATM 'machinery' must be, and are, thoroughly tested and tried. In practice, this results in controllers becoming very familiar with particular ways of working, and reluctant to embrace (risk?) change. In effect, their potential for adaptability and flexibility may be reduced. If and when some new technology or process is introduced, it may prove necessary to go through the process of working-up the team, i.e., enabling them to try out the new facilities, to become familiar and proficient in its use off-line, as it were, before 'going live.' In so doing, the experienced team members will adapt to the new system/technology, but may also find faults or defects in operation, process and procedure.

For other sociotechnical systems, the situation may seem different. A fighter interceptor aircraft with a crew of one, or two, say, forms a seemingly small sociotechnical system, until it is realized that, while airborne at least, the crew is in close communication with other fighters, ground controllers, airborne warning and control aircraft, naval ships, ground command and control, military forward controllers, etc., etc: so, very much (part of) a highly dynamic sociotechnical system. . . . When designing, developing and constructing the aircraft, on the other hand, it may seem to engineers in particular that the human connections are rather localized; after all, most of the internal subsystems are technological: engines, powered flying controls, generators and alternators, power distribution, fire-and-forget air-to-air weapons, counter-measures, and many, many more.

An alternate view of the fighter interceptor aircraft is that it performs *only* as a sociotechnical system. Without the crew, it is an inanimate object; with the purposeful crew, it becomes a synergizing part of a team of such aircraft, and it becomes dynamic, open, interactive, adaptable, and even innovative. So, another way of looking at the technology is that it extends the capabilities of the crew beyond their human limits. With the aircraft, the crew can travel faster, climb higher, see further, 'throw' further, see threats earlier, and so on. Looked at in this way, the technology can be seen as making the whole aircraft better able to contribute to the social group of air defenders; so confirming the view that the fighter interceptor is a sociotechnical subsystem, although one, perhaps, with many technological sub-subsystems. . . .

Since the systems methodology concerns itself with realizing purposeful systems to solve problematic situations, solution systems will seldom be other than social or sociotechnical systems: few machines exhibit purpose, conceive of a mission, and pursue that mission purposefully of their own volition; in the future, intelligent robots will, no doubt, do so; but not today. The machines that we build at present tend to be artifacts; tools made by humans to be used by humans. It is the humans who have the purpose, while the artifacts are at best purposive, i.e., an observer may ascribe purpose to them. A hammer, for example, has no purpose. An observer may suppose it is for hitting things — he or she ascribes purpose to it. A workman may use the hammer to achieve his purpose of driving a nail. However, he may also use the hammer to prise open the lid of a paint-pot, or to stop a piece of paper blowing away. Together the man and the hammer form a small sociotechnical system, in which the adaptability of the man extends the use of the hammer. (Hence, perhaps, the old saying: 'If the only tool you have is a hammer, soon everything starts to look like a nail.')

Organizational considerations

The various subsystems comprising the solution system may be created at the same time in the same location, perhaps as a single project; where the solution system is largely managerial or technological, that would not be unusual. On the other hand, the whole may be developed as a number of parallel projects, with each project, perhaps, creating one of the major subsystems. This may be more appropriate where the whole solution system is large and diverse; where the subsystems are significantly different in nature from each other; where the technology of each major subsystem is significantly different; and so on.

The projects may be independent, although that carries with it the risk of so-called specification 'creep,' such that the dynamic emergent properties, capabilities and behaviors of the various subsystems no longer constitute a matched set; this may not become evident until they are brought together for final test and integration. To guard against that, it may be prudent to retain a coordinating function, which regularly audits the emerging subsystems to see if they are conforming to specification. Sometimes deviation from specification can be rectified; sometimes it is unavoidable. In the latter case, there may be a need to rejig the whole design; tests using the dynamic test environment may guide the decision. Where people are part of the various subsystems, they may adapt without further ado, or may need further instruction or retraining.

Integration considerations

When the subsystems have been formed, by whatever means, they may be brought together, or integrated, with the dynamic test environment acting as the watchdog on the success of proceedings. Even at this late stage, and in spite of all previous efforts, it is not unknown for errors to emerge; some may be ignored, if minor; some may be repaired 'on the fly' and incorporated into the specifications; others may be 'showstoppers,' and may require significant redesign, if not abandonment of part or the entire project. This can arise in technological solution systems, for example, where all functions seem to be operating correctly, but the timing and coordination of interactions between functions is incorrect, erratic, etc., and cannot be corrected: it is to discover this kind of error, in part, that dynamic testing is important particularly in the final stages.

There are those who do not favor such final integration and test prior to deployment. One alternative approach is to transport the various developed parts of the whole to their operational site, and to integrate them there, going straight to operation. It can be tempting, since it seems on the face of it to save time and money. It may also prove risky, as witness the Hubble space telescope, where pre-launch test of the optical system would have revealed defects that in the event were not discovered until switch-on in orbit: such approaches may turn out to be false economies.

Instead of fully testing the whole SoS prior to setting it to work, it is also possible to 'work-up' a solution system in situ, and in 'live operation.' Here, all the various subsystems are delivered to their various sites and interconnected so that the whole operates. It may not work optimally, however, so there may follow a series of 'adjustments' to individual subsystems and to their intercommunications, to enhance whole system operational performance. Much of this is unlikely to involve technology in the first instance, since that tends to be relatively inflexible; instead, the people-elements adapt their behavior, refine their processes and procedures, adjust, or even change tactics, etc.

For complex sociotechnical systems, then, there may be a two-stage process for integrating the various subsystems:

- In the first stage, technological facilities are integrated progressively, joining perhaps two parts initially, and ensuring they work together correctly, before adding a third, and then a fourth, and so on. This progressive integration helps to avoid the situation where all the technological parts may be brought together and connected in one go: thereafter, isolating the source of faults and defects may prove difficult-to-impossible; hence, the progressive approach is designed to aid fault detection and isolation. This integration and test is best undertaken where there are facilities for effecting any redesign, repairs and reprogramming. . . .
- In the second stage, all of the subsystems, social, sociotechnical and technological, are deployed to their operational situation, comfortable in the knowledge that the technological parts at least should work as required. The human elements — the people systems — may now be encouraged to interact with each other and with other systems in the operational environment. The whole solution system will then take some time to adapt to the situation, as the human elements accommodate the changing situation, and as the whole, open, interactive SoS adapts — as it was designed to do.
- This is, in effect, post-delivery optimization — some would call it operational systems engineering — see Figure 12.1, and is common practice in, for example, multilateral forces, where force elements are brought together from different nations and expected to work as a unified whole. It also happens, on a smaller scale, when design teams are formed at the start of some major new project; team members may be drawn from many different sources, and may have many different backgrounds, experiences, methods, techniques and procedures. See also Case C: The Total Weapon System Concept.

A third example might be that of the communications infrastructure for a major military intervention such as Kosovo, where there was a complex mix of military and nongovernmental organizations (NGOs), including the Red Cross and the Red Crescent. Many different organizations were continually arriving and leaving, such that there was — and could be — no established communications infrastructure. Instead, the structure was in a permanent state of flux: this *was* the optimum state, as the infrastructure adapted its links, protocols, switches, directories, etc., to the turbulent need.

The proven solution system may be a one-off, or reproduced many times. Typically, a command and control system, a new political party, or a new enterprise might be a one-off, while a new car, tank, ship, aircraft or avionics system might be destined to be replicated many times; the latter to be installed in all of the aircraft in a fleet, or on a production line. Conceivably, then, there may be different sociotechnical solution systems, each with the same technological content, each pursuing different purposes and performing different missions: the human element in each sociotechnical system employs the same technology for different purposes.

Operational systems engineering — continual optimization in operation

Once proven, the SoS may be put into operation, interacting with other systems in some dynamic operational environment. The operational SoS, as we may now call it, will not remain constant, however: the problem it has been designed to solve may be continually morphing, and the SoS will have to adapt and evolve to match; the environment may be changing, and new technology may become available which, if introduced, might make operational behavior, support, or resourcing better, less expensive, less polluting, more secure, less energy dependent, etc.

The SoS may 'enter service,' i.e., become operational, without the design having taken account of later changes to the original problem; this is not uncommon practice where the SoS has major technological components that have may have taken some time (a decade or more would not be unusual) to develop and construct. In the short term, the human element of the delivered SoS may be able to accommodate the difference between the SoS as it is, and the SoS as it should be to address the morphed problem. In any event, there will probably be need to 'improve' the SoS, to catch up with the changing situation/problem. Unless, that is, there are insufficient resources in the kitty for the improvement; or, perhaps, the owner of the new SoS decides to continue with the SoS as it is, considering it 'good enough.' In essence, the owner may choose to resolve the problem, rather than solve it as originally intended with the concomitant expense, and the potential loss of capability during upgrade. . . .

If the SoS is able to continually reinvent itself, through redesign, reconfiguration, etc., then the SoS may survive indefinitely — although not necessarily in the same form as it started. If the SoS can adapt and evolve as fast, or faster, than the environment in which it exists, then it has the prospects of lasting indefinitely. If not, like the dinosaurs, it may die out. . . or be supplanted by a better-adapted species.

Component of the Whole SoS

The whole solution system consists essentially of three elements — see Figure 12.2. In practice, these are unlikely to appear as three discrete subsystems. Instead, they will be 'woven into the fabric' of the overall solution system:

■ The mission/operating system, that which interacts directly with other systems in the operational environment.
■ The viability management system, which ensures that the mission system is capable of pursuing the mission. Conceptually, the viability management system provides a platform, or base from which the mission/operating system may operate securely and safely.
■ The resource management system that ensures the availability of all resources, to maintain the mission/operational system, the viability management system and itself, the resource management system

The notion of the three complementary elements of the whole SoS is universal. Apollo would not have reached the Moon and back without the ground facilities around the Earth for communication, telemetry, recovery, etc. The Apollo crews would not have been proficient without crew training facilities, both for working together within the craft, and for undertaking extravehicular activities in space to inspect the exterior, make repairs, etc. The complete mission system would not have been

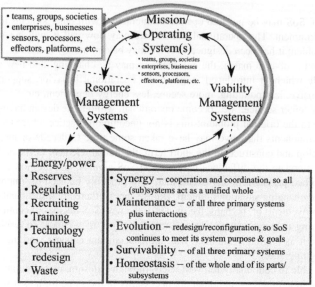

Figure 12.2 The major components of the whole SoS. In addition to the Mission/Operating system, i.e., that which directly addresses the task of solving the problem, the systems methodology generates the Resource Management System, to ensure that the Mission Operating System is continually/continuously resourced, and the Viability Management System, which both maintains the whole system capabilities and evolves/adapts them in line with developing needs and threats. All three components combine to assure that the Mission Operating System continues to pursue its mission and purpose, conceptually endowing the SoS with unlimited life.

viable without maintenance, servicing, telemetry and facilities to test for defects and failures. The public may have seen the rocket and the space vehicles — the mission/operational system — but the other elements of the 'triad' became highly visible when things did not go according to plan.

Considering an individual person as an archetypal operational SoS model:

- the brain, senses and motor controls might be viewed as the Operating System, collecting information, strategizing, planning and executing 'missions';
- the body with its skeleton, organs, nerves, immune system, etc., might be seen as the Viability Management System, providing a dependable base from which to mount operations;
- and the Resource Management System might be comprised of:
  - the food and liquid intake, processing and waste disposal elements,
  - blood circulation, carrying oxygenated blood to the tissues and muscles (energy distribution),
  - returning venous blood to the heart and lungs (waste management), and so on,
  - all needed to support the Operating System and the Viability Management System — not forgetting the Resource Management System itself.

In creating new systems, it is not uncommon to concentrate on the mission/operational system, to the detriment, perhaps, of the other two. The whole SoS may be created under a number of different project headings such that the mission/operating system is started first, followed later by the viability management system as a series of disjointed projects, and later still by the resource management system. It can make better sense, however, to develop and introduce all three as a unified whole.

For example, when introducing a new fleet of passenger aircraft, it is important to introduce at the same time all the support facilities for maintenance and servicing, for training the maintainers, for training the crews (including simulators, part-task trainers, etc.) and the requisite logistics support system to ensure availability of consumables, facilities, spares, etc. Where a new SoS is being added to an existing catalog of systems, it may appear to make economic sense to rationalize the components of the SoS such that they are compatible with those already in use. On the other hand, it may prove much quicker and less costly to keep the whole SoS support and resourcing independent of other systems.

Summary

The processes of creating the solution system and its parts are conceptually straightforward, provided the SoS design has been successfully and comprehensively achieved. To confirm this, it is prudent to confirm the design by simulating the requirement specifications for the various subsystems and their interactions, and by setting the simulated SoS to interact with an open adaptive, representative interactive environment.

Once confirmed, subsequent creation involves synthesizing the various subsystems and their mutual interaction infrastructure, and then bringing the realized parts together and testing them dynamically and interactively. The acid, or decisive, test as always will be the ability of the realized, dynamic, interactive, adaptive SoS to eliminate all of the symptoms associated with the original problem. Where the problem has morphed since the original analysis on which the design is based, there may be a need to update the design and SoS before putting it to work; else, concessions may be made and any capability shortfall made up later.

The way in which subsystems are synthesized will vary according to the nature of the particular subsystem. Solution systems and their subsystems are generally either human activity systems (HASs), such as teams, groups, societies, communities, enterprises, etc., or sociotechnical systems where people are supported/enabled by more or less technology, with continuous/continual interaction between the two. Within the technological elements, there may be discrete sub-subsystems that are entirely technological; the technological sub-subsystem will be open, interactive and may even be adaptive. . . .

The whole SoS is realized by bringing together the various synthesized subsystems, causing them to interact, and setting the whole to work in its operational context, as an open adaptive, interactive whole. There are arguments in favor of performing the integration of the various subsystems in a nonoperational context, using a simulated operational context; this has the advantage that any errors, defects or faults that emerge may be addressed more readily and with less risk in this nonoperational environment. It may also be possible to work up team performance offline, in this manner, to avoid the problems of deploying an immature solution system.

The systems methodology conceives, designs and creates whole solution systems, including not only the operation system, but also the viability and resource management systems. If these are manifested as the whole SoS, then the SoS will have the ability to continually redesign itself, to adapt to changes in the problem, or even to quite different needs. Such a system may be intended for just one mission, or for a lifetime — where, if the SoS is sufficiently adaptable, the duration may be indefinite.

Assignments

1. You are concerned with the urgent delivery of a complex, and very expensive, technological system to be launched into near Earth orbit, where it will be used for remote sensing of Earth resources and for observing the Sun. All the various subsystem elements have been developed and tested separately: all meet their specifications. Time does not permit full test and integration of all the parts prior to meeting the next launch window, in one month's time, and the next feasible launch window is a year later. You have a choice, then, between proceeding with full test and integration and missing the window, or installing the parts directly into the launch vehicle, to meet the launch window. Or, perhaps there is a third way? List the factors you would take into consideration in making your decision, and justify your choice, as though you were presenting to a congressional appropriations committee — who will wish to hear neither detail, nor 'techno-babble!'

2. Conceptually design a self-healing human activity system, i.e., one that is capable of repairing itself. Explain and justify your design. Conceptually design a self-healing technological design for a system of your choice. Explain and justify your design. In both instances, explain how you would synthesize and prove your designs, and where they might be applied in the real world.

13
The Systems Methodology — Elaborated

Géronte: It seems to me you are locating them wrongly: the heart is on the left and the liver is on the right.
Sganarelle: Yes, in the old days that was so, but we have changed all that, and now we practice medicine by a completely new method.

Molière (J.-B. Poquelin) 1622–1673)

Ideal World vs Real World

The elaborated systems methodology concerns itself, in the first instance, with establishing an ideal world solution to the problem: in the event, it could be decided that the ideal world solution, the optimum, or 'best in the circumstances' solution, might not be appropriate — see Systems Methodology: conceptual model on page 172. Instead, it may be more appropriate to move the goal posts, i.e., dissolve the problem by changing the situation which generated it. Alternatively, having discovered what the full SoS might look like and cost, and considering the degree of turbulence that would be involved in introducing it, it may be more appropriate to:

■ do nothing, and live with the problem — sometimes the cure can be worse than the condition;
■ use the difference between the ideal world solution system, as presented by the systems methodology, and the real world to act as an agenda for change.

This second option may be pursued by examining what would have to be done to the present system to make it more effective, i.e., more like the ideal world system. A program of change may then be introduced effectively to modify or upgrade a current system, or systems, rather than replace it lock, stock and barrel. The ideal world system would, then, serve as a marker, target, or template; i.e., something to be aimed for and hopefully achieved over time. In management speak; the two different approaches may be characterized as revolution vs evolution, where revolution makes major change in a short time, while evolution makes more gradual change over a longer period.

Systems Engineering: A 21st Century Systems Methodology   Derek K. Hitchins
© 2007 John Wiley & Sons, Ltd

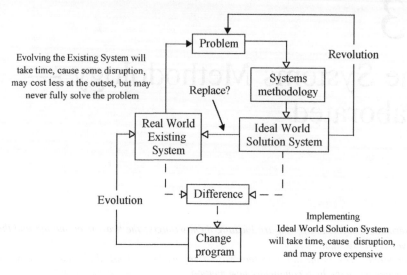

Figure 13.1　Evolution vs revolution. In the first instance, the systems methodology aims towards designing an Ideal World Solution System (SoS), which may either replace the existing system — revolution — or act as a goal design, towards which the existing system may be progressively adopted, upgraded, etc. — evolution.

The comparison is exemplified at Figure 13.1. As the figure shows, the arguments for and against evolution vs revolution can be finely balanced in terms of likely timescales, degree of turbulence, interim loss of capability while systems are either upgraded or replaced, and so on. In the final analysis, the decision is likely to be swayed as much by the willingness of decision-makers to allocate resources as by the urgency and severity of the problem. In any event, the overall systems methodology encompasses any and all of the approaches above: dissolution, resolution, or solution: and, even rationally choosing to do nothing . . . .

The Systems Methodology — as Products

The systems methodology produces a complete set of products for the SoS. These may be seen in the behavior diagrams from Chapters 7–12, and are implicit in Figure 13.2.

The Systems Methodology as a Whole

Chapters 7–12 have introduced the systems methodology as a succession of parts, with each part represented as a so-called behavior diagram:

■　SM1: addressing complex issues and problems
■　SM2: exploring the problem space
■　SM3: exploring solution system purpose
■　SM4: developing a concept of operations (CONOPS)

- SM5: design the solution system
- SM6: optimize the solution system design
- SM7: create and prove the solution system

The Systems Methodology — as Process?

Further elaboration in the form of behavior diagrams cannot be conducted here because the resulting diagrams would be too large and unwieldy. For a different perspective of the systems methodology as a whole, consider Figure 13.2. The figure shows the principal activities/processes as they might occur in going from problem space to operational Solution System (SoS). Examining the tree of processes, it is suggestive of a strategy for solving problems and creating viable, effective solutions.

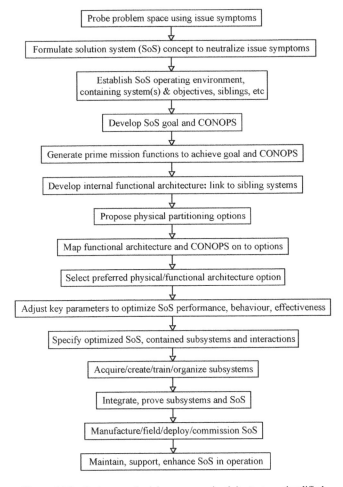

Figure 13.2 Systems methodology process / activity tree — simplified.

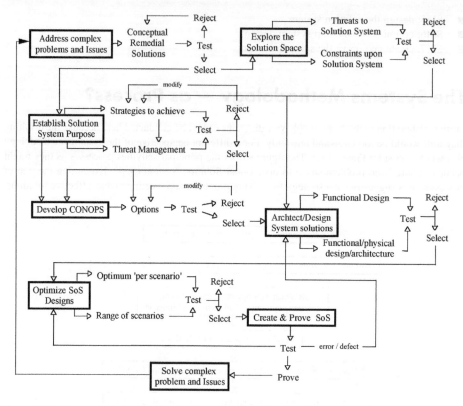

Figure 13.3 Route map, indicating the circuitous route that following the systems methodology may invoke, with testing of deductions, proving of results, etc., resulting in iterations and in the discarding of nugatory interim results.

However, the tree is not a process model, *per se*; it should not be thought, for example, that the overall process is likely to be linear. Instead, there are likely to be many recursions and iterations. See Figure 13.3, which gives some small indication of the reasons. Many of the activities and processes will produce results that have to be tested for validity, credibility, cultural acceptability, legality, regulatory conformance, etc., not forgetting their direct relevance to solving the original problem. Some tests may fail, suggesting a 'return to the drawing board,' a revision, or discarding of some potential solution or characteristic.

As Figure 13.3 suggests, for example, it would be quite possible to go around the 'conceptual remedy' loop, top left in the diagram, many times: indeed, it is by no means certain in every case that a conceptual remedial solution will be found that will stand up to scrutiny. The various iterative loops may be different in nature, too. In the case of threats, it is incumbent upon practitioners to identify threats — which may require considerable experience of the solution space, plus no small measure of insight — but at the same time not to 'over-egg' the situation by including low probability/low impact threats that may be difficult and/or expensive to address. There could be many CONOPS, each of which may need testing and evaluating, perhaps using dynamic simulations, perhaps using verbal 'walks-through;' Getting these early steps right is essential in setting the stage for effective design of the solution system; the figure shows that there can be many others, too.

Outer loop – inner loop design

A quite different kind of iteration is shown in Figure 13.4. In this case, the whole SoS design is completed at the top level, resulting in a number of complementary, open, adaptive, interactive subsystems that, together, constitute/synthesize the whole SoS. In many cases, this process/procedure will result in sufficient subsystem information to enable their creation. For some systems, however, and for some subsystems, it may be 'useful' to iterate once more, to conduct 'Inner Loop' design: that is, for one or more subsystems to be identified and elaborated in terms of its purpose (part of, and contributing to, the overall SoS purpose) CONOPS (similarly, part of and contributing to the overall SoS CONOPS), design/architecture (again, linked to the overall design and architecture of the SoS — of which it becomes an elaborated segment . . .). Such iteration would employ the same tools, techniques and methods as will have been already used, but would be more specific, would elaborate designs more fully, and so on.

Readers may recall an earlier statement that it is only possible to optimize at whole system level; optimizing a subsystem in isolation may subsequently de-optimize the whole. Conceptually, inner loop design might result in optimizing a sub-subsystem design, to the detriment of overall SoS optimization. This potential risk will not materialize if the inner loop design of subsystems and sub-subsystems is pursued with them connected and dynamically interactive; so, design elaboration requires that the subsystem or sub-subsystem to be elaborated is still an open interactive dynamic part of the dynamic, interactive, whole. This is best done, as before, in dynamic, open system simulation, where the design of the sub-subsystem can be 'adjusted' (i.e., parameters varied and selected) to sustain optimum performance of the whole.

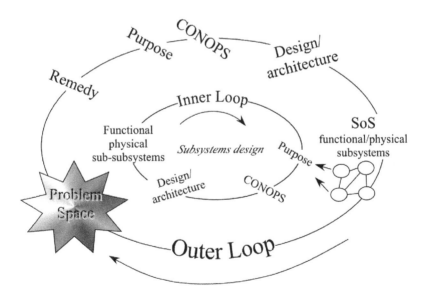

Figure 13.4 Outer and inner design loops — conceptual diagram. The outer design loop addresses the problem space and produces a solution system design of open, adaptive, interactive functional/physical subsystems. Any or all of these subsystems may then go through the same methodological process — the inner design loop. Exceptionally, it may be useful to incorporate yet another loop inside the inner design loop . . . .

Outer loop, inner loop and systems engineering

Outer loop and inner loop system design, Figure 13.4, goes some way to explain the apparent divergence of opinion about systems engineering as seen by systems thinkers and operational customers on one hand, and procurement agents and engineers on the other. It is not unusual for major customers, particularly in defense and aerospace, either to undertake the outer loop of the overall design process themselves, or to employ systems companies to do it for them. These major customers (or their system company 'customer's friend,' as they are sometimes called) may then issue specifications of requirement for some of the subsystems, and call for engineering organizations to respond, that is in effect to undertake the activities of the inner loop.

Where inner loop activities are conducted in line with the systems methodology, i.e., with each subsystem being considered as an open, interactive adaptive whole within the environment set by the other sibling subsystems and the operational environment/context, then systems engineering is clearly in operation, and the systems approach is being followed.

Where a subsystem is considered in isolation from all other subsystems, with the analysts, architects and designers unaware of, or taking no account of, the overall system, the other, interactive subsystems and their mutual operational environment, CONOPS, etc., then it would seem unreasonable to describe the activities as systems engineering; the systems approach is not being followed.

At some stage, there may arise the need to design and fabricate an artifact, with its associated interfaces, interactions, properties, capabilities and behaviors, etc. That stage may sensibly be called engineering, for that is what it is, but it may also be considered as an essential part of the overall systems engineering process — for that is what it also is. It is also practicable to adopt the systems approach when conceiving and designing artifacts, which encourages the soubriquet of systems engineering.

Discussions about perceptions of engineering vs systems engineering may also obfuscate the situation — many of the subsystems will be social and/or sociotechnical, where standard engineering practices will be inappropriate, but where the systems methodology, in the hands of good systems practitioners, will comfortably fit the bill.

The Systems Methodology — in Parts and Phases

A degree of elaboration of the systems methodology may be presented using a different form of illustration. See Figures 13.5–13.8: the figures form a contiguous set:

- Figure 13.5 illustrates SM Step 1 Explore Problem Space, and Step 2 Explore Solution Space.
- Figure 13.6 illustrates SM Step 3 Focus SoS Purpose, Step 4 Develop SoS High Level CONOPS, and part of Step 5 Design SoS.
- Figure 13.7 illustrates the second part of SM Step 5 Design SoS, and Step 6 Optimize SoS Design.
- Figure 13.8 illustrates SM Step 7 Create and Prove SoS.
- Together, the four figures present one view of the elaborated systems methodology.

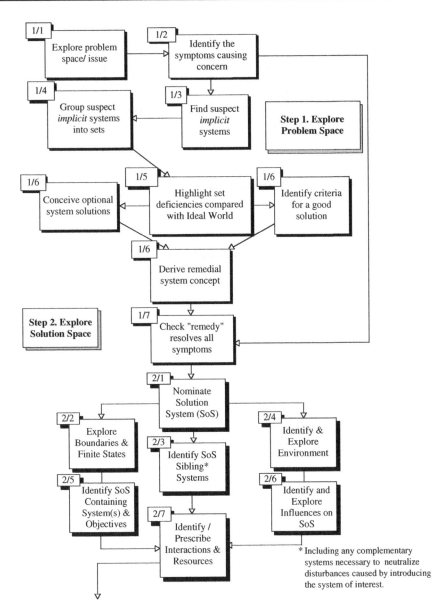

Figure 13.5 Systems methodology high-level process view, Steps 1 and 2.

Note that Figure 13.8, Activities 7.2 and 7.3, incorporates some possibly unexpected inputs, concerned not so much with the nature of the SoS, but more with the nature of the environment in which the systems methodology manifests itself. The systems methodology is a complex system in its own right; it is open, adaptive, interactive, and may even be vulnerable to threat and attack.

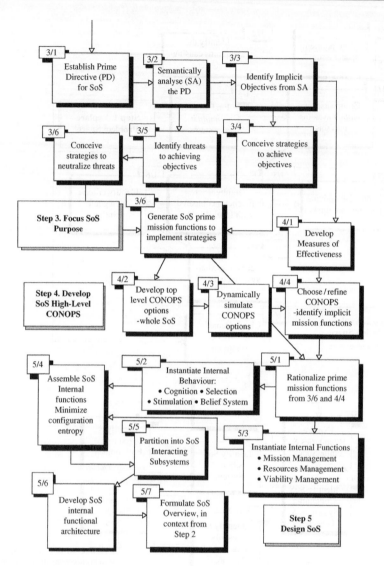

Figure 13.6　Systems methodology high-level process view, Steps 3, 4 and (part of) 5.

This becomes particularly relevant to achieving the intended outcome — a viable, enduring solution to the original problem — during the creation phase, where the SM, the team conducting the process, and the resources they require to drive the process forward may be threatened by competing projects, may be denigrated by competing reductionist processes as time consuming, expensive, etc., all of which will divert effort away from prosecuting the systems methodological process, and may prejudice a satisfactory outcome.

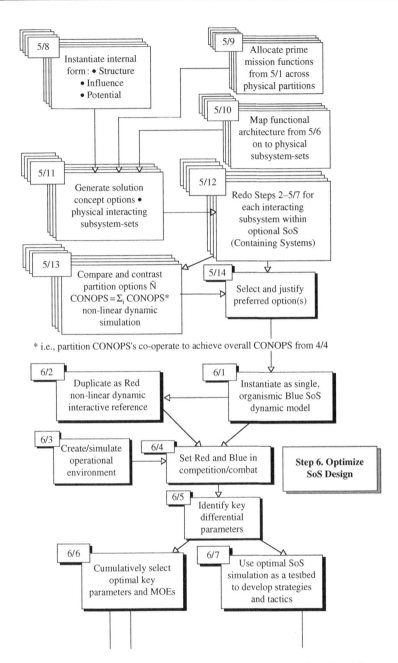

Figure 13.7 Systems methodology high-level process view, Steps 5 and 6.

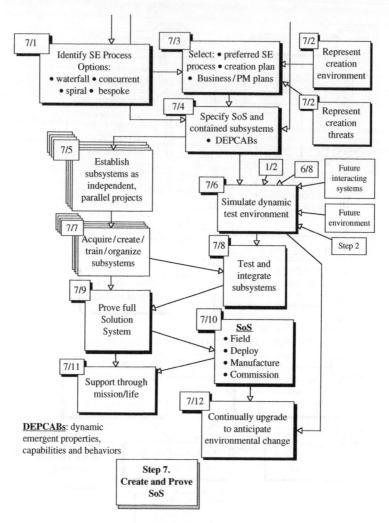

Figure 13.8   Systems methodology high-level process view, Step 7.

Summary

The elaborated systems methodological process is, in the first instance, targeted at creating an ideal world, or optimum solution system (SoS). However, in the overall systems methodological approach to problem solving, this may be only a preliminary stage: see Figure 6.4 on page 175, where the three of the four possible outcomes are shown: dissolving, resolving and solving the problem. The fourth possible outcome, having reviewed the first three, would be to choose to do nothing; that is, to 'live with the problem.'

The full systems methodology, then, encompasses all of these optional outcomes. The systems methodology process/procedure (as exemplified in the behavior diagrams of Chapter 12, and the process or activity models of this chapter), on the other hand, concentrates on designing and creating the ideal world whole SoS, or best solution in the circumstances, since that serves as a marker by which to judge the best course of action.

If the ideal world, or optimum, solution is chosen, then the process is straightforward — create, prove, introduce, work-up, and continually evolve the whole. If a less than ideal solution is chosen, one that still solves the original problem but which perhaps has less than ideal properties, capabilities and behaviors, then the process is equally straightforward. A less-than-ideal solution may be quicker or cheaper, or both, in the short term, for instance, and may even be a wiser choice in a particularly turbulent or uncertain environment, where a SoS's useful life may be relatively short.

The systems methodology may be viewed in part as a process, or series of activities — of course, that is only a small part of the story, since addressing the processes and activities requires experience, expertise, tools, techniques and methods. However, the process/activity charts tell a story, indicating the broad strategy and sequence of activities, with the generic reference model providing a series of waypoints or markers, navigating forwards from the problem space, and backwards from the solution space to provide a design that is intended to satisfy both problem and solution.

It should not be thought, however, that the path through these activities is linear, from start to finish. Instead, the route finds its way through a landscape of infinite possibilities and probabilities, with tests, rework and iterations along the way. At a number of points, options are generated and pursued before some of them will be dropped. A wide spectrum of threats may be generated and examined, but not all of them addressed, perhaps because of low probability, low impact, or high cost. And so on. Barely concealed then within the elaborated systems methodological charts of this chapter are blind alleys, elaborate iterations, etc., many of which should be explored to maintain rigor in the method — sometimes a seemingly blind alley turns out to be an innovative, even radical, new idea.

The outcome of the 'top level' design activity is a set of complementary, open, interacting and adaptive subsystems. Some of these may be social systems, or human activity systems, as in an enterprise, a military formation, a disaster relief organization, or a team. Others may be sociotechnical, as in a command and control system, an assembly plant, a passenger transport aircraft, a lean volume supply system, an F1 (Formula One) racing car team, etc. There may be need/advantage in iterating the systems methodology for these subsystems, particularly the sociotechnical ones, with each subsystem being considered as the whole, with its own requisite dynamic emergent properties, capabilities and behaviors (DEPCABs). Each subsystem would be addressed in the same way as the whole, i.e., as an open, adaptive system, interacting with its siblings and with other systems in the solution space: so, still fully connected and interacting, using the same dynamic simulation approach.

In this context, it is reasonable to view the top level of design as the outer loop, and this second 'go around the system' as the inner loop. Inner loop activities will elaborate subsystems into open, interactive sub-subsystems, some of which may turn out to be specified as exclusively technological.

Assignment

You apply for a position as a Systems Design Manager, to be responsible, with a team of specialist systems engineers and designers, for establishing the avionics design of a new air defense fighter aircraft. You are surprised to be given the job. On arrival in your new position, you set to work, trying to find out all you can about this new fighter and its avionics system, which you know comprises navigation, communications, flight management, primary and secondary radars, weapons and weapons management, processors, data highways, etc., etc. You are able to find plenty of technical data about the various technological subsystems that are to form the avionics system, but you can find no trace of any concept of operations (CONOPS) for the fighter in its air defense role.

Do you:

1. Press on regardless, without a CONOPS, and proceed with the processes of integrating the various subsystems, reasoning that the CONOPS is for operators anyway, not for systems engineers
2. Concoct a typical CONOPS by talking with air defenders, particularly those who have operated previous fighters in the air defense system, and who know how it worked with earlier types of fighter aircraft
3. Ask the customer, a government procurement department, for the approved CONOPS — knowing that there is none
4. Set up an expensive and time-consuming operational analysis, using one-on-one and many-on-many combat simulations, in an effort to find out what the CONOPS should be?
5. Something else. . . ?

Consider the above (real world) dilemma, discuss the options, choose and justify your course of action. N.B. You do not need to know anything about fighters, air defense or avionics to undertake this assignment.

14

Setting the Systems Methodology to Work

I like work: it fascinates me. I can sit and look at it for hours. I love to keep it by me: the idea of getting rid of it nearly breaks my heart.

Jerome K. Jerome, 1859–1927

Previous chapters have presented the systems methodology in abstract form: whereas the tasks to be undertaken have been nominated and interrelated, there has been no mention of the manner in which the undertaking of the many and various tasks involved might be established, organized and set to work, to achieve the goal — the solving of some problem. This chapter examines how the systems methodology as a whole may be 'overlaid' on to, or mapped into, real world organizations and structures.

Systems Methodology in Phases

There is a natural, inalienable sequence embedded within the systems methodology: essentially, some things have to be done before others, and some things can only be done based on what has gone before (see The Time dimension on page 167). It is possible, then, to consider the systems methodology as comprising a number of contiguous phases, such that the output/outflow from one phase becomes the input/inflow to its successor. So, there might be:

1. A problem-solving phase that addresses the issues, explores the problem space, and emerges with one or more conceptual remedial solutions.
2. A phase of examining how conceptual remedy might work, how it might go about achieving its objectives, how achieving those objectives might be threatened in the real world, and what might be done to neutralize the threats.

Systems Engineering: A 21st Century Systems Methodology   Derek K. Hitchins
© 2007 John Wiley & Sons, Ltd

3.   Following on, a phase to identify a range of functions that the putative solution system should be able to perform, both to achieve its objectives of solving the original problem, and of sustaining itself as a viable system.

4.   The many and various functions can then be related and interrelated so as to develop an intrinsic functional architecture, which can be mapped into physical partitions to accommodate dynamic solution space constraints of space, economy, and time — the optimum design.

5.   In the light of this optimum, ideal world solution system design, a phase may follow in which choices are made as to whether to solve, resolve or dissolve the original problem — so, potentially, changing the direction and outcome within the applied systems methodology . . . a solution system will be required to resolve the problem, and a solution system — perhaps of a different nature — may be needed to dissolve the problem.

6.   According to the choices made in 5, there may follow a phase of specifying, developing and making physical parts, while at the same time recruiting/acquiring and, if necessary, training people as members of human activity system (HAS) teams.

7.   The various complementary parts forming the whole SoS may then be interconnected and 'encouraged' to work together — this might be termed a work-up phase, during which the performance of the whole is progressively improved until the whole exhibits requisite emergent properties, capabilities and behaviors

8.   The SoS will then enter its 'operational phase,' during which it will continue to solve/resolve/dissolve the problem for which it was created, and will continually repair and redesign itself as the problem evolves/morphs.

9.   Should the problem for which the SoS was created cease to exist, the SoS may then be 'phased out,' or, alternatively may be 'repurposed.'

There may be many variations on the theme expressed above: solution systems may be introduced piece by piece, perhaps to aid affordability or to minimize organizational turbulence; and so on. However, one way to organize the implementation of the systems methodology is simply to form a team of suitable individuals, to train them to work as a team, equip them with the necessary tools to address each phase in turn, and encourage them to practice. . . .

Systems Methodology as HASs

There can be many ways to arrange and organize the systems methodology for practical application. Figure 14.1 illustrates the point; while there may be a 'natural sequence,' it does not follow that the organization need mirror that natural sequence precisely, although it is likely to follow it in some degree. Moreover, not all situations require the full extent of the systems methodology to be 'applied;' sometimes it will be necessary only to address the problem and produce remedial solution concepts, for instance.

On other occasions, a group of people may be presented with a solution concept, already selected by some customer or arbiter; it may be the responsibility of the group to work from that point onwards, without investigating the remedial solution options. (Such a group may elect, judiciously, to investigate the problem and the options anyway, so that they understand the nature of the problem they are being asked to solve. . . .)

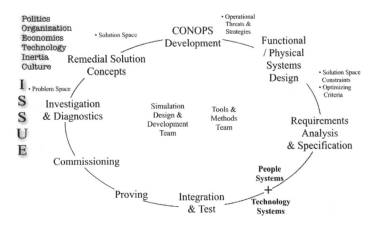

Figure 14.1 Systems methodology 'in the round,' showing principal activity centers, which may be manifested as individuals or as teams of people forming human activity systems (HASs). So, Investigation and Diagnostics could be the team title for a HAS formed of analysts with knowledge of the particular domain(s) of interest. Similarly, CONOPS Development might be the title of a team concerned with elaborating and testing various concepts for the operation of the solution system with a view to selecting and validating the most appropriate CONOPS to solve the problem in the circumstances . . . .

Systems Methodology as Tools

An alternative view of the systems methodology illustrates it as a pipeline of successive processes, many of which may benefit from the use of context-free systems tools, methods and techniques. Such tools enable systems methodological practitioners (systems engineers?) to handle and accommodate amounts of information that might otherwise swamp them, or at least prevent them from seeing the forest for the trees.

Figure 14.2 shows a tool-rich view of the systems methodology, using a selection of tools that have proved useful; there may be many others. The figure starts top left with an Issue and a Problem Space, which is addressed in the example using the Rigorous Soft Method, which may provide one or more conceptual remedial solutions — always supposing there to be any solution. The systems methodology then proceeds as illustrated, following the arrows at top and bottom of the diagram. In the center, the Generic Reference Model and the Automated ISM (Interpretive Structural Modeling)/N2 Chart are shown as being employed in several activities/processes.

Similarly, the use of a proprietary dynamic simulation tool, such as STELLA™ is indicated; there are several others. However, it should be noted that such tools are not used as their manufacturers might have expected; in particular, within the systems methodology, the dynamic simulation tool simulates whole, open, adaptive, interactive systems rather than phenomena, and the same tools are used to undertake optimization through cumulative selection — for which task, future tools yet to be developed may be better suited.

Finally, at bottom right, the solution system emerges — in full design, at least, since this is as far as the figure progresses. Note that information from the problem space — dashed line — is passed along the process line to prove the solution system design, which must be able to solve the original problem.

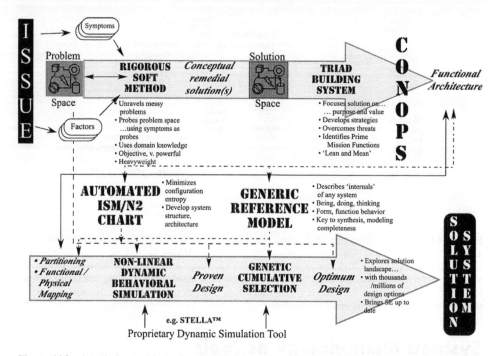

Figure 14.2 A typical assemblage of tools and their points application within the progress of the systems methodology. The tools are all context-independent, but have in common their facility for reducing information and configuration entropy, leading from a disordered, perhaps even chaotic, issue, to an ordered, specified solution system design.

Organization for Applying the Systems Methodology and for Systems Engineering

There are many ways in which an organization for the implementation and application of the systems methodology may be set up and managed. In the past, and particularly in the 1960s and 1970s, such organizations were commonplace in the so-called systems houses — these were independent companies who explored problems and issues and created solutions to those problems for customers. Systems houses treasured their integrity and objectivity, and owed allegiance to no company or brand name that might predispose them to use facilities or equipments that were less than appropriate. So, they rarely manufactured — instead they either selected and bought-in suitable equipments, or put build specifications out for tender to engineering companies to make what was needed.

Figure 14.3 shows one such, in which the organization was managed under five main headings, as shown. It is noteworthy that the objective of operations and requirements analysis, some 40 years ago, was to find the 'real' operational requirement.

This was because it was common practice for customers to believe that the system they wished to procure was to solve one problem (expressed in a customer's so-called operational requirement), but for the systems analysts and systems engineers to find, upon proper investigation, that it should

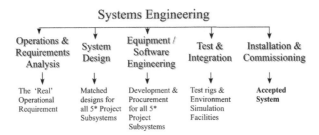

Figure 14.3 Systems engineering organization, *circa* 1965–1975 (Hitchins, 1992). The systems methodology 'process' proceeds from left to right. See text for the '5* systems.'

have been to solve quite another, and in an unanticipated way. Operations and requirements analysis undertook the task of exploring the problem space, of finding conceptual solution remedies and then of identifying the putative solution system and its CONOPS, usually by extensive simulation of competing solution concepts in their future operating environment. It was not uncommon for the systems engineering team to come up with a solution to a problem that was quite unexpected by a customer — often one that not only solved the problem effectively, but also cost considerable less.

System design, as illustrated, produced matched designs for five project subsystems:

1. The primary/operational system — the system that the customer expected.
2. The in-service maintenance system that the customer would require to keep the primary system operational.
3. The in-service training systems needed to train maintainers and operators, including crew simulators, part-task trainers, maintenance and servicing rigs, etc.
4. The in-company engineering system needed to develop 1–3.
5. The in-company maintenance system needed to keep the engineering system operating.

There were, and are, relationships between the various systems. For an aircraft or ship project, for instance, a crew simulator would have the look, feel and behavior of the corresponding operational aircraft or ship, so that the crew felt that the training was entirely realistic. During test and integration, there was also a need for simulation, such that the displays and controls, for instance could be tested and their behavior realistically assessed. There was evidently a relationship between the two kinds of simulation.

Similarly, there was a relationship between the facilities needed to maintain the equipments that had been bought-in or manufactured, and were to be integrated, and the same equipments when they were in service. For instance, in-company test and maintenance might involve factory test equipment (FTE), while in-service maintenance might involve automatic test equipment (ATE); although different in some respects, there was a degree of commonality that would have made it sensible to develop ATE from its precursor FTE.

So, as long ago as the mid 1960s, if not earlier, the practice of creating whole solution systems, where the operational, or mission element was only a part, was in evidence. It happened only rarely, however, because procurement agencies of the time were determined to maintain competition and to spread aerospace and defense contracts around, regardless of overall cost or integrity of the whole solution system.

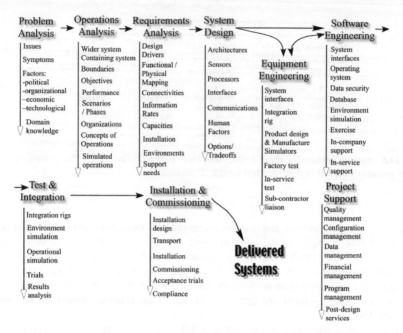

Figure 14.4   Systems house project organization, showing tasks, skills and activities; *circa* 1965–1975 (after Hitchins, 1992).

Not evident from Figure 14.4 is the systems methodology — the 'how' of systems engineering that runs as a weave across the teams' vertical weft. So, in addition to such figures about organization, there has to be some process/project/program plans and charts which show what activities are to be done in what sequence, what inputs such activities will need, what tools may be used, where the output from the activities is expected to go, and so on. This chart effectively runs laterally across that of Figure 14.4, so that skills, capabilities, etc., that are required to undertake tasks are seen as being available as required.

SM GANTT charts

Figure 14.5 shows a notional GANTT chart which might be used for planning an application of the systems methodology. Time progresses along the x-axis. The burden of work passes progressively through the various teams: Problem Analysis, Operations and Requirements Analysis, Systems Design, and so on. Although work passes from team to team as they address the various systems methodological phases, it does not follow that the teams are comprised of different practitioners; on the contrary, it is likely that some practitioners will move with the burden of work, from team to team, according to their skills, expertise, domain knowledge, etc. The resources section at the bottom could be filled in for a particular project, enabling the beginnings of project cost assessment. Note also in the chart that milestones — often marking major deliverables — will occur towards the end of each phase. Program evaluation and review technique (PERT) Charts can be employed in similar fashion to establish detailed patterns of work.

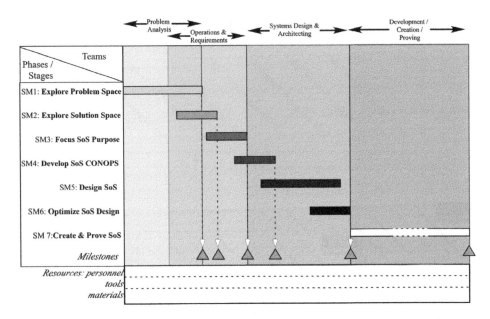

Figure 14.5 Notional GANTT-type planning chart for an application of the systems methodology.

Teams of teams

For large and/or complex projects, a team of teams may be appropriate – see Figure 14.6. Such an arrangement would not be uncommon for projects such as Apollo, in the past, and for major civil engineering, aerospace and defense project today. There may be top level team, which addresses the issues and the problem at the top level, generates conceptual remedial solutions, develops a number of potential CONOPS — perhaps several for each remedial solution concept — and progresses towards a putative set of open, interacting, adaptive, complementary viable subsystems. Each of these subsystems may be a complex system in its own right, and so properly merits its own systems team, applying the systems methodology. Indeed, the resultant arrangement would be a team of teams. . . .

However, the teams cannot sensibly work independently, else the various subsystems may not remain complementary. At the top level, each of the subsystems will be prescribed in terms of its dynamic emergent properties, capabilities and behaviors, its interactions and interfaces, etc., or to use the apposite engineering term, 'fit, form and function.' Also, the various subsystems must be compatible with the CONOPS of the whole, such that when the various parts are brought together, they contribute severally and together to the prosecution of the CONOPS.

In practice, a high degree of harmony may be required between all the teams and this is may be achieved by including members of first-level teams in the corresponding top-level team, and *vice versa* — remembering, of course, that teams of more than half a dozen find consensus and decisions increasingly hard to reach. The development of CONOPS, the functional design, the functional/physical design, etc., may all be the subject of continual variation, as concepts and designs of parts impinge on the whole and as the constraints of the whole impinge on the parts. For example, were the emerging design and development of one of the subsystems to vary significantly

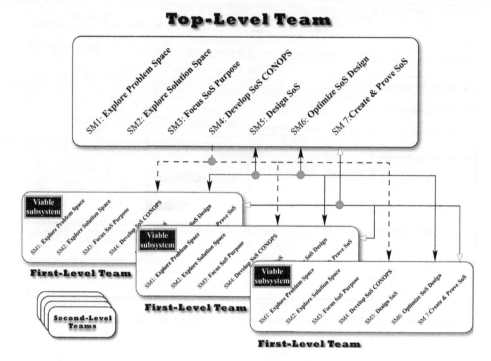

Figure 14.6 Systems methodology as a team of teams.

from expectations, it might be necessary to redesign or reconfigure the offending subsystem; or, if the situation could not be ameliorated, then other subsystem may have to be 'adjusted' to accommodate the imbalance, resulting in an all round redesign.

An early indicator of impending difficulties may be found by budgeting; that is, by periodically assessing and evaluating key parameters. For instance, in a largely technological solution system, it may be prudent to check forecasts against budgets for mass, volume, centre of mass, shape, moment of inertia, capacity, endurance, energy consumption, dissipation, waste, processing power, communications rate, etc., etc; i.e., many of the various emergent properties, capabilities and behaviors of the subsystems, which contribute to the similar emergent properties, capabilities and behaviors of the whole.

Team of teams and inner/outer loops

Figure 14.6 presents an alternative view of outer loop – inner loop design as previously seen on page 293; in the figure, the perspective is one of hierarchical, rather than 'inner loop contained within outer loop.' A moment's consideration, however, will convince that they are really the same — either viewpoint is rationally tenable; each equates to the other. The figure shows a top level and several first-level teams; there could be further levels, effectively forming the downward extending roots of a tree of systems teams.

Team of teams and system of systems

As we have seen, the developing design of a complex system may support the formulation of a number of complementary subsystems: these may be functionally bound subsystems, or they may be viable subsystems, i.e., systems which may stand on their own in their environment.

The difference between functionally bound subsystems and viable subsystems may be relatively trivial, or it may be significant. Functionally bound systems may exist within a whole that has perhaps complex, but nonetheless unified, resource management: similarly, viability management may be complex, but singular. Example of such organizations might be the human body, with its many, tightly functionally bound subsystems, all of which are resourced via the one complex system which provides energy, removes waste, replaces aging cells, etc. Similarly, for the body, the management of viability is undertaken for the system as a whole (the whole body) with the organs being addressed as composite parts of the whole, rather than as separate subsystems. An alternate example of this arrangement might be an enterprise operating from one location, with various departments and sections, and with resources and viability being managed locally for the whole.

There can be many arrangements, according to the design outcome, with some subsystems tightly functionally bound, while others may be viable; in some instances, all of the subsystems may be viable. This last may come about in a different manner, with the formation of a whole by the bringing together of a variety of extant operating systems under one 'umbrella' to form an association of systems. Such an association need not be a system — it may not satisfy the simplest definition of the term 'system.' However, it may be capable of becoming a system, perhaps post-association, with the various parts and their interactions being configured and adjusted to make the parts complementary, cooperative and coordinated, usually under the guiding hand of a directing, controlling (orchestrating?) entity.

Figure 14.7 shows how a number of subsystems within an overarching containing system may be progressively conceived, designed and developed, where the subsystem emerging from the design process are viable systems. Examples of such system are legion, and include Apollo, where the mission subsystem was comprised of a number of craft (command module, lunar excursion module, etc.) that, although fitting together like a 3D jigsaw puzzle, were also capable of independent operation as viable systems in their own right. A second example might be of an aircraft fitted with short-range air-to-air missiles for self-defense. The whole aircraft, with its crew, is a viable system. The missile, once launched, is a viable system too, with its own purpose and control, its own sensors, effectors, resources (propellant, refrigerant, payload. . .), etc.

As the figure suggests, there may be several teams operating, perhaps at the same time, perhaps not, but all would be operating within the one systems methodology. The top-level team is concerned with solving the root problem, and in pursuing the systems methodological process; it will establish a preliminary overall design, which may include a number of potentially viable subsystems. The top-level team may then spawn, or otherwise encourage the formation of, first-level systems teams, and the systems methodological process may then be repeated, but with all the teams working in coordination. The CONOPS for the whole system need not be the CONOPS for each viable subsystem part, but the conceivers and designers of those parts need to be aware of, and understand, the overall CONOPS, so that the CONOPS for the parts both enable and contribute sensibly to, that of the whole.

Similarly, the purpose, behavior and mission features of the parts support, enable and contribute to the purpose, behavior and mission management features of the whole. As Figure 14.7 shows, the mission management features of the whole will plan (choreograph) and coordinate (orchestrate) the various subsystem activities such that their combined functional behavior results in the requisite dynamic emergent properties, capabilities and behaviors (DEPCABs) of the whole.

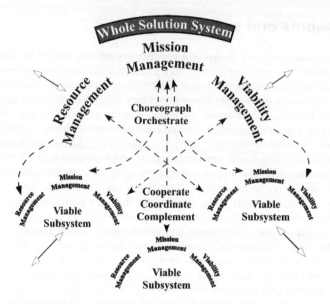

Figure 14.7   The whole solution system manifests resource, viability and mission management. Mission management for the whole choreographs (plans) and orchestrates (activates/coordinates) the various subsystems, each with its own discrete management capabilities, to coordinate the execution of missions. The whole will be neither fully resourced, nor viable, unless its subsystems parts are both resourced and viable.

However, since some of the subsystems may be viable systems in their own right, resource management for each viable subsystem may be independent — at least, in some degree. At a fundamental level, resources come in two varieties: throughput resources, that the system takes in, processes, and passes on; and resources to maintain, replace and build up the internal features of the system, to establish, maintain and improve its capabilities. Provided that the DEPCABS of the viable subsystem remain as required, this second variety of resources may be independent of such similar resources for other viable systems in the complementary set making up the whole.

An example will make the point. A global lean volume supply system may be comprised of many parts, major viable subsystems in their own right, formed into a fan-in pipeline of subsystems with the last 'pipe' in the line being final assembly, output, marketing and selling to the consumer. The whole system is choreographed and orchestrated by this lead system, the one that interface with the market. However, some of the various plants in the pipeline may be in different countries, may employ people of different nationalities and cultures, and may employ different machinery. Resources that maintain, build up and replace these parts do not have to be the same for each of the various viable subsystems, provided that the performance and behavior of the whole is reflected in that of the parts. So, while the throughput and quality of the manufactured and assembled parts is choreographed and orchestrated from the center, the people and machinery used to create and process those throughput resources need not be, provided their behavior is as prescribed and expected.

Similar considerations apply to viability management (S-MESH); provided the viable subsystem remains viable, the means by which it achieves this end may be of less concern in some respects. Synergy is the cooperation and coordination between the various parts to produce some desired external effect — the same (emergent) effect may be produced using different parts (people and machinery in this case). Maintenance similarly implicates chiefly 'local' people and machinery. Evolution may arise locally, but may require direction from the containing system center, such that the evolution of the parts contributes sensibly to, and is compatible with, evolution of the whole. Survivability of the viable subsystem part may be a local consideration, where there are distinct local threats, but it may also be a whole system consideration where, for example, interactions between the parts are threatened. Homeostasis, as ever, is a sensitive consideration. Viability for the viable part requires homeostasis, notable of local resources and capabilities. Viability for the whole, on the other hand, requires a continuing homeostatic balance between inflows and outflows, not only for each viable part, but also for the whole.

Returning to Figure 14.7, it can be seen that the various systems design teams at top level and first level may apply the same systems methodology, and may progress the top level design hand-in-hand. At the same time, the first level design teams will be identifying and defining the implication of the top level design team's findings as they 'flesh out' the specific CONOPS, mission management, resource management and viability management features of their respective subsystems, such that the DEPCABs, rather than internal substance, of viable subsystem parts, is realized.

Summary

Setting the systems methodology to work may be as simple in concept as bringing together, training equipping and practicing a team of suitably qualified and experienced people, but there are nonetheless a variety of viewpoints on the subject. The systems methodological process may be perceived as a series of contiguous phases, for example, with the output from each phase forming the input to the following phase. An alternative, but related view sees the process as punctuated by the use of different tools, each appropriate to one or more phases, such that the information that emerges from one phase and its tool/method/technique is in the right form and format to serve as input to the next phase, tool/method/technique, making the whole systems methodological process 'seamless.'

There has been, in the past, a variety of successful way of organizing the systems methodology process (the 'how' of classic systems engineering) such that it is conducted by a number of teams within a systems engineering organization. Examples are given of such arrangements from the 1960s/70s, as employed by so-called systems houses, which were objective, independent systems engineering companies set up to employ the systems approach to solving complex problems in the defense, aerospace and civil engineering arenas, including hydroelectric energy, atomic and nuclear energy, and many more.

These successful examples from the past serve as a guide for organizational paradigms of the present and future. The chapter continues with a look at modern organizational paradigms, with the employment of teams and teams of teams, applying the systems methodology to systems and systems of systems.

Assignments

1. You are Vice President, Corporate Organization, of a manufacturing organization that has recently acquired five major companies: two for manufacturing parts, one for storing and transporting parts, one for marketing, and one for distribution. Each of the acquired organizations

was a viable, profitable business in its own right prior to acquisition. The President of your organization sees the future as being largely concerned with agile, lean volume supply systems, and it is with this goal in mind that he has acquired the new subsidiaries. He is aware, however, that he does not yet preside over a system of systems. Indeed, he feels unable to regard them as even an association of systems, since each of the acquired subsidiaries remains fiercely independent under its own operational control. The President goes so far as to call the subsidiaries 'a dissociation of systems!'

- The President has decided to set up a working party, with members from each of the five subsidiaries. The terms of reference for the working party are, broadly, to consider how the various subsidiaries may be 'harmonized' (his term) into a unified, lean volume supply system.
- You are to head up the working party, but in the first instance, the President requires you to determine the agenda for the working party: you have one day.
- You retire to consider what needs to be addressed, discussed, agreed, planned and executed in order to reconfigure the separate subsidiaries into an agile, lean volume supply system . . . after nervously scanning the recruitment advertisements in the trade magazines, you settle to work and produce the draft agenda. . . .

2. You are a member of the top-level systems team addressing the problem of sending a team of astronauts to land on Ceres, one of the largest of the asteroids orbiting the Sun. There is already a team of system teams in existence, with first-level teams for ground environment, communications, imaging and graphics, and for the various parts of the mission system — launcher, command modules, etc. Quite how physical contact with the asteroid is to be accomplished is still under discussion as part of the top level CONOPS. . . .

- You, however, are tasked with examining a different subject: software. It has become the practice to form a central, software development center, with responsibility for developing all software, including that for each of the discrete parts of the mission system. This centralized approach was introduced many years ago within your organization to standardize and improve the quality and reliability of software
- This centralized approach has created some difficulties in the past, since each of the mission subsystems, with its respective first- or second-level systems engineering team, has no direct responsibility for its own mission software. There is a proposal to rectify the situation by 'embedding' the software for each viable mission subsystem within that system.
- So, the systems engineering team for the command module, for instance, would be directly responsible for conceiving, developing, testing and proving the software, just as they would any other piece of the command module.
- Remembering that the CERES Mission Program already operates as a systems engineering team of systems engineering teams, your task is to consider the pros and cons of centralized versus distributed/embedded software development in this context and to present a considered recommendation — with justification — as to which way to organize software development, integration, test and proving.

Case D: Architecting a Defense Capability

Note

The following Case Study synthesizes an entirely mythical requirement for a defense capability. It originated as an INCOSE tutorial to illustrate how the Systems Methodology might be used in a practical case, but has no basis in fact. Note the date on the following, supposed, 'news extract.'

SM1: Explore Problem Space

Extract from the Washington Defense Business Herald Times, Declassified Briefing Note, April 1st, 2004

Mojave Maneuvers

The rise of terrorism is a cause for concern, not least because it is almost impossible to say where they, the terrorists, will strike next. However, it would be foolish to concentrate on the terrorist threat to the exclusion of conventional warfare.

Proliferation of nuclear capability seems to be ongoing and inevitable. As with terrorism, however, to overly-concentrate on the threat of nuclear warfare would be to offer a potential enemy a so-called "free ride" in the conventional warfare arena.

It is not as though the West has conventional warfare "sown up." There are major arenas around the world where the US, for instance, would find it difficult to operate. One such is the desert, and it may not be without significance that we see DARPA,

---

the US Defense Advanced Research Projects Agency, hosting a race between robot vehicles across the Mojave Desert. Why, one asks, would they be so interested in such an activity as to offer significant prizes? They are not renowned for their altruism.

So, US forces are faced with a shortfall in capabilities when it comes to land warfare over large open areas: deserts, tundra, Great Plains, savannah, etc. There is plenty of room for potential enemies to raise, operate, maneuver and hide sizeable forces. Interestingly, a number of such areas are in regions not too friendly to the US.

The US has a particular problem when it comes to casualties, too. The US public does not like 'body bags,' and they soon lobby their politicians if even one casualty arises. While 9/11 may have changed circumstances somewhat, casualties are still a major issue.

Desert conflict can be cripplingly difficult on man and machine: the Second World War showed that in N. Africa, where Rommel and Montgomery faced off. Rommel was the proponent of the *blitzkrieg*, while Montgomery was more by way of a set piece battle exponent. Neither party had it all their own way. Seemingly, neither strategy was dominant, at least not in that conflict.

It would be comforting to think that such arenas would find employment for existing weapon systems. The evidence suggests otherwise however. Our tanks and personnel carriers do not like desert operations: they overheat as the filters clog with sand;

they consume enormous amounts of energy to keep their occupants cool or warm; they get stuck in deep sand, or bogged down in, uh, bogs, and need to be pulled out.

Communications can be difficult, too, with thermal inversions playing havoc with radio communications. Heat shimmer and mirages can upset visual sights. Radar has problems, too, when it has to be operated from vehicles on the move in undulating country; even the best radar may not work too well when at the bottom of some desert *wadi*.

Altogether, it has to be said that the problems facing the military in such hostile circumstances are more akin to those facing a naval task force that a conventional army land force. Perhaps the army 'think-tanks' should catch up with their naval colleagues and compare notes!

There will always be a money issue when it comes to defense. One positive aspect of an otherwise bleak and forbidding 9/11 experience is that the arguments against defense spending are more muted than before. On Capitol Hill the question seems to be more about the risks of not spending, than of the expense *per se*.

When asked about the need for a new kind of open land force capability, Paul Weinhard did not confirm the need. Significantly, perhaps, he did not deny it either.

The smart money, then, is observing the significance of the events in the Mojave Desert, and is forecasting an announcement of a new defense capability requirement within the next administrative period. Just what that new capability will be is anyone's guess. Our guess is that the winners of the Mojave competition will have a head start on the competition, and that robotic vehicles operating in deserts may have something to do with it.

Darren Stevens
WBHT Defense Business Analyst

Examining the report by influential analyst Stevens reveals the following:

Extract from briefing note	Symptom of issue
1. There are major arenas around the world where the US, for instance, would find it difficult to operate.	■ Perceived US military limitations in open land warfare
2. DARPA, hosting a race between robot vehicles across the Mojave Desert	■ Implied robot vehicle solution
3. Shortfall in capabilities when it comes to land warfare . . . open areas: deserts, tundra, plains, etc. in regions not too friendly to the US	■ Perceived US military limitations in open land warfare
4. The US has a particular problem when it comes to casualties, too. The US public does not like 'body bags'	■ US political issue with casualties
5. Seemingly, neither strategy was dominant, at least not in that conflict	■ Uncertainty over desert operations strategies
6. . . . existing weapon systems. The evidence suggests otherwise however	■ Perception that existing weapon systems unsuited to desert operations
7. Communications can be difficult, too . . . visual sights can be upset . . . Radar has problems	■ Communications, visual sights, radar - among problem systems
8. . . . problems facing the military in such hostile circumstances are more akin to those facing a naval task force	■ Perception of military land situation being akin to naval operations at sea
9. . . . money issue . . . risks of not spending	■ Perceived threat likely to overcome financial inhibitions
10. New defense capability requirement within the next administrative period	■ Political urgency to attain new capability

N.B. Take each symptom in turn. Identify possible causes, using pejorative terms, e.g. 'poor,' 'lack of,' 'inability to,' etc. From list of possible causes, develop causal loop models, but drop the pejoratives, to create an 'ideal world' representation of processes and systems.

The first symptom, perceived US military limitation in open land warfare, is addressed in Figure D.1. The figure is laid out as prescribed for the Rigorous Soft Method, page 195, *et sel*, with the symptom and a laundry list of possible causes top right. From these, a CLM is developed, bottom left, which discards the pejorative terms used in developing the laundry list. The CLM is complex, in this particular instance, as it addresses not only the perceived military limitation, but also economics of defense spending and political concerns.

The CLM reveals a number of implicit systems, bottom right, which may be in imbalance, remembering that the CLM represents some ideal world in which the perceived (or, perception of) US military limitation would be resolved. . . .

It might be thought that the CLM of Figure D.1 is a sufficient Rich Picture in its own right, and it certainly classifies as rich; however, there are other symptoms to consider . . .

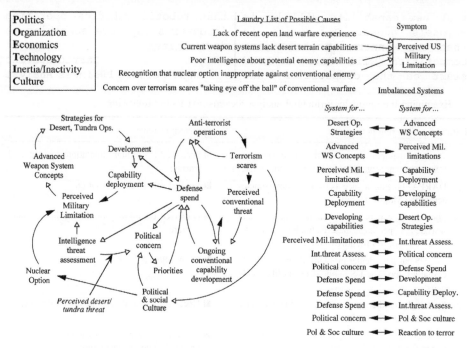

Figure D.1 Perceived US military limitations. Causal loop model (CLM) showing how symptoms might be resolved in an 'Ideal World.' In the CLM, hollow arrowheads encourage/enable, solid arrowheads discourage/oppose.

Figure D.2, by contrast, examines the problems facing an open desert land force, and highlights the difficulties for military personnel operating under such conditions, as well as the technological difficulties. With concern over potential future 'body-bag' counts and the relative frailty of people in such conditions, it would seem sensible, indeed, to consider some robot force — except that, hopefully, no one would want to trust to the decision-making qualities of an autonomous, armed, mobile robotic force. . . .

Between them, Figures D.1 and D.2 address many of the symptoms from the original briefing note. The POETIC acronym offers a useful way to examine the models:

- Politics and Economics. Figure D.1 indicated politics as a major focus of concern. Intelligence advises that there is a an international threat that the US military seems unprepared to meet; meeting that threat would cost money, which implicates defense spending, which is already high as a result of international terrorism. The potential use of nuclear weapons in the proposed desert and tundra regions of the world is seen as politically, as well as environmentally, unacceptable. Of course, defense spending funds Intelligence, so — no 'suitable' Intelligence, no funding?

- Organization, Technology and Culture. Figure D.2 addresses the perceived current shortfall in defense technology, particularly in the light of the social, cultural and political intent to minimize casualties in any conflict.

- Between the two CLM-based analyses, most of the symptoms have been addressed, and the problem space has been explored . . . without any obvious solution emerging.

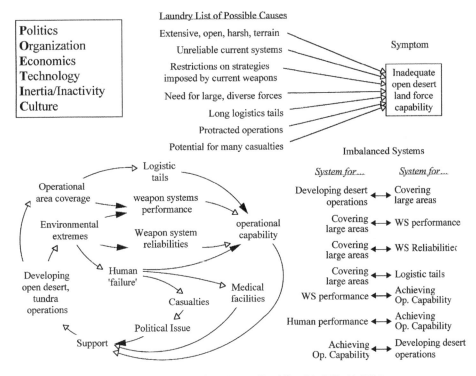

Figure D.2 Perceived Open Desert Capability. Ideal World CLM.

Figure D.3 shows some of the factors from the briefing note that emerge as contributing to a remedial solution system concept. They include:

1. Uncertain warfare strategy. There is ongoing concern about the relative merits of 'positional warfare' and maneuver warfare (e.g., *blitzkrieg.*)
2. Competitions to race robotic vehicles across the Mojave Desert suggest that the ideal strategy, one of being able to adopt either positional, or maneuver warfare, or hit-and-run, etc., has been inhibited by the inability to operate quickly and without loss of life over large, inhospitable tracts.
3. Current technologies clearly have limitations in extreme environments
4. The concern over casualties is ongoing, and could threaten US abilities to deploy military force effectively. A solution with few, or even zero casualties would be attractive. . . .
5. There are very large, open, sparsely populated areas around the world
6. The idea that land operations could be likened to naval operations, with these wide, open areas equating to the oceans, is intriguing.

The various implicit systems in imbalance can be brought together into an N^2 chart see Figure D.4. The process of developing the chart consists of a number of steps:

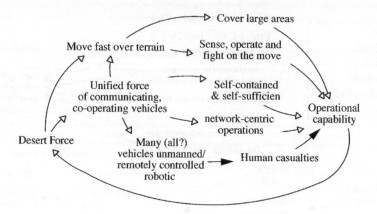

Figure D.3   Factors contributing to a remedial system solution concept. Note that the use of robotic vehicles, although it might reduce human casualties, is seen as also potentially prejudicing operational capability.

```
                                        1    2    3    4    5    6    7    8    9   10   11   12   13   14   15
US Culture    Rct terror     1         [B]  1
              P&S Culture    2          1  [D]  1
US Politics   Pol Concern    3               1  [F]  1         1
              Int Thrt Ass   4                    1  [E]  1
              Perc Mil Lim   5                         1  [C]  1
              Def Spend      6          1                  [G]           1
Feasibility   Hum Perf       7                         1       [A]       1
Constraints   Des Cp Strat   8                                    [O]  1
              Area cover     9                                    1  [N]  1
              Capab Deploy  10                                    1   1  [M]  1
              Dev Capab     11                               1            [H]  1         1
Future Weapon Adv WS Conc   12                                           1  [J]  1
Systems       WS perf       13                                      1   1  [L]            1
Characteristics WS reliab   14                                          1            [I]  1
              Log Tails     15                                                   1   1  [K]
```

Figure D.4 N^2 chart drawn from the analysis of Figures D.1–D.3. Interrelationships between entities were first developed using Interpretive Structural Modeling. N.B. The chart was produced by a proprietary tool, which abbreviated the various entity titles. So, 'Rct Terror' is Reaction to Terror. P&S Culture is Political And Social Culture. DesOpStrat is Desert Operational Strategy. Adv WS Conc is Advanced Weapon System Concepts. Etc.

1. Consolidate and rationalize a list of all the implicit systems in imbalance from Figures D.1 and D.2.
2. Develop a reachability matrix for the implicit systems. (See Warfield's interpretive structural modeling (ISM) on page 191.)
3. Convert the reachability matrix to an N^2 chart: seek out and repair asymmetries and evident omissions, recursing to correct the originating laundry lists and CLMs as necessary.
4. Recluster the N^2 chart, minimizing configuration entropy.
5. Partition the N^2 chart into 'subsystems' of the overall system, such that the interfaces between subsystems are minimized.
6. Embrace each subsystem within a suitable title.

The resulting partitioned N^2 chart can be converted to the more readily understood systems interaction diagram (SID) of Figure D.5. The SID is formed by recalling that the entries in

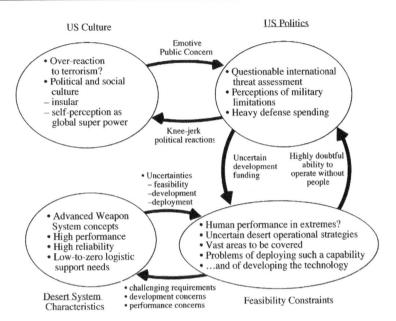

Figure D.5 Systems interaction diagram (SID) showing the major, open, adaptive, interactive subsystems within the whole problem space. (N.B. Due to the nature of the technique, the SID may seem critical and blunt.)

the N^2 chart derive from the original CLMs, which were 'ideal world', i.e., they represented what should have been happening in some ideal world, but which was not so doing: else, the corresponding symptom would not have emerged. In effect, then, the SID re-inserts the pejoratives that were dropped in forming the CLMs, but at a higher level (strategic level, as opposed to tactical level?)

Figure D.5 is reasonably comprehensive, yet compact. Each of the ovals incorporates aspects of interest and concern for the respective subsystem: the interactions between the ovals indicate what each is giving to the other — again in pejorative terms. The use of pejoratives may make the SID appear blunt, even harsh, but it is not intended to offend — simply to get down to 'bare bones.'

The 'Real' issue?

It may seem prudent to revisit the Issue: initially, it seemed to be about the need, or otherwise, for a new land force capability. Looking at Figure D.5, the Issue now appears to be rather broader: could it be about US culture and politics, about the political aspirations of some to 'police the world' and of others to virtually withdraw from the international scene? There appears to be a cultural/political dilemma, in which there is a perceived (political) need to employ force to maintain a democratic 'way of life,' yet (some of) the people comprising the democracy are unwilling/uncomfortable about deploying force, particularly if it risks casualties.

Figure D.5 does not indicate a solution to the problem, although the reader may infer one. The systems methodology, and systems engineering, can proffer answers by dissolving, as well as resolving and solving. Looking carefully at the SID, it may become apparent that the problem could be dissolved:

- More and better intelligence might convince the politicians that there is no real threat (see under US Politics)...
- ... or, they could simply disregard the intelligence as lacking credibility.
- They might also consider, looking at the SID, that if the task of setting up a force were really so complex, expensive and difficult, then perhaps no one else would attempt it — so, no viable threat. (See under Feasibility Constraints.)
- The US could choose not to police the world (at least, not in this particular instance — see under US Culture.)

The chart also suggests, somewhat cynically no doubt, that politicians could seek to 'shape' US culture, by somehow reducing the public's emotive reaction to casualties — at least some of which may be triggered by political rhetoric and knee-jerk opposition and media reactions to international situations. From a political standpoint, it may be difficult to prosecute a protracted, but necessary, conflict if the public at large is opposed to any casualties. And such opposition is a relatively recent cultural phenomenon, developing notably since World War II.

The other side of the coin is also presented in the SID — the prospects for the development of advanced technological platforms and weaponry designed to operate under extreme conditions, in sensitive environments, and perhaps even robotically or remotely controlled. From a defense business standpoint, there is a prospect of research and development funding on a grand scale if the politicians can be persuaded, and the public pacified....

However, it is not impossible that the idea of some robotic force roaming the deserts and tundra of the world, taking on 'the bad guys,' is an entire fiction. Such a fiction would serve to reinforce belief in US military and technological domination, both at home and abroad. Nor it should not be forgotten that the contemporary emphasis on fighting terrorism may draw attention — and funding — away from the high-technology developments and advanced systems that have been the hall mark of US space and defense systems.

Cynical? Maybe, but then, look who wrote the original article!

Figure D.3, together with the bottom half of Figure D.5 do suggest the essence of a potential solution, or resolution of the problem, by the direct method of conceiving, designing and creating a solution system. Following sections will skirt around the political quagmire and address the emerging solution system instead.

SM2: Explore Solution Space

A convenient way to accumulate the information generated by the various systems methodology activities is to generate the information in tabular form, using the stage/task numbers used — in this instance — in Figure 13.5: see The Systems Methodology — in parts and phases on page 294.

Systems methodology stage/task numbers	Response
2/1 Nominate System of Interest: (SOI)	Mobile Land Force 2010
2/2 Explore Boundaries and Finite States	Complementary set of interacting, all-terrain, fighting vehicles operating multiple UAVs after the manner of a land-based aircraft carrier task force. States: training, standby, operational, covert, recovery, turnaround and repair

2/3 Identify SOI Sibling Systems	Air transport, air insertion, air recovery, satellite intelligence, satellite communications/navigation, logistic support, repair facilities, vehicle recovery
2/4 Identify & Explore Environment	Vast areas of desert, plains and tundra, wide temperature variation, plains with rocky outcrops, frozen lakes, bogs, little vegetation, savannah, delicate ecological balance . . . .
2/5 Identify SOI Containing System and Objectives	US Global Peace Command. To neutralize enemy incursions into UN designated global deserts plains and tundra.
2/6 Identify and Explore Influences on SOI	Political desire to operate without loss of US lives. UN desire to operate without loss of any lives. US belief in advanced hi-tech weapon systems. US defense business interest in developing advanced, nonlethal weapons.
2/7 Identify and Prescribe Interactions and Resources	Air Transport and Insertion; re-supply; intelligence; doctrine; Rules of Engagement; fuel, weapons and consumables; unmanned air vehicles; trained operators; repair staff, logisticians, communications, satellite navigation, etc., etc.

SM3: Focus SoS Purpose

Prime directive and semantic analysis

The third phase of the systems methodology employs the TRIAD building system: see page 225. A suitable PD for the solution system in this instance might be: 'To neutralize enemies in open desert, plains and tundra regions around the world.'

This PD may be semantically analyzed as follows:

- To neutralize . . .
- enemies in . . .

- open desert, plains and . . .
- tundra regions . . .

- around the world.

- To render ineffective . . .
- those opposed to the US as identified by UN directive ABC existing and/or operating in . . .
- open, desolate, largely uninhabited tracts . . .
- and Arctic plains with permanently frozen subsoil, lichens, mosses, and dwarfed vegetation . . .
- wherever sanctioned by the UN.

1. The final element indicates that the intent would be to create a force that would operate only under the aegis of the United Nations.
2. The definition of desert, plains and tundra lack specificity.

The semantic analysis allows the identification of implicit objectives, as follows:

From the Semantic Analysis	Implicit Objectives
■ To render ineffective . . .	To deploy swiftly
■ those opposed to the US as identified by UN directive ABC existing and/or operating in . . .	To operate over wide areas radically different environments, temperatures, going, etc.
■ open, desolate, largely uninhabited tracts . . .	To move rapidly to scenes of incursion/activity
■ and Arctic plains with permanently frozen subsoil, lichens, mosses, and dwarfed vegetation . . .	To identify legitimate enemies specifically To engage and deter, or overcome To operate within a UN mandate at all times
■ wherever sanctioned by the UN.	

Measures of effectiveness (MOEs)

Once the implicit objectives of the Mobile Land Force 2010 are established, it is apposite also to establish Measures of Effectiveness (MOEs) for the whole force. These may be regarded in three categories: performance: availability (of performance;) and survivability (of performance.)

■ Performance. (a) Time to scene; (b) time to neutralize; (c) degree of neutralization; (d) Blue casualties; (e) Red casualties; (f) operation costs and cost-effectiveness; (g) cost exchange ratios; (h) casualty exchange ratios; (h) outcome
■ Availability. Generally calculated as 'operationally effective time ÷ (operationally effective time + down time)' Addresses, therefore, both the reliability of the whole, of its machinery and technology, of its operators and of their support in terms of surveillance, reconnaissance and intelligence — without which they cannot be operationally effective.
■ Survivability. Generally addressed under three headings: avoidance of detection, self-defense, and damage tolerance. May include damage repair.

Strategies to achieve objectives, overcome threats

In their turn, these implicit objectives suggest strategies for their achievement; the identification of suitable, credible strategies requires knowledge and experience of the future operational environment:

Implicit objectives	Strategies to achieve objectives
To deploy swiftly	Air transportable
To operate over wide areas radically different environments, temperatures, going, etc.	Air deliverable
To move rapidly to scenes of incursion/activity	High-powered, high-speed, all-terrain vehicles
To identify legitimate enemies specifically	Unmanned air vehicles (UMAs) for remote identification and engagement where appropriate
To engage and deter, or overcome	Vehicles to operate and fight on the move as an integrated unit, for speed, area coverage, avoidance of detection

	Fleet formation management to reduce enemy threat — open and tight, etc.
	Some vehicles to be self-steering, but under control of personnel in nearby vehicle/command post.
To operate within a UN mandate at all times	Command and control to be linked directly to UN Operations HQ

Looking at this elaboration from implicit objectives to strategies to achieve objective, it is evident that concepts are developing based on the original issue and its analysis, e.g., the use of UMAs is aimed towards minimizing interference with delicate ecosystems, and with preventing casualties to own forces. Similarly, the notion that some vehicles might be self-steering is aimed towards reducing the personnel complement, and hence to reducing casualties and logistics support. Fighting on the move and fleet formation strategies echo the notion that the force might be likened to a naval task force at sea, rather than a conventional land force.

In the following elaboration, threats are implied to achieving the implicit objectives, and strategies are posed to overcome those threats. Note in the elaboration that UMAs are proposed to engage opposition; psychological operations (psi-ops) are included as part of a strategy to neutralize and deter, rather than engage in warfare; that some novel weapons are proposed to achieve this 'passive' neutralization; that some heavy duty conventional weaponry is included, but as a backup if persuasion fails.

Implicit objectives	Strategies to overcome threats to achieving objectives
To deploy swiftly	Pre-deployed cadre forces in area
To operate over wide areas radically different environments, temperatures, going, etc.	Some WS/vehicles specialized for hot, wet, cold, ice, etc. conditions
To move rapidly to scenes of incursion/activity	Use of armored UMAs to accelerate ahead of ground force
To identify legitimate enemies specifically	Use of nonlethal force to neutralize:
To engage and deter, or overcome	Equipped: psi-ops, loudspeakers, leaflets, stun weapons, nonlethal anti-riot weapons
	Equipped: fuel-air and thermobaric weapons (to warn as well as neutralize) + short-range electromagnetic pulse (SREMP)
To operate within a UN mandate at all times	Equipped: canon, anti-tank missile, etc., anti-sniper lasers, enhanced remote ethnic/nationality laser identification

The threats are implied rather than stated, in the above elaboration. So, the threat to rapid deployment is that enemy forces on the ground may attack the force before it can assemble — hence the proposal that a cadre force should be in place, to provide defensive cover during force assembly. Other strategies in this section also address an implied threat.

Nonlethal weapons

Nonlethal weapons are proposed in this section to overcome the threat of being unable to engage the enemy without killing, which is to be avoided wherever possible. The use of psychological operations (psi-ops) is an obvious choice in such circumstances, as are stun weapons and nonlethal anti-riot weapons (rubber bullets, tear-inducing gases, etc.) Thermobaric weapons have been included since these have the potential to clear an entrenched enemy, perhaps in underground caves, or locally built-up areas. To operate nonlethally, they would be combined with psi-ops, warning any enemy that the weapon was going to be deployed, was being deployed, had been deployed, was seeping into the area, and was about to be initiated. . . .

SREMP, the short-range electromagnetic pulse weapon, does not exist at present; it would require development and trialing before being deployed. However, in principle, it could be invaluable. A high-energy plasma-stroke would release a pulse of electromagnetic energy which would render all nearby vehicles, electronics and electrical systems unusable; flora and fauna, including humans, would be unaffected by the EM radiation. So useful would this capability be in such operations in ecologically delicate areas, that it is seen as well worth developing and deploying.

A 'remote ethnic/nationality laser' also does not exist, but seems to be feasible and could reduce the risk of inadvertent casualties. A nondamaging laser beam would be reflected from suspect targets and the reflection analyzed for spectral absorption: such absorption characteristics can detect that individuals have been eating different kinds of spicy foods, for instance, are wearing perfumes, deodorants, after-shave, are sweating heavily, etc.

Functions from strategies

The various strategies may be collected, and converted into functions that the SoS should be able to 'perform;' these will become many of the overall system's Prime Mission Functions. Note that the systematic elaboration process means that each of these Prime Mission Functions can traced back to the objective it is helping to achieve, and hence to the Prime Directive.

(This does not imply that the process is 'handle-turning:' on the contrary, conceiving and substantiating appropriate threats and strategies is challenging, and require domain knowledge and experience. Different teams with different levels of expertise and experience would produce different results — all hopefully generated systematically. Moreover, different results will impact significantly on the cost and time to develop for the whole Solution System, and particularly on its effectiveness in operation. This part of the design process is potentially highly creative and innovative: such innovation has been the hallmark of systems engineering since the beginning.)

A collection of strategies and the consequent implied functions follows (SM3/6). This is, in effect, embodying each strategy in the solution system: if a strategy is a particular way of doing something, then the implied function is the embodiment of that way of doing things in function, activity, process, structure, capability, etc.

Strategy	Implied function
Air transportable	Dedicated air transport
Air deliverable	Palletized low-altitude insertion
High-powered, high-speed, all-terrain vehicles	Highly specialized vehicles
UMAs for remote identification and engagement where appropriate	UMA launch, control, operate and recover—on the move!
Vehicles to operate and fight on the move as an integrated unit, for speed, area coverage, avoidance of detection	UMA turn-round, repair, refuel, rearm, re-equip—on the move!

Fleet formation management to reduce enemy threat — open and tight, etc.	Formation control: reacts to threat, disperses, closes, rearranges vehicle formations in face of threat from terrain and enemy
Some vehicles to be partly robotic, but under control of personnel in nearby vehicle/command post.	Mobile repair units, operating on the move
	Short-range only communications for security, 'silent' running
Pre-deployed cadre forces in area	Cadre forces maintenance, communications and intelligence
Some Weapon Systems and vehicles specialized for hot, wet, cold, ice, etc. conditions	Special vehicle support
Use of nonlethal force to neutralize	Lethal weapons training/practice
Use of armored UMAs to accelerate ahead of ground force	Nonlethal weapons training
Equipped: psi-ops, loudspeakers, leaflets, stun weapons, nonlethal anti-riot weapons	Fuel-air and thermobaric weapons training/practice
	Human target identification
Equipped: fuel-air and thermobaric weapons (to warn as well as neutralize)	Sniper location
Equipped: canon, anti-tank missile, etc., anti-sniper lasers, enhanced remote ethnic/nationality laser identification	Improved non-cooperative identification
(To operate within a UN mandate at all times)	Real-time control of Rules of Engagement

SM4: Develop SoS High-level CONOPS

There are many different concepts of operations for Mobile Land Force 2010: there are several ways in which the preferred CONOPS might be developed and presented. (It is presumed in the following that competing, optional CONOPS have been considered, modeled, and compared, leaving the preferred CONOPS as the choice. In practice, this would be a significant task.)

Figure D.6 shows the preferred CONOPS as a sequential process model, using the presentation style of a causal loop model. The CONOPS comprises several interconnected process-parts:

- ■ UN command and control, top left, who have the authority to trigger the main force element.
- ■ Once triggered, the main force element loads into its dedicated aircraft, launches and arrives in area where the force is inserted.
- ■ Part of the strategy involves positioning a cadre force in the area, maintaining surveillance and developing local intelligence; the main force formates with the cadre force, which provides defensive air and ground cover in the event that opposition is already attacking.
- ■ The combined force then reconnoiters, using their own UAVs and satellite radars, to identify incursors — the supposed opposition.

■ Incursors are engaged and neutralized
■ The force continues to seek and locate further incursors, and to ensure that those previously neutralized remain so.
■ Once 'tranquility' has been restored, the main force is extracted, leaving the cadre to maintain watch
■ Throughout, the logistics supply has been continual — see bottom arc of model.

CONOPS options may be simulated using a nonlinear dynamic modeling tool, to investigate time scales, logistics, the effects of operations on reserves, maintenance, and vice versa. The results of such operations analysis may invoke changes to CONOPS. For instance, a typical query might be: 'should transport aircraft be on airborne quick reaction alert (QRA) to minimize delays? The response would trade responsiveness for cost. . . .

The preferred CONOPS, illustrated at Figure D.6, indicates the need for additional Prime Mission functions, including:

■ Intelligence gathering, development, corroboration, deployment and briefing
■ Reconnaissance — satellite, high-altitude, standoff, and UAV-based
■ Command and control — UN/US Global Peace Command, Mobile Land Force 2010 and cadre force
■ Force extraction
■ Base resupply/repair
■ Force self-defense

(Additionally, if the cadre force is considered part of the whole, which would not be unreasonable, there will be a variety of functions associated with forming, equipping, inserting, supporting, maintaining, directing and communicating with the cadre force. For exercise purposes, these will be subsumed into the full set of Prime Mission Functions.)

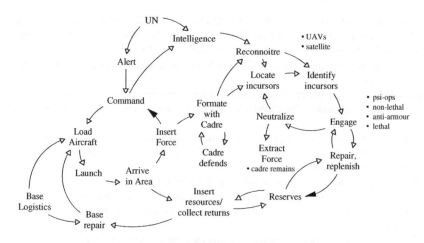

Figure D.6   High-level CONOPS for Mobile Land Force 2010.

SM5: Part 1. Design Solution System — Functional Design

The addition of CONOPS-generated Prime Mission Functions (PMFs) adds to the previous list to make a composite list of PMFs:

- Cadre forces maintenance, communications and intelligence
- Special vehicle support
- Lethal weapons training/practice
- Nonlethal weapons training
- Fuel-air and thermobaric weapons training/practice
- Human target identification
- Sniper location
- Real-time control of Rules of Engagement
- Dedicated air transport
- Palletized low-altitude insertion
- Highly specialized vehicles
- UMA launch, control, operate and recover—on the move!
- UMA turn-round, repair, refuel, rearm, re-equip—on the move!
- Formation control: reacts to threat, disperses, closes, rearranges vehicle formations in face of threat from terrain and enemy
- Mobile repair units, operating on the move
- Short-range communications for security, 'silent' running
- Intelligence gathering, development, corroboration, deployment and briefing
- Reconnaissance — satellite, high-altitude, standoff, and UAV-based
- Command and control — UN/US Global Peace Command, Mobile Land Force 2010 and cadre force
- Force extraction
- Base resupply/repair
- Force self defense

This list of PMFs has been derived exclusively from operational considerations: it does not include the host of internal functions essential to establish and maintain any mobile, yet unified, fighting force; these can be derived using the Generic Reference Model (GRM) as a template — see below.

SM5/3 instantiate internal functions

The table below is used to instantiate the 'internal features' of the Mobile Land Force 2010. The table incorporates those features that will be used to 'manage', or 'orchestrate' the Prime Mission functions, already generated above.

Internal architecture generation table				Instantiate function management	
Mission management management of . . .		Viability management management of . . .		Resource management management of . . .	
GRM	SOI	GRM	SOI	GRM	SOI
Information	Intelligence Reconnaissance Communications Center Image Center	Synergy	Formation management; command and control system	Acquisition	CPRM Base re-supply Training

(Continued)

Internal architecture generation table				Instantiate function management	
Mission management management of . . .		Viability management management of . . .		Resource management management of . . .	
Objectives	CPRM	Survival	Formation management Self-defense system	Storage	Logistic support vehicles Ready use stores
Strategy and plans	Command and evolution control (C2)		Performance recording systems	Distribution	Mobile distribution fleet
Execution	C2 Weapon management UMA management	Homeostasis	Climate control	Conversion	Mobile support Air transport
Cooperation	C2	Maintenance	Mobile maintenance teams	Disposal	CPRM

CPRM: Contingency planning and resource management.

SM5/2 instantiate internal behavior

The following table instantiates those 'internal features' that will go to determine the behavior of the Mobile Land force 2010. Many of these functions and features will interact with, and influence, the functional processes of Mission Management.

Internal architecture generation table				Instantiate behavior management	
Cognition management management of . . .		Selection management management of . . .		Stimulation management management of . . .	
GRM	SOI	GRM	SOI	GRM	SOI
Tacit knowledge	Desert, plain and tundra combat experts OJT	Nature	Psychological monitoring Counseling	Motivation	Training Command and control
World models	Doctrine, maps, satellite imagery, cultural perceptions, Intel briefing	Experience	'Simulate before activate'	Activation	Command and control Exercising
Constraint	Doctrine, Discipline, Training, Rules of Engagement				

OJT: On-the-job training.

SM5/4 assemble SoS internal functions — minimize configuration entropy

The assembly process employs Warfield's interpretive structural modeling (ISM) to identify and establish the relationships between the entities, using the trigger question — 'Does entity A help to achieve entity B, or is it the other way around?' This allows the development of an intent structure, which is converted to the N^2 chart, highlighting the essential interactions between the elements. The N^2 chart is then 'clustered,' either by eye or, as in this case, using a clustering algorithm, to reveal the incipient architecture in the system, which is dictated by the many and various interaction pathways between the entities.

The N^2 chart, Figure D.7, reveals some question marks, indicating where the process may have missed some important connections. It seems likely that there would be a relationship between climate control and CPRM, for instance, since climate control will necessitate the continuing utilization of resources. Similarly, command and control should have a relationship with rules of engagement, since the one governs — or should govern – the behavior of the other.

SM5/5 partition into solution system interacting subsystems

Having corrected the omissions shown in Figure D.7, the N^2 chart may then be re-clustered to produce the form shown in Figure D.8 below, where the clusters are different because of the additions/corrections made in response to Figure D.7.

The clustering process brings together related functions, which may then be grouped by partitioning the figure as shown, and giving sensible titles to the emerging subsystems. The groupings are straightforward, revealing four principal subsystems: combat, C^3I, logistics and transport, and human resources.

```
First Moment
Clim Contr      1     J                        ?
Wpns Man        2        S   1
Int/Recce       3          R 1
UMA Man         4        1 1 Q 1
Self Def        5            1 P 1
Formate Man     6              1 0 1
C and C         7                1 N 1                      ?
Engag e Sim     8                  1 M 1
Image Centre    9                    1 L       1
Mobile Sup     10                      H 1 1
CPRM           11     ?                 1 I   1
Log Supp       12                       1 G 1 1
Comm Centre    13                    1   1 1 K   1   1
Air Transp     14                         1   F 1
Base Resup     15                       1 1 E 1
Training       16                         1 D 1
Perf Rec       17                     1       1 C 1 1
Psych Mon      18                             1 B
ROE Man        19              ?               1   A
```

Figure D.7 Initial N^2 chart of the internal functional architecture for MLF2010, showing suspected missing relationships. The names of the various entities have been abbreviated by the CADRAT™ tool, of which this is a printout.

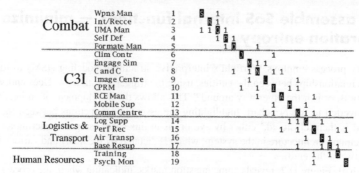

C3I: Command, control, communications and intelligence

Figure D.8 Re-clustered N^2 from Figure D.7, partitioned to show interacting subsystems. The titles of the various functions have been truncated by the CADRAT™ tool, of which this is a printout.

These four subsystems proffer the basis for functional architecture. It is noteworthy that, using the approach illustrated here, architecture *emerges* — essentially it has emerged from the nature of the problem and the issue, since each of the functional elements relates back to the LMF 2010 objectives, which in turn came from the Prime Directive for the whole solution system. The PD, of course, was formed so as to 'neutralize' the symptoms emerging from the problem space, i.e., the issue symptoms.

A closer examination of the source of the structure shows that it facilitates the necessary interactions between the functional elements. It is, in essence, the infrastructure that enables the synergistic interactions between related functions that will create emergent properties, capabilities and behaviors in the eventual whole, operational system.

SM5/6: Develop SoS internal architecture

The subsystems identified in Figure D.8 can be drawn up into a functional architecture, Figure D.9, which shows all the interconnections/interactions from the clustered N^2 chart, and at the same time shows subsystem affiliations.

The many Prime Mission Functions (PMFs) already identified may be 'attached' at suitable points to this internal architecture. In the conceptual design, weapons, both lethal and non-lethal, will be delivered by UMA, which may be launched singly or in formation, under the direction of command and control. UMAs may also defend the mobile land force, as well as be used for reconnaissance and to support the gathering of intelligence. So, all weapons, UMAs, self-defense, etc., will connect with the corresponding internal management function.

SM5/7: Formulate Solution System Overview

The functional architecture of Figure D.9 is incomplete: it does not interconnect to the outside world. Moreover, by virtue of its derivation, it applies to the whole MLF 2010. It does not, for example, refer exclusively to the mission elements that are 'inserted' into hostile territory. An abstracted view shows the wider world and the environment, with the MLF as an entity interacting with other entities — see Figure D.10; Figure D.9 may be an intermediate abstraction; and there is at least one further degree of elaboration, in which the interconnected architectures of various vehicles, infrastructure networks, command and control links, etc., can be seen in some detail.

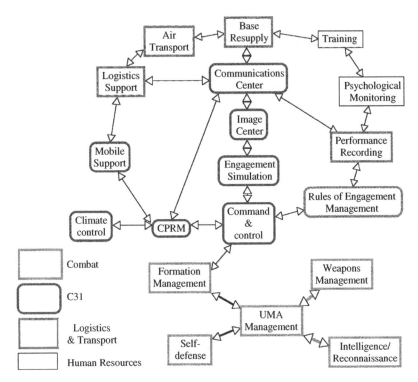

Figure D.9 Mobile land force functional architecture, from Figure D.8.

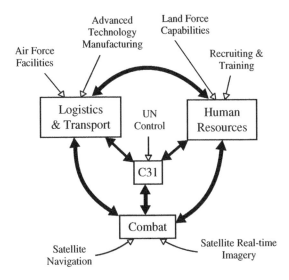

Figure D.10 Abstracted MLF 2010 architecture, showing the force interacting with other systems. . . .

SM5: Part 2. Design Solution System — Functional/Physical Design

Only the GRM (Function) and the GRM (Behavior) models have been used so far: that leaves the GRM (Form) model outstanding. This model identifies power, energy, subsystems, cohesive and dispersive influences, etc. In particular, many of the features in the Viability Management Model — S-MESH — find physical 'effectors' in the different parts of the Form model.

- Maintenance, for instance, may emerge as tangible self-test facilities, maintenance technicians, spare parts, replacement personnel, on-the-job training, test facilities, etc.
- Survivability may emerge as tangible armor, camouflage, physical redundancy, anti-missile missiles, damage repair crews, etc., etc., according to the design.

Sufficient information about the putative solution system has been gathered in previous sections to posit optional physical solution concepts. The concept is emerging of a highly mobile and transportable land force, one that may be air-inserted near the scene of activity. This Mobile Land Force (MLF) makes extensive use of unmanned aircraft (UMAs), beyond anything currently in service. UMAs will engage any contacts/incursors/enemy at a distance from the main force elements.

The concept is not dissimilar to a naval task force formed around an aircraft carrier. Carrier-borne aircraft carry out reconnaissance, defend the fleet, and mount attacks ahead of fleet. An aircraft carrier can be highly vulnerable, however, and demands that considerable effort be expended to defend it.

1. The MLF could be formed around a land-based 'carrier' able to launch and retrieve UMAs while on the move. Other fighting vehicles/aircraft would be needed to defend the land carrier.
2. On the other hand, the MLF could comprise a variety of vehicles, some able to launch UMAs, others able to retrieve, with yet others able to control, so that there would be a functional split in the control and operation of UMAs. Other fighting vehicles would defend the core UAV force.
3. Or, the MLF could comprise a number of semi-autonomous vehicles, each able to move, fight, launch, control and retrieve own UAVs, so that the vehicles would be all substantially identical. This would make for a more flexible force, although each vehicle might be more complex since it would have to perform all functions.

Intermediate task!

1. Consider each of the optional solutions, and one other of your own.
2. Develop a distinct CONOPS for each, highlighting any differences.
3. Are you able to identify the relative strengths and weaknesses of the options.
4. How well? How sure are you?
5. Are you able to choose which you prefer?
6. How firm is your choice?
7. Could you prove (e.g., to your boss) that your choice is the best?
8. Could there be much better, quite different options? How would you know? How could you find out?

In addition to vehicle configurations, there are Measures of Effectiveness (MOEs) to consider:

- What if the UMA's cannot get airborne, cannot see, or get shot down? Can the terrestrial vehicles find and engage the enemy, and defend themselves in the absence of UMAs?
- What about survivability?
  - Avoidance of detection?
  - Stealth features might be important, particularly to minimize radar and IR signatures, equivalent echoing areas, etc.
  - Camouflage could prove important, not only to avoid detection, but to blend in with the environment. So, a UAV might be made to look, say, like an indigenous bird so that it drew neither human nor faunal attention.
  - Terrain screening. It will be in the nature of the terrain — plains, tundra, desert, and savannah, perhaps even wetlands — that there will be no trees and little or no vegetation for hiding. There may be hills and outcrops to facilitate hiding. In some situations, it might be possible/necessary for vehicles to submerge beneath lakes and in bogs as the only cover.
  - Passive radars would reduce the prospect of being detected and targeted, along with greater reliance on detailed satellite reconnaissance.
  - Undetectable CNI radios would be advantageous — they would act as the network for network-centric operations, providing not only communications but also accurate relative navigation and identification. If, in addition, they operated in a part of the radio spectrum that absorbs radio waves, then the transmissions would not be detectable at ground level beyond a very few miles. Such absorption bands, normally the bane of communicators, could prove invaluable in this situation.
- Self defense?
  - New close-in weapons system (CIWS) requirement:
    - Proposed as a new guided-energy weapon, capable of multiple simultaneous engagements,
    - Also doubles as an attack weapon, with the energy beams being reflected/steered off mirrors on UMAs and directly on to targets.
  - The same mirrors and attack optics may be used in reverse to observe potential enemies visually, making this new weapon system conceptually multi-role, with surveillance and reconnaissance capabilities.
- Damage tolerance?
  - Light-weight active armor,
  - Multiple redundancy at vehicle and systems levels;
  - Self-healing systems;
  - On-the-move damage repair teams . . . if the MLF is to fight on the move, which is the more rational approach in open territory, then damage repair, and maintenance, must also be done on the move, so that mobile teams, with their own form of transport, may be needed.

SM5/11 Identifying an Option

The more we look at the potential design in this way, the more options can be seen, and design decisions have to be made. Looking at Mobile Land Force 2010, it is not unlike the challenges facing early Apollo. Looking back at Apollo, there were several levels of system design. The top level included:

- all of the major modules;
- how they fitted together;
- how they acted as one;
- how they could separate, act independently, dock, etc.;
- particularly, how they served the CONOPS;
- in fact, more like their emergent properties, capabilities and behaviors;
- not very much about the internals of any of them;
- all of that is second level design — one for each major module, each with its own CONOPS — part of overall CONOPS.

Similarly, system design for MLF2010 consists of:

- air transport vehicles, those loaded with the 'transportable land elements' (TLEs);
- TLEs forming the ground element of MLF2010;
- UMA/RPVs forming the interdiction, air attack, air defense and close air support elements;
- as TLEs emerge from the transport aircraft, so ULAs emerge from the TLEs;
- it may also be that some options require the transport aircraft, with appropriate systems and crews, to assume additional roles as forward air controllers, remote pilots for UAVs, and a communications relay;
- the air transport element looks set to be viewed as an integral part of the system — as any feasible CONOPS would dictate anyway.

The limiting form factor in this example is likely to be the capacity of the transport aircraft:

- what weight it can carry;
- what size vehicle it can upload and insert;
- so, how many vehicles can it carry at once . . .
- . . . and, what kind of aircraft is it anyway?

As with Apollo, the delivery vehicle provides the overall limitation in terms of weight, size and shape. The delivery vehicle also has to be appropriate to the operating environment. To cut the Gordian knot, i.e., to identify a particular option, we will boldly assume a bespoke, V/STOL transport, 2500 nautical miles single-hop range when fully loaded, with a carrying capacity at maximum range of 35 tons. This means that there will have to be more than one Land Force 2010 base around the world to provide 'instant global cover.'

Our solution option carries three transportable land elements (TLEs) of 10 tons each, in tandem in the cargo hold. TLEs will be the manned and unmanned vehicles that will travel over land, permafrost, savannah, bog, etc. The remaining 5 tons of the 35-ton capacity are made up from:

- command and control + CPRM;
- remote vehicle control stations for:
  - TLEs;
  - UAV/RPVs;
- intelligence suite;

- communications, including satellite communications (Satcom) and a focus for network centric operations;
- logistic supplies;
- repair bays.

A full force might comprise 20+ such aircraft, with 60+ TLEs deployed at once, each with multiple UMA/RPVs active simultaneously, all on the go. . . .

Transportable land element (TLE) design concept

At 10 tons, the TLE vehicle is much lighter than a tank: the concept employs vehicle agility and camouflage, rather than heavy armor for survivability. The TLEs, or ground vehicles, are not intended primarily to engage an enemy. Instead, they carry a wide range of UMA/RPVs that can deliver weapons. Operators are not intended to come into contact with enemy. Hence, Blue casualties should be minimal . . . in principle.

Each of the TLEs is externally similar. Each has a skirt which can be used to hover, so that the vehicles can get out of bogs, ponds, quicksand, cross water, ice, rocky desert, etc. Under the skirt are retractable drive wheels/half-tracks for normal road/off-road use.

The TLEs have no windows, doors, or visible apertures. The sides are silvered like a mirror, which renders the vehicles virtually invisible from ground level.

The top of the TLE displays a live 'photocopy' of the road being passed over. This concept is based on the hand-held scanner used to scan pieces of paper by passing the scanner from top to bottom of the page. For the TLE, the scanner is fixed under the front of the vehicle, so that it scans the ground as the vehicle moves. A copy of the scan is presented on top of the vehicle, which therefore looks like the road when viewed from above. In this way, the vehicle can be virtually invisible, both while stationary and when on the move. Because of its ability to blend in with any terrain, the TLE will be called Chameleon.

Since such suggestions of invisibility may seem fanciful, the design concept may be substantiated with models. Figure D.11 shows three Chameleons on a rocky plain. One Chameleon is plain, with a matt grey paint finish, and is highly visible. The second is mirrored on the sides and on top, with the result that the top reflects the bright blue sky so that, while the sides may be hard to see, reflections from the upper surface 'give the game away.' The third is silvered on the sides, but the top is a photocopy of the ground that the vehicle has just passed over, so the third vehicle is hard to distinguish from above, as well as from the sides.

Of course, shadows will be difficult to conceal, as the figure shows, and dust may be thrown up as the vehicles move. . . .

Figure D.12 shows a single Chameleon a short distance away, again on a rocky desert plain. Only the shadow cast by the sun is visible. The front face of the vehicle is also in shadow, but that is part of the photocopy of the ground, and its brightness may be adjusted. Conceptually, such adjustments could be effected using a UMA as an 'eye in the sky' to observe and minimize any different in brightness between photocopy and surrounding ground.

The difficulties being experienced by those trying to create effective robotic vehicles suggests that the Chameleon should be driven using full internal controls in the first instance. However, each Chameleon should be designed so that it can also be remotely controlled from another Chameleon, or from the transport aircraft. Each Chameleon will be supplied with stereo TV, and high-definition radar, so that the driver may operate the vehicle effectively either from within, or remotely, using the integrating mesh network to share both the sensory information and the vehicle controls.

Figure D.11 Transportable land element (TLE), or Chameleon. From left to right: unmirrored; all-mirrored; sides mirrored, with top showing 'photocopy' of ground beneath vehicle.

Figure D.12 Single Chameleon on the move, showing concealment effectiveness.

Figure D.13 'Naked' prototype Raptor UMA/RPV. Each Raptor can be 'dressed' to represent a large bird of prey indigenous to the action territory, with appropriate feathers and coloring.

Ongoing developments may make the vehicles steer and drive autonomously in the future, and in the full concept, there may be only one or two manually driven Chameleons, with the other vehicles formatting on each other after the manner of a swarm, shoal or flock.

UMA/RPVs and weapons

The Raptor Concept. Two categories of UMA/RPV are envisaged: the Raptor and the Dragonfly. As the name implies, the Raptor is larger, shaped like an indigenous bird, and can roam some distance ahead of the main Chameleon Swarm – see Figure D.13 for an artists representation of the prototype Raptor before it has been 'dressed' to look like a bird of prey indigenous to the area of action. The Raptor is powered to sustain flight, but is intended to make best use of thermals and air currents. The Raptor:

- Uses nano-technology
- Uses biological muscle parts;
- Wings are solar panels;
- Legs are radio antennae;
- Eyes are video cameras;
- Tail is 'flat' radio antenna;
- Is able to soar and fly on its own;
- Can also be guided;
- Can carry 'Dragonflies,' weapons and mirrors;
- communicates via a network not unlike that of a third-generation mobile phone system, carrying photographs and video 'back' to command and control, voice in both directions, and control signals to direct flight and manage weapons.

Figure D.14 Dragonfly UMA/RPV hovering over mud-brick desert dwellings, looking and listening for sings of activity. . . .

The Dragonfly concept. See Figure D.14, which shows a prototype Dragonfly hovering over a group of mud brick buildings; the Dragonfly is listening and looking for signs of activity which will be transmitted in real time either directly to command and control, or via a nearby Raptor, serving as a relay.

Dragonflies are much smaller than Raptors, and their wings beat rapidly to maintain hovering flight. Like the Raptor, they too use nanotechnology, but in addition, they use biological muscle taken from the humming bird to power the rapid wing beat. Dragonflies may be launched from Chameleon directly, but having relatively short range, they are more likely to be launched from an airborne Raptor when it is near to a suspect target.

Each Chameleon carries several Raptors and each Raptor carries several Dragonflies, and weapons: some Raptors/Dragonflies are used for reconnaissance, with encrypted video links back to Chameleon and to the transport aircraft/command post. Raptors and Dragonflies can be launched on the move: a concealed roof panel opens, the UMA rises on a platform, resealing the Chameleon, and the UMA lifts off. The process takes less than 5 seconds. Recovery is the (automated) reverse procedure.

Weapons. Raptors and Dragonflies may be equipped with specialized payloads for psi-ops, for negotiation, and to spread confusion. Dragonflies, for instance may be fitted with miniature loudspeakers speaking pre-recorded messages to dissidents. Using the same equipment, command and control personnel may engage in two-way negotiations, using the Dragonfly as a relay.

Reconnaissance Raptors glide and soar, staying aloft for hours, scouring ground for clues, using thermals to maintain altitude and solar energy to recharge their batteries.

Both Raptor and Dragonfly can carry weapons:

- nonlethal personnel anti-riot weapons;
  - sleep-gases, stun devices;
- area anti-technology weapons;
  - SREMP, the nonlethal short range electromagnetic pulse weapon
- area blast weapons;
  - fuel-air and thermobaric weapons to access buildings, caves, etc.;
- point impact and blast weapons;
- energy weapons, rockets, canon, etc.;
  - directed energy weapon housed in the TLE;
  - energy beam is reflected off a mirror positioned on a Dragonfly;
  - the Chameleon operator sees the target through telescope, remotely adjusts the mirror on the UMA, fires, assesses damage — so addressing the full SATKA cycle, i.e., surveillance, acquisition, tracking and kill assessment;
  - this is a development of the fighting mirror proposed for the strategic defense initiative (SDI), but here used in a tactical role, and at relatively very low energies.

Swarming and formation control

A number of Chameleons operating together constitutes a Swarm. Many of the Swarm may be robotic, maintaining their distance with others in the Swarm automatically, as do bees in a swarm, or birds in a flock, etc. Potentially, a Swarm might be an attractive target to an enemy, so the TLEs employ 'formation control,' which maintains the general Swarm concept, but which allows for different distances between vehicles according to terrain, threat, etc.

Figure D.15 shows a typical Swarm reforming after having broken into three groups to negotiate a rocky outcrop. The central passage presented a potential trap, hence the tactic of dividing, both to minimize any casualties and to counter attack any concealed opposition from behind. In such a situation, the central group might comprise robotic vehicles only, with any manned vehicles skirting around the outcrop, yet able to 'stage manage' any conflict that might arise. This is consistent with the aim of minimizing casualties.

As the Swarm reforms, it is approaching a narrow defile, where it will adopt a defensive 'arrowhead' formation, in anticipation of hostile forces further up the gulley. Formation Management, identified in the Functional Architecture design, may exercise such formations and formation control automatically.

The chameleon TLE — internal design concept

It is possible at this early stage in whole system design to propose a schematic layout for the internal design of the Chameleon TLE. Figure D.16 shows an outline. There is a central area, Operations, with seats and screens for any human operators. A vehicle management system occupies the front section: this will allow for manual or automatic steering, using built in TV, infrared and short-range radar sensors. Separate transmission units are located on either side, with the main power unit sitting athwart, and low down, to keep the centre of gravity within sensible bounds. The power

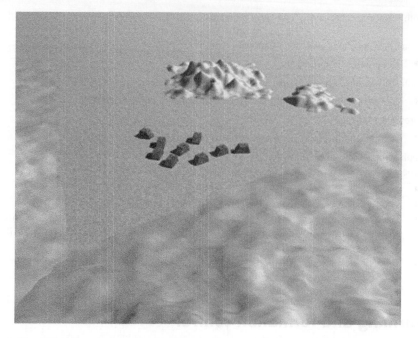

Figure D.15 A Chameleon Swarm reforming after negotiating rocky outcrops. Note, the Chameleon are shown with neither silvering nor 'photocopy' top surfaces; else, they would be virtually invisible.

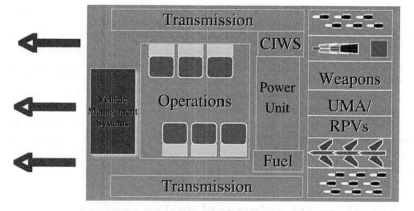

Figure D.16 TLE chameleon internal layout schematic.

unit powers the transmission, provides electrical power and also blows the air into the skirt for hovering operations — the Chameleon can employ the hover blower and conventional transmission at the same time, to minimize impact on the environment. Not shown, the Power Unit is also able to blow up flotation bags stored under the skirt; these would be used to rise to the surface after being submerged to avoid detection.

The rear of the vehicle is dedicated to weapons of various kinds, including the close-in weapons system, canons and ammunition, SREMP, Raptors, Dragonflies, together with their arming, launching and retrieving mechanisms, etc. Raptors and Dragonflies are self-arming, with automatic takeoff: appropriate stores are coupled to air vehicles on a rack-mounted chain belt system — the whole is selected, armed, and controlled by operations, either from within or remotely.

The VSTOL transport aircraft/operations HQ/logistics support

The design option that has been explored in these sections was determined in no small part by establishing the limit in size for the transport aircraft. An internal schematic for the aircraft main cargo bay is shown in Figure D.17

As the figure shows, three Chameleon TLEs are loaded in tandem, nose to tail. They may be unloaded on the move, using parachutes to extract them on palettes. Alternatively, the VSTOL transport aircraft is able to land almost anywhere, and allow the Chameleons to drive out, down the rear ramp, even while the aircraft is taxiing at speed.

Also in the cargo hold are the local operational command HQ facilities, including intelligence, contingency planning and resource management, together with a communications center.

Using Raptors as radio relays, operators in the parked aircraft may monitor any action, and may remotely steer, direct, negotiate, etc., all with minimal risk. Alternatively, they could do the same while airborne, and perhaps well to the rear of any area of action. If some of the Chameleons are manned, then the aircraft crew can act in support, liaise with UN/US Global Peace Command, gather and analyze intelligence, etc.

There is potential for the VSTOL transport aircraft to take a more active role: it could be fitted, for instance, with sideways looking radar and infrared (SLR and SLIR), with self defense weapons for use both on the ground and airborne, and possibly with attack weapons in support of the ground elements.

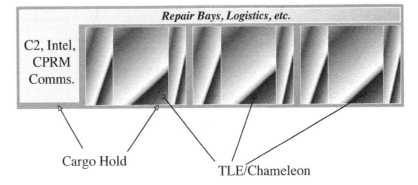

Figure D.17 Transport aircraft cargo bay schematic, showing the full load of three Chameleons, plus repair bays, logistics support fuels, lubricants, weapons, batteries, etc., and the aircraft-based control center.

SM5/12 Re-do Steps 2–5/7 for each Interacting Subsystem within Optional SoS (Containing System)

So far, we have described only one option, which we selected somewhat arbitrarily as an example, and we have shown it to satisfy its CONOPS, and meet the design objectives set out earlier. Now, it would be prudent to 'go around the inner loop' of system design, i.e., to re-do SM2 to SM5/7 for each of the subsystems, in the context of the whole system. This would be best done for all subsystems at the same time, to enable interface and interaction negotiation and adaptation. For example, the system design of Raptor must be compatible — in detail — with that of both Chameleon (which stores, arms, launches, controls and retrieves it), and with Dragonfly (several of which can be housed within Raptor, launched from Raptor, and possibly even retrieved by Raptor).

So, each subsystem — VSTOL aircraft (command post), Chameleon, Raptor, Dragonfly, network, etc. — will have its own prime directive, semantic analysis, objectives, threats, prime mission functions, CONOPS, etc. Figure D.18 shows a diagrammatic concept of operations. The emphasized loop, upper right, shows the continuing process of patrolling, gathering intelligence, identifying potential targets, deciding what action to take, allocating targets, etc. A recognized air and surface picture (RASP) is shown as being formed in the center of the loop: the RASP is an active map of combat area formed using sensed information from all sources — satellite, transport aircraft command post, SLR, SLIR, intelligence, Chameleons, Raptors and Dragonflies. The RASP is formed by all, and available to all: using the Raptors in particular, the RASP is able to move with the action by effectively forming a patchwork from the separate 'pictures' that each of the Raptors transmits.

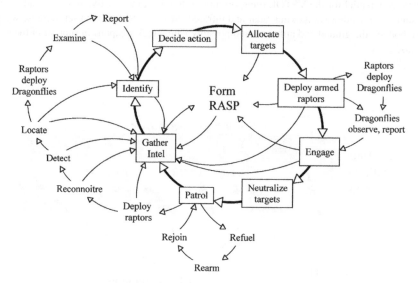

Figure D.18 Operations concept: a more specific and detailed elaboration of the engagement phase of the MLF 2010 CONOPS. RASP: recognized air and surface picture, where 'recognized' means 'identified'. The RASP is a picture of ongoing operations, showing situation, targets, resources allocated to targets, etc.

Command and Control

The loop to the upper left in Figure D.18 shows the process of detecting, locating, identifying, and tracking potential targets. As with the CONOPS, so may there be many Operations Concepts. Note that the Operations Concept at Figure D.18 is incomplete; it does not show, for instance, what happens to Dragonflies post-engagement. Are they expendable, or are they reacquired: if the latter, then by a Raptor in the air or by a Chameleon on the ground *and* on the move?

The design of each subsystem will emerge as an elaboration of the whole solution system: in particular, the design elaboration of each subsystem will interface and interact with that of the other subsystems, ensuring fit, form and function of the subsystem and of the whole.

In this 'inner loop' design process, some whole system functions will not map neatly on to particular subsystems: these will be functions of the whole that cannot be reduced. Examples include formation management, forming the RASP, network management and the coordination of SATKA (surveillance, acquisition, tracking and kill assessment), and target allocation, which has to be organized and 'deconflicted' at whole system level, to avoid UMAs from different Chameleons interfering with each other's actions, either by wasting resources through targeting the same hostile, or by targeting each other.

All of these indivisible, whole-system functions are elements within the vital, coordinating, executive command and control (C^2) system — referred to by some as C^3I — command, control, communications and intelligence — and by others as C4ISTAR — command, control, computers, communications, intelligence, surveillance, target acquisition, and reconnaissance. (Indeed there is a veritable alphabet soup of acronyms and letters attempting to bring various aspects of command and control within their remit.)

It is a feature of wholes, such as MLF2010, that there will be functions of the whole that cannot be properly ascribed to, or mapped on to, the parts and subsystems. Command and control is one of these — in nonmilitary systems, different words would be used to express much the same notion as C^2, e.g., executive control, central government, etc. Command and control is the brain of the whole, the part that observes, assesses, plans, formulates strategies, makes decisions, executes the plan, controls operations and maintains the integrity of the force in the face of hostile action.

Figure D.19 shows a typical command and control 'loop,' or continuous flow of information and activity. The loop shows how information is gathered from various sources — radar, video, sideways looking infra red, intelligence, etc. — and brought together to assess the situation in which the mobile force finds itself. Some of these sources provide their information automatically, once the whole system is operational, and some have to be 'polled.' Situation assessment itself may be a human operation, since experienced operators are likely to be superior to, and more reliable than, machinery in conducting this task.

The C^2 process then identifies threats and opportunities — which may be a human activity supported by machinery to manipulate and present information in revealing ways. Having identified either a threat or an opportunity, or both, it is prudent to review constraints before developing a strategy to address the threat or opportunity, since some otherwise attractive strategies might not be feasible in the circumstances. Any strategy should also be consistent with doctrine (the 'book of how to conduct operations properly, effectively, ethically and morally') and the contemporary rules of engagement, which will have been conveyed from higher levels of command. . . .

Not shown in Figure D.19, but a potential element of C^2, will be the ability to simulate operations before actually executing them: so, a strategy may not only be chosen, but it may be tested

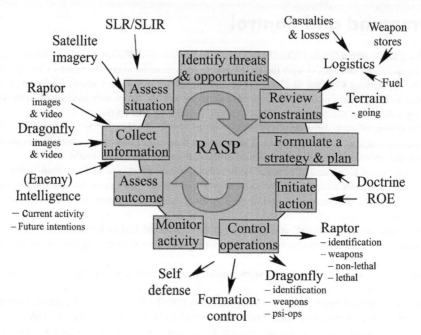

Figure D.19 MLF2020. Typical command and control operations loop. RASP is recognized air and surface picture. ROE is rules of engagement. SLR/SLIR is sideways looking radar/Infrared.

in simulation prior to execution. Such simulation may be carried out by a TLE commander, or it could also be carried out in the command center of the transport aircraft, or even back at US Global Peace Command, with UN involvement — this would be made possible by the communications network which underpins all of the cooperative and coordinated activities and information sharing.

C^2 then proceeds to initiate and control operations, generally using the UMAs to identify and interact with the 'enemy,' and the CIWS for self-defense of the TLEs. The result of operations changes the situation and the cycle repeats continually, with multiple operations perhaps happening in parallel.

If function management choreographs and orchestrates emergent properties, capabilities and behaviors, as has been proposed earlier, then C^2 is the orchestra conductor. In MLF2010, C^2 will be a sociotechnical system of the whole, rather than another of the subsystems that it 'orchestrates,' and the major human element will require training, exercising and experience — without which the whole MLF 2020 will be ineffective.

Fractal C^2

Although C^2, like executive management and central government, may be indivisible, it may be looked upon as fractal: i.e., there is a degree of C^2 self-similarity in the various systems and subsystems. As central government 'spawns' regional government and local government, and as executive management spawns divisional management and departmental management, so C^2 of the whole MLF 2010 will spawn C^2 of the transport aircraft/command center, the TLE/Chameleons, the UMA Raptors and Dragonflies: each viable subsystem will have its C^2 system — a miniature version of the whole.

Each TLE will require a command and control system to monitor the sensors, plan the route and drive the vehicle: the C^2 system will activate, arm and launch Raptors, control them in flight, direct them towards suspect targets, steer them away from danger, receive and present their surveillance and reconnaissance information, enable their 'attack modes,' recall and recover them. Each Raptor will have a command and control system to activate its various subsystems, to enable it to steer, climb, detect thermals, soar, travel between thermals, direct its sensors, transmit information, receive and respond to control signals, arm and launch weapons, activate and dispatch Dragonflies, report back to its parent TLE, etc.

Design Summary

In other contexts, there would have been a great many options to examine, model, simulate and compare, and many of the comparisons would have been detailed and numerical. We have identified many of our singular option's properties, capabilities and behaviors in general terms, and we could similarly describe other options — and verify them against their CONOPS.

In the process, we have identified a variety of new, novel and/or updated requirements: e.g., Chameleon camouflage, raptor surveillance, new weapons, etc; this is to be expected from employing a systems methodology, which should present opportunities for useful innovation and creativity. However, we have yet to be specific about vital parameters: radars have been mentioned without identifying transmitter power outputs, receiver sensitivities, capacities, ranges, effectiveness . . . etc. As the design progresses further, such definition will be needed if the eventual design is to be optimized.

SM6/1 Instantiate as Single Organismic Blue SoS Dynamic Model

Figure D.20 shows a layered GRM version of the MLF 2010 design. (See the Instantiated layered GRM on page 138.) Note that all the vehicles/subsystems are treated as one system. This is valid, since we have been following the systems methodology that synthesizes whole, organismic system solutions. In particular, we have mapped the whole functional architecture on to the functional/physical partitions, so designing them to be open, interactive, adaptive subsystems. In the case of the MLF 2010, the network-centric nature of the design ensures that the parts behave as a unified whole, that is organismically — provided, of course, the network is transparent to all the necessary subsystems interactions, sensor information, communications traffic, automatic target allocation negotiations, etc., etc.

However, different design options would introduce different values for many of the parameters. For example:

■ battle damage might be greater with fewer, larger, concentrated vehicles. On the other hand . . .
■ . . . battle damage repair might take much longer with more, widely dispersed vehicles;
■ similarly, rearming and refueling on the go would be quite different for different options;

- different approaches to camouflage might render the TLE — which we have dubbed the Chameleon since our design option enables it to blend into its background — more visible, and hence more vulnerable;
- using active radars within the TLEs, or homing beacons to direct returning UMAs might reveal TLE location to an enemy making good use of electronic surveillance measures (ESM), with the concomitant risk of enemy action

Altogether, given that we have somewhat arbitrarily chosen a particular option, it might seem unreasonable to find that we still have an inordinate number and variety of options to address. This is because we have yet to optimize the design.

SM6/1 and 6/2 Instantiate Blue MLF2010 and Red Opposition

One method of optimizing the design is outlined in The GRM and the Systems Approach on page 135. Essentially, the idea is to instantiate the design of the MLF 2010 and to set the simulated model to action against an opponent. In the absence of any identifiable opponent, as in this instance, it is convenient to duplicate the MLF2010, and set it to fight itself. Keeping this Red model as a dynamic, interactive reference, the Blue Model can then be optimized using cumulative selection

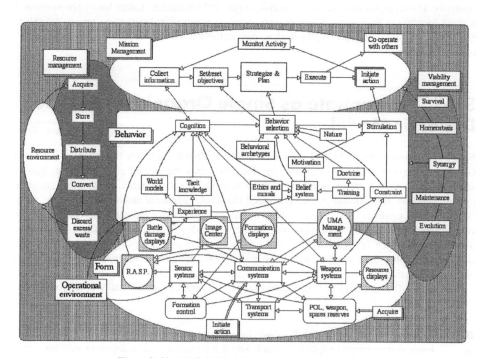

Figure D.20   MLF 2010 as an instantiated, layered GRM.

methods — see Optimizing the whole on page 266. This approach has the distinct advantage that it optimizes and balances both offensive and defensive capabilities at the same time.

Space does not permit the full optimization process to be illustrated. However, the choice of optimizing parameter is an important consideration. There are choices, the commonest being cost-effectiveness, cost exchange ratios and casualty exchange ratios. In light of the original issue, with its concerns about minimizing own force casualties, it would seem sensible to emphasize casualty exchange ratios.

However, cost will be important, too. Parameters will be included in the optimization process that will be used to predict both the effectiveness and the cost of various contributing technologies. For example, there may be a parameter (often presented as a graph) relating power, range and performance of a weapon, a vehicle, a radar, etc., with its cost; in making incremental changes to the projected subsystem performance as part of the cumulative selection process, the model would also make concomitant incremental changes to its predicted cost.

Using this approach, the cost of each of the myriad of options generated by cumulative selection can be accumulated, and used as the denominator in the effectiveness-to-cost ratio. The dynamic simulation will also generate the numerator, so permitting cost-effectiveness to be calculated and used to 'steer' the path of cumulative selection towards the most cost-effective solution, where effectiveness in this instance will have to include minimizing loss of manpower.

The outcome is a matched set of values, specifications, etc., for all the parts as they operate together to form the whole.

Having optimized Blue against Red, it would be prudent to try the optimized MLF2010 against a variety of different opponents in a variety of different scenarios and environments, to check that there are no 'holes' or deficiencies in the optimized system's emergent properties, capabilities and behaviors.

Summary

The result from applying the systems methodology Steps 1–6, including both the outer and inner loops, will be a matched set of optimal parameters—specifications, schematics, architectures, values and measures—for the Blue MLF2010.

This is a great advance on conventional methods. Matched specifications show each subsystem making its best contribution to overall mission effectiveness while the various subsystems are operating and interacting with other systems under operational conditions: this is organismic synthesis, creating open, adaptive solution systems! It is, moreover, a proper application of the systems approach.

The systems methodology, if carried through into SM7: Create and Prove SoS, will result in a solution system with dynamic emergent properties, capabilities and behaviors. Interestingly, the simulation of MLF2010 exhibits emergent behavior, too, plus intelligence — unexpected for a simulation.

For instance, the (simulated) force realizes when it is losing an exchange and will change tactics, withdraw and reform, change Rules of Engagement, etc. So, too, of course, will its opponent in the simulated engagement, with the result that the patterns of behavior in a conflict may prove complex, even counterintuitive, with each side responding to changes in behavior of the opposition. (This, after all, is what happens in practice – see Case C: The Total Weapon System Concept.)

The solution system parameters contribute to an optimum solution: not too little, not over the top, but just right for successful operations, and consistent with the chosen MOEs. The dynamic simulation approach also determines optimal support, maintenance, and logistics, too. The whole can be shown to resolve the original problem, if SM6:Optmize SoS Design is omitted, and to solve the original problem if SM6 is incorporated.

Conclusion

The Case Study has employed/applied the systems methodology in solving, resolving or dissolving a complex, albeit contrived, problem. Ways of dissolving the problem became apparent early on in the process, when the problem space was explored; as might be expected, these ways were political and social, and the systems methodology could have been employed towards such ends. However, the study elected to pursue the direct route towards solution and resolution of the problem, which was to conceive, design and simulate/build a Mobile Land Force. The resulting design was innovative and creative, particularly considering that the systems methodology was 'applied' by an individual, rather than by an experienced team; or, perhaps that individuality was the reason for the creative outturn.

Reviewing the Case Study, it is interesting to speculate how it would have differed if a team of practitioners, or perhaps a team of teams of practitioners had been employed, i.e., with a team for each of the major subsystems as well as a team for the whole MLF 2010. Certainly, the results would have been more credible, specific and detailed where domain experts had been involved in the process. Would the degree of innovation, with the involvement of teams of experts, have been more or less than for an individual?

One noteworthy aspect from the application of the systems methodology is that a set of research projects has been identified for a number of innovative lethal and nonlethal weapon systems, plus two revolutionary UMAs. This aspect distinguishes systems engineering (and the systems methodology as the 'how' of systems engineering) from reductionist approaches to design and problem solving, which generally seek to respond to requirements stated by some customer, rather than — in this case — identify what the optimum set of requirements might be. . . .

Were this systems methodology carried through to the creation and proving of the MLF 2010 — or MLF 2020 as it would be by the time it was completed — then some at least of the subsystems might be available in prototype form. It is useful to consider staged prototypes, as these may provide essential subsystem dynamic emergent properties, capabilities and behaviors, needed to make the whole MLF 2020 function as a unified whole.

For instance, a prototype Raptor might consist in the first instance of a miniature, radio-controlled glider fitted with a pair of third-generation mobile phone parts to provide the essential elements of network communication, photography and video transmission: prototype Dragonflies might be strapped under the Raptor wings. Dragonfly might be prototyped in similar fashion, perhaps using radio-controlled miniature helicopters, so that early versions of the MLF2010/2020 might operate with prototype UMAs during early exercises and deployments. Similarly, SREMP, CIWS and Fighting Mirror would all engender research programs, and might appear initially in prototype form as a result of such research.

The purpose of the Case Study was not, of course, to design a Mobile Land Force, but instead to demonstrate the efficacy of the systems methodology. It is for the reader to decide whether, in this respect, the case study has proved itself.

Part III
Systems Methodology and Systems Engineering

15

Systems Engineering — the Real Deal

If you can keep your head when all about you
Are losing theirs and blaming it on you
If you can trust yourself when all men doubt you
But make allowance for their doubting too;
If you can wait and not be tired by waiting,
Or, being lied about not deal in lies,
Or, being hated don't give way to hating,
And yet don't look too good, nor talk to wise . . .

From 'If' by Rudyard Kipling 1865–1936

Distinguishing Systems Engineering From the 'Look-alikes'

The essence of systems engineering was identified on page 120.

> The essence of systems engineering is in: selecting the right parts, bringing them together, orchestrating them to interact in the right way and so creating requisite emergent properties, capabilities and behaviors of the whole. Essential systems engineering is executed such that the parts and the whole are operating dynamically in their environment, to which they are open and adaptive, while interacting with other systems in that environment.

This refers principally to manmade systems and it includes systems of all kinds: social, sociotechnical, organizational, technological, process, complex, straightforward. . . . We have also deduced earlier that manmade systems are generally social or sociotechnical — exclusively technological 'systems' turn out to be artifacts that become parts of a system only when they are used. So, it is not helpful to think of a hammer as a system; on the other hand, a man wielding a hammer may constitute

Systems Engineering: A 21st Century Systems Methodology   Derek K. Hitchins
© 2007 John Wiley & Sons, Ltd

a system for driving nails. Such a system can indeed exhibit emergent properties, capabilities and behaviors: speed, accuracy, and accidental collateral damage spring to mind . . . while a hammer on its own does nothing.

Similarly, it is not helpful to think of an automobile as a system; without the driver and passengers, it is an artifact, contrivance, or — if you like — an incomplete whole. By itself, it exhibits no purpose, it does nothing, and it is inactive, inert. Add the driver, spouse and children, perhaps, and we have a sociotechnical system which travels, seeks out destinations, achieves goals, observes traffic regulations — most of the time — creates a comfortable environment with constantly changing scenery, steers, accelerates and decelerates, etc; none of which the car-as-an-artifact can do on its own. And which, come to that, precious little of which the driver could do as well on his/her own. Bring the two parts together and the whole exhibits properties, capabilities and behaviors . . . that are, by definition, emergent, since none can be attributed exclusively to either of the rationally separable parts — automobile or driver.

Is this intending to imply that those who conceive, design and create artifacts are not systems engineering? No, that would be a non sequitur, since — if an artifact were inevitably part of a sociotechnical system — then, to be working on an artifact would to be working on (part of) a sociotechnical system; which, might still be systems engineering.

It does, however, suggest that there may be something special in the manner, approach, method, or way of working to conceive, design and create an artifact-as-part-of-a-sociotechnical system, that may classify that way of working as systems engineering. The key to this special manner, approach, method, or way is characterized by:

- The systems approach, that is, any system-of-interest (SOI) is considered (conceived, designed, changed) to be active, interactive, in contact with, and adaptable to, its environment.
- An omnipresent human element: the SOI either is, or has strong associations with, people as actors, users, operators, controllers, workers, directors, assessors, evaluators, decision-makers, etc.
- The SOI is never considered on its own as an isolated entity.
- The SOI is regarded always in its dynamic context, i.e., open to its environment, adapting, inter-acting, functioning, doing, emitting, absorbing, acting, coordinating, cooperating, combating, reacting, and responding . . . .
- Synthesis, as in the bringing of parts, functions and processes together in the right way, but . . .
- . . . not reduction, as in creating separate parts and either omitting, overlooking, or obstructing, process and functional interaction pathways between the parts.
- Holism, creating the whole solution to some problem, and holism, too in the sense that a system is a whole — and, again, must include the person/operator/user, to synthesize the sociotechnical whole.
- Organicism, causing the parts of the system to operate as a unified whole — essentially through cooperation, complementation and coordination of and between the parts.

'Functioning' is necessary since functions mutually interact to create emergent properties, capabilities and behaviors. So, systems engineering is less about building an edifice vertically from blocks, like a brick wall, and more about enabling functions and processes to interact laterally across a supposed hierarchy, or network, of parts: changing the nature of such lateral interactions can change a system's emergent properties and their degree — as Case C, The Total Weapon System Concept, illustrated. Systems engineering is dynamic, not static. In design, in the absence of extant system substance, the essential dynamic context may be afforded by simulating function and behavior.

Distinguishing Systems Engineering From the Engineering of Systems

The two titles are seductively similar — upon first meeting them, you may be encouraged to think that they are the same thing. However, they can be quite different. Engineering of systems is, as the title makes clear, primarily 'engineering.' It is therefore concerned with the application of science in the design, planning, construction and maintenance of machines, buildings, and other manufactured things.

Human — part of the system, or user of the artifact?

So, the engineering of systems concerns itself with the artifact, the tool, the weapon, the platform, etc., but generally considers the user, operator, etc., to be outside its remit, except perhaps for an interface that may be needed between man and machine. Engineering has a range of established methods, tools and practices which, because of the nature of engineering work, tend to be founded in Cartesian Reduction, and which operate on the principle that the whole equals the sum of the parts — no more and no less.

Cartesian Reduction is based on the notion that it is possible to break down a complex problem into a number of simpler problems, to find explanations of the simpler problems, and hence to aggregate these explanations into an explanation of the whole. Cartesian Reduction has a good track record — it has worked well for centuries, and many of our finest achievements can be attributed to adoption of the reductionist philosophy. However, as we have seen in earlier chapters, it does not always work — some problems resist attempts to be broken down into simpler problems. Some systems have properties of the whole that cannot be exclusively attributed to any of the rationally separable parts: while this is true of a few simple systems, it is particularly true of complex, dynamic systems; for these, systems engineering — the real deal — will prove necessary.

Linear vs nonlinear

An approach to engineering systems, then, is to create 'building blocks' with fixed and predictable characteristics which can be put together and which are intended not adapt to each other. In electronic design, for instance, the blocks might be circuits that can be described in 'black box' terms as two-port networks, with defined inputs and fixed transfer functions providing predictable outputs. The term 'transfer function' applies generally to linear, time invariant, single input, single output systems.

In electronic engineering, for example, considerable effort is made to ensure that, while signals/data may be transferred between transfer function 'blocks,' the transfer functions do not themselves adapt to such interactions. Electronic blocks might be coupled optically, to prevent any electronic connection/reaction, or impedances might be driven towards zero or infinity, with the same intention. In practice, such desired linearity and loose coupling are achieved only approximately, partly by ensuring that devices are not overdriven.

Engineering in general at present, then, is largely linear, and technological devices resist adapting. It is possible, in this context, to configure a series of blocks, each with its unique transfer function, such that the output from one forms the input to the next, and to describe the overall, end-to-end transform mathematically as the product of the individual transforms — and be right. So, the intent, and to a large extent the achievement, is to make the whole equal the sum of the parts. And, using this approach, it is possible using the building block method to 'build up' larger, more complex functions.

Systems engineering, on the other hand, as identified with the systems methodology, owes little to linearity. The various parts — systems, subsystems, sub-subsystems — are by definition open to their environments, interacting with other systems in that environment and adapting to the interactions, so that the parts and the whole may be nonlinear — as would be the case if the parts were organs in a body, people at a meeting, or predators and prey in Nature or in business.

On the other hand, there is nothing in the systems methodology that *requires* nonlinearity. It is straightforward to apply the systems methodology and to conceive, design, configure, simulate, etc., assuming linear cause and effect throughout — observing, of course, the physical laws — Newton's Third Law, Le Chatelier's Principle, the Conservation Laws, etc., as they refer to action, reaction, homeostasis, balance, regulation, and so on. In effect, the linear system may be thought of as a special case of nonlinear systems, with 'a nonlinearity coefficient of unity.'

Top-down vs bottom-up

Comparing the two paradigms, then, systems engineering and engineering of systems, it is apparent why systems engineering is sometimes referred to as 'top-down,' while engineering of systems may be thought of as 'bottom-up.' In other words, systems engineering starts from some high-level purpose and CONOPS of the whole, and successively elaborates to reveal increasing functional detail — which is top-down.

Engineering of systems, on the other hand, builds up desired functions through the linear combination of simpler parts, or building blocks — which is bottom-up. Elaborating top-down inevitably shows the interactions between the parts, since they emerge with the elaboration. Synthesizing bottom-up, however, requires the interactions and their pathways to be identified, forged, and made to work; if any is missing, incorrect or inadequate, then the whole may not perform/behave as expected.

In practice, top-down elaboration can lead the inexperienced practitioner to elaborate functional designs that could not be readily constructed: so, 'top-down' may require a 'good understanding of the bottom.' This is why, perhaps, many good systems engineers were originally engineers, or physicists, and why 'top-down and bottom-up should ideally meet somewhere in the middle.'

Flavors of Systems Engineering

Systems engineering — the real deal, that is, as circumscribed above — comes in many 'flavors,' or varieties. Some of these are shown in Figure 15.1.

Unprecedented, one-off systems

Systems engineering was developed initially, it seems, to address the big, blockbuster, unprecedented system, such as Apollo and national power generation of electricity using atomic energy.

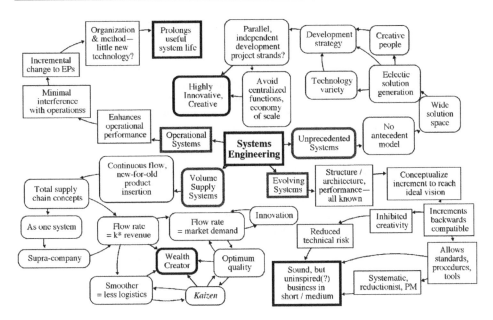

Figure 15.1 Mind map showing four different systems engineering archetypes: operational; volume supply; evolving; and, unprecedented systems engineering.

At some point, each of these was an unprecedented system; there had been nothing like it before, nothing to compare with, no experience of similar things, and so on.

Given this unprecedented nature, the solution space was wide open, and innovation was both essential and inevitable. Creative people were a necessary component of this flavor of systems engineering; people with vision who were prepared to make difficult decisions, take calculated risks. Similarly, since these unprecedented systems were challenged with achieving unprecedented goals, it was inevitable that they would be using innovative technologies — 'at the cutting edge.'

Such innovation and unpredictability required new ways of organizing and managing, particularly where such blockbuster projects were large and extensive. Program management came into vogue with such projects, so that the extensive whole could be divided into connected projects which were able to progress — or not — independently.

This 'separation into connected projects' provided a degree of control over progress, as glitches in one program 'strand' need not affect developments and progress in other strands. Where one strand fell behind, or was not progressing as expected, then more money, effort and resources could be brought to bear, or that strand could even be started afresh, perhaps using different technologies, different concepts of operations, etc.

Case D presented an example of an unprecedented system, and the degree of innovation that emerged was characteristic of such situations and solution systems.

Evolving systems

Solution systems may be evolved, changed slowly over time, often by making a series of changes/alterations/amendments. The designs of cars and airplanes evolve over time, with seemingly new designs emerging as different marks or variants of the original version. Such changes are unlikely to impact on basic purpose, structure, architecture, and function: instead, they may add capability ('bells and whistles'), change superficial appearance, correct deficiencies in earlier variants, and so on. Significantly, too, the human element of these sociotechnical systems changes too: not, of course, in physical terms, but certainly in terms of expectations. The driver of a new car in 2007, for instance, has significantly greater expectations than had the driver of a new car in 1957, or 1907. Correspondingly, the driver of the new car will have greater capabilities by virtue of experience — he/she will be able to drive faster, brake later, accommodate a wider range of road and off-road conditions, traffic congestion, etc., than predecessors.

Evolution may be planned. Designers and manufacturers of personal computers, digital cameras, software, brown goods (hi-fis, TVs, video recorders, etc.), white goods (washing machines, refrigerators, microwave ovens, etc.), may design a comprehensive product, but introduce it initially in a 'cut-down,' minimal form. Subsequent variants will be marketed that appear to have additional capability, such that the initial version, although still working adequately perhaps, may seem obsolescent. Consumers buy the new variant and discard, or set aside, their initial version. This process may be repeated with a periodic issue of newer versions, until eventually, the full comprehensive product of the original design is available to buy. At that point, the product range may need some substantial revision — or replacement, and the process will start again.

Evolution may not be planned. Sometimes the problem or environment changes, such that the initial solution system no longer works as well as it once did, and must be changed/improved. Sometimes, new concepts, new methods, or new technology come along, such that a system can perform more effectively, occupy a smaller volume/weigh less, cost less, be more reliable, etc.

In relation to such evolving systems, systems engineering faces a less daunting range of tasks compared with unprecedented systems, for instance. Generally, with evolving system solutions, the nature of the problem changes little, the architecture and structure of previous versions is largely unaffected, and indeed much of the internal and external relationships may be governed by the same protocols, interfaces and procedures. Case B: The Practice Intervention examines one small, but significant step in the evolution of an organization.

Systems engineering of evolving systems may involve less innovation, and less risk, if only because there is less scope for creativity. And the results, as Figure 15.1 shows, are likely to be a sound, if uninspired, business in the short to medium term. The prudent manufacturer of evolving solution systems is likely to have a number of parallel strands to the business, creating a balanced portfolio of products, if only as a hedge against changing fashions . . . .

In considering evolving systems, it may be fruitful to consider whether it is the artifact, tool, weapon or contrivance that is subject to systems engineering, or perhaps something else altogether. There is another system involved: the system for persuading consumers to purchase new products and services, even although their current products and services may suffice. And that system is, in its turn, part of a supply system, which is in turn part of some capitalist/consumer economic system — all of which interact with each other, and all of which stabilize, like any open system, at a high energy level.

Extant operational systems

Systems engineering may be active with operational systems, changing them while they are in operation — as opposed to, say, unprecedented systems, where much of the systems engineering seems to precede operation. Case C: the Total Weapon System Concept, which looked at the Battle of Britain, showed operational systems engineering in action. The objective is generally to improve performance, effectiveness, reliability, responsiveness, etc., without disruption of operations. Since these features are all emergent, we may say that operational systems engineering aims to enhance whole system emergent properties, capabilities and behaviors. Often such improvements can be the result of reorganization, rationalization, and changes to practices. Sometimes new facilities and equipments can make the difference.

The hallmark of operational systems engineering is that the system in question improves in some way and becomes 'fit for purpose.' Operational systems engineering may be continual, and may extend the life of a system that would otherwise become obsolete or obsolescent.

Volume manufacturing/supply systems

Last of the systems engineering flavors shown in Figure 15.1, volume supply systems operate potentially at Layers 4 and 5 of the 5-Layer Systems Engineering Model: global lean volume supply systems are significant wealth creators, on a grand scale. Case A: Japanese Lean Volume Supply Systems addressed the global supply, particularly of automobiles, but the principles and practices developed by some of the top Japanese companies can be universal.

Systems Engineering 'Strategies'

The application of the systems methodology may be organized, undertaken and managed in a variety of ways, each of which characterizes the particular project and which features largely in the respective project management plan:

■ Waterfall, in which each phase of a project (e.g., problem solving, concept development, requirements analysis, specification, design, development, implementation, etc.) is completed before proceeding with the next
■ Spiral, or Helical, involving the use of prototypes to elicit better understanding of the requirement and to explore potential solutions
■ Evolutionary, in which new systems are incremental 'improvements' on previous tried-and-trusted systems
■ Simultaneous or Concurrent, in which:
  ● activities which might otherwise be sequential are overlapped to reduce time to completion
  ● designs anticipate the needs of development, production, assembly, etc.
■ Sashami — like waterfall, but with some overlap between phases
■ Regression — which allows each phase to linger throughout the project
■ Chaos — breaks the rigid notion of phases and allows them to come and go as the project evolves.
  ● Once rework breaks out, most projects *de facto* tend towards chaos.
  ● Managers may become unduly concerned about this state, which may offer the best performance in complex situations
■ Goal-oriented, in which the objective is to create the emergent features of a goal system by synthesis and without reduction.

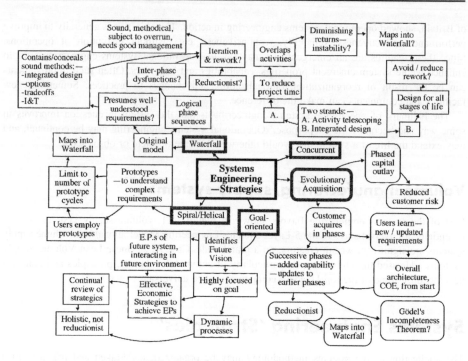

Figure 15.2   Archetypal systems engineering strategies.

Figure 15.2 shows several of these archetypal project 'strategies' in a mind-map, which suggests that the four shown, at least, map on to the waterfall archetype.

The waterfall

Figure 15.3 shows the waterfall archetype in context, as it might apply to a largely technological subsystem. The upper part of the diagram shows an outline of the process of going from problem space to solution space, as guided by the systems methodology. At the end of the whole system design optimization, there will be a matched set of design/requirement specifications for the solution system as a whole and for those viable subsystems contained within the whole.

Each of these viable subsystems may then go through a repeat of the systems methodology, in which each is perceived as open to the environment of the whole, is interacting with the other viable subsystems and external systems, too. This, as has been illustrated earlier, may be thought of as the inner loop, and one archetypal way of organizing is to employ the so-called waterfall model. The model comprises a number of steps or stages, each of which is completed before moving on to the next step or stage.

So, 'Understand the (Operational) Requirement' would be complete only when each of the (possibly very many) operational requirements had been understood, explored, dissected and related to the other operational requirements to form a coherent, complete set. At that point, and not before, the project would move forward to the next step or stage, and — conceptually — all the individual requirements would 'flow down' to the 'Create Solution Design' step or stage.

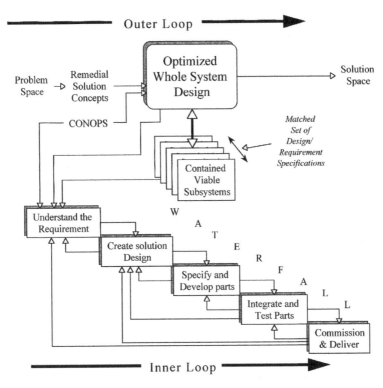

Figure 15.3 Systems engineering archetypal 'waterfall' in context. The whole system design is shown at top, as part of the outer loop. This process identifies a number of open, interacting viable systems, all part of the whole system. Each of these may become the subject of an interconnected inner loop project, which may be practiced in a variety of ways: one of these ways, the classic waterfall, is shown . . . .

As Figure 15.3 shows, this neat waterfall may not proceed entirely to plan. There is likely to be interplay between successive stages, as queries develop. Less obvious from the figure, following stages may be involved before their stage work starts. For example, system designers who are in the process of specifying the functional requirement for a piece of hardware or software, co-opt the practitioner who is about to receive the specification. So, a system designer specifying some operational software, for example, would enlist the support and guidance of the operational software specialist in developing the specification — that way, glitches and misunderstandings, omissions and the accidental specification of the unrealistic or absurd are avoided.

The figure also shows that other problems may arise to prejudice the neat concept of continual flow down the steps of the waterfall. During the process of integrating and testing parts, for example, problems may emerge which reveal that, in spite of best efforts, the parts do not fit together/work together correctly: this may necessitate some redesign. Worse, problems may not emerge until commissioning and delivery, in which case — supposing concessions were inappropriate — it may be necessary to go 'back to the drawing board' or even to revise the original operational requirement — or at least, the understanding of that operational requirement. This would be an extreme situation, but is not unheard of . . . .

However, the archetypal waterfall model is characterized by its dedication to careful, measured progress. This dedication is exemplified by typical systems engineering 'slogans:'

- 'Level at a time,' indicating that elaboration of functional requirements should be executed progressively, with all first-level elaboration/decomposition being completed, before any attempt was made to elaborate further
- 'Breadth before depth,' indicating that diving down into detail in any one area should be avoided, as it would waste time and effort overall, and may bias the design.

The waterfall archetype epitomizes care and caution, and for these reasons, it can prove to be relatively slow. It also works best when the nature of the solution system, e.g., the technology to be employed, is well known and understood. That may not always be the case, particularly in respect of unprecedented systems. In such cases, the spiral archetype may prove more suitable.

The spiral

The spiral archetype employs a series of prototypes to 'try out' potential solution systems, usually under controlled conditions, to find out how well they operate, in what way they may be deficient, and what needs to be added, amended, improved, etc., to create the full, effective solution.

The archetypal spiral model is shown in Figure 15.4, where it can be seen as a part of the inner loop of the systems methodology, being applicable to a viable subsystem of the whole.

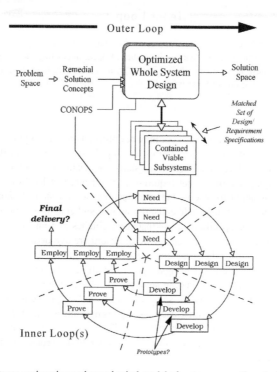

Figure 15.4   The systems engineering archetypal spiral model of system conception, design and development.

Examples of the need for, and utilization of, prototypes, abound. London Underground developed a prototype of a new underground railway carriage, and put it to work in between other regular carriages, to see how passengers reacted to the smoothness of ride, the internal noise levels, the comfort and resilience of furnishings and fabrics, etc. As a result of what London Underground learned in the process, they improved on the designs for the production version of the new carriages.

As the figure shows, it is possible to go around the 'need; design; develop; prove; employ' cycle several times, although patience may become strained if the process repeats too often. Note that each cycle is equivalent to one trip down the waterfall. So, where prototyping is concerned, it may prove a much longer, drawn out process than the simple waterfall. However, the spiral approach is also one of cautiously feeling one's way, to ensure a sound solution system in the end.

It is not unreasonable to regard the Apollo program as being based around a series of prototypes, since the craft used for each mission, and the knowledge and experience of the mission support crews and the astronauts, improved with each mission, until they finally landed Neil Armstrong on the Moon — and retrieved him and his colleagues. With spiral, the person can be part of the system . . . .

Concurrent

Pressure to bring products to the market more quickly, and to reduce costs, have brought the concurrent archetype, sometimes known as the simultaneous archetype, to the fore in recent years, at least in industry. There are two main strands of logic to the concurrent model; see Figure 15.2

■ Telescoping various sequential processes such that the output from one is fed as the input to the next before the first has been completed.

- This is at odds with the waterfall archetype, where each set of activities would be completed before moving on to the next set.
- However, there can be cases where it is reasonable to anticipate the next stage, notably when a sequence of activities has little impact on, or relationship with, other lateral sequences.
- It may also be that the output from one stage is known and understood before the whole of that stage is completed, documented, etc., and so can be passed on without risk.

■ So-called integrated design, where the design process anticipates the needs of manufacturing and production, maintenance, upgrade in operation, and even replacement. This anticipation obviates the need for costly subsequent redesign, reduces manufacturing costs, operational costs and replacement costs.

- This is an entirely sensible thing to do, and need not be particular to concurrent/simultaneous engineering.
- There is no reason why such anticipatory design cannot be conducted within either the waterfall or spiral archetypes.
- Note: the term 'integrated design,' as anticipating downstream needs in this way, is not the only use of the term.

Concurrent engineering, in the context of telescoping, can certainly reduce timescales and hence costs, but it may also introduce risk; if a downstream activity starts work early, on the basis of information from a preceding, as-yet-unfinished activity, and should that information subsequently turn out to be spurious, then the downstream work will have been wasted, and the activity may have to be reworked. Worse, further downstream activities may have set to work based on the just-generated false information, such that a snowball effect may propagate through the project.

While concurrent engineering may be the contemporary favorite, it is dependant upon pre-existing good knowledge and experience of the solution system type and technology, manufacturing of that technology, and operation, maintenance and upgrade of that technology: so, not for every application . . . .

Chaos

The chaos archetype does away with the rigid phased approach altogether and lets activity follow on activity freely, while retaining the logic, of course, of which activity should follow which. A re-examination of Figure 15.3, and in particular the potential feedback paths which may be activated when something does not go according to plan, may suggest that in some cases the waterfall model may 'descend into chaos' — and this has been known to happen. And, as Figure 15.2 showed, since each of the many archetypes can be mapped on to the basic waterfall archetype, there is a distinct possibility that any or all of them can 'descend into chaos'.

Experience shows that the project plan is continually revised. Project managers employ the program evaluation and review technique (PERT) to manage this continual change, which is brought about by a host of different occurrences, which may be lumped together as 'friction:' changing situation, unexpected events, shortages and surpluses, etc., etc.

'No campaign plan survives first contact with the enemy' (General Helmuth Graf von Moltke). The same may be true of systems engineering plans. However, in the case of both warfare and systems engineering, the objective is not to stick rigidly to some detailed plan, but to pursue a sound strategy through changing situations and circumstances, and in the face of unforeseen and unexpected events. In practice, and contrary to management expectations, chaos may prove to be the best approach in terms of time, quality and cost. For systems engineering, the strategy is the systems methodology.

If the project plan has been well designed, and if the practitioners who are undertaking the various tasks and activities set down in the plan are experienced and motivated, then it may well be that neither stages nor management is necessary: the system will self-organize, with individuals tackling tasks, cooperating with each other and yet pursuing the innate logic (strategy) of the plan, in terms of what needs to be done first, what cannot be done until later, which activities necessarily relate to others, and so on. Individuals may cooperate on more than one activity at once, may part finish one activity and move on to another, either coming back to the first activity alter, or perhaps leaving to another individual. The group may even modify and adapt the plan to meet changing circumstances. And so on.

To an observer, the whole will appear chaotic (in the everyday, rather than mathematical, sense). Surprisingly, however, it can be shown that this chaotic approach can be quick, inexpensive and most effective, sacrificing nothing to quality. It may make managers and 'anal-retentive control freaks' uncomfortable, though . . . . Not to mention those who insist that there is only one way to do these things.

The observation that chaos is, to some extent, inevitable, is borne out by experience: even employing the waterfall archetype offers no guarantees against it. So, where does this leave management forecasts and budgets of time and money for the conduct of complex systems engineering projects? As a hostage to fortune.

Functional, Project and Program Management

Systems engineering, as we have seen, is concerned with creating solution systems to solve, resolve or dissolve problems: the systems methodology offers a way of conceiving and creating solutions. Since the systems methodology is comprised of many tasks that may be undertaken by many people of different skills, it can be helpful to manage and organize the procedure: there are two archetypal approaches to organization and management of systems engineering:

- *Functional management.* Set up 'function groups,' equipped with appropriate tools, that can undertake successive phases of the creative process. There might be a group for exploring problems and conceiving remedial solution concepts, another for developing concepts of operations and operational concepts, and so on. A particular problem-to-be-solved then passes through each of the various groups in sequence, and several problems can be addressed in parallel within any group.
- *Program and project management.* Set up discrete project teams that can undertake all of the creative processes associated with a single problem and its single solution

At the time of writing, the second of these approaches is in vogue. In the second half of the twentieth century, however, functional management was fashionable, particularly in systems houses, where particular disciplines could develop experience and expertise readily by working on a variety of different subjects in succession. For example, an operations analysis section might undertake an analysis of a naval engagement one month, a hydroelectric scheme in the next two months, then a defense logistics problem, and so on. As a result, the section members would become expert in, and develop tools for, all kinds of operations analysis: their capability and performance would improve.

Functional management presented a problem to company accountants. Different individuals in a functional section might work on different problems at the same time; indeed, the same individual might be working on several problems for several customers at the same time. Not only did this make cost accounting more complicated, but it also encouraged systems practitioners to be too interested in their specialization, and to have less interest in getting fast results — none of which impressed the accountants.

So, management gurus urged a switch to project-based working, as it was more 'accountable, controllable and efficient.' It may not always be so effective, however: practitioners inducted to a new project team may find themselves in unfamiliar territory, having to learn how to do the job from the ground upwards. They may then spend years on the one project, and in so doing may fail to keep up with developing skills, methods, techniques and technology, so that — at the end of a long project, loyal project team members can find themselves outdated and unemployable. Despite this, at the time of writing, project management has become so ingrained in western business culture that many systems practitioners are unaware of any alternative to project-based working.

Systems engineering, then, is often undertaken in association with a project or program. Perhaps because of the switch away from functional management and towards project management, the constitution of a systems engineering project has changed, too. Industry, for example, may now expect to be told by customers what they require — often in detail as a set of so-called requirements, and often in respect, not of the whole solution system, but of only a part. In systems methodological terms, industry may be expected to undertake SM7: Create and Prove Solution System.

If that were the sum total, then systems engineering would have been truly emasculated. However, some customers — not all — undertake a full systems methodological process, perhaps in-house, but often using various parts of the appropriate industry to conduct different phases of the overall systems methodology. So, customers may commission concept studies, feasibility studies, project definition studies, operational analysis studies, prototypes, and so on — effectively following a systems methodology, albeit in a somewhat piecemeal fashion.

The seemingly piecemeal approach is, perhaps, an indication of caution or uncertainty — an unwillingness to commit funds to major constructions until certain that their designs will work be affordable and sustainable. Unfortunately, the approach can increase overall costs greatly, as funds are bid for and released for each phase in turn, spinning out and delaying the onset of any actual creation. Moreover, the delays militate against the project outturn being effective, since the problem/threat/opportunity that it was intended to address may have changed or evolved out of all recognition.

Nowhere is this more evident than in defense, where new platforms may be introduced into operational service at least twenty years after they were first conceived and designed. Aircraft, ships and tanks may be introduced, for example, to meet a Cold War threat that no longer exists . . . . Air defense systems may be introduced, piecemeal, as discrete platforms: a ground radar system here, a new interceptor there, an airborne early warning platform later on; only once they are all operational might attention focus on binding the various parts into an organismic whole — or system of systems, perhaps.

Unfortunately, it has often proved impossible to do this retrospectively, since the various parts, not having been designed with this in mind, have lacked the built-in facilities ('hooks') for lateral connectivity to enable coordination, cooperation and complementation. Without those interaction hooks, it has proved impracticable to interconnect the various platform technologies, and to interact other than verbally via the human operators: this may be too slow to meet the needs of modern high-speed warfare, and precludes the development of whole system capabilities such as dynamic target allocation, and airborne RASP formation; i.e., whole-system emergent capabilities.

Where a customer is, in effect, managing the creation of the whole solution to a problem, he/she/they may be said to be practicing program management, where a program consists of a number of interconnected projects. This is consistent with the full systems methodology, which may result in the creation of a set of interacting viable subsystems, the conception, design, development and creation of each of which may be considered as a separate but related, interconnected project. It is also consistent with the employment of teams of teams, perhaps with a central team overseeing the whole solution, and with 'subordinate' teams overseeing the viable subsystems.

Program management and project management address more than the application of the systems methodology, however. They are generally concerned with acquiring resources (funds, people, skills, tools, facilities, partners, etc.), applying those resources to conduct their systems processes and procedures, meeting economic and time budgets, in ensuring the overall quality of products and processes, and in ensuring an outturn to satisfy the customer and the contract. Both program and project management are concerned with setting and achieving business goals; systems engineering (and the systems methodology) is one way in which they can go about achieving their goals.

There is a perceived potential for conflict, therefore, between systems engineering, with its integrity ethic, and project management with its business ethic. There should be no need for such perception, however, which seems to arise from a narrow view of systems engineering as applying only at Layer 1: Product/Subsystem and Layer 2: Project. (See The 5-layer systems model on page 113). If Layer 3: Business/Enterprise is considered, then a different perspective emerges, in which systems engineering concerns itself with designing the project as part of the business or enterprise, such that it meets constraints of cost, time and quality while at the same time creating a sound solution system in the approved manner, i.e., building the systems approach into the project plan, etc. To see how this might work in practice see Layer 2: Project Systems Engineering on page 117, and Figure 15.12 which shows, inter alia, how the design of a project may be undertaken cooperatively. The detailed project plan, however formed, should apply the systems methodology . . . .

'Eine Kleine Systems Engineering Archaeology'

M'Pherson's system design framework

Figure 15.5 and the two following figures date back to 1980/1981, and show the state of understanding associated with, and development of, systems engineering at that time (M' Pherson, 1981). Philip M'Pherson has a background first in the Royal Navy, later in the UKAE, concerned with stability in nuclear reactors, before becoming a leading academic. Even cursory examination will indicate that systems engineering at that time was significantly in advance of its current state as practiced in many industries, and it is quite different from 'systems engineering management,' promulgated by the US DoD just a few years later in such documents as MIL-STD-499A. So different is it, that the US DoD version bore virtually no resemblance to the original in terms of philosophy, methodology, practice, performance, products, or outcome; it was, however, also referred to by many as 'systems engineering.'

An alternative look at systems engineering, and at the system design process in particular, is shown in Figure 15.5. This is a remarkable diagram, worthy of study: it contains a wealth of information about the organization and practice of systems engineering design.

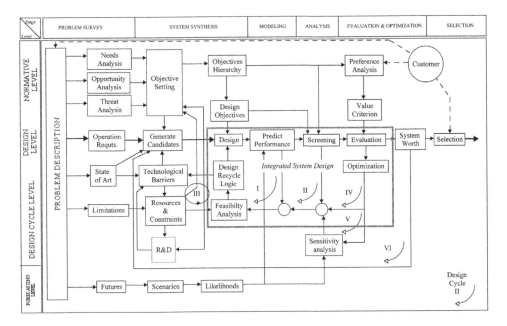

Figure 15.5 M'Pherson's system design framework.

The diagram is developed within a rectangular matrix. The x-axis shows a sequence of activities constituting system design: problem survey; system synthesis; modeling; analysis; evaluation and optimization; and, selection. The y-axis shows various levels: normative level; design level; design cycle level; and, forecasting level.

Within this matrix, the central theme running from left to right goes from problem description; through operational requirements; design; performance prediction; screening; evaluation; system worth; and selection. At top right, the customer can be seen as involved in, and concerned with, problem description, performance analysis and selection of the preferred solution design. The whole is, as it should be, context, scale and system type independent.

Roman numerals indicate six loops within the system design framework, showing that the process is recursive:

I.   Predicting design performance by modeling
II.   Screening designs that perform well, to determine if they meet objectives
III. Testing the feasibility of designs in terms of technology and other constraints
IV.   Optimizing otherwise acceptable designs
V.   Testing the sensitivity of optimized designs against different scenarios and situations
VI.   Evaluating the worth of a designed solution, for comparison with others.

Five of the six loops are shown within a rectangle marked 'Integrated System Design' which, in this instance, means that all of the relevant disciplines work together to create a composite design. It could also mean that the designs will anticipate the needs of manufacture, operations, maintenance, etc., but none is specifically mentioned: 'integrated system design' seems to be generally about systems, rather than specifically about engineering.

M'Pherson's system design framework is system design at two levels: first, there are the obvious design processes I to VI; second, the whole framework is a designed system, revealing as it does a methodological approach to design. The framework covers some of the same ground as the systems methodology, although the framework shows tasks and their interrelationships and interactions, without indicating how they might be accomplished. Nonetheless, it starts at the problem and progresses through the various stages and recursions to create an optimized solution system design.

The system design framework addresses both sociotechnological systems and technology systems: and, while it does not appear to address social or societal systems specifically, there seems to be no reason why it should not, simply by omitting the technological elements.

M'Pherson's design hierarchy for a complex system

Philip M'Pherson also addressed the relationship between systems engineering and project management. Figure 15.6 shows his design hierarchy for system design during the design phase, other phases being development, production, testing and insertion. Each phase has a management hierarchy of its own, and the various phases will need system-wide integration and overall coordination. Note goal coordination in the figure, necessary since different subsystem designers/decision-makers primarily pursue local goals; and interaction between subsystems, as decisions made for one subsystem affect other subsystems. The design hierarchy adopts the systems approach, with each subsystem being designed as an open, adaptive system, interacting with, and adapting to, all the other subsystems.

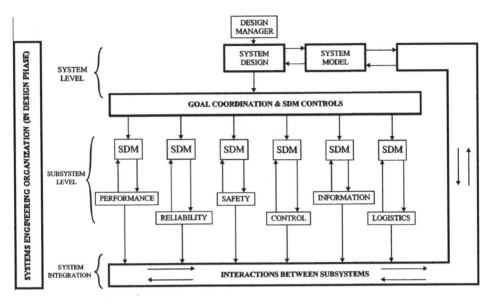

Figure 15.6 Design hierarchy for a complex system. SDM is subsystem design manager/decision-maker. Note that different SDMs are concerned with different aspects of the whole system, including logistics, safety and information as subsystems: the sum of these different, interacting aspects is intended to represent the whole system. Reproduced from M'Pherson, P.K. (1981) *The Radio and Electronic Engineer*, 51(2), Paper 1971/ACS 82, with permission from IET.

M'Pherson's systems engineering organization in project management

M'Pherson also illustrated the relationship between project management and systems engineering — see Figure 15.7, which should be viewed in conjunction with Figure 15.6. Figure 15.7 shows the relationships and interactions between project management, resource management, systems engineering, customers, etc., together with the progression of a system through its various phases until it is finally delivered to a new owner as an operating system.

The figure shows elements of the project that are 'systems engineering functions:'

- Project coordination;
- Customer liaison;
- Phase coordination and integration;
- Inter-phase integration;
- Phase resource management...
- ... and, of course, subsystem design/decision-making.

On the other hand, project management functions include the project manager as director of the project, with phase managers for each phase. The project manager interacts with project coordination, and with the customer. (Not shown is any relationship between the customer and

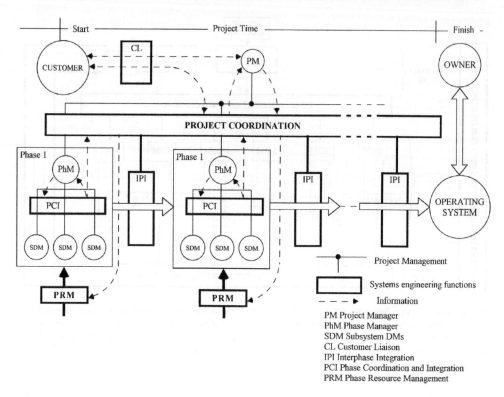

Figure 15.7   Systems engineering organization in project management. Reproduced from M'Pherson, P.K. (1981) *The Radio and Electronic Engineer*, 51(2), Paper 1971/ACS 82, with permission from IET.

the eventual owner of the operating system: they could be the same, or the owner may employ an agent to be the customer and to acquire the operating system on the owner's behalf. In any event, it is likely that the eventual owner will have identified the problem that the solution system is to solve, and he may have influenced the concept of operations, and aspects of the design to make the operating system compatible with other systems already in, or proposed for, the owner's portfolio.)

SEAMS

M'Pherson's insightful diagrams suggest that the various phases of creative activity may generate a large amount of information. Moreover, each of the SDMs, the subsystem decision-makers, as M'Pherson dubbed them, will exist within a web of constantly changing information as their design decisions affect other subsystems and vice versa. Information about the eventual operating system is also becoming more specific and detailed as time progresses: this is entropy reduction, the underlying thrust of systems engineering design.

Design (architecting?) can become bureaucratized, too, with specifications defining methods of design, and specifications showing what the output from a design activity should contain.

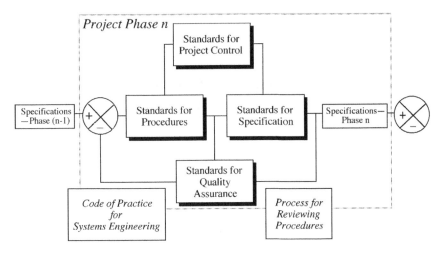

Figure 15.8 Systems engineering standards and specifications.

Figure 15.8 illustrates the point, showing a project phase, such as those in Figure 15.7: the input from the previous phase is a specification, and the output to the following phase will also be a specification; these may be called 'work products' of each phase (Hitchins, 1992).

Within each phase, there will be standards for procedures (analysis, design, elaboration, decomposition, coupling, interfacing, integration, etc.), standards for specification ('skeleton' specifications showing layout, content, etc., but with blanks instead of specifics which will be filled in during the phase), standards for project control and standards for quality assurance.

This degree of bureaucratization, intended to streamline and simplify, could be restrictive in some instances, so there will be a process for reviewing the procedures and, permeating all, a code of practice for systems engineering. The code is designed to ensure that the standards and practices are consistent with the systems approach and systems engineering philosophy, that synthesis is employed in place of reduction, that elaboration, decomposition, etc. are executed top down and in knowledge of the containing system's objectives, etc.

With so much information, and so many standards and practices, it may be difficult to maintain coherence. An individual subsystem designer, for instance, may require access to only a restricted subset of a vast store of knowledge and information about the problem, other subsystems, the CONOPS and the operations concepts, which methods, techniques, practices and specifications are appropriate to the task in hand, etc., etc. One way to address the issue of potential information overload is to create and provide a 'systems engineering environment;' an information system which stores, records, handles, distributes and generally manages all kinds of information, including particularly information about the subsystem, and information about acceptable ways of analyzing, understanding, designing, testing, proving, etc.

SEAMS, a notional systems engineering analysis and management support environment (Hitchins, 1992), is shown in Figure 15.9. The figure shows a project management structure, systems engineering management structure and workstations for systems analysts and designers, all interconnected via a local area network to a server bank. The figure also shows that there may be connections to remote sites, so that teams and teams of teams may all work on the same systems engineering project.

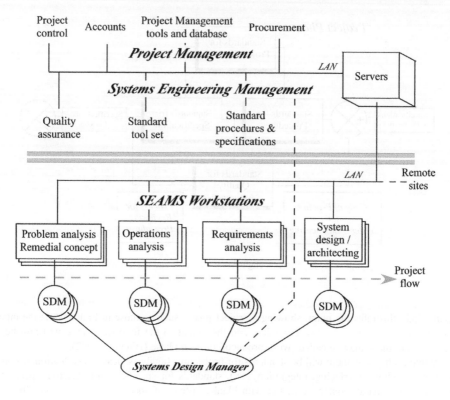

Figure 15.9   Systems engineering analysis and management support (SEAMS) Environment.

Does SEAMS exist, some 15 years after it was first proposed? Apparently not, at least not in the form of Figure 15.9: SEAMS went the way of the systems houses that were conceiving such environments as they developed and evolved systems engineering — squeezed out by the major aerospace manufacturing companies. Looking at M'Pherson's diagrams, it may seem to the knowledgeable reader that systems engineering may not have advanced much since 1981 . . . .

Systems Architectures

Some systems engineering concepts, tools, methods, etc., have advanced since those shown in the preceding 'systems engineering archeology' section: one such is architecture. So much so, that some prefer the term 'systems architect' to 'systems designer,' and there does seem to be some correspondence between ideals and goals of the civil architect and those of the system designer.

However, systems architecture is less well understood. At its most basic, systems architecture is the pattern formed by linked clusters of systems and subsystems. Since such clustering and linking can occur in many different ways, there are many different patterns, so many different system architectures . . . .

There ought to be a science of systems architecture, systems architectonics, which would indicate the most appropriate architecture for systems in different situations, to assure the best system solution; no such science appears to have been formulated. Nevertheless, looking at system architecture is one avenue for understanding the whole system and its emergent characteristics.

The effect of systems architecture

The question arises, then, is there a relationship between architecture and system behavior, performance, resilience and vulnerability? Since architecture is delineated by the connections between parts of a system, clearly the interruption of those connections could prevent interactions, could prevent the parts from operating as a unified whole and could impair performance. By the same token, if there were multiple connections, such that the severing of any one did not impair subsystems interactions, then the parts would continue to operate as a unified whole. So, in principle, redundant linkages coupling parts of a system could make it more resilient.

Systems architectural strengths and weaknesses

Basic notions of systems architecture arise around ideas of binding and coupling. Where a number of subsystems all mutually interact, there is said to be tight functional binding. Where such tight functional 'blocks' are interconnected, there is said to be loose coupling.

Figure 15.10 shows the concepts of functional binding and coupling in the form of an N^2 chart, which contains N rows and N columns, resulting in N^2 squares. The leading diagonal identifies entities, or subsystems. All other rectangles represent potential interfaces, through which data, information, signals, communications, etc., may flow. Archetypal architectures may be represented on an N^2 chart, where they form characteristic patterns.

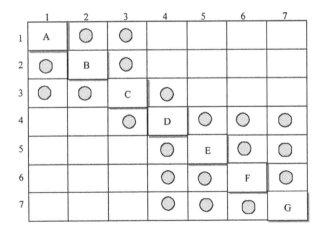

Figure 15.10 N^2 chart showing binding and coupling. ABC and DEFG are tightly functionally bound blocks, as all their mutual interfaces are connected. The two blocks are coupled via interfaces 4, 3 and 3, 4.

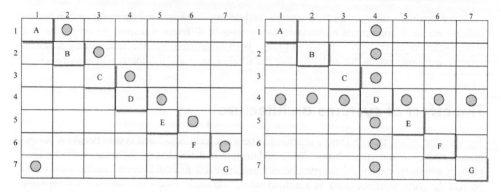

Figure 15.11 N^2 charts showing architecture archetypes. At left: pipeline, or ring, architecture, with ABCDEFGA linked in order. Information, energy, or substance from A is communicated to B, from B to C, and so on around the ring. At right: nodal, or star, architecture, with all subsystems linked to D. Should D fail, then none of the others would be connected, and the system would be reduced to separate parts . . . i.e., no longer a system.

In Figure 15.10, ABC forms a tightly functionally bound block, since all of the mutual interfaces between the three subsystems are active. This is likely to represent some degree of functional stability, or at least resistance to change, as the three subsystems are mutually interdependent and supportive. Similarly, DEFG forms a tightly bound functional group. The two groups are coupled by bidirectional (duplex) interfaces between subsystems C and D: coupling is not indicative of stability, simply of connectivity and interaction.

Figure 15.11 shows two more archetypal architectures. At left is typical ring-pipeline architecture, with the subsystems and interfaces arranged in a unidirectional (simplex) circle, or loop. Police, judiciary, courts, prisons, parole and police, are organized in ring-pipeline: recidivists go around the ring system repeatedly. Volume supply systems may be formed as a pipeline, with fan-in at the start from various suppliers, and fan-out into the market place. Alternatively, if recycling is considered, they may form a ring pipeline . . . (see Keiretsu on page 157.)

A characteristic feature of pipeline architectures is the so-called waterfall, which can be seen in the left-hand diagram; information or substance moves from A to B, from B to C, from C to D, and so on. Such waterfall structures are sometimes used, for example, in major infrastructure systems, power distribution networks, etc., where they can be vulnerable to cascade failures, in which a failure in one system triggers failures in all of the others in quick succession. Power surges, or 'spikes,' have been known to trip extensive, interconnected, load sharing and power distribution networks over vast areas. Similar architectures, and vulnerabilities, are to be found in digital encoders and decoders, where accurate timing signals are required to coordinate sequences of activities across a number of boards, modules or subassemblies.

At the right is a typical nodal architecture, otherwise called a star-connected architecture. One subsystem, D, serves as the single point for all inputs from, and outputs to, the other subsystems. Such architecture is indicative of a vulnerable system, since failure of the singular nodal system will reduce the whole into separate, disconnected parts. Nodal architectures are to be found in hierarchical organizations, where a single superior controls a number of subordinates individually. The disadvantage, apart from the vulnerability aspect, is that the superior has to manage all of the interfaces with all of the subordinates, which may involve a high and sustained level of work.

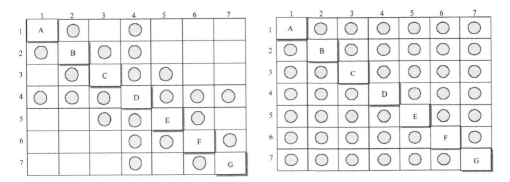

Figure 15.12 Archetypal architectures. At left, a duplex circular pattern is combined with a node, D, and a nodal bridge (5, 3 and 3, 5) to form a composite architecture. At right, the system is fully networked, with all interfaces active. There are no apparent nodes, and the whole is tightly functionally bound.

Star architectures may also be found in multi-processor systems, where a central processor partitions processing tasks into separate parts, peripheral processors then processes the parts before returning the results to the central processor for re-integration.

Figure 15.12 shows two further architectures identified by interface patterns. At left, the architecture has three elements:

- a node at D;
- a duplex waterfall, (i.e., down the waterfall and up again), often indicative of a chain of command with orders being passed down, while acknowledgements and reports are passed up;
- a node 'bridge.'

The node bridge is formed by interfaces at 5, 3 and 3, 5: the bridge would accommodate the elimination of node D, without the system breaking into separate parts. Information, for instance, could follow the waterfall route ABCEFG, using the additional interface 5, 3. This type of architectural structure may be invoked where an alternate war headquarters (AWHQ) might be needed as a backup to the main HQ in military operations. In the diagram, an AWHQ might be located at C or E, equipped with the necessary command and control (C^2) facilities (maps, communications, tote boards, IT, etc.) and manned with a backup C^2 team, ready to takeover the reins should the need arise. Similar architectures enable military land systems to move from place to place, with main HQ being temporarily incommunicado during the move, while the alternate HQ holds the reins.

The right-hand N^2 chart in Figure 15.12 represents a fully networked system, with everything connected to everything else. This might be seen as the architecture of the Internet, of network enabled capability, network-centric operations, and many more. According to viewpoint, it may be seen as a non-nodal architecture, a multi-nodal architecture, or tightly functionally bound architecture.

One way to achieve such total interconnectivity is to use network communications systems, either on the ground as telecommunications or in the air as radio/wireless networks. Bluetooth and Wi-Fi are current commercial communications facilities that support wireless network operations. Military operators have employed data-link systems for many years with such capabilities, only recently dubbing them 'network centric' facilities.

The ideal radio facility is likely to involve network packet radio, with combined facilities for communication, navigation and identification. The radios would be expected (at least) to frequency hop at high rates, and at random frequency and time intervals, with sophisticated encoding, error correction, message redundancy and encryption making it near impossible for any third parties to jam, intercept, or exploit information. Given such a system, perhaps as an enabler for network-centric operations, it might be thought that the system would be free of nodes, and relatively invulnerable. In communications terms, it might be, but the whole system is rather more than the communications network. It would not be unusual to find that the participants in some networked enterprise had created organizational nodes, thereby rendering the non-nodal aspect of communications superfluous.

For example, a naval task force might deploy several fighter aircraft for air defense of the force: it might be evident to an observer that one of the ships, say, was directing the various fighters. In such a situation, that ship would be an organizational node, and a prime target to an enemy, defeating a principal purpose of the non-nodal communications system.

A similar example shows some of the difficulties facing network-enabled capabilities. The physical architecture of an air combat group is constantly changing as aircraft move around relative to each other, and as aircraft run out of consumables and have to return to base for refueling, rearming, repair, etc. This constant change presents a challenge to the overall system, such that group members running low on resources are not tasked to do the undoable, new and returning members are identified, welcomed into the group with a full information update, and tasked appropriately to their abilities, on-board resources, etc.

The potential benefits are, however, significant. Network enabled capabilities, or network-centric operations (both terms are used) have the potential to create a range of emergent properties, capabilities and behaviors for the whole group that has hitherto not been possible. Tying the separate parts of the group into a coherent, cohesive, organismic whole will speed up and coordinate interactions, and enable many whole group functions that would otherwise not be feasible. See Case D for ideas on what might become feasible . . . .

So, an architectural viewpoint enables an overview of the whole system: architecture is both holistic and extensive. Architecture can have a significant effect on operational capability, vulnerability, cooperation, coordination, complementation, synergy . . . etc. And, not mentioned above, but demonstrated in the various case studies, architecture is characterized by minimal configuration entropy: each of the N^2 charts above is presented in its minimum entropy configuration, which reveals the characteristic patterns we call architecture. (Scrambling the N^2 chart of Figure 15.10, for instance, would obscure its neat, ordered pattern.)

Architecture analogies

The concept of architectural analogies has been introduced earlier (see Models of Systems Architecture on page 108). Using the human body as an architectural template is not a new idea. Indeed, it has been mooted (Hitchins, 2003) that we humans inevitably, perhaps subconsciously, design systems that are analogous to our own, human, design. However, the human body is so delightfully powerful, in design and performance terms, that we could do worse than emulate human structures in our manmade system designs. Apart from other considerations, this may make the systems we create more intuitive in their behavior and operation.

As an example of direct analogies between human and manmade systems architectures, consider Figure 15.13. The diagram at left shows a simplified representation of the central nervous system. In the lower part, a stimulus excites sensory neurons. The excitation is passed by interneurons, relay

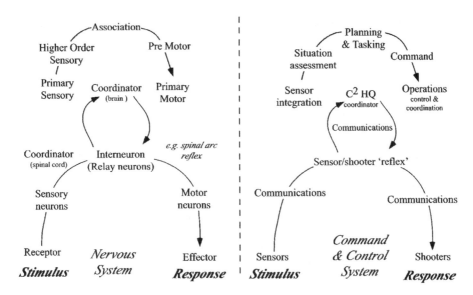

Figure 15.13 Architectural analogy: left, simplistic representation of the nervous system, showing the responsive spinal arc reflex associated with 'knee-jerk;' right, tactical command and control showing the responsive 'sensor–shooter' linkage. See text.

neurons in the spinal column, to initiate a motor response. This is the spinal arc reflex, intended to create rapid response. If a person walking barefoot, for instance, were to tread on a thorn, the pain from the thorn-prick would activate the spinal arc reflex, causing the foot to lift off the thorn before damage was done. This is the so-called knee-jerk. Without the spinal arc reflex, pain signals would pass up to the spinal cord to the brain, be interpreted and processed to result in motor signals that would pass back down the spinal cord to arrive too late at the foot to prevent damage from the thorn entering the foot. . . . The pain signals from the thorn still go to the brain so that the individual senses the pain and is aware of what is happening/has happened. . . .

The diagram on the right of Figure 15.13 draws an analogy between the human central nervous system and a tactical command and control situation, where a force is engaging an enemy. Information about the enemy is the stimulus that activates sensors, and that information is transmitted to the C^2 HQ, where it is combined and correlated with other information to create a situation 'picture,' enabling a situation assessment. The assessment may reveal threats and opportunities, which the command and command team will plan to act upon. The plan is then put into action with the operations control team coordinating the actions of forces under operational control. All of which takes time, during which the enemy may have moved, taken evasive action, etc.

In an analogy to the spinal arc reflex, it is possible to pass sensor information to the force 'shooters,' i.e., artillery, tanks, snipers, etc., along with authority to fire on the enemy according to current rules of engagement (e.g., 'you may fire only if fired upon'). This sensor/shooter reflex makes for a highly responsive force. Note, too, that in both cases, the main coordinator (brain vs. C^2HQ) can receive information from remote sensors (eyes, ears, nose, vs radar, TV, etc.) and can therefore anticipate threats and risks as well as react to events.

Once the analogy between the central nervous system and command and control has been drawn, it may seem rather obvious. After all, the human body is an information decision action (IDA) system, and the CNS is its command and control system. So, does the analogy extend to other IDA systems: boards of directors, business executives, lean volume supply systems, etc?

If the analogy does extend, it will be because all of these various systems share the same functional/physical architecture, where the term functional/physical implies that there is correspondence between functions and physical partitions of the system. In other words, discrete functions are not spread across two or more physical partitions, which would involve interfaces, intercommunications, and potential delays, interference, and exploitation.

Contained and encapsulated systems viewpoints

Systems architecture can be seen from different viewpoints — see Figure 15.14. At top left is a version of the 'poached egg' diagram (see Representing and modeling systems on page 76). It shows three levels of hierarchy explicitly, and implies two more. Three nominal level systems are marked: these are at the current point of view. Within the nominal level systems are nine contained subsystems, three in each: by implication, each of the contained systems will contain further sub-subsystems, and so on, (not quite) *ad infinitum*. The nominal level systems are contained within an overall containing system that exists within an environment, where it interacts with other containing systems — there is then an implied level above these containing systems that contains them in their turn. Hence, there are five implied levels of hierarchy — at least.

The diagram at the right of Figure 15.14 shows a perspective view of this containment hierarchy. However, hierarchy may be something of a convenient viewpoint, a useful artifice that simplifies the view of architecture. Each level within the hierarchy can be viewed as encapsulating the layer beneath it. If, for example, the properties, capabilities and behaviors of a nominal level system

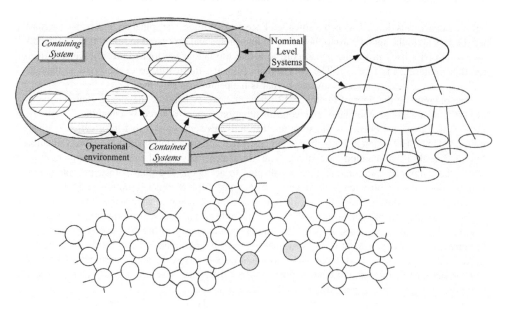

Figure 15.14   System hierarchy, system elaboration.

can be fully, described, then the 'contents' of that system may be of no immediate interest. We can 'encapsulate' it, put a cover over it and label the cover with the properties, capabilities and behaviors of the system . . . .

Not all representations of systems are so amenable however. The bottom diagram shows a 'flat view,' i.e., one in which there is no evident hierarchy and in which all subsystems are shown. Moreover, there are more subsystems than in the upper two diagrams, making it more difficult to identify any particular system. Note, in the lower diagram, that all of the circles representing subsystems are connected. So, does the lower diagram represent a system, several systems, or no system?

Insofar as there appear to be a number of clusters of subsystems, it may be possible to identify potential systems, but it is far from clear. Do the shaded circles, for instance, represent interfaces, or are do they form part of some infrastructure system — an energy supply, or waste disposal system perhaps?

The architecture of complex systems may be difficult to discern, and may be dynamic, i.e., constantly changing. The architectures of enterprises and businesses may prove highly interwoven and in a constant state of flux. Fan-in/fan-out pipelines may be more appropriate, for instance, than vertical hierarchies, when considering business architectures.

The lead company, the one that interacts directly with the market, assembles parts and subassemblies provided by first tier suppliers into fully assembled products for the market. First-tier suppliers have the same arrangement with second-tier suppliers, who provide them with parts and sub-subassemblies; and so on, up the chain.

In operation, there may be alternate suppliers for some parts, so that one supplier supplies some while a second supplier supplies the rest: this provides a hedge against delivery shortfalls. Should the quality of either supplier's products fall off, then the lead company may try to help remedy the problem, but may also choose to reduce the quantity taken from the failing supplier, and to take more from the other supplier instead. The failing supplier will start to lose money, while the successful supplier will start to make more . . . there may subsequently follow a change to the fan-in pattern, as a failing supplier is replaced.

Supply of products is market-demand led, and demands are passed up the chain, spreading in a reverse of the materials fan-in, i.e., an information fan-out. There will be another fan-out, this time of products, as they enter the market and fan out into different parts and sectors of businesses and societies. Money will then flow in reverse, fanning-in to the lead company as revenue and profit, with which to sustain the whole structure, architecture, industry, etc.

Purposeful Systems Architectures

As the systems methodology showed, functional architectures emerge unaided as an intrinsic part of the system conception and design processes. Physical configurations, however, may be governed by the constraints of the solution space. System implementation is concerned with maintaining an effective functional architecture by mapping it into a suitable physical configuration without in any way impeding functional interactions and functional behavior.

Preceding sections have shown that system architecture presents an extensive, whole-system viewpoint. The architecture archetypes also reveal that architectures present emergent properties, capabilities and behaviors: for instance, a nodal architecture may signify that the whole system is vulnerable to single point failure. Architecture, too, indicates connectivity and potential cohesion, since it shows the extent of connection — a fundamental aspect of any system being, of course, that all of its parts are interconnected.

In 'branched' architecture, it is possible to label the branches with an indication of what is to be found at the leaves: we are used to the idea in the context of a library or bookshop, where various rooms may contain books on different topics or of different genres. Inside each room, bookshelves are labeled with the categories of the books to be found on the particular shelves.

Architecture of various kinds can be so labeled. Databases, instead of holding data, can instead hold labels, or pointers, to the data, which may be stored remotely. This leads to a different aspect of architectures — whether they are 'push,' or 'pull.' In a command and control system, for instance, the main command and control center needs access to potentially vast mounts of information. Some of this date is relatively fixed: geography, topography, etc. However, much of it is ephemeral: intelligence, logistics, operational readiness states, enemy activity, etc., etc. While such ephemeral data may be changing on a second-by-second basis, the staff in the HQ may need to access a particular piece of data only occasionally — perhaps on an hourly, or daily basis: when staff do access that data, however, it must be current.

One way to address the issue is to send large amounts of information from the periphery to the central database every few seconds: this is 'push;' the data is pushed towards the user of that data, even though it may not be used, but may be overwritten very many times before the user accesses the relevant part of the database. An alternative approach is not to put data into the central database, but instead to enter the source address for that data. This is 'pull;' no data is sent from the periphery to the central database unless specifically asked for, which happens when the central database user accesses the particular data he/she is interested in. If the design is sound, the central database user will not be aware that he/she is accessing remote data; the data will appear on screen just as though it were stored locally.

The difference between push and pull architectures can be significant: pull can require several magnitudes less in terms of communications capacity. On the downside, if communications should be lost, a pull system is isolated, while a push system will still have a full set of information in local storage, although that information will be aging . . . . In practice, a combination of the two approaches may be advisable.

Push–pull architectures are becoming more common in areas such as social services, policing, etc., where there may be some reluctance to share data fully between agencies. In some countries, for example, there are laws preventing the unwarranted disclosure of personal data. A pull system permits security checks on the authority of an enquirer prior to passing data.

There are many other purposes that architecture might support — see Figure 15.15 for a selection, appropriately presented in the form of an attribute enhancement structure. At the top of the structure, the ultimate purpose of purposeful architecture is seen as being to 'Support the Mission.' Other architectural contributions include:

- Physically grouping entities that are functionally interactive minimizes inter-function time delays that might otherwise prejudice that interactivity.
- Architecture enables synergy — without all the parts of a system being connected, cooperation, coordination and complementation would not happen — but in addition, the disposition of parts may make synergistic interaction possible. For example, a CCTV sensor has to be positioned so as to cue a security team to respond in time to detain an intruder, or better yet to prevent an offence from happening . . . .
- Reconfiguration may enable a complex system to tolerate damage, perhaps by switching in reserve sensors, by coordinating and sequencing the use of different weapons against 'difficult' targets, by rerouting communications around an area of damage, and so on.
- Given knowledge about configuration and current fitness, a system may reconfigure its architecture, substituting reserves to failing primary facilities, rerouting information, etc. Some

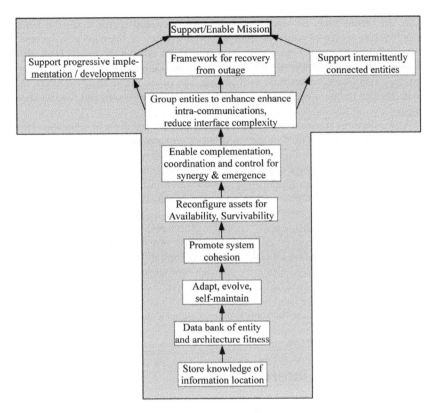

Figure 15.15 Attribute enhancement structure showing possible purpose of system architecture: developed using Warfield's interpretive structural modeling (ISM) in Chapter 7. Reproduced from M'Pherson, P.K. (1981) *The Radio and Electronic Engineer*, 51(2), Paper 1971/ACS 82, with permission from IET.

facilities that require routine or exceptional maintenance can be temporarily taken out of service, and a substitute inserted. Potentially, such systems may be self-healing.

■ Acquiring and sharing knowledge about entity and architecture fitness. This enables the whole system to be aware of its operational status and, if need be, mount support activities to repair damage, find faults and mend defects.

■ Architecture can delineate groups and subgroups: similarly, architecture can 'ring fence,' conceal, restrict access . . . .

■ Architecture aids interoperability and interoperation, including openings for parts to join the system and to cooperate and coordinate with corresponding parts easily.

■ And many, many more . . .

Layered architectures

Layering presents a different view of architecture. The archetype of layered architecture is the medieval castle, with its outer wall, baileys or courtyards, successive inner walls, with their baileys, and finally the central keep, to which the castle 'owners' would retreat if all else failed. The layered architecture presented would-be assailants with a series of 'hurdles' to overcome. They had to breach or climb over the outer wall, upon which they entered the bailey. Here, defenders may outnumber the small number of men able to scale the wall at any one time. If the invaders prevailed, they then faced another wall, then another, and so on, before finally reaching the keep which was intended to be impregnable, and which often had its own well and food stores, to last out a siege.

Layered architectures are amenable to mathematical analysis and dynamic simulation (Hitchins, 2003). They invoke a concept of 'leakage,' with whatever seeps through permeating to the next layer: each layer stops much of the attempted flow through; given sufficient layers, it should be possible to ensure minimal leakage overall. Such layered architectures are used in modern defense, in the way that battle formations are drawn up, in defense against air attack, etc: had the Strategic Defense Initiative (SDI) antimissile system come to fruition, it may have presented a layered defense architecture.

Modern security systems often employ layered architectures, such that a would-be intruder or burglar has to overcome a series of obstacles or barriers. The obstacles are designed both to obstruct and delay, such that the would-be intruder gives up in frustration, or runs out of time, or both. They might also require the intruder to carry so much equipment with him to deal with different obstacles that the exercise becomes logistically impracticable. And, for some applications, it is not unknown for security systems to have 'teeth,' in the form of razor wire, electric fences, automatic machine guns, flame throwers, anti-personnel mines, gaseous emissions, etc.

ISO open systems interconnect

The International Standards Organizations's (ISO's) Open Systems Interconnect (OSI) model is the standard model for networking protocols and distributed applications. It is a 7-layer model with the following layers:

1. Physical — the physical medium for the transmission of signals
2. Data Link — the point-to-point protocol
3. Network — the network protocol
4. Transport — ensures transport control, the correct transmission and ordering of packets of information
5. Session — defines the format of the data sent over the connections
6. Presentation — converts (interprets) data used for transmission to data for an application
7. Application — provides network services, such as mail, file transfer protocols, etc.

The Open Systems Interconnect model employs a layered architecture, or protocol, sometimes referred to as a nesting protocol, as each layer effectively 'nests' in the one above. (This is also true of the 'The 5-layer systems model' of systems engineering on page 113.) So, two remote users of some application can interact with each other through their application, with the Open Systems Interconnect protocol performing invisibly, or transparently, as when someone with a personal computer browses pages on the Internet.

N.B. The term 'open' in the OSI has a specific connotation: it refers to the protocol being open to all potential users, and managed openly, too: so, open in the sense that nothing is being arranged, organized or managed in private by any self-interested groups, for instance.

Command and control architectures

Some architecture may be thought of as fractal, in the sense of self-similar. We are familiar with this in industries and enterprises. A corporation may have a board of directors. The company may consist of a number of major divisions; each of which may have it own board of directors. Each division may comprise a number of business units, each of which also has its own board. . . .

Military command structures can be fractal (using the term somewhat loosely: a fractal can be explored to an infinite degree of resolution, as in the Mandelbrot Set.) An army will have a general and his staff. The army may be made up from several divisions, each with a commander and his staff, and so on through brigades, battalions, etc. Command and control is similarly fractal, with strategic C2 HQs, tactical C2 HQ, etc.

Figure 15.16 shows a typical command and control arrangement, with two levels of command: on HQ at an upper level, with three units reporting from a lower level. At each level, there are staffs for operations, intelligence, plans, logistics and engineering. At each level, members of the respective C2 team interact with each other (team intra-links), but are likely to talk between levels only within their own discipline. So, as highlighted in the figure, logistic staff at the upper level will interact with logistics staff at the lower levels (interlinks between teams), but generally with no other staffs. In effect, this means that there is an overlay of communications matrices both vertical and horizontal, one for each of the various 'disciplines. Note, too, connections to lateral and vertical formations not shown: command and control may be the ultimate 'open system!' (Note: conventionally, command and control includes neither sensors nor effectors, although of course they are integral to the overall IDA system).

Military forces may be deployed 'in fractal form,' too. Conventionally, airpower is deployed by application: squadrons of bombers are organized and managed together; squadrons of air defense fighters; squadrons of close air support, and so on.

Suppose, however, they were arranged into small cells, with each cell containing bombers, fighters, transport, C2, communications, etc. Then, each cell could, in principle, be autonomous; each could attack, defend and move on its own. This is the notion behind MOSAIC (Moveable,

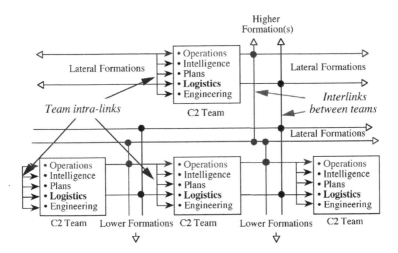

Figure 15.16 Fractal command and control architecture. Reproduced from Hitchins, D.K. (1987) 'MOSAIC—Future Deployment of Air Power in European NATO': *Advances in Command, Control and Communications Systems*, Peter Perigrinus, London, with permission from IET.

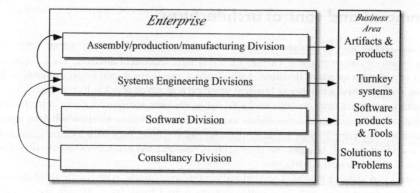

Figure 15.17   Enterprise architecture promoting inter-divisional synergy.

Semi-Autonomous Integrated Cells, Hitchins, 1987). In the MOSAIC concept, air power would be deployed as a number of such cells, able to configure themselves into various patterns: lines of defense, lines of advance, etc. — and able to orientate those lines to face a shifting enemy. MOSAIC offered the ultimate in airpower flexibility and agility. It also offered a high degree of survivability, with self-defense an damage tolerance particularly in evidence. The architecture of a MOSAIC force would be highly flexible.

Enterprise architectures

Enterprises also exhibit architecture; as in the other example of architecture, different architectures promote different emergent properties, capabilities and behaviors of the whole system. Figure 15.17 shows an enterprise that consists, not of competing divisions as quite frequently occurs, but of complementary divisions. In the figure, each division is a separate business, with its own business area and goals. However, the divisions have been set up so that they may interact to their mutual benefit:

- Consultancy Division has been established essentially to address customers' problems, to undertake feasibility studies, to establish requirements for new systems and products, and so on. In addition to their business in the market place, they also act internally, with systems engineering divisions as their customer. In effect, they are able to perform the front end of the systems methodology, including problem solving, solution conception, modeling and simulation etc.
- Software Division likewise deals in the market place, bidding for software projects, creating and marketing software tools and products. Systems engineering divisions can be a customer, too, since they may use the tools, and may require system software to be designed, developed, tested and proved as part of some system-to-be-delivered.
- Systems engineering divisions are set up to design, develop and provide turnkey systems, i.e., complete solutions to customers' problems, including information technology, operators, managers, and so on. Amongst the parts of these whole systems, systems engineering divisions will specify artifacts and products, which the hardware division can be called upon to design,

develop and manufacture. Systems engineering divisions might consist of one each for land–sea, land–air and sea–air, or one each for Global Environmental Systems, IDA Systems and Aerospace Systems, for example.

Figure 15.17, therefore not only represents an organization, business or enterprise; it also represents a system, with the various parts cooperating, coordinating and complementing each other: this is not always the way in which divisions operate in large organizations, where divisional rivalry can prevent the development of synergistic interactions. Getting the architecture right may be the key to success.

Human Activity Systems

Why human activity systems?

None of the foregoing in this chapter has been specifically about either technology, or people: in general, systems engineering includes people and technology in the whole-system and its subsystems. If there is one factor that characterizes systems engineering it is the inclusion of humans as systems and subsystems, as system users, as system operators, but always in the system. Conversely, excluding the human from the system, or considering the human as an adjunct to the system, does not reasonably qualify as systems engineering.

There seems to be a fundamental reason for this. Systems engineering is founded firmly in the organismic paradigm, rather than the machine paradigm (see Functionalism and the Organismic Analogy on page 18). And, using the systems approach requires that the 'thing' being studied be viewed in its context, open and adaptive to its environment, which is philosophically organic, rather than mechanistic.

However, there seems to be more to it. If we look at the nature of many manmade artifacts, we can see them as extensions to human capabilities, rather than something that is quite different. For instance, we use radar to help us see better, further, through bad weather and at night; it is an extension to our limited human vision. We use vehicles to travel faster and more easily than we are able to on foot. We use spanners to help us undo nuts that are too tight for our fingers, and we use our hands to wield the spanner. We use catapults and missiles to throw things further than we can throw rocks by hand. We use radio to communicate over greater distances than we can by shouting and waving. And so on. In almost every case, our manmade artifacts are designed to extend our limited human capabilities — and the artifacts that we find the easiest and most natural to use are often those that are designed just like that — a natural extension to our human capability.

Readers can conduct a mind-experiment to underline this point. Imagine you are on a golf course, and you are about to hit a golf ball. You know that you must keep your eye on the ball. You prepare yourself; you draw back, pause, swing and strike the ball, keeping your eye on the spot even after the ball has gone. Curiously, throughout that mental exercise, you will not have visualized a golf club.

You are not a golfer? Try a different sport — the same thing works for tennis, squash, etc: you will not visualize the racquet, only the ball. We become so adept at using these implements/artifacts/clubs/racquets, etc., that they become a natural extension to our bodies – which we do not visualize either. . . .

Even where the human has no innate capability, such as flying, it seems that some designs are more intuitively organic than others. Pilots of the legendary Spitfire during World War II observed that strapping into a Spitfire was more like putting on a familiar suit than climbing into a machine: the aircraft felt like an extension of their bodies, such that flying was intuitive and easy. They described the machine as 'very forgiving,' meaning that pilot flying mistakes were easily rectified. . . .

Including the human in the system, then, may seem natural and inevitable to systems engineers. Of course, we cannot design humans, but we can identify functions that the system may have to perform, and we can elaborate functions into related activities . . .

The human–machine divide

There may come a point in the problem–concept–design–create process, however, where it may prove necessary to allocate some functions, or parts of functions, to people, and other functions, or parts of functions, to technology. This, in turn, implies that some subsystems may be human activity subsystems (HASs), others may be technological subsystems, and some may be mixed, sociotechnological subsystems.

Deciding whether functions and activities should be human or machine is a decision not based wholly either in science or in engineering. In the West, there is a penchant for automating processes and procedures to reduce the need for manpower and hence running costs: in the East, we may find precisely the opposite philosophy. In many eastern countries, the objective may be to create as many jobs as practicable, rather than to put people out of a job.

One approach to deciding whether or not to automate is to start by creating a system powered entirely by humans. This is feasible for IDA and information systems, data systems, many media systems, management systems, manufacturing and assembly systems, (cars, airplanes, computers, white goods, brown goods, etc.). A human-only system (HAS) may consist of a number of teams, each addressing different functions, so that teams cooperating and interacting manage the whole job.

While in some cultures, training up these teams and progressively improving their overall performance would be 'operational systems engineering,' in other cultures, the process might extend further. It may be possible to introduce technological support to the teams, such that either the teams do not need so many people or, alternatively, each team can accommodate a higher throughput rate with the same number of people. *In extremis*, complete processes may be automated, as in chemical and nuclear processing plants, for instance. The motive in such cases may be to safeguard workers as much as to reduce costs.

Often, then, the apportionment of functions between people and machines is a cultural thing. Managers may convince themselves that automotive machines are essential, but in many cases, they are not — at least they are not essential to undertake the task in hand; they may be used to reduce operating costs, if that is an objective. After all, the ancient Egyptians built the Great Pyramid, still today the largest freestanding stone building in the world, in 20 years; and the only technology they had was ropes. Necessity, it seems, was the mother of invention. . . .

HASs, it has been suggested above, may be comprised of teams of trained individuals performing coordinated tasks — and that is often the way. Air Traffic Management is a case in point, where teams of controllers sit at their respective consoles communicating with each other and with aircraft only via the console, which becomes their eyes, ears, voice and even their noses, in the sense that that good controllers can 'smell' when trouble is brewing. (This is Recognition-Primed Decision making –see Instantiated layered GRM on page 138.)

Self-organizing HASs

It may be practicable for such HASs to self-organize. Consider a situation in which the objective is to develop expert command and control (C^2) teams. One approach is to select experienced people, who need not know each other, and to situate them in a C^2 center, part of a simulator which can represent a variety of combat situations with interactive capability, i.e., if C^2 makes the 'right' decisions — whatever that may mean — they will win the day.

The process of self-organization starts by announcing to the collection of individuals that they are to take control of the (simulated) situation: the simulation commences. Radar screens will come alive, showing various force deployment. Tote boards will activate, showing resources in various locations. Telephones will ring, and people at the other end will relay messages and provide intelligence.

Because they are experienced, and because some of them will be better at some aspects of C^2 than others, individuals will sit down at the screens, and start to interpret what appears to be happening. Others will start to put an intelligence picture together from the scraps of information they are given over the telephone.

The pace of simulated activity is then stepped up, so that the players start to be swamped. This will encourage them to form into teams, so that they are better able to cope. Moreover, the numbers in each team will be matched to the need set by the simulated activity rate. Once the teams have formed, stabilized and performed, the rate can be increased again, and the nature of the problem they are dealing with can change, too. Over time — several days — the teams will develop, become flexible, work faster and be more effective.

At that point, the teams are either ready for operational deployment, or a further stage can be invoked, in which the teams are asked to identify technological aids to improve their performance, in terms of being faster, accommodating more information, predicting enemy behavior, etc. This technology can then be provided and the simulations recommenced at even higher rates of change of situation. In this way, the team can define for themselves what they need in the way of technological support, and can try it out to see if it really helps them perform.

Using this self-organizing approach, the size and composition of teams will be self-organizing, and any technology will be suited to its task and to its users, which together will perform some useful function(s).

Training

For self-organizing teams there would need to be an underlying assumption that some, if not all, of the participants were experienced. Where that is not the case, individuals and teams may need to be trained. Since teams perform functions, there is a direct analogy with designing and developing technological subsystems to perform functions. For many manmade systems, it may prove advantageous to train the team where it consists of both people and machines to perform, perhaps, sets of functions.

For instance, suppose the objective were to create an intelligence team. The objectives of the 'intelligence function' can be established, together with a concept of operations (CONOPS). The objectives might be:

- to evaluate current enemy strengths and dispositions;
- to observe current enemy activities;
- To deduce enemy current and future intentions.

Each of these objectives can be achieved by performing a series of functions. These involve gathering information, reconnaissance, surveillance, etc., to establish a picture of the enemy that, almost by definition, will be incomplete. As this so-called intelligence picture builds, it may be combined with other information about own strengths, weaknesses and disposition, to enable a situation assessment. From this, threats and opportunities may be perceived, which will lead on to plans for proactive and reactive actions.

A team of people may be formed to process different kinds of information that might contain intelligence, and to synthesize the intelligence picture from it, hopefully finding correlations between information from different sources to lend credibility. Different sources might include reports from agents (HUMINT), photographic intelligence (PHOTINT), electronic surveillance measures (ESM), signals traffic analysis (SIGINT), and so on. Others in the team will draw these separate sources together into a hopefully coherent picture. Yet others, working from this picture, will try to assess the enemy's disposition, knowing that the information they have will be incomplete. And so on. The whole process becomes a fan-in pipeline, with different team members addressing different elements of the fan-in and different stages in the pipeline.

To help them in the process, the team may be provided with technological support. Clearly, those gathering raw intelligence will need the appropriate technology for taking photographs, intercepting radio messages, etc. But, in addition, the information they glean from these various sources may be entered into a database and time-stamped, to establish currency/latency. So, intelligence information may be stored, correlated, accessed and presented using information technology (IT) as an aid — or not, since people can do this work without technological aids, and have done so for generations.

So, to perform the functions and to achieve the objectives, the team may need training at the individual level, to address technological specializations, and training at the team level so that they can work as a unified whole to achieve their objectives. Good IT design may help them be faster, smarter, and to have better memories, both as individuals and as a team. On the other hand, of course, an IT system containing a full set of intelligence could be vulnerable to attack, disruption and exploitation.... It may be unwise to become totally dependant on IT in some situations — apart from other considerations, it may break down, or power may be interrupted at the critical moment....

Unlike most technological designs, human activity systems adapt readily and swiftly, and can become much better at performing their functions over time. It is important, then, that any technological support with which they are provided does not hinder team development. Generally, then, their machines should be able to develop and evolve with the team members.

Exceptionally, such development and evolution may be a bad thing. If, for instance, the system to be considered was one for controlling a nuclear power reactor, then the correct procedures may be stored and managed through an IT system. In such situations, and after the power generation system has been in operation perhaps for decades, it is quite possible for the team of operators to have developed a distorted view of how the reactor control and safety systems operate, and in consequence, they may no longer be well placed to modify procedures and processes. In such instances, the control procedures in the IT system together with the locks and interlocks built in to the control and safety systems should not be able to evolve and adapt with the operators. Instead, successive generations of operators and maintainers should be carefully trained and re-trained so that they do not 'evolve' their understanding in the wrong direction.

Societal Systems

Societies are aggregations of people, often with a shared culture. We humans are naturally gregarious, so societies abound: we are also inclined to accept the familiar and to be suspicious of the

Figure 15.18 Neolithic dwelling in the village of Skara Brae, Orkneys. (*c.* 3200–2200 BC) There are stone 'beds' to left and right, and a 'dresser' facing, with a central, stone-rimmed area for a fire. The whole was covered over with whalebones (trees will not grow on Orkney because of the wind) and covered with turf. Low alleyways linked the dwellings to form the village (photograph by the author).

unfamiliar — it is probably a 'survival thing' — so societies also disintegrate. Indeed, we humans are not only highly adaptive; we are also highly instinctive, although we seem peculiarly intent on denying it.

Living in societies, in large social groups, probably started off as families and extended families living together for mutual protection, for cooperative hunting and gathering, etc. Visitors to the Neolithic village at Skara Brae in the Orkney Islands cannot but be struck by the coziness of the dwellings, Figure 15.18, huddling together on this most inhospitable of windswept shorelines. The evidence from there suggests that there was no 'head' of the village, no hierarchy of elders, etc. Instead, individuals took the lead in whichever pursuit they excelled. So, one man might lead at hunting, another at fishing, another at tool making, and so on. Carved and shaped stone artifacts found on the site suggest that there may have been cultural beliefs and practices that we find difficult to even guess at.

From the beginning, then, humans in social groups have been concerned with safeguarding resources, protection from danger, nurturing offspring, etc. As societies grow, they develop 'organs of state' which seem necessary to maintain the society as a functioning entity, i.e., to prevent it disintegrating. These often include:

- some form of social hierarchy;
- bureaucracies to manage, make decisions, implement them, supposedly on behalf of the society as a whole, and to communicate, control and regulate throughout the reaches of the society;
- an economic system, not necessarily involving money, but enabling individuals and groups to buy and sell, trade, barter, etc., in a dependable, sustainable manner;
- systems for providing the society with resources — food, water, energy, clothing, shelter . . . ;
- systems for defending society from external and internal threats, including policing, military and possibly conscription;
- rules of behavior that govern, particularly, how people may interact with, and behave towards, each other — these are often set out in spiritual or religious terms, and their impact pervades the culture;

- punishment for those offending against societies rules;
- places for social meetings — feasting, entertainment;
- seats of learning;
- etc., etc.

Societies with their social organs, institutions, groups, clubs, clans, tribes, etc., can be highly complex and interwoven. Moreover, it may prove difficult to identify any clear purpose for a society: on the contrary, different sectors within society and different individuals within sectors may have individual and even contradictory views, opinions, goals, etc. For millennia, people have tried to organize, manipulate, enthuse, empower, suppress, etc., different societies and different sectors within societies. Such efforts have sometimes been called social engineering, and it has a checkered history.

Studies of ancient societies suggest that there many different ways in which societies can exist and survive. The ancient Egyptian civilization is a prime example, lasting as it did for some 4000 years. The civilization was far from consistent — there were periods of stability interspersed with so-called Intermediate periods of disorder, even invasion. Like most civilizations, there was a Golden Age, which many looked back on through rose-tinted spectacles (which had not, of course, been invented). Their Golden Age, the Old Kingdom (2686–2181 BC), was a time of peace and plenty, but with the ever-present threat from the desert, and the potential failure of the annual Nile inundation, upon which they depended for growing food.

Their king was a god, with the power to intercede with other gods to ensure successful Inundations: see Figure 15.19 (cf. Figure 2.4 on page 46). Promoting a belief in gods, and their power to influence the natural world, gave individuals satisfying explanations of everyday phenomena that

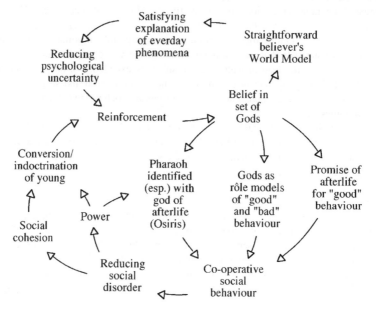

Figure 15.19 Early social engineering: promoting the ancient Egyptian pharaoh as an omnipotent god to give assurance to the people and to promote peace and social cohesion.

would have otherwise have dumbfounded them. The pharaoh also became an icon, propagating ideas of good and bad behavior, promoting the rule of law and ideas of 'fairness,' and 'justice.' These not only promoted social cohesion, but, at the same time, consolidated the pharaoh's supreme position. The belief system worked because it was believable, and it became the focus of sustained societal cohesion for millennia, offering cogent lessons on the establishment and maintenance of social order for today's would-be social engineers.

Social engineering

Since societies began in prehistory, there have been attempts at social engineering by secular and religious leaders, by reformists and radicals. Ancient Egypt engineered 'good behavior' by propagating the idea that there was an afterlife for those who behaved well during this life, and by defining what was meant by 'good behavior.' The Greeks introduced democracy as a way of social engineering, although their society was maintained by a slave labor underclass. The Romans obliged all subjugated states to adopt Roman culture, which was not unsuccessful. Napoleon dreamed of world domination, so did Hitler . . . and then there was Soviet communism — social engineering on the grand scale, which collapsed spectacularly after a relatively short, if bloody, time.

Looking back through history, it seems that attempts to control social behavior en masse result in counterintuitive results, sometimes sooner, always later. With societal systems forming such complex, interwoven, dynamically shifting patterns, predicting outcomes may not be practicable: not that power-crazed dictators seem to care too much about 'how things will turn out.'

Since World War II, particularly, there has arisen in the West notions of 'freedom,' 'liberalism,' and 'human rights' in increasingly secular societies. It might seem that these notions run counter to ideas of social systems engineering. Meanwhile, some followers of Islam have become appalled at the behavior of 'liberalized' western people, individually and en masse. The global stage seems set for yet another societal paradigm shift.

Is there a connection between systems engineering, the systems methodology, and social engineering? In Afghanistan, for instance, there are current attempts to restore infrastructure damaged by decades of warfare. The country is divided into more than thirty provinces, there are many warlords, a significant proportion of the population is nomadic, infant mortality is some 17% a the time of writing, women have been suppressed by the Taliban so there are insufficient nurses, doctors and teachers, and yet the nation somehow needs to recover. The Taliban, however, believe that they are right, and that they are true followers of the faith . . . .

The UN approach (concept of operations?) to restoration at national level is to conduct Peace Operations in stages:

- *Peace-making*. May involve the use of force to oblige warring factions within a country to come to the negotiating table and negotiate peace
- *Peace-keeping*. Maintaining peace while basic infrastructure and services are restored to provide food, water, shelter, power and energy, communications and infrastructure, so that the nation can function again under civil control.
- *Peace-building*. Building up the organs of state, media, industries, etc., to establish a vibrant, self-sustaining economy

The objectives of Peace Operations seem so reasonable and so helpful. However, things that can go wrong do go wrong. Can systems engineering offer anything? Perhaps, but not as presently set up. The following example will show the problem.

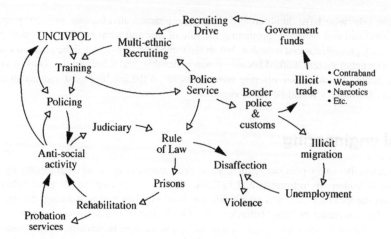

Figure 15.20   Restoring the Rule of Law. (Section of a larger model synthesizing Peace Operations in Afghanistan.) UNCIVPOL is United Nations civilian police.

As part of Peace Operations, it is seen as important to restore the rule of law, see Figure 15.20. Over decades of warfare, civil courts have been destroyed, police have been dispersed or killed, prisons have been wrecked and abandoned, and so on. There is, basically little left of the original systems. So, how to start afresh?

In Europe or the US, money would be made available to recruit and train police, lawyers and judges, build police stations, courthouses and prisons, etc. Businesses would come forward, eager to take the money and to do the work for gain and profit.

However, when operating in a country such as Afghanistan, it is by no means clear that simply making the money for rebuilding 'available' would have the desired effect. Would local tradesmen and craftsmen respond? Would they be prepared to take 'foreign money?' If so, how long would they take to respond and what strings would be attached? Would their fellow countrymen attack tradesmen for 'dealing with the infidel?' And, what legal system is to be established? Will it be secular law, Shariah Law, or something else? Will potential police, lawyers and judges come forward to be educated, trained, developed, etc? If not, what is to be done, and even supposing the population does cooperate, how long will it all take to establish a working system?

The interwoven nature of the problem is hinted at in Figure 15.21, which shows at top left the disruption caused to industry by violence, which in its turn causes unemployment, loss of tax revenue for government and disaffection, leading potentially to more violence. At right, a peacemaking force is shown suppressing the violence and allowing time to reestablish the national infrastructure on which both industry and agriculture depend. Note the need for de-mining: many areas are unsafe with thousands of unexploded mines, which need to be cleared, requiring extensive help from the international community.

Other sections of the simulation model include:

■   Narcotics, a deadly but valuable crop. The opium poppy is easy to grow, and commands a high price. Without it, the farmers believe they will starve. With it, the West faces a continuing flood of illegal and damaging narcotics

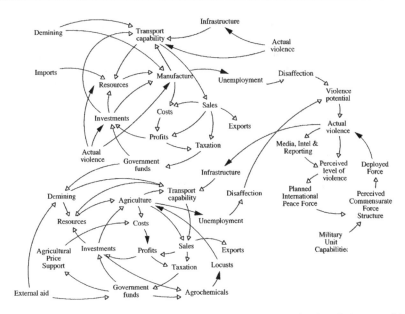

Figure 15.21 Peacemaking for restoration of agriculture and manufacturing (section of a larger model synthesizing Peace Operations in Afghanistan).

■ Population health and medical care — how to restore the nation's health, how to create a healthy workforce able to rebuild the nation....

■ Shelter... much of the housing stock has been destroyed.

■ Energy and power... both generation and distribution on a national scale.

■ Water and sanitation... water is a precious commodity, for drinking and for agriculture, some of which is presently imported.

■ Exploitation of natural resources... Afghanistan has traditionally exported minerals mined from its extensive reserves. It could do so again....

It soon becomes clear that building dynamic simulation models of war-ravaged societies is far from straightforward, and it is not clear how those societies will behave, either in their war-torn state, or as attempts are made to restore their battered nation. It is possible to build models of Afghan societies as extended Learning Laboratories, i.e., to allow simulated experiments with different kinds of international aid, in different sectors, and in different ways. While such models might allow a broad comparison between alternate strategies for reconstruction, they boast neither precision nor completeness. Much research will be needed before dynamic simulation models of societies, cultures and social behavior become viable.

Meanwhile in the West, social engineering goes on at an accelerating pace, driven by political factions and ideologies. *Homo sapiens*, it seems, impatient for biological evolution, has imposed social evolution on societies, nations, and internationally. What may seem to be enhanced freedom and human rights from one perspective may be seen as progressive de-civilization from another.

For instance, did anyone consider the impact on social relationships of the contraceptive pill, which was introduced 'to enable married women to plan their families.'? Its original purpose having now been lost in the sands of time, it is being prescribed to prepubescent girls as a safeguard — not quite in line with the intention, perhaps, but entirely predictable.

And, has anyone looked forward to see the potential impact on society of civil union between homosexual partners, the so-called gay marriage? While such legal unions may seem entirely reasonable from the viewpoint of personal freedoms and human rights, they may further undermine the institution of marriage, with its implicit purpose of providing a stable, nurturing environment for offspring. Already it is becoming clear that, on average, the children of one-parent families perform less well at school, are more likely to take to crime, are more likely to be unemployed, and are more likely to propagate one-parent families . . . .

Whereas police were introduced in the nineteenth century to *prevent* crime, today they are largely powerless to act unless a crime has already been committed. Police rarely act to deter crime or criminal behavior, for fear of breaching human rights laws. This, in turn has led to 'atomized' societies in cities living in fear, with little or no social cohesion . . . so changes to the law designed to enhance freedom of the individual have resulted in the individual retreating into his or her abode, adopting a fortress mentality, and being fearful for his or her life and property. And all of this has occurred as a result of social engineering over a space of only half a century.

All of which suggests that there is a real need for politicians and others to adopt the systems approach when conceiving new laws and regulations, and when proposing to unleash new influences on society. We had better introduce some sensible systems methodology before it is too late . . . .

Summary

The chapter characterizes systems engineering, and what sets it apart from other approaches to creating artifacts, products and systems. Functional, project and program management are examined briefly, followed by a look at various 'flavors' of systems engineering, including unprecedented, evolving, operational and volume manufacturing systems engineering. Archetypal systems engineering strategies are outlined, including waterfall, spiral, evolutionary, concurrent/simultaneous, Sashami, regression and chaos strategies, together with relative advantages and disadvantages.

A look at past 'versions' of systems engineering and project management from the 1970s and 1980s shows the state of the disciplines at that time to be somewhat in advance of present practice in many respects. There have been developments since that time, however, including new tools, methods, techniques, and, particularly, an appreciation of architecture. Several architecture archetypes are presented, and their characteristics examined.

'Designed' human activity systems are discussed, showing that such human systems can be designed and created by breaking down 'functions-to-be-performed' into subfunctions and activities, and training individuals and teams to undertake the tasks in a cooperative and coordinated manner. On the other hand, it seems reasonable to suppose that HASs may be potentially self-organizing, given that some of the members are experienced and motivated to achieve appropriate objectives and goals.

Human activity systems can also be seen as presenting structure/architecture, and as exhibiting emergent properties, capabilities and behaviors. Many purposeful systems may be viewed as sociotechnical, in the sense that the human drives the purpose of such composite systems, while the technology extends and expands upon limited human capabilities. This conceptually places the human at the focus of systems design rather than, as in some engineering methods, somewhat on the periphery of, or an adjunct to, a technological artifact.

The chapter concludes with an examination of societal systems and social engineering, deducing that much research is required if we are to be able to sensibly apply the systems approach, systems methodology, and systems engineering to complex social problems. This conclusion is drawn from brief examinations of ancient Egyptian society and present day, war-torn Afghanistan, which suggest that we are some distance from being able to understand and predict social behavior in either of such situations . . .

Assignment

1. You are asked to create a 5-minute presentation, to be delivered by the company chairman, on the subject of systems engineering, in which he will explain to a group of knowledgeable laymen what systems engineering is about, 'how it works,' why it is unique, and what its pros and cons might be. The chairman is very particular: the presentation is to consist of no more than five slides in PowerPoint; each slide is to contain a simple-but-meaningful picture or diagram and no more than fifty words. Having no choice, you accept the assignment enthusiastically . . . .

2. If the engineering of systems brings together linear parts such that the whole equals the sum of the parts, then there should be no emergent properties: every property should be attributable to one or other of the rationally separable parts. On the other hand, interconnecting and activating a number of such linear parts will form architecture. There may be nodes, redundant links, etc., such that the whole may exhibit vulnerability, resilience, damage tolerance, etc: these are emergent properties. How is it possible for an assemblage of linear parts to present emergent properties? Discuss.

16

Systems Creation: Hand of Purpose, Root of Emergence

Yet I doubt not thro' the ages one increasing purpose runs
And the thoughts of men are widened with the process of the suns

Alfred, Lord Tennyson 1809–1892

The Hand of Purpose Flowing Through Human and Machine

Manmade systems are purposeful: the purpose they serve is that of their designers/makers/owners/operators/users. It is possible to follow the route from the mind of the owner/operator, through the whole system, to observe purposeful activity and to see the purpose achieved, so giving closure.

This simple circuit is perceptible at a number of levels: the first level is within the human individual. We are purposeful creatures — much of the time. We are also singularly unaware of how good we are at achieving our purposes. Consider, for instance the simple act of a man throwing a dart at a dartboard. First, it should be appreciated that man alone has the facility do such an everyday thing. Even our closest cousin the chimpanzee is poor at throwing and catching by comparison with quite young human children.

The would-be dart thrower steps up to the oche (throwing line), leans forward, regards the desired point of impact on the board, raises his or her arm and throws, seemingly in one smooth movement. However, analyzing the whole action indicates that there is a myriad of muscular actions that must take place in the right sequence and with the right degree of vigor. And, prior to the throw, there must have been a rather smart calculation to allow for the distance to the board, the gravitational drop, the weight and flight characteristics of the particular dart, etc.

Research suggests that we have the ability to establish mental 'templates,' sets of neurons that 'remember' the actions that we take, in sequence, and that can be called upon to repeat a complex series of actions, both in imagination and in reality, without having to 'go back to basics' each time. So, the dart player learns to throw the dart by trial and error, and then refines his or her performance with practice, until the neuron 'template' in the brain has stored a well-honed and dependable pattern relating to throwing darts. There must be many such patterns, relating to how we stand up,

Systems Engineering: A 21st Century Systems Methodology   Derek K. Hitchins
© 2007 John Wiley & Sons, Ltd

and manage to stay upright; how we walk and run; how we use a QWERTY keyboard without looking (touch-typing), and many, many more.

Humans become so adept at behaving purposefully that we can lose sight how we do it. You think not? Close you eyes and try to imagine yourself walking by operating the muscles in your back, arms and legs in the correct order needed to walk — not forgetting all the corrections necessary to retain balance. Give up? It's impossible. However, we may be able to work out what is going on in our brains and to use that understanding to help create purposeful social and sociotechnical systems, incorporating artifacts with which we can achieve our human purposes.

Figure 16.1 shows a notional model of purposeful behavior. The diagram might refer to an individual, with the functions being bodily functions, or it might be a complex sociotechnical system with some functions being performed by people and others by artifacts. Indeed, it might be repeated, creating two models one above the other: the upper one could refer to a human operator; the lower figure might refer to a complex artifact being controlled by the human operator. The two figures would be interlinked to show how the intent of the human operator activated and coordinated the various functions of the complex artifact.

Figure 16.2 develops this notion of the two models as one. At the top, the (human) mind perceives a situation, and determines upon some intent. According to psychologists (Klein, 1989), this intent is likely to engender a mental simulation of how we propose to achieve the intent that indicates whether our initial plan is credible. If not, we mentally change the plan and retry, until we find a plan that satisfices, i.e., one that is 'good enough.' All of this, it seems, is performed in the twinkling of an eye.

We then activate the plan, that is, orchestrate/perform the planned activities/functions, etc., all the while observing the developing situation. If things are not going according to plan, we may revisit the plan, perhaps more than once, homing in on an acceptable solution (satisficing). Finally, having realized our intent, we achieve mental closure: we have reached our goal, and achieved our purpose.

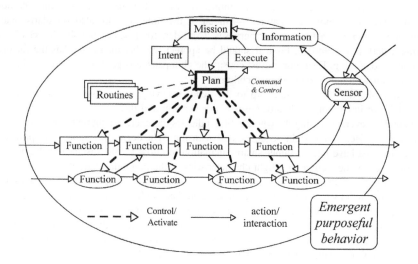

Figure 16.1 Orchestrating functional synergies to create emergent, purposeful behavior. Intent gives rise to a plan, comprised largely of practiced, rehearsed routines that may be adapted in real time. The plan is executed by activating functions/processes according to the routines in the plan, giving rise to synergistic functional activity in pursuit of the Intent; i.e., emergent purposeful behavior.

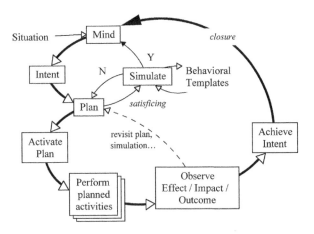

Figure 16.2 Notional model of dynamic control in achieving intent, i.e., purposeful behavior.

Figure 16.2 could refer to something as simple as a golfer at the driving range, practicing his or her swing. It could refer to an aerobatic display, with the pilot controlling stick and throttle as he/she undertakes a series of preplanned maneuvers, making corrections for crosswind, buffeting, etc., as the display progresses. In this case, the model could refer to just one stick movement, an aerobatic maneuver, or to a complete sequence of maneuvers. . . .

Yet again, the figure could refer to a concert pianist playing a concerto, listening to the orchestra, observing the conductor, and initiating all the various routines in sequence that constitute the piano score. . . .

In each case, the figure shows 'the hand of purpose,' i.e., the closed loop that leads from intent to closure when intent is satisfied. It incorporates both function and control of function. . . . And, to a significant extent, the creative phases of the systems methodology and of systems engineering are about identifying, establishing and maintaining this 'hand of purpose' as it goes from human intent, through human and artifact in such a way as to create requisite emergent purposeful behavior, and hence closure.

Which ought to be easy. . . but which, as following sections will show, may not be, as the loop representing the hand of purpose may be come tangled, obscured and confused with other loops and features attendant upon any complex system.

Preserving Interfunctional Connections in Functional-to-physical Mapping

Figure 16.3 shows the notion of functional-to-physical mapping. In the upper diagram, a set of functions is shown, with inter-function flow and function control: this is part, only, of a model such as Figure 16.1, with the elements of purpose, intent and planning omitted for clarity.

The lower diagram of Figure 16.3 shows the functional architecture compartmentalized to create structure, but with all the interflows and controls maintained exactly as in the upper part. This is simple functional to physical mapping, resulting — in this instance — in three compartments: these

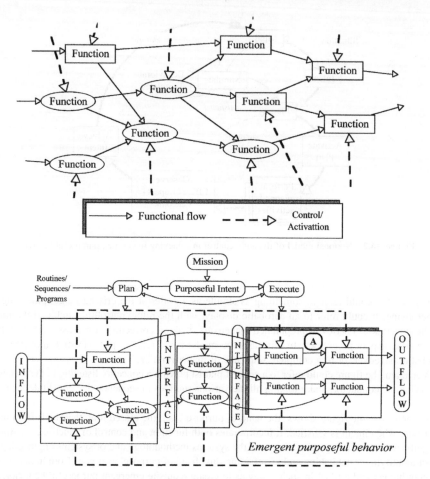

Figure 16.3 Preserving functional architecture in functional-to-physical mapping. The upper figure shows a notional (prime mission) functional architecture, with functional interactions/functional flow. Also shown dotted, are the control/activation lines that coordinate functional activity, and which give rise to synergy. In the lower figure, the elements of the upper figure have been partitioned into a simple physical structure, carefully preserving both the functional features and the control/activation features, without which requisite behavior will not emerge. The right hand block, A, is shown as a closely coupled subsystem of the whole, but will be reconsidered as a viable subsystem in Figure 16.4, below.

might be candidate subsystems, subassemblies, etc. Compartmentalizing creates internal boundaries, necessitating interfaces or accesses to traverse the boundaries.

Figure 16.3, lower diagram, restores the missing Mission, Purposeful Intent, Plan and Execution elements. Note, too, that the control 'lines' have been organized: these are analogous to the central nervous system, providing sequences of controlling signals to the muscles and organs of the body. The whole is beginning to look more like the model of some manmade device.

The right-hand compartment in Figure 16.3 has been marked 'A.' As shown, it is part of the whole, which whole may be presumed to incorporate overall function management, i.e., mission

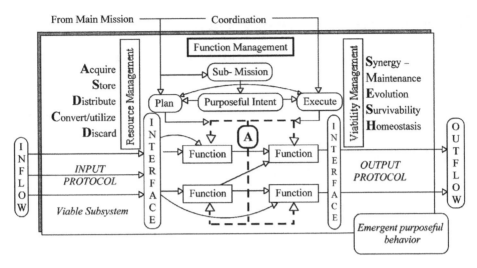

Figure 16.4 A viable subsystem structure model, showing at lower center the functional block, A, as also shown in Figure 16.3. The viable subsystem is able to exist, maintain itself, resource itself, and function on its own. However, when transparently connected to the other subsystems of the whole, as in Figure 16.3, it will be functionally identical with block A in that figure, and will contribute equally to the emergent purposeful behavior of that whole. Note that viable independence has greatly increased the complexity of the whole, and has introduced the need for communications and interchange protocols and interfaces. However, viable independence may also have reduced the vulnerability of the whole, allowed the whole to change form more freely, and presented opportunity for specialization and independent evolution. . . .

management, resource management and viability management, for all three compartments. . . . Suppose, however, that compartment A was to be created, not as an intimately coupled subsystem, but as a discrete, separate, viable subsystem, i.e., able to exist on its own. . . .

Compartment A is revisited in Figure 16.4, this time as a viable subsystem The functions of compartment A are shown, marked 'A,' center bottom of the figure: note that the interflow lines and control lines are all unchanged. However, being a viable system, 'A' now possesses its own, discrete mission management, resource management and viability management, also shown in the figure. This viable system still forms part of the original whole, but it is no longer closely coupled. It sets up its own missions — as 'sub-missions' of the whole systems' mission(s). It executes its own plans, although these are coordinated with the plans and execution of the whole. It manages its own resources. It manifests its own viability. And, because it may be physically separate from other parts of the whole, it may need interflow/intercommunication protocols to allow it to operate remotely, yet at the same time be part of the whole.

Viable subsystems are common. They may be divisions in a business, brigades in an Army, fighter aircraft, ground radars or command and control in an air defense system, navigation systems in a commercial airliner's avionics suite, vehicles in an Apollo mission, etc. Viable subsystems have the characteristics of systems: being subsystems, additionally, they also complement other subsystems in contributing to the whole. (Viable subsystems may undertake missions and pursue purposes in addition to those inherent in being part of the whole: such additional features will not be examined here.)

Emphasizing the Process View

Looking at Figure 16.4, it may become clearer how it is that the 'hand of purpose,' if not overlooked, may at least be played down, during the processes of creating, particularly, viable subsystems of some greater whole. It is a linchpin of systems engineering that this hand of purpose is *not* played down, but is kept visibly to the fore during detailed design, development, engineering, integration and test. It is particularly evident during integration and test, and when fault finding, where the procedure is generally to follow the 'hand of purpose' through the system, finding where it has either not been connected, or has come adrift. Such misfortunes are common during development, and may appear as timing errors, buffer overloads, absence or distortion of signal, loss of control, excessive control, overpowering, under powering, etc., etc. They are also common during operation, with failures, damage, etc., interrupting the 'hand of purpose,' which then has to be reconstituted by repair, replacement, reprogramming, retraining, etc., as appropriate.

Figure 16.5 shows an alternative view of Figure 16.4, emphasizing the systems approach. The viable subsystem, or SOI (system of interest), is shown as an open system in its environment, to

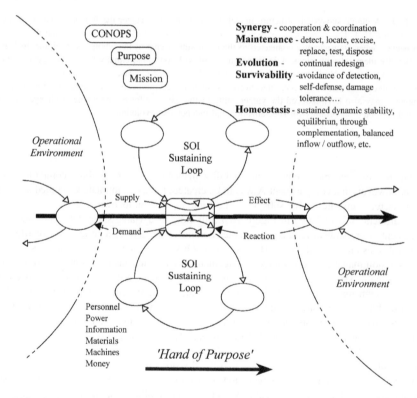

Figure 16.5 A viable subsystem process model. The systems approach regards the viable subsystem of Figure 16.4 as an open system in context, with resource management and viability management sustaining the 'hand of purpose,' i.e., purposeful processes. The viable system will have its own CONOPS, derived from, and contributing to, the CONOPS for the whole of which this viable subsystem is a part. The 'hand of purpose' will pass through all of the prime mission subsystems. . . . N.B. Functional Block A from Figure 16.4 presents, center, as a series/parallel arrangement of processes, A, which together 'perform' the requisite functions.

which it may adapt. The 'hand of purpose' passes from left to right through the subsystem: this is the operational aspect of the system, the prime mission function, processes and behavior. The functional block, shown as 'A' in Figure 16.4, is seen in Figure 16.5 as a series/parallel arrangement of coupled processes, which achieve the functions. For example, in a volume supply system, the function 'assemble' would present as a series of assembly processes in which various parts were offered up to a chassis or substrate, or to each other, in sequence until the assembly was complete. The function view is relatively static: the process view is more dynamic. (Coordination of processes might be necessary, invoking an information system to activate the relevant sequence of coupled processes at the appropriate time.)

Similarly, a function 'track target' would become a series of processes: a function 'strategize,' or 'plan,' would present as two different series of processes; and so on. In general, any function can be presented as a dynamic set of interacting processes — see Understanding Open System Behavior on page 12.

Sustaining loops in Figure 16.6 'enable,' or resource, the prime mission functions with manpower, materials, money, information and power. They also sustain the viable subsystem *per se*, establishing a dynamic stability or equilibrium at high energy levels, and maintaining order (negative entropy) by ensuring connectivity, cooperation and coordination, neutralizing threats from inside and outside of the viable subsystem, and by progressively and continually redesigning the viable subsystem as the need and opportunity arise — see Organismic Control Concepts on page 20. In this Weltanschauung, training systems, simulators, design systems, maintenance systems, financing systems, etc., are part of the viable subsystem, are designed as part of it, created as part of it, and — if appropriate — delivered and operated as part of it.

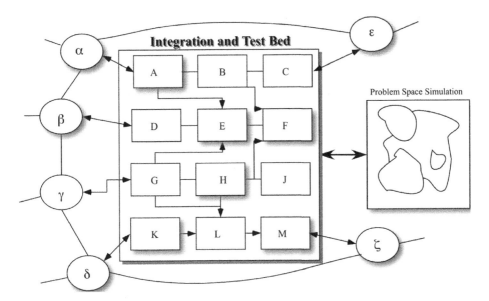

Figure 16.6 Integration and test bed. The whole starts as simulation. The central box contains the modules of a (viable?) subsystem, interconnected and interacting. The subsystem is one of many making up the whole system. The ovals (α–ζ) represent other interacting systems and subsystems that affect behavior of the subsystem, and of each other. The subsystem interacts with the problem space, in concert with other subsystems of the whole, to observe and measure the effectiveness of the whole. Progressive integration sees simulation modules (A – M) being replaced by engineered modules or parts, currently AEFHKM (shown shadowed), until the integration is complete . . . .

System design of the viable subsystem follows the Inner Loop of the Systems Methodology: see Outer Loop– Inner Loop Design on page 293, *et seq.* In so doing, both the structural view of Figure 16.4 and the process view of Figure 16.5 are relevant.

Elaborating the design of the viable subsystem will be facilitated by simulating it dynamically in its open systems context, interacting with other open subsystems as a complementary part of the whole, which is also interacting dynamically with other open systems in the environment. This is best achieved using the process view, since this lends itself more readily to dynamic simulation. Using the simulation approach, the effects on behavior of the whole can be envisaged/predicted by reconfiguring processes, selecting/substituting processes as situations develop, enabling processes not only to interact, but also to adapt/modify each other, reacting to outputs, improving complementarity, cooperation and coordination, responding to threats and failures, etc.

It is this process view, in which the so-called 'hand of purpose' runs through the various subsystems, integral and viable alike, that characterizes systems engineering. The interactions between these various functions and processes are the source of emergent capability and behavior — see Does the GRM Capture emergence? on page 141. This is why it is vitally important to maintain visibility of the lateral processes running through the various subsystems throughout the development, engineering, integration, and proving processes.

In proceeding from design to specification, to development, to engineering, to integration and test and into operation, the process view is, then, held to the fore. The various structures that are designed, developed, manufactured and maintained may be compared, albeit somewhat fancifully, with a trestle railway bridge, designed to support the train as it crosses a ravine, where the train, with its engines and carriages become the prime mission processes which can proceed only provided the trestle bridge is robust and enduring. The bridge creates and provides a supporting capability, which 'enables' the train to operate. In much the same way, the viability and resource management features of the viable subsystem provide an enabling capability, which the prime mission processes use to further the whole system purpose.

This conceptualization is consistent with a neat, insightful approach attributed to USAF scientists at Wright Patterson AFB during the 1980s. They regarded the systems within a fighter aircraft under three headings: mission, resource, and platform. Platform systems (equivalent to viability management systems in the context of this book) were subsystems concerned with establishing and maintaining the aircraft as a platform, i.e., abilities to take off, fly, land, maintain and defend itself. Resource systems were subsystems for acquiring various consumables such as fuels, lubricants, spares, etc., to maintain the platform, and acquiring and managing deliverables such as weapons. Mission systems were concerned with achieving specific missions, such as navigation, reconnaissance, targeting, jamming, weapon aiming, weapon delivery, evasion, recovery, etc. So, the platform and resource systems provided a capability, but without purpose; the mission systems, on the other hand, provided the crew-driven purpose that was supported and enabled by the other two.

Design, Integration and Test

The continuing emphasis on the dynamic aspects of the system-to-be-created presents both needs and opportunities for dynamic simulation. Figure 16.6 shows one approach, which offers significant advantage in maintaining the 'hand of purpose' and of developing requisite emergence.

Investigation of the problem space may have employed dynamic simulation to understand the problem, and to locate the sources of the various symptoms that characterize the problem. Design of the whole system will have employed dynamic simulation to address the interactions that take place within the whole, and between the whole and its environment, with other systems in that environment, etc. These simulations may be brought together to create an overall simulation after the style of Figure 16.6, with the problem, the solution system, and the environment all interacting. In the figure, the large central box may represent the whole solution system, or one of several subsystems that together represent the whole.

Within the box, the simulation modules A – M correspond to the physical tangible modules of the solution system, A – M. Initially, the whole is run in simulation, although it may be possible to connect the solution system simulation to extant, real world sources and sinks (α, β, γ, δ, ε and ζ, shown shadowed in the diagram). In this state, the (simulated) solution system should 'solve the problem,' i.e., should eradicate the problem symptoms, ideally without creating counterintuitive and adverse 'side effects.'

Creating correspondence between the modules of the simulation and the modules to be engineered into the final solution facilitates the development of specifications of behavior for both: this in turn emphasizes the process view, and maintains the flow of purposeful behavior through the solution system.

Designed and specified modules are engineered and/or trained as appropriate — there is nothing in Figure 16.6 that requires a module to be technological — any module could be an individual performing a function, a person operating a machine to perform a function, or a machine performing a function.

As modules, however composed, become available, they are introduced into the test bed in place of their simulated version, interconnections for inflow, outflow and coordination are made, and the whole is run to test the effectiveness of the composite arrangement. If all has gone according to plan, there will be no discernible difference between the behavior of the simulated module and the behavior of the real-world, engineered/trained module. Where differences arise, it may be possible to restore order by including adjustment facilities in the module design. . . otherwise, investigation will be necessary to sort out which is wrong — the simulated or the real-world module. It may also be possible to accommodate deviations from expected behavior within other modules. . . but, if all else fails, it may be back to the drawing board for whoever simulated, specified, engineered and/or trained the offending module.

Introducing tangible modules one at a time is prudent, if time consuming. The alternative, replacing many of the simulation modules with their real world counterparts simultaneously, is fraught with risk. If the resulting whole does not operate according to expectations, it may prove difficult to find the source of the problem, if only because of the complex of actions and interactions between the many parts as they react and adapt to each other.

Summary

Manmade systems have the user's/operator's intent coursing through them; this has been dubbed 'the hand of purpose.' It is the unifying theme that runs through systems and subsystems as they cooperate, coordinate and interact to create emergent properties, capabilities and behaviors of the whole system: it is also their instigator, activating and orchestrating those whole system emergent properties, capabilities and behaviors.

Systems designs may be viewed either from a structural perspective, or from a process perspective, where coupled serial/parallel process sequences constitute functions and functional behavior. Both views are important during the creation phases of detailed design, specification, development, engineering, integration, proving and operation.

Systems engineering emphasizes the process perspective, creating, elaborating and continually testing the design using dynamic simulation of the system or subsystem as an open system in its environment, part of a containing system whole, interacting with, and adapting to, other systems and subsystems. This is the systems approach, and is fundamental to systems engineering.

Assignment

You work at senior level within a dynamic business. The business presently operates from a single site, so that the five operating divisions of the company, each with about 500 employees, share all the common services, etc. You are tasked by the CEO with drawing up plans to relocate one of the five divisions to a separate site some 45 miles distant, where there is significant room for expansion. As a start, the CEO, who sees the move as a systems problem, wants to see a notional site plan showing the division, with its three 'contained' groups, and all of the supporting facilities it will require to exist as a viable operating unit, remote from the main site, but still firmly linked to it.

For the transferred division to remain viable, it will have to address homeostasis, dynamic equilibrium, in terms of revenue, manpower, energy, resources, facilities, security, etc. The move will clearly disturb manpower in the short term, if only because some employees will be reluctant to travel the extra 90 miles per day to reach the new site. Consider homeostasis for the transferred division, and for the four divisions remaining at the main site, and present your considerations in no more than five PowerPoint slides to the CEO

Case E: The Police Command and Control System

The Problem Space

Civilizations since the earliest time have developed and been organized around class structure: it has been the hallmark of society. The earliest civilizations had three classes: agricultural workers and herders generally formed the lower class; merchants and shopkeepers formed the middle class; and rulers, nobles and landowners formed the upper class (Haywood, 2005). Societies as a whole were pyramid-shaped, with a large lower class, a smaller middle class and a much smaller upper class.

Civilizations through history, including present-day western civilization, have continued to be based on three classes, or tiers, although with the advent of mechanized farming, the lower classes migrated from the land to the cities to become an industrial lower class, working in factories and on shop floors, mass producing yarns, cloths, implements, automobiles, consumer goods, etc. — see Figure E.1. The lower class was renamed 'the workers,' while the middle class comprised managers, teachers, shopkeepers, etc., much as before. The upper class remained the upper class, accumulating wealth and living enviable lives.

That envy led to social anger, with workers (no longer 'gatherers,' so perhaps reverting to 'hunters') angry that others benefited from their work, while they benefited little. Class structure, which had stood as the cornerstone of civilization for five thousand years was now challenged. Marxism and socialism emerged. The scene was set for an overthrow of the class structure: those driving the overthrow were aware that they were creating social upheaval, social disorder: they expected a new social order to emerge. Were they, instead heralding continuing and accelerating social disorder — even, perhaps, the breakdown of their civilization, which may have depended fundamentally upon the hated class structure?

Systems Engineering: A 21st Century Systems Methodology   Derek K. Hitchins
© 2007 John Wiley & Sons, Ltd

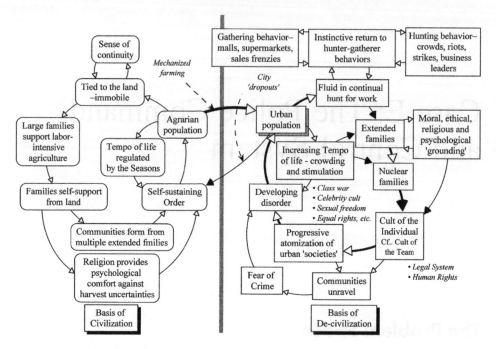

Figure E.1 The Advent of De-Civilization in the West. At left, a largely agrarian population provided stable, extended family-based social structure, tied to the land, with a history of some 5000+ years. The industrial revolution brought mechanized farming, forcing agricultural workers to go to cities seeking work, at right. No longer tied to the seasons, the tempo of city life increased inexorably, fanned by consumerism, competition, crowding and mass entertainment. No longer tied to the land, worker populations became more mobile, and instinctive hunter–gatherer behavior re-emerged. Extended families fragmented. Social bonds broke. Societies 'atomized,' with individuals, older and younger, living in fear. De-civilization had arrived . . . .

Social engineering — irresponsible liberalization?

Two books characterized and gave impetus to social change in the 20th Century. One was Adam Smith's *An Inquiry into the Nature of the Wealth of Nations* (Smith, 1904). Smith (1723–1790) proposed that wealth could be created for all by the promotion of free trade between nations, and the elimination of restrictive business practices. The other was Karl Marx's *Das Kapital* (Marx, 1887), who perceived that the mass of workers were being exploited by the capitalist upper classes, and that social upheaval would be necessary to restore balance and equity. For most of the 20th century, Marxist ideas seemed to predominate over those of Adam Smith . . . .

The second half of the 20th century has seen a progressive liberalization, with relaxation of oppressive rules, regulations and laws. Successive governments competed with each other as social engineers, although they would probably object to being so described. Unfortunately, the long-term impacts of such social engineering were not envisaged — see Social Engineering on page 389, *et seq.*

Much of the second half of the 20th century in the UK has been spent breaking down the so-called class system that supposedly inhibited society and social development. Conceptually, there was a wealthy, landowning upper class, an industrious white-collar middle class and an honest-but-downtrodden, blue-collar, working class. And there were supposed divides between the classes, which inhibited movement up the ladder — everyone was supposed to 'know their place,' and stick to it: at least, that was the caricature.

In attempting to break down the perceived class structure the social engineers revised state education into a homogeneous, so-called comprehensive system, taxed inheritance to deplete the wealth of the land owning 'gentry,' and revised the House of Lords, the upper house of parliament, so that it no longer consisted primarily of lords and wealthy landowners with inherited privilege. In a recent petty, but highly emotive, act of parliament, the social engineers, or socialists, banned foxhunting with hounds on the pretext that it was cruel to the fox: many, including those opposed to foxhunting, will have seen it, however, as a further instance of using (abusing) the law to destroy the last vestiges of an upper class, believed by many city dwellers to be those enjoying traditional foxhunting activities in the country. Throughout, no one noticed that the fox was continually cruel to the chicken. . . and that the fox population is now, paradoxically but predictably, in decline. (Why in decline? Perhaps because the hunts generally managed to catch only the old, unfit foxes, which left more food for the next generation. Perhaps, too, because hunts needed foxes, so foxes would be effectively protected between hunts: no hunt—no protection.)

As might have been expected, social engineering has not had the expected outcome. Class structures have not gone away; they have simply changed, with a new class of celebrity forming a tinseled upper set, populated by the instantly wealthy, the publicity seekers, the flamboyant and the outrageous. Spurred on by the media, the general population is encouraged to celebrate the decadent, the disgraceful, the immoral, the bizarre, the incompetent, and — only rarely — the praiseworthy. 'Celebs' become society's role models, to be admired and emulated by the young and impressionable. The experimental comprehensive education system has not proved to be a success, not so much dragging the average level of education down across the nation, as homogenizing, of bringing all towards a median level, with little room for the very bright to succeed, while the dull remain, uh, dull.

With these major social changes has come social turbulence, which initially expressed itself as civil unrest and violence between supporters of rival factions, often football clubs. Political reaction has generally been to provide more, widespread mass entertainment, presumably to keep the youthful, more vigorous section of the population fully occupied in harmless pursuits such as outdoor rock concerts, all-night raves, greatly increased varieties of television entertainment, blockbuster movies of greater and greater disasters, and so on. Mass entertainment has become the soma of Aldous Huxley's *Brave New World*: 'All the advantages of Christianity and alcohol; none of their defects.' Meanwhile state religion, Christianity, Marx's 'opium of the people,' has been sidelined, disregarded and mocked: he would have approved greatly.

The net effect of this liberalization and barrier removal should have been greater freedom for the individual, greater prospects for self-determination, the establishment of a 'level playing field,' with equal opportunities for men and women, able and disabled, young and old.

However, it has to be said that the sum effect may be viewed as a progressive breakdown in civilization. The fabric of society, at first made threadbare, is showing signs of disintegration. There has been a widespread breakdown in the family, the basic building block of society. The institution of marriage is in decline. Parenting appears to be a largely forgotten skill, with children no longer being nurtured and taught how to behave socially in the family, before being launched on an unsuspecting public. Art, often an indicator of civilized culture, has gone from weird to bizarre, with the famous Turner Prize for Art recently being awarded to an unmade bed, with semen-covered bedclothes strewn over it. Fashion has gone from the elegant to the degenerate, with new trousers on sale ready soiled and worn, with tears and knee patches . . . .

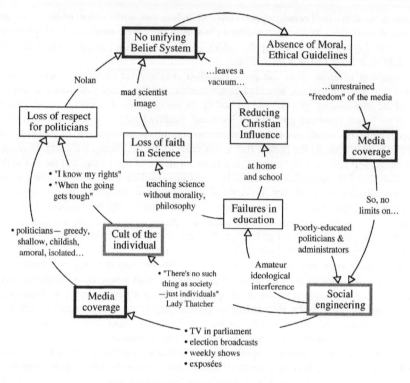

Figure E.2   Unrestrained social engineering.

Political correctness: the new secular religion

In place of rewards in heaven promised for good behavior on Earth, there has emerged 'political correctness' — the new, secular religion. There is certainly a need for something — see Figure E.2, which purports to show the effects of social engineering when combined with a concept of 'human rights' which is described in the figure as the Cult of the Individual, i.e., there is no such thing as society (famously stated by Margaret Thatcher, although she later tried to recant), individuals are the basic element, individuals have certain inalienable human rights as laid down in law, and so on.

Political correctness may be seen as (excessive?) zeal in the observation of these Human Rights laws, and in the interpreting them in such a way as to defy common sense: one UK police force recently refused to publish photographs of escaped criminals for fear of infringing their human rights!

This may suggest, correctly, that the police services in England and Wales are 'treading on egg shells.'

■   Prisoners in jail successfully sued the government for making them go 'cold turkey' when first imprisoned, instead, presumably, of either supplying them with illegal drugs in prison, or providing some legal substitute.

- Police questioned an elderly Christian couple for an hour for stating openly that homosexuality was immoral: the couple was subsequently awarded £10 000 ($20 000) by a court for the distress caused to them.
- Police refused to give chase to a perpetrator escaping on a motorcycle because he was not wearing a crash helmet as required by law: if, while pursued, he should crash and be injured, the police would have 'infringed his human rights!'

In Figure E.2, 'Nolan' refers to the Nolan Report on Standards in Public Life. His report identified seven Principles of Public Life:

1. *Selflessness*. Holders of public office should take decisions solely in terms of the public interest. They should not do so in order to gain financial or other material benefits for themselves, their family, or their friends.
2. *Integrity*. Holders of public office should not place themselves under any financial or other obligation to outside individuals or organizations that might influence them in the performance of their official duties.
3. *Objectivity*. In carrying out public business, including making public appointments, awarding contracts, or recommending individuals for rewards and benefits, holders of public office should make choices on merit.
4. *Accountability*. Holders of public office are accountable for their decisions and actions to the public and must submit themselves to whatever scrutiny is appropriate to their office.
5. *Openness*. Holders of public office should be as open as possible about all the decisions and actions that they take. They should give reasons for their decisions and restrict information only when the wider public interest clearly demands.
6. *Honesty*. Holders of public office have a duty to declare any private interests relating to their public duties and to take steps to resolve any conflicts arising in a way that protects the public interest.
7. *Leadership*. Holders of public office should promote and support these principles by leadership and example.

These principles apply to all aspects of public life. The (Nolan) Committee has set them out for the benefit of all who serve the public in any way.

It is an interesting commentary on the depths to which politics has descended that such principles should need to be stated: the occasion was brought about by a number of politicians behaving in a manner that called their integrity, and the integrity of parliament, into question. It is not unreasonable to suppose that the seven principles were set out specifically because they had *not* been observed, perhaps indicating the depths to which standards of behavior had fallen.

Politics in policing – tough on crime, tough on the causes...

At the same time, and for much of the last two decades, politicians have been fond of repeating the political mantra: 'tough on crime and tough on the causes of crime.' Unfortunately, no one seems to be quite sure what the causes of crime might be, so the mantra reduces in practice to 'tough on crime. . . ' As a result, the prison population has risen steadily, until the UK now has more offenders in prison per capita than any other European nation, prisons are bulging to breaking point, and a program of building new prisons is underway. So, not much impact on the *causes* of crime, then. . . .

Crime has been a thorn in the flesh of successive governments since World War II. When annual crime statistics are published, they are invariably up or down on the previous set of figures. If they are up, government congratulates itself on its success. If they are down, the opposition howls in derision, and more draconian sentencing of prisoners is introduced, or the government embarks on some enterprise such as 'three strikes and you're out,' or 'zero tolerance,' and so on.

The facts are, however, that government actions seem to have little effect. The graph under the heading of Fractals on page 39 shows the reported crime statistics for a county of England over several years. Analysis of the graph shows that it is made up from two elements: an underlying, steady increase in reported crime levels, year on year: and, a highly variable element on top of this underlying 'ramp.' The variable element turned out to be fractal, i.e., weakly chaotic. This, it is reasonable to deduce, means that government actions are having little or no effect on the crime statistics, which vary chaotically, and would do with, or without, government action. The statistics on which the graph is based also show a positive correlation between reported crime clear-up rate and the numbers of police and sergeants on reactive duties. Curiously, adding in the numbers of inspectors, chief inspectors, superintendents, etc., significantly reduced the correlation. . . .

The Solution Space

Policing in a democracy

Policing in a democracy is a contentious subject. If a democracy is 'rule of the people, by the people,' then perhaps policing should also be 'of the people, by the people.' To some, even the idea of policing is inconsistent with that of democracy and personal freedom. However, most people see the need for laws, if only to protect the weak from the strong and powerful, and consequently they see the need for police to 'enforce' the law. 'Enforce' is, of itself, a contentious word in the context of policing, implying as it does that the police may use force: so contentious that the term Police Force is now considered inappropriate, to be replaced by the more politically correct 'Police Service.' The ever-present threat of a 'police state,' looms large in some people's minds, presenting to them a distasteful image of armed police on every street corner, herding and brutalizing the cowed, downtrodden population. . . and there are such police states in the world.

Victorian London was a hotbed of crime and disorder, with cutthroats, cutpurses, highwaymen, robbers, pickpockets, prostitutes, druggies, drunks, beggars, an antiquated legal system, and bulging prisons. ('So, what's new then?') In 1829, Sir Robert Peel established the Metropolitan Police of London, the first such force in the world. Cognizant of concerns about freedom and democracy, he also set out his Principles of Law Enforcement, as follows:

1.   The basic mission for which the police exist is to prevent crime and disorder (NB: *prevent*, not catch after the event).
2.   The ability of the police to perform their duties is dependent upon public approval of police existence, action, behavior and the ability of the police to secure public respect.
3.   The police must secure the willing co-operation of the public in voluntary observance of the law to be able to secure and maintain public respect.
4.   The degree of co-operation of the public that can be secured diminishes, proportionately, with the use of physical force.

5. The police seek and preserve public favor, not by catering to public opinion, but by constantly demonstrating absolute impartial service to the law, in complete independence of policy, and without regard to the justice, or injustice, of the substance of individual laws; by ready offering of individual service and friendship to all members of society without regard to their race or social standing.
6. The police should use physical force to the extent necessary to secure observance of the law or to secure order only when the exercise of persuasion, advice and warning is found to be insufficient.
7. The police at all times should maintain a relationship with the public that gives reality to the historic tradition that the police are the public and the public are the police; the police are the only members of the public who are paid to give full-time attention to duties which are incumbent on every citizen in the interest of the community welfare.
8. The police should always direct their actions toward their functions and never appear to usurp the power of the judiciary by avenging individuals or the state.
9. The test of police efficiency is the absence of crime and disorder, not the visible evidence of police action dealing with them.

Reading through Sir Robert's words, they are as fresh and relevant as when he wrote them in 1829. However, these Principles do not seem to be observed today.

Changes in policing

So what happened? Well, technology happened, for one. The invention of the car and the radio convinced police in the 1930s that they could become remote from the people, could rush to the scenes of incidents, could be alerted and directed by radio and so they didn't really need to be out there, with the public, part of the public: rapid response also meant that less police were needed, saving on costs. . . and playing with the new toys was such fun! Principle Number 7 went out the window, and never returned. With it, Principle Number 1 had to go since, with few police on patrol, there was little advance intelligence of what was going to happen; with no officer on the spot, the opportunity for 'nipping things in the bud' no longer arose. Prevention, the fundamental mission of the police, gave way to catching after the event.

The idea gradually arose that patrolling the streets was for 'wooden-tops' (uniformed patrol officers), and that 'real policing' was catching criminals: not to mention rushing around in fast cars and armed response vehicles, arriving at the scene in only minutes, but often hours after the crooks had flown.

Principle Number 5 disappeared with the advent of multicultural societies (an oxymoronic political term, since a number of adjacent cultures tends to become a number of self-segregating, separated societies, rather then one unified society). Police forces, particularly in the cities, failed to recruit members of the various ethnic cultures, and failed to understand the cultures they were attempting to police. With Number 5 went Principles 2, 3 and 4.

But perhaps the final nail in the coffin for crime prevention was the employment of statistics; this scuppered Number 9 overnight. You cannot easily measure the effect of deterrence. There is no immediate evidence that an officer has calmed things down, perhaps prevented a scuffle or worse, simply by his or her calming presence. Nor is it evident when a criminal decided not to perpetrate a crime because he saw a policeman near the scene and thought better of it. Since this deterrent effect cannot be directly measured, it cannot be entered into a statistic, so in effect, it does not exist. Hence, you cannot measure the efficiency of the police as Sir Robert, so correctly, defined it through 'the absence of crime and disorder' — at least, not directly using statistics. But then, police officers at that time were primarily Peace Officers, not crime fighters.

As part of the general liberalization and enhancement of 'human rights,' the law change, too. Lawmaking became a political game, with political parties competing to introduce new and better laws. The death penalty for murder was abolished. More subtly, the so-called vagrancy laws were abolished, too. No longer could a policeman require someone to 'move on,' or show 'visible means of support.'

Hawks, doves, liberals and terrorists

Modern policing operates, if at all, in this confused, contradictory moral vacuum, where laws, some of them bad laws, seek to direct and control personal behavior. In George Orwell's prophetic novel, *1984*, he postulated the existence of the Thought Police, who control and regulate what we, the public, are allowed to think. The Thought Police are with us today, inhibiting freedom of speech and action, and obliging people to 'think differently.' People, it seems, do not think differently — they just keep their thoughts to themselves in this 'thought police' era.

Within government there are, as always, hawks and doves. The hawks would, if they could, introduce draconian punishment for even relatively minor crimes: they would restore the death penalty; they are 'against' almost everything; and, they are generally xenophobic. The doves, on the other hand, would rather issue community work orders to criminals than imprison them, would rather encourage with carrot than beat with stick — a mixed metaphor, perhaps, but apposite. Liberals seek to reduce all constraints on personal freedoms, and are 'more concerned with the environment' than their opponents. . . .

A balance is to be drawn between fear of a police state on the one hand, and fear of crime and violence on the other. Concern about excess policing is winning the day, with fewer police visible on the streets, an ever-increasing ratio of public to active police, while many other police are consumed by documentation, paperwork and mind-numbing legal procedures. . . with a population of around 60 million, England and Wales have some 130 000 police, of whom only a very small proportion are visible on the streets at any time.

Terrorism changes the scene: not only terrorism, but terrorism generated from within England's much vaunted multicultural society: home-grown terrorism, even if externally sponsored and inspired. Gradually, it dawns that there is a problem, and that the erosion of Peel's Principles, together with unthinking social engineering, has opened the nation up, made it vulnerable.

Social atomization, fear of crime, and the reactive spiral

In many communities, there are people who are afraid (Peak and Glensor, 1996). Fear of crime may be irrational, since the statistics indicate that the likelihood of being on the receiving end of criminal behavior may be relatively small, but the fear remains, repeatedly and incessantly stoked by the media. Fear is prevalent amongst the elderly and infirm, and amongst the young and vulnerable. Much of this fear is transmitted by bullying, and by reports of youths roaming the streets in gangs, particularly at night, causing mayhem. Certainly, there has been a sea change over the last 50 years, with youths seemingly ever ready to offer verbal abuse and physical violence at the drop of a hat. And this behavioral phenomenon has spread to young females, too, fueled by alcohol and binge-drinking.

The reaction of many, especially those living on their own, is to retreat into their fortress home, apartment or condominium of an evening, and to avoid social contact. This is leading — has led — to social atomization in cities, where individuals exist, go about their business, but have no relationships with other individuals outside, perhaps, of daily work.

Such people cannot be said to live in a community: they are isolated. The most effective policing, however, comes not from established police in uniform, but from communities, which effectively police themselves, characteristically through 'neighboring,' as the police call it. Isolated individuals do not take part in neighboring, do not recognize and are not recognized by neighbors. (Kelling and Coles, 1996)

Police working in such supposed communities and societies have become largely reactive, i.e., they react to alerts. Sometimes they arrive after the incident is over, the villain is gone, etc., which does not inspire confidence in the police. As a result, it is common practice for members of the public not to report every crime to the police. Estimates vary, but typically, it seems that only some 25% of crimes are reported. Police respond to reports of crime, and produce a 'reported crime clear-up rate,' which is typically about 20–25%. So, the overall rate for solving crime is about 25% of 25%, which is 6.25% — far from brilliant.

Police numbers are established with this reported crime clear-up rate in mind. An associated phenomenon is known in police circles as 'the reactive spiral.' Suppose that a particular police service is doing rather well, and is catching and prosecuting more perpetrators than previously. One byproduct of this is increased public confidence in the ability of their police: the public is more likely to report incidents to the police, believing that it to be worthwhile since the police are doing so well. As a result, the better the police are at reacting to reports and incidents, the more work they attract, and the more manpower they require — this is the reactive spiral in operation.

A diagram showing the reactive spiral is at Figure E.3. If police resources are unlimited, the reported crime clear-up rate will increase, there will be more reports of crime, more demand on

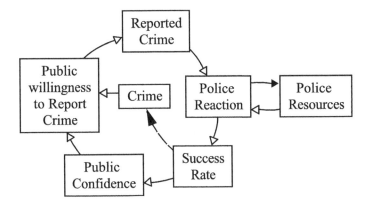

Figure E.3 The reactive spiral. Success in solving crime boosts public confidence, increasing their willingness to report crime, which increases the work of the police. Further success in crime solving will lead to spiraling demand for ever more reactive police. The spiral tops out due either to limited police resources or to suppression of crime.

police resources, more crime cleared up, and so on. Only two factors prevent the spiral from rising:

- Limited police resources. This is the usual limiting factor, and the spiral increases the demand on police until they are fully stretched, and tends to settle at that level.
- Reducing the propensity for crime. Since so little of crime is presently reported, the spiral could rise some way before the general level of crime is significantly affected. Occasionally, prolific offenders are caught and imprisoned, which temporarily reduces the crime rate, but their places are often swiftly taken by replacement 'wannabees,' so restoring the status quo.

The first bullet suggests that the reason why crime clear up is so low is simply because governments, central and local, are unprepared to pay for more policing: we get the level of crime we are prepared to tolerate as a society, and that level seems to be quite high. . . .

Reactive policing, then, is something of a failure, both in principle and in practice. Even if the police could react at the speed of light to reports of crime, they would still pick up only a small fraction of crimes. Criminals in the US soon learned to oblige their victims not to report until the perpetrators had left the scene.

Progressive reduction in the state of stability

Over time, and particularly over the last fifty years, social turbulence has increased in the UK — social turbulence, is, after all, a consequence of liberalization, which permits and encourages people to do different things, travel more, be more vociferous in opposition to 'authority,' and so on. Turbulence is an aspect of the stability of society: state of stability is a useful measure, which may be formulated over time to assess social dynamics, see Figure E.4. At the top of the figure, the population is shown in three 'compartments:' ordered society; disordered society; and, prison. Population turnover, i.e., people entering the area (county, state, country) from other areas, counties, states or countries, for whatever reason, increases the numbers in disordered society, at least initially.

Ordered society represents, hopefully, the bulk of the population, where people lead orderly lives, go to work, bring up their children, go to the cinema, concerts, sports meetings, college, etc., belong to clubs and associations, take family holidays, etc. . . . Socialized humans are creatures of habit and routine, and tend to lead ordered lives. They are also, by nature, gregarious and form groups: families, associations, clubs, teams, companies, etc., etc. These also represent order.

Disordered society is that proportion of society that is not settled into an orderly routine and that is not socially bonded to others within the society. These are the people who do not observe social norms, either because they are unable to do so, or because they are isolated from society and no longer influenced by, and grounded within, its norms. It is from this disordered section of society that, in general, crime emerges, leading to antisocial activities and to fear of crime. Fear of crime, as we have seen, can have the effect of atomizing parts of society so that people living on their own, and older people in particular, feel afraid and marginalized — this has the effect of drawing them from ordered, into disordered, society, see Social capital on page 50.

The reactive spiral is shown at the right of Figure E.4, with detected crime (rate) enhancing the reported crime (rate). Detected crime results in some small proportion, currently about 2%, of perpetrators going to prison. This is police reaction, or reactive policing at work. Police may also take direct action, as indicated, to combat crime, to curb antisocial behavior, and to reassure the

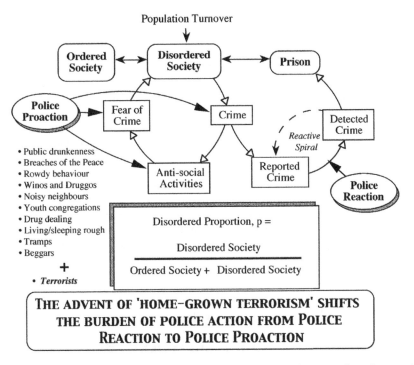

Figure E.4 Measuring society's state of stability. Home-grown terrorism adds a new dimension to antisocial activities . . . .

public that they have no real need to fear crime. However, as the reactive spiral explains, so many police may be absorbed in reacting, that there are likely to be few left over for other activities.

A parameter, or measure, the state of stability for the area, county, state or nation, may be developed as shown in the figure. The state of stability p is the ratio of disordered society to ordered society plus disordered society. Those in prison (some 70 000 at the time of writing) are not included, since they are not, for the time being, in society.

The state of stability p is relatively meaningless as a one-time measure. Its value is in how it varies over time. Moreover, it would be wrong to think that a fully ordered society, with a value $p = 0$, is desirable: on the contrary, as the police are aware, such a society would be effectively dead. With no turbulence at all, society would be missing one of its 'vital signs!' On the other hand, a rising value of p over time may be an early indicator of societal issues. Similarly, a falling value of p over time may suggest either that social capital is increasing, or, if p falls too far, that society is losing its dynamism. (Oppressed societies would be expected to exhibit a low value, p, by virtue of the heavy hand of bureaucratic control.)

Graph 2.5 on page 39 indicates that, for a particular county in England, the crime trend was increasing steadily over some twenty years, suggesting that the state of stability was falling. Over the same period, the prison population has risen, along with the fear of crime. Reactive policing using a minimum of police may satisfy the liberals, and may pacify concerns over 'police states,' but it is evidently ineffective in accommodating a progressively more turbulent and atomized society.

The terrorism issue

In particular, and as a number of countries have experienced, reactive policing is no match for terrorism. 9/11 in the US and July 7th in the UK were wakeup calls — our policing systems are inadequate. Particularly, they are inadequate where, as in the UK, the terrorists turned out to be UK citizens, born and raised in the UK, but seemingly recruited and subverted by agents of a foreign country and an alien culture.

As any military man will know, it is impossible to mount a perfect defense. But a policing system that can, in effect, only react once a crime has been committed is of little use when trying to prevent terrorism — particularly when fanatics seem to be willing, even eager, to die while committing their acts. Then, reactive policing is no defense at all.

Looking back over the last fifty years of increasing liberalization, advances in human rights legislation, etc., it may be seen that, admirable though these advances may be, taken individually, they have resulted in creating a soft underbelly within democracy. Having rationalized that global war and international conflict are outdated, Western democracy is now so liberalized as to be in danger of being unable to defend itself, and its way of life. The pathway to hell, it seems, truly is paved with good intentions. . . .

Remedial Solution Concepts

Proactive policing, gearing and reactive demand

If reactive policing is not working fully, which seems to be the case, then combining it with proactive policing will undoubtedly help — see Figure E.5. In the diagram, reactive policing is seen as largely dealing with crime. Proactive policing, which involves the police in taking an active line, operates in a variety of ways to enhance social stability and build social capital.

As the figure shows, there are two arms to this proactive approach:

- To address and help to reduce the fear of crime through neighborhood schemes, self-policing schemes such as Neighborhood Watch, increased victim support, counseling to those in fear, and visible policing. The general public has, for decades, demanded to see more police on the streets — requests that police have felt unable or unwilling to satisfy, through lack of manpower.
- To encourage and support social groupings through education, cross-cultural association, sports, youth clubs, etc., and to positively assist with employment, e.g., of ex-convicts who need to become part of ordered society quickly so as to avoid the temptations of returning to crime.

Proactive policing does not replace reactive policing: rather, it complements it by encouraging a more cohesive society, reducing the process of social atomization and increasing social bonding. Proactive policing offers a so-called gearing effect: whereas reactive policing tends to be one-on-one, with one police officer versus one offender; proactive policing tends to be one on many. So, one policeman can encourage many dissociated people to 'join in.' This effect is referred to as 'gearing,' and can help to minimize the number of police needed in an area, if only by progressively reducing social turbulence and hence reactive demand.

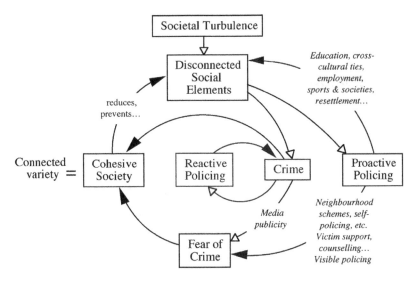

Figure E.5   Proactive and reactive policing together. In this causal loop model, open arrowheads 'encourage,' close arrowheads 'discourage.'

Terrorism changes the picture...

However, neither reactive policing, nor the kind of proactive policing outlined above, is a match for terrorism. If terrorists have determined to perpetrate crimes against the general public, and if those terrorists are virtually indistinguishable from that general public, then the only approach that affords any prospects of success is one of combined anticipation, detection and deterrence.

■   Anticipation seeks to predict the terrorist events and activities and so to prevent them happening by positive interception — before the event, not after.
■   Detection seeks to observe terrorist recruiting, training, development, resourcing, preparation, rehearsal and activity.
■   Deterrence seeks to prevent terrorist action by being on the spot before and during the intended action and either warning the terrorist off, or catching the terrorist as he/she acts, or both.

Density of police on the ground needed for deterrence, e.g., of terrorist activity, is high. If, as may well be the case, there is no definite intelligence to warn of the location or timing of an attack, then deterrence will require police to be out on the streets in large numbers, forming a mobile, shifting network of visible individuals looking out for untoward behavior and ready to act. For such a network to be effective, policemen would have to be on street corners, patrolling residential areas, talking to people in areas of dense population such as railway stations, airports, shopping malls.

So close would policemen have to be to each other that, on alert, several could converge on a spot literally within seconds — minutes would be too slow. All of which may sound unreasonable — unless, that is, one were to recall Metropolitan London in the Victorian era, where just such an arrangement was in place for crime prevention. Just such a dense network of police 'on the spot' was needed to cope with the level of street crime, petty and not so petty, that was the hallmark of Victorian London. In many ways, it is not the problem that has changed, so much as social perceptions, the law, the level of crime and the style of policing to which we have had to become accustomed, willy-nilly, since World War II.

Cost-effectiveness of deterrent, crime prevention policing

It might be thought that the cost of such policing would be prohibitive, and would render the notion instantly unacceptable. Not necessarily.

At present, the UK is spending significant amounts of money on prisons, prison warders, probation services, courts, lawyers, proceedings, etc; it is difficult to calculate a total figure, but it is mounting all the time. Estimates suggest that it costs £100k/$200k to keep a prisoner in jail per annum: with 70 000 prisoners in UK jails, that amounts to £7billion/$14billion. That does not account for costs to the state of the 'headless' family: if the breadwinner is imprisoned, spouse and children are likely to need state support for food, education, healthcare, etc. It gets worse: children brought up in an environment where one or more parents, or older siblings, are criminals and 'do time' are, themselves, more likely to turn to crime, so creating a crime family dynasty. Once sent to prison, the so-called crime academy, offenders are more likely to offend again. This may be one reason why the numbers in prison are rising, seemingly inexorably. . . £7billion/$14billion may be the tip of the iceberg.

If deterrent, crime prevention policing can stem and reverse the tide, then the cost of deterrent policing may be offset by the savings to the state from reduced cost of keeping ever more people in prison, operating probation services, supporting families, building new prisons, etc., etc. If deterrent, crime prevention policing can minimize the risk of terrorist attack, and reduce the fear of crime that presently paralyzes sections of the population, then the value to society would be inestimable.

Of course, there is a problem; governments of all persuasions seem remarkably reluctant to accept such 'creative accounting,' i.e., offsetting the cost of policing against savings in prisons, etc. Nevertheless, there are great savings to be made — more than sufficient to offset the cost of peace operations, Robert Peel style.

Recruiting proactive police

How might one recruit larger numbers of police to be on the streets at all times? This is not Victorian London, and nervousness about a police state must be addressed. Robert Peel understood the problem, and the answer is, in part, given in his seven principles. In particular, the police are the people, and the people are the police. So, the increased police presence that would be needed can only be derived from the communities and societies to be policed. As was always done in medieval times, the people should nominate their peace officers.

In a modern twist, however, this notion may be turned to society's advantage. Conceptually, it would be possible to call for volunteers, especially from young men and women who have finished schooling and are keen to go to university or further education. They could form part a 'voluntary county service:' police recruits would be selected from among the volunteers and offered full training, with full pay, as police officers. After training, they would go through a probationary period, accompanied by an experienced officer, and would spend up to two years policing their own communities. During this time, they would have the opportunity to save money and they would subsequently attend university free of tuition fees. On leaving university they would have a guaranteed position in the police force, but only if they so wished.

The advantages of such a scheme are more than the provision of much needed police numbers. Since these young men and women would be drawn from the communities that they policed, they would satisfy Robert Peel's principles, and could hardly be said to be part of some 'police state.' Moreover, their time in the police service would instill within them a self discipline, self-respect and understanding of community dynamics that would enable them to build social capital in their communities, even when no longer members of the police service.

Peacemaking, peacekeeping and peace building: Levels 1, 2 and 3 policing

The UN approach to Peace Operations is to address them in three phases: peacemaking, peace-keeping and peace building, see Social Engineering on page 389. Conceptually, the UN regards these three phases as sequential, i.e., non-overlapping.

The idea of Peace Operations may be applied to policing. In police terms, peace is order, and the task of a peace officer is the maintenance of order — or, as it used to be quaintly referred to, 'maintenance of the Queen's Peace.'

- Peacemaking is the activity of restoring order, where disorder has broken out. So, reacting to crime, responding to incidents such as brawls, domestic violence, football riots, etc; all of these and many more would be peacemaking.
- Peacekeeping is maintaining order, preventing crime, nipping an altercation in the bud before it flares into violence, deterring disorder and criminal behavior, preventing terrorist acts, etc.
- Peace building is to build social cohesion and social capital, to reduce the fear of crime, to encourage cross-community and cross-cultural cooperation, to support people in need and at risk, and so on.

In policing, unlike UN Peace Operations, the three activities occur at the same time, rather than sequentially:

- Peacemaking corresponds with reactive policing. As such, it is fast response, and often invokes the use of police toys: fast cars, flashing lights, sirens, etc.
- Peacekeeping corresponds with proactive policing, including deterrent, crime prevention policing. In priority terms, it follows on behind peacemaking, which has first call on manpower. Unless, that is, terrorism is taken into account. . . then, peacekeeping has first call on police availability
- Peace building corresponds with building social cohesion and social capital. Peace building is a longer-term activity, with projects perhaps taking months or years to come to fruition.

Although the three peace activities may not be sequential, they evidently call upon police manpower with different priorities. This encourages the concept of 'levels of policing,' as follows:

- Level 1 policing. Essentially peacemaking, with fast response to incidents, emergency calls, etc.
- Level 2 policing, corresponding to peacekeeping operations. Visible security patrols, operational intelligence gathering, intervening to calm disputes, reassuring vulnerable people, deterring terrorists, catching perpetrators in the act, etc.
- Level 3 policing, corresponding to peace building operations.

Concepts of Operations

Policing models

The various concepts presented above can be drawn together into a notional organization — see Figure E.6. The figure shows an organizational diagram in which the three levels of policing are represented. The objective of the police is to improve the state of stability, which is shown at the top of the diagram. The organization is intelligence driven, with intelligence having been derived from a number of sources, including — but not limited to — that garnered by patrol officers as they do their rounds, talk to people, build profiles of individuals, places and things, and look for deviations from the established profile — 'deltas' — that indicate unusual behavior.

India (intelligence) officers support patrol officers in real time by radio. India officers are able to advise patrol officers about the profile of the area/person, what to expect, known associates of individuals, escape routes from areas that the patrol officer is approaching, etc.

The area sergeant prepares a so-called menu of task for each patrol officer. How and in what order, the patrol officer undertakes the tasks is at his or her discretion, since they may at any time be diverted for a Level 1 response; this explains the term menu, indicating that the patrol officer may, to some extent, pick and choose the order, so long as all of the tasks are addressed by the end of the period of duty — if possible.

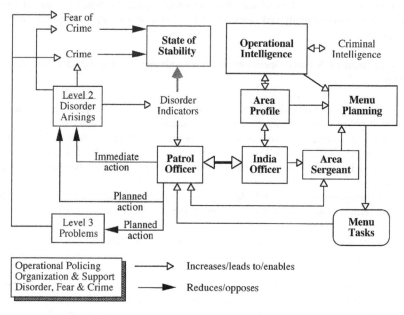

Figure E.6 Conceptual organization of Levels 2 and 3, for intelligence-led police operations. N.B. India Officer equates to intelligence officer. Menu is a term denoting the range of tasks allocated to individual officers, which they can address when not required for instant-response, Level 1 activities.

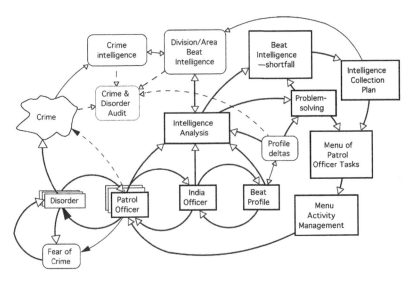

Figure E.7   Functional architecture of intelligence-led policing.

Intelligence-led proactive policing

If there is to be a significant increase in the number of patrol officers, consistent with terrorist crime prevention as well as with crime fighting, then the opportunity arises to make better use of operational intelligence. This will be especially important, since the patrol officers are unlikely to welcome 'strolling about aimlessly,' as the general public might see it.

One approach to organization might be as in Figure E.7. Two kinds of intelligence are indicated: crime intelligence and beat intelligence. Crime intelligence is gathered largely from investigating crimes, from infiltration and from 'contacts.' Beat intelligence is picked up by patrol officers as part of their daily activities: while they may pick up information by chance, patrol officers will also be tasked to find out particular things — the intelligence collection plan — as requested by the analysis of intelligence, which will highlight shortfalls and profile deltas, which are suggestive of change and disorder. Intelligence might also be needed about places, buildings, utilities, resources, and a host of other things with which the patrol officer has to deal. Patrol officers will also be 'prefects' for groups of people, perhaps 100–200 in number, with special responsibilities for knowing about their people (relationships, occupations, associates, etc.) and for looking after them.

As the figure shows, the various aspects complement each other to create a system for detecting and solving problems, for reducing disorder and fear of crime.

Designing a Solution System

An overall system for police peace operations might look as shown conceptually in Figure E.8. A police command and control system allocates police resources (police, vehicles, dogs, helicopters,

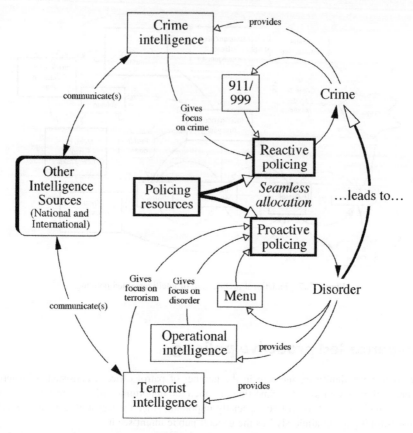

Figure E.8 Overview of conceptual policing system. At the heart of the system is command and control (C^2/C^3I), including a computer assisted dispatch facility for allocating resources — generally police officers — to incidents, tasks, problem solving, etc.

etc., etc.) seamlessly between reactive and proactive policing. It may be the intent that an individual policeman or woman might be working on Levels 1, 2 and 3 policing throughout the course of a day, i.e., peacemaking, peacekeeping and peace-building.

Figure E.9 shows part of an N^2 chart for intelligence-led police Peace Operations. The chart shows dispatch, towards top left, allocating resources in real time to deal with incidents, and to allocate levels 2 and 3 assignments. In the normal course of events, patrol officers advise the dispatcher which of their menu task they are undertaking, and the dispatcher may organize support if needed. India officers (intelligence) liaise with dispatchers and provide real time intelligence to patrol officers on Level 2 Operations. In a well-supported system, information fed back from patrol officers would be largely automatic, using video cameras to transmit situation, scenes of crime, and other data automatically and securely. Two-way electronic traffic would allow the peace officer on patrol to receive graphic layouts of streets, buildings, utilities, etc. from the India officer, using a police digital assistant, a PDA, not forgetting images of suspects, known terrorists, etc., to which the officer could refer if in doubt.

				Other data-bases	Terrorist Intel	Inputs ↕				Outputs ⟵ ⟶				
Resource avail'		Status				Status						Status		
	Incident alert (999)	Reactive demand												Data
Allocation		Dispatch				Liaison				Despatch support	Assign-ment	Assign-ment		Activity data
		Task Lists	Menu M'ment							Task Lists	Task Lists			
				Op. Int.	Crime data	Int. OUT							Intel Shortfall	
		Officer safety			Crime Int	Officer safety								
		Liaison			Crime data		India Officer	Updates			Real-time advice		Profile analysis	
				Intelligence Categories			Intel' Database	Profiles				Planning Overview	Trends, Warnings	
Reactive demand								Incidents		Reactive demand				
Reactive demand								Updates	Disorder	Proactive demand	Problem solving			Data
		Situation, Needs.			Real time observation			Peace-making		Level 1	Support	Support		Arrests
		Task-in-Hand			Real time observation				Peace-keeping	Support	Level 2		Profile analysis	
		Task-in-Hand			Real time observation				Peace-building	Support		Level 3	Profile analysis	
Resource needs										Super-vision	Super-vision	Super-vision	Area superv'	Profile analysis
			Content, priorities									Resource needs	Menu planning	
Resource effec'ness		Situation Data		Situation Data	Situation Data	Situation Data	Situation Data					Status	State of Stability	
												Activities	Crime & Custody	

Figure E.9 Police peace operations N^2 chart for intelligence-led operations.

Summary and Conclusion

Policing in a liberal democracy is a delicate business, requiring sensitivity and a 'light touch.' From being originally concerned with crime prevention, the burden of policing has switched to crime fighting. The law has similarly shifted, in line with increasingly liberal notions of personal freedom and human rights, from an emphasis on crime prevention to one on crime fighting. Police resources are kept low, seemingly, to avoid any suggestions of there being a 'police state.' Consequently, crime levels are higher than they might otherwise be with greater numbers of police.

This is not a situation with which the general public is happy. Few relish the fact that the police are generally powerless to prevent a crime, but are there only to catch perpetrators: this brings little comfort to victims; indeed, it ensures that there are victims. The public has insisted for many years that they want to see visible policing, with uniformed officers patrolling the streets. Politicians ignore such requests, choosing instead to limit police numbers and face down challenges that they are not coping with rising crime statistics.

The advent of homegrown terrorist bombings changes the situation. Detecting and preventing such terrorist activities will always be a high priority, but it is unrealistic to believe that detection and prevention can be perfect, when the terrorists are seemingly well-adjusted members of their local communities, born and raised in the country, well educated and with rewarding jobs. . . .

That suggests that, in addition to efforts to anticipate, detect and prevent terrorist activities, it will be necessary to prevent such activities as they happen, out on the street, in railway stations, airports, bus queues, shopping malls, etc. And for that, there will be a need for a different approach to police organization and manning. In a sense, it means a return to the original philosophy of policing — that the police are out on the streets in strength primarily to *prevent* crime, either by deterrence, or be catching perpetrators 'in the act.'

Instead of police numbers being determined by some arbitrary formula related to 'reported crime clear up rate,' they would be determined instead by the need to set up a dynamic network of officers patrolling streets, residential areas, business parks, centers of congregation, etc. As in Victorian times, any patrol officer coming across something suspicious could call for assistance and it would arrive in seconds from his or her adjacent patrol officers in the network.

Greatly increased numbers of police on the streets, performing 'intelligence-led policing' would, of course, cost more money than is presently spent on policing. However, a phenomenal and increasing amount of money is being spent on locking up offenders and keeping them in jail for longer and longer periods. Not only are there costs in housing prisoners securely, but in building new prisons, in enhanced probation services and, less obviously, in state support to the 'headless families' left to fend for themselves while the breadwinner is locked up.

Putting more officers on the streets would both deter would-be criminals and catch petty criminals before they graduated to become bigger 'operators.' So, instead of having a continually rising prison population, it would fall, more than offsetting the costs of the increased policing. Increasing police numbers would also reduce the fear of crime, a major source of social atomization especially among the young, old and vulnerable elements of society. Fear of crime, it seems, is a major source of social disorder — yet, like crime deterrence, it does not appear on police statistics, and so is largely ignored by liberal politicians.

Organization of a larger, intelligence-led police service could sensibly employ the same concept as that proposed by the UN for Peace Operations, i.e., comprised of peacemaking, peacekeeping and peace-building; for police Peace Operations, these three would operate contemporaneously. Peacemaking would concern itself with reacting to incidents and disorder. Peacekeeping would concern itself with crime prevention and deterrence, including terrorist activities. Peace-building would concern itself, on a longer timescale, with building social capital in the form of cross community and cross-cultural ties, associations, youth clubs, neighboring schemes, etc.

With this concept of operations in mind, the case goes on to present a notional intelligence system architecture and a full system architecture for Peace Operations. The resulting policing system is consistent with liberal attitudes and developments in human rights: but, it also addresses fear of crime and of terrorism, it deters so reducing the levels of perpetrated crimes, and it proactively rebuilds social harmony and social capital. And it seems to come potentially cost-free. . . .

Would the resulting policing system constitute a police state? Well, there certainly would be many more police on the streets, night and day. And the police would maintain profiles of people, places and things, so that they could detect deviations from normal behavior. On the other hand, the police service would remain county-based, and policemen and policewomen would necessarily be recruited from local communities, many of them for relatively short periods between leaving school and going to university. Since they would, in effect, be policing their own communities, it is difficult to see them as agents of a police state. Although, some would always see police in that light. . . .

Case F: Fighter Avionics System Design

For want of a nail, the shoe was lost;
For want of shoe the horse was lost;
For want of a horse, the rider was lost;
And for want of a CONOPS, the battle was lost.

With apologies to Benjamin Franklin,
1706–1790

The Problem Space

It is the late 1960s. The Cold War is in vogue. Mutually assured destruction (MAD) is the political philosophy of the day. The nation faces an ever-present threat of air attack and must defend itself against the enemy who, Intelligence reports, has the potential to attack in force at high and low altitudes, using a variety of weapons, including nuclear bombs . . . parity must be maintained.

However, it is all very expensive. So, instead of weapon systems aimed at specific targets and specific enemy vulnerabilities, perhaps it may be possible to be a bit smarter and produce airborne weapon systems that can undertake a variety of roles, such as interdiction and strike, counter-air, ground attack, reconnaissance, etc.

A conceptual solution is mooted in which a new aircraft will be designed primarily to undertake the low-level ground attack role. With the addition of suitable sensing and photographic equipments, the same aircraft should be able to undertake reconnaissance missions, and with suitable radar, it should be able to detect and engage targets at sea and ashore. So, the conceptual solution is a multi-role aircraft for operation against enemy targets at low level, over land and sea.

Perhaps the same aircraft can be adapted for use as an air defense fighter-interceptor? Now that really would save money . . . .

Systems Engineering: A 21st Century Systems Methodology   Derek K. Hitchins
© 2007 John Wiley & Sons, Ltd

Prescribed Solution

So, (the defense department of) government established a requirement for an air defense variant of the multi-role aircraft. Government already had in development advanced interceptor radar with many of the features that would be needed by this new variant. Government also had a new, semi-active homing air-to-air missile that would work with their new radar. Good planning, it seemed, had set the stage for a swift, economic and successful development of the new, air defense variant of the multi-role aircraft.

To ensure that its production costs were kept in check, government decreed that the air defense variant was to use 80% of the equipments and facilities already designed for, and being fitted to, the other variants. This certainly made sense economically, and seemed to make sense in systems terms: the two-man crew could use the same controls and displays, perhaps with different legends; most of the sensors would be the same, for altitude, attitude, airspeed, mach number, etc., etc; wings, engines, powered flying controls, etc. — surely, none of these need to be changed.

So, the stage was set for a truly cost-effective, new air defense fighter. All that was needed now was to engage a group of designers to integrate the new radar and the new missile with the aircraft and systems already on the multi-role aircraft.

Except... the new radar and the new missile were being developed under fixed-price government contracts, within which there were no allowances for the developers to communicate, let alone cooperate, with anyone wishing to integrate their advanced technological artifacts. So, they would not be talking to the aircraft designers and system integrators — not, that is, unless a large amount of money was made available and not unless government accepted that, in consequence, both the radar and missile programs could be delayed. As it transpired, the radar would experience development problems extending over many years in any event....

Designing the Solution System

Blissfully unaware of such difficulties, the system design team assembled, and organized along the lines shown in 'Organization for applying the systems methodology and for systems engineering' on page 304. The avionics system design, test and integration contract had been awarded, not as expected to the aircraft manufacturer, but to a systems house: this mirrored, in part, what had happened to the original multi-role aircraft. The aircraft manufacturer was not pleased: they immediately formed a system design team of their own to 'shadow' the systems house team man for man. No design meetings could be held, no decisions could be made, they declared, without their corresponding team members being present... not too happy, then, and determined to recapture their self-declared, former preeminence as systems designers.

Flies in the ointment

The first problem facing the assembling systems design team was getting information. To understand their design and integration task, they needed to gain information about all the many and various systems and subsystem that were to be pulled together into the avionics suite for a modern air defense fighter. That was a long list. There were the usual instruments, of course, an automatic flight control system, a defensive aids suite, a new head up display (HUD), the new missile, and of course the new radar. There were secondary radar identification systems, short-range air-to-air self-defense missiles, and even a proposed night visual identification system.

The design of this Night-Visident System, and even the way it was to operate, were to be determined. Its purpose was to enable visual identification by the crew of intruder aircraft, so that they could be warned off, or even engaged, but only once the interceptor crew were satisfied as to their true identify — shooting down the wrong aircraft during the Cold War could prove 'regrettable.' But, how to illuminate the intruder without blinding the crew — neither a searchlight nor a simple flashgun device would do . . . perhaps some short wavelength radar? Clearly, this was a research problem, not a systems design issue.

Many of the new subsystems that would be form part of the AD variant avionics system existed as self-contained systems. There were weapons management subsystems that were designed to interface with equipments other than those on the current list of parts. There was a long-range navigation aid that came as a box with its own displays and power supplies. Fine for a transport aircraft with a navigator and a plotting table: for a fighter, the information provided by the aid would have to be fed into the avionics system, compared with other sources navigational data, and a 'best estimate' value presented automatically on the pilot's and operator's displays.

At the heart of the navigation suite was an inertial navigation system, based around a so-called stable table: a gyroscopically stabilized platform with three orthogonal accelerometers mounted on it. Integrating the accelerometer outputs once gave velocity in each of the three directions: integrating a second time gave distance traveled in the three directions; provided, that is, the stable table was correctly and precisely initialized, and related to the appropriate map in three dimensions, too. Which took a considerable degree of time, and which was essential for a bomber penetrating covertly, deep into enemy territory. It was, however, something of a liability for a fast-reaction fighter that may have to scramble in seconds in response to an alert.

So, the avionics system was initially a collection of separate parts. To make the parts work together, interface units had to be designed and constructed, which would transfer signals and information between the various parts, and a central processing system would be needed to correlate, calculate, etc. And to build the interface units, it was vital to understand precisely how the various equipments worked, how relevant signals could be 'extracted,' what the characteristics of those signals were, particularly in terms of errors, etc., etc. Today, some might refer to such an avionics system using the tautological mantra: 'a system of systems:' then, it was to be an integrated avionics system.

Fighting the aircraft — the missing CONOPS

All of which was a mile away from the immediate concerns of the new avionics system design manager (SDM), just retired from the air force, where he had spent some twenty years in air defense — which might conceivably have had something to do with his getting the new job. While his team busied themselves working on the various technicalities of subsystem integration, and trying to extract blood from the various stones representing the aircraft company, the radar company and the missile company, he addressed a much bigger problem.

The new multi-role aircraft had been designed as a stable, low-level bombing platform at which it was turning out to be very good. However, the very things that made it good as a bomber were the same ones that militated against its air defense role. For a start, the engines were underpowered for an interceptor. Typically, a fighter has excess engine power-to-weight ratio, such that it can accelerate in a vertical climb when fully fuelled and armed, and reach altitudes in excess of, say, 50 000 feet or more. Low altitude bombers do not need that kind of power, and are not provided with it.

Similarly, fighter interceptors are expected to be agile, so that they can turn inside the turn radius of their quarry, and evade air-to-air missiles. Potential agility is indicated, partly, by wing loading: effectively, the force that the wing can sustain in turning flight. Wing loading for bombers is relatively low — they are not required to be that agile.

For fighter interceptors, both excess power-to-weight ratio and high wing loading are de rigueur. The new air defense variant had neither. There was no way that the new fighter could fight — at least, not in the conventional fighter interceptor, 'get stuck in and join the melee' sense. On the other hand, the aircraft had very good sortie duration, much longer than many conventional fighters; long duration meant long time on station, enhanced air patrol and surveillance capability....

The SDM visited old air force chums in government to raise his concerns informally, and see if there were any options. They were as concerned as he. After much soul searching, they realized that the only way in which the new AD variant could possibly operate would be to keep the enemy at arm's length. To achieve the necessary kill rate without being shot down, the air defense variant would have to serve as a 'missile platform,' standing back and launching missiles at a number of targets simultaneously or in rapid succession; the 'smarts' would have to be in the missiles. The air defense variant would not survive if it attempted to 'mix it' with enemy fighter aircraft, which would almost literally run rings around it. Unless....

There was another aspect to consider: one air defense variant on its own would be highly vulnerable, but how about several variants, say two or three, operating together and watching each others backs? The operations analysts worked out a number of scenarios in which two or more variants worked as a team. It seemed to work, but to be effective it required close cooperation and coordination between fighters. And that would require some system redesign to introduce a data link system to pass enemy target data automatically between the interceptors. This would enable the fighter 'group' to automatically allocate targets between them (vital to prevent missiles being wasted by two variants firing at the same target), to triangulate on radar-jamming aircraft, and even to permit one variant to be the 'eyes' of another that had experienced radar failure, and so enabling the latter to still fire his missiles at the right target.

None of this had been considered in the government's air defense variant requirement. Indeed, it became obvious that there was not, and never had been, any credible concept of operations (CONOPS) for the new 'air defender,' either operating singly or in a fleet. Was the air force being sold a pig in a poke?

Meanwhile the systems design team was developing the design of the avionics system so that the two crew members would be able to 'fight the aircraft:' i.e., in addition to flying it, they would also be able to fight with it. A major and continuing concern was the division of labor between front and rear cockpits.

Many fighter interceptors in service are single-seat — the pilot does everything: takeoff, vector to target, detect, locate, track, engage, defend, return to base, land. With a two-man aircraft, the many and various tasks can be shared between the front and the rear cockpit. Moreover, since navigation is less of an issue in a relatively short-range fighter that generally stays over national territory, the rear seat crewmember need not be a navigator: instead, he or she may be a radar operator, weapons manager, multiple-interception planner, etc., or better still, 'mission manager.' Pilots, of course, just love being relegated to the position of 'front-seat jockey!'

The appropriate air force operations branch of government reviewed the system design periodically as it developed. Experienced fighter pilots and navigators populated the branch on tours lasting two to three years, with posts alternating between a pilot and a navigator. Each had their own ideas of how best to fight the aircraft. And that created unforeseen problems.

When a single-seat air force pilot filled the post, he influenced the design towards the front cockpit, with the pilot doing the bulk of the work. When an air force navigator filled the post, he influenced the design towards the rear cockpit, so that the navigator/radar operator/mission manager had much more control.

In retrospect, these problems also arose because of the lack of a CONOPS. No one, not even experienced air defense aircrew, knew how to fight this new hybrid animal.

The problem was highlighted by the capability of the new radar. It was of a new type, with excellent lookdown capabilities: it could look down from high altitude and see moving targets beneath it without the confusion of echoes from the stationary ground. It also had the ability to track-while-scan, i.e., it could track a number of targets while the radar dish was still scanning back and forth.

Bearing in mind that the new variant could not 'mix it' with other fighters, and that it would have to stand back from any melee, the system designers realized that the new radar would allow the crew to set up a sequence of interceptions, rather than engage just one target at a time. It was possible to:

- track a number of targets;
- plot an intercept line that ran past the targets in sequence, or perhaps even in parallel;
- launch weapons at the appropriate points in the sequence 'run;'
- maintain a watch on other target tracks and plots;
- connect other targets into the sequence, to 'keep it going,' until they ran out of weapons or fuel.

Conceptually, at least, this might start to offset the disadvantages that the new variant inherently possessed: it was to be a way of fighting, but it seemed feasible.

To test out that feasibility, the design team approached an air force unit dedicated to developing, trialing and teaching tactics: it seemed the sensible thing to do. The unit was populated by pilots and navigators with considerable experience at operating aircraft already in air force service, and therefore designed some 20–30 years earlier: they had no knowledge of more recent technology, and no idea how to take advantage of it tactically. In essence, they only knew how to fight their old aircraft using only old technology . . . .

Government reluctance

With a possible way of fighting the aircraft in mind, the system design team set about designing the system to enable the 'hand of purpose,' i.e., to develop the functions and processes that would allow the crew to perform their tasks, examine their sensors, manage their weapons, engage their targets, defend themselves against attackers, fight their aircraft and cooperate with others.

One of the most important capabilities needed, in view of the variant's proposed 'standoff' fighting philosophy, would enable the crew to engage a number of targets simultaneously. The variant would have two kinds of air-to-air missiles: a shorter range, infrared seeker, fire-and-forget missile, nominally for self-defense; and a longer range semi-active homing missile that required its target to be illuminated by the interceptor's radar before and during missile flight. The shorter-range missile was no problem — it could be locked to a target's IR signature, launched and left to 'do its thing.'

The semi-active homer was problematic. The missile system was designed with an earlier kind of radar in mind. Earlier radars had a single dish, which stopped scanning when the radar locked on to a target, and the dish pointed thereafter at the target. Coupled to the dish would be a 'CW illuminator,' a continuous wave radar transmitter that shone a beam of energy at the target. It was the reflection from this beam that the semi-active homer missile detected, locked on to and followed. The target had to be illuminated for the whole flight time, else the missile would 'lose sight of' the target and miss.

The variant's new radar, however, was a track-while-scan (TWS) radar, with all the advantages that afforded for multiple target tracking and multiple target engagement. Its CW illuminator was physically locked to the scanner dish, so that the CW illuminator wagged back and forth with the main dish. If the whole dish assembly, including the CW illuminator, had to be locked to the target sightline to engage the missile with its target, and for the duration of missile flight, then the radar would, for that time at least, no longer be a TWS radar — it would not be scanning at all.

The solution seemed straightforward, conceptually at least. If the new variant was to be able to engage several targets at range simultaneously or in rapid succession, then the TWS had to be in operation, and the targets and missiles had to be illuminated too, but only part time as the scanner swept the CW beam across them. Otherwise, it would be necessary to have separate scanners, one for steerable multi-beam CW illumination and the other for the TWS radar. The first solution required modifications to government's specially developed missile: the second solution required modifications to government's specially developed, advanced radar.

Government would countenance neither solution; no modifications would be allowed to their fixed-price, fixed-specification development contracts. Which shot rather a big whole in the only feasible fighting philosophy that the system design team had developed. So, still no credible CONOPS . . . this would be like a man on crutches, wearing boxing gloves, taking on a group of ninjas wielding samurai swords!

Systems concepts — triangulating ghosts

Back to the drawing board. The threat was alleged to include jamming aircraft, which may prevent the TWS radar from tracking targets: instead, the radar screen would show a line on the azimuth angle corresponding to the direction of the jammer.

Two cooperating variants could exchange the azimuth angle at which they respectively perceived the jamming target. Plotting own position and cooperating variant's position with the respective azimuth angles resulted in a cross on the map display: this was triangulation on the jammer and it could be accomplished automatically, provided the two variants had a data link to enable data exchange between them.

Suppose, instead, there were two jammers, or three . . . for N jammers there would be N^2 triangulation points, or crossovers, only N of which would be real; the remaining $N \times (N-1)$ would be 'ghosts.' The systems design team got to work and came up with several ingenious schemes for picking out the real target from the ghost target. They all depended on automatic exchange of data between cooperating aircraft using some form of data link. There was no data link in the original requirement, however: without a CONOPS, who would have perceived the need?

Systems concepts – LANCE

The SDM came up with an idea (allegedly while taking a bath with his big toe up the spout of the cold tap). Since the TWS radar could track a number of targets simultaneously, it could provide the data to draw target vectors on the radar map, with each vector showing the respective aircraft's track, and the vector's length corresponding to its speed. Similarly, it was possible to draw a vector in front of the symbol for own aircraft, but this time showing the maximum relative range (MRR) of the missile. MRR is a useful measure: when a missile is fired, its motor provides an impulse, which takes it rapidly to high speed; thereafter, with the motor burnt out, the missile coasts to its target, slowing down as it does. As it slows down, the launch aircraft starts to overtake it. So, there will exist a point at which the missile reaches its furthest distance from the launch aircraft — this is the MRR. Beyond the MRR, the launch aircraft starts to overtake the missile, which may still be a long way ahead. The MRR represents the farthest distance/earliest point at which an intercept can be made.

Suppose now that a vector is drawn on the electronic map protruding from the interceptor variant's symbol, showing the MRR for the missile, and the number of seconds from launch to reach that maximum relative range, T_{MRR}. Draw a small circle at the end of the vector to mark the spot. Suppose, too, that the vectors in front of each target track are scaled to show where each target will be in T_{MRR} seconds: put a small cross at the end of each vector to mark the spot.

If a pilot flies his aircraft so that the missile vector's circle lays over a target aircraft's cross, then the coincidence represents the point where both missile and target will be in T_{MRR} seconds; i.e., the earliest engagement point.

One of the team named this idea LANCE — line algorithm for navigation in combat environments. The analogy of a lance protruding in front of the aircraft was apt. All the crew had to do was to fly the aircraft so that the tip of the LANCE lay over the tip of a target track vector, and press the button! The concept was ideal for the air defense variant, since it would enable the variant to keep its distance, and to stand back as far as possible from any melee.

LANCE was a natural — government liked it, air force fighter crews like it, it seemed to have everything going of it. Except. The aircraft manufacturer, still no-doubt smarting because he had not won the avionics system design contract, incorporated a prototype version of LANCE on the development rigs in the factory, and decided that the display was 'too unstable,' and therefore impractical. Perhaps it was . . . .

Six months later, virtually the same idea was incorporated in another nation's fighter aircraft under a different name. Not too unstable for them, then? Now, there's a thing.

Conclusions

It is interesting to note, from this cautionary tale, just what systems engineering was, and always is, about. It was operating at two levels:

- First and foremost, it was trying to understand the problem facing the crew in achieving their purpose, and to provide them with the facilities and services they would need, not just as isolated crews, but as members of a group or fleet of defending aircraft.
  - This was not engineering — this was 'system operations design,' for want of a better term — working out how the whole system, with the crews at its foci, is going to achieve its mission, fulfil its purpose.
- Second it was configuring and integrating the various technological facilities, including sensors, weapons, displays and controls, so that the various mission functions could be choreographed, activated, orchestrated and controlled by the crew, working in cooperation with other crews as necessary.

- This was not engineering, either — this was systems design, tracing and forging the 'hand of purpose' from crew member, through controls, subsystems, interfaces, subsystems and — generally — back to the crew. Or, if you prefer, enabling the crew to orchestrate and conduct operations and prime mission functions

The system design team, many of them excellent engineers, physics graduates, systems analysts, operations analysts, etc., were obliged to try and establish a CONOPS. They did as well as they could, but should this have been their job? I refer to the bastardized quotation at the beginning of the chapter.

And the aircraft: how did it fare in service? Well, let us put it this way. It was not the best fighter the air force ever had. It went through a number of marks and improvement phases during which some of the original shortcomings were recognized and rectified — but not all. As usual in such cases, the motivation, training and excellence of the aircrews managed to overcome many of the shortcomings, even if they were effectively fighting with one hand tied behind their backs.

And, come to think of it . . . the whole project was systems engineering with one hand tied behind its back, too.

17

SoS Engineering Principles and Practices

'Wilkes,' said Lord Sandwich, 'you will die either on the gallows, or of the pox.'
'That,' replied Wilkes blandly, 'must depend on whether I embrace your
Lordship's principles or your mistress.

John Wilkes, 1727—1797

Creating, Developing and Evolving a SoS

We have already met the concept 'system of systems' (SoS): see System of Systems on page 94. (In this chapter, the abbreviation SoS indicates system of systems: in earlier chapters, SoS indicated solution system as an outcome of the systems methodology. The different contexts should hopefully prevent any confusion . . . .) While there is much confusion about the term, with various pundits expounding contradictory views, it seems that the term refers in general to some new whole, which is formed by bringing together a number of complementary, extant, operational systems. The reference gives an example of an integrated, city-wide, transport system formed from extant transport systems such as underground/metro, rail, bus, taxi and other services.

Other examples would include:

- A volume supply system formed by bringing together existing businesses for making parts, designing, assembly plants, marketing, etc, each of which may be a viable business in its own right.
- An air defense system formed by integrating separate operational elements: ground radar system; surface to air missile systems; fighter interceptor aircraft; dispersed airfield support systems; airborne radar system; early warning sensor system; etc.
- A university formed by bringing together a number of existing colleges, choosing the disciplines of said colleges to be both complementary and contributing to a distinct discipline for the whole university, e.g., humanities, social sciences, physical sciences, etc.
- An army, formed from various operational units of infantry, cavalry, artillery, signals, intelligence, engineers/sappers, etc., to address a particular situation.

Systems Engineering: A 21st Century Systems Methodology   Derek K. Hitchins
© 2007 John Wiley & Sons, Ltd

In each example, the new whole is relatively large, and tends to be comprised of enterprises, i.e., of social or sociotechnical systems, which were viable systems before becoming part of the new whole, and may remain so after. A viable system is one that is able to exist, survive and operate in a suitable environment, acquiring its own resources, performing its functions and, as an open system, maintaining dynamic equilibrium.

When a number of extant systems is brought together, it is unlikely that the various parts will be fully complementary, nor will they cooperate and coordinate their actions in an ideal manner: after all, prior to being brought under one central umbrella, each of them had its on purpose, its own CONOPS, its own 'way of doing things,' or culture. Once brought together, there may be a need to harmonize their operations, to make the various parts 'more complementary,' to enhance cooperation and coordination between them — or not.

It may be impractical to render radical changes if they interrupt the operations of the various parts. Besides, how much change is needed? Are bridges and interfaces to be established/enhanced between the enterprises? And, in establishing such interfaces, is there likely to be resistance? It is as well, too, to bear in mind that these are open, nonlinear systems interacting in an open, often fluid, environment. It would seem to be essential to employ the systems approach to exploring the possibilities and limitations in this 'joining together' process.

Limitations in SE for SoS

The conception of any new system is, or perhaps should be, in response to a perceived problem. In this respect, a system of systems should be no different. Joining systems into an SoS triggers questions:

- What is the purpose of the 'bringing together? What is the purpose of the new created whole? What problem or shortcoming is it designed to preclude or overcome?
- Why join together these particular extant operational systems? What makes them suitable?
- What is to be gained in the process?
- What might be lost?
- How is the composite system of systems expected to operate:
  - will there be a central control and coordination bureaucracy, or. . .
  - will each part continue to 'do its own thing,' and ignore the need to cooperate and coordinate?
  - will there be functions and departments to be created 'at the new center? What will they be for? Are they an unforeseen cost? Can they be justified?
- How is the integration to be achieved — what degree of intervention is appropriate/too little/too much?
- And, most importantly, will the anticipated integration solve the problem, and at what cost?

These, and many more, are the kinds of questions and issues that the systems methodology is designed to address. So, should the systems methodology, with its seven steps (Chapter 6 *et seq.*), be applied to SoS as to any other problem?

In principle, yes — but with provisos. The systems comprising the whole are extant and operational. Whatever changes might be proposed are unlikely to be appreciated if they seriously interrupt operations — unless the value of any such disruption can be shown to outweigh the cost.

Importantly, too, the new whole should have a singular, distinct, high level purpose, to which the composite parts can contribute severally and together.

It may be an objective of classic systems engineering to create an optimum solution to a problem, but it is unlikely that bringing existing operational parts together will constitute an optimum SoS solution. Inevitably, there are going to be imbalances: capability and capacity mismatches, incompatible process rates, etc. For the extant operational systems, there may be limits to any changes and to the rate of such changes that should not sensibly be exceeded. Effectively, this suggests that it may not be possible to configure the parts into an optimum solution system of systems....

This in turn suggests that a high degree of judgment may be called for, to determine how far and how fast to go towards full integration. The situation is unlike engineering integration — with SoS, the constituents are largely people systems (HASs) and people-managed sociotechnical system, both of which exhibit such human characteristics as:

- *Motivation* — inclinations to conform to social norms (achievement motivation and compliance motivation)
- *Dominance/submission* — tendency to lead or be led
- *Territorial imperative* — strong sense of territory ownership
- *Territorial marking* — visible signs/symbols of ownership
- *Personal space* — ego-centered space, physical and emotional
- *Family loyalties* — unquestioning adherence to relationship
- *Tribal loyalties* — unquestioning adherence to relationship
- *Dyadic reciprocity* — interactions between communicants
- *Natural predisposition* — inherited tendency to respond
- *Cultural predisposition* — learned tendency to respond
- *Group polarization* — tendency for group discussions/decisions to move to the extremes—'risky shift'

Consider just one of these factors, tribal loyalties, in the contexts of the four examples bulleted at the start of the chapter, and the reader may start to realize that this integration task may not be straightforward. Each of the extant operational systems is likely to have developed its own culture, making the people who work there members of the 'tribe.' Such people will not readily become part of the much larger tribe, the SoS tribe of tribes: they do not know the other people, and have little in common with them. So, there will be mutual suspicion, even open hostility at being 'taken over,' and at the very least, barriers to bring down, as well as interfaces to build up. One key to success is motivation.

Of all the human characteristics on the list, perhaps this is the most significant and 'helpful.' It may prove vitally important to motivate the members of the extant operational constituents to be active, willing achievers in the new whole. After all, with a much wider whole, there should be greater prospects for advancement, a higher promotion ladder, enhanced prospects for training, etc., etc.

Strategy for SoS Engineering

Systems engineering for Systems of Systems (SoSs) is operational systems engineering, i.e., conceiving, designing, redesigning and reconfiguring 'on the go:' it is not static, 'green field' systems engineering — See Working Up the System — Operational Systems Engineering on page 234 for a perspective on operational systems engineering.

'Spinning plates'

Operational systems engineering is constantly adapting the system to improve effectiveness and performance, increase efficiency, etc, and to keep pace with changes in the environment and in other, interacting systems. In the example referenced in the previous paragraph, the strategy adopted for operational systems engineering was one of 'spinning plates.' (The spinning plate analogy refers to the jugglers who spin a number of plates balanced on the tips of canes: to keep them all spinning the juggler has to constantly attend every plate in turn, spinning-up those that start to wobble.)

This is certainly one approach, and it has been adopted in the automobile industry. (Womack and Jones, 1990) In one notable example, the 'plates' turned out to be queues of parts forming at particular machines in an assembly plant waiting to be assembled. Queuing indicated that the work rate for the machine was too low, or conversely that two or more machines were needed in parallel to increase the mean rate. Each time actions were taken to reduce one queue of parts (which constitute work in progress (WIP), cost money, and reflect in the cost of the end product), another queue would appear in front of a different machine, and then another. However, by continually addressing queues in this way, the overall amount of WIP gradually reduced, and along with it down went unit production costs (UPCs).

Continual redesign

For complex, nonlinear, social and sociotechnical systems, an alternative to keeping up with this constant change is continual redesign. Continual redesign continually addresses the problem space, detecting and addressing changes in situation, and seeking revised and better conceptual remedies. It goes on to conceive and design remedial solutions, CONOPS, etc., to identify risks, threats and opportunities, and to include strategies and functions in the developing design to address them. A functional architecture emerges as part of the design process, which can be compared with the design of the existing system of systems. When differences between the redesigned functional architecture and the extant functional architecture exceed some threshold, it is time to act — to change the extant system to be more like the redesigned version.

Continual redesign ideally employs systems thinking in the form of behavior models — dynamic simulations of open interacting systems. It is one basis for evolving intelligent systems, where such systems are deemed intelligent when they can constantly adapt to changes in their environment, performing, remaining effective, avoiding or neutralizing threats and, ultimately, surviving in the process.

A sensible, practical approach to systems thinking and therefore to continual redesign is to employ so-called learning laboratories (see Systems thinking on page 17, Systems thinking and the scientific method on page 73 and the Battle of Britain Simulation on page 238.) The learning laboratories allow the analyst to experiment with system design solutions dynamically, in context, and while they are in operation — just what is needed for SoS continual redesign. Learning laboratories can be made simple to use and to experiments with, even if the models of dynamic opens systems and their nonlinear interactions may be complex.

With such simplicity of use comes the opportunity for managers, CEOs, VPs, commanders, and the like to explore possibilities themselves, rather than delegate to others, to share concepts, ideas, strategies and plans with staff at all levels and to focus motivation: if a SoS is to be successful, it seems likely that it will be because the workforce is motivated to cooperate and contribute in harmony towards a common goal, to understand and accept the need for change, to embrace it, and to effect that change swiftly and cleanly.

All of which depends on developing sound and effective behavior models for the open SoS, with its constituent open, interacting parts, open to its environment, and interacting with and adapting to other systems in that environment.

SoS Architectures

SoS pipeline architecture

The architecture of a SoS may be suggested in the title — a system of systems. There may be rather more to it, however. Figure 17.1 shows a notional architecture for a 'pipeline' SoS; see System of Systems on page 94. Pipeline architectures pass material and information through a series of systems, each system processing and transforming the throughput in sequence before passing it on. Lean volume supply systems are pipeline systems, where the throughput is material for manufacturing, processing and/or assembly; so, too, are policing systems, where the throughput is offenders and lawbreakers.

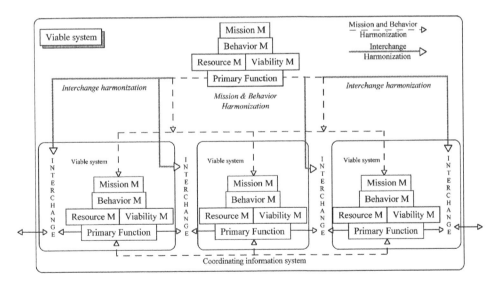

Figure 17.1 System of systems (SoS) 'pipeline' architecture. Each of the contained (sub) systems is a viable system. The three subsystems exchange material and information. Activity initiations, rates of operation, activity coordination are effected with the aid of a (primary function) coordinating information system. Primary functions of the whole system, shown at the top of the figure, include mission and behavior harmonization of the subsystems, and interchange harmonization to enhance cooperation, coordination, compatibility and synergy.

Figure 17.1 shows three elements in a pipeline; the three contained, viable systems at the bottom of the figure. The three are formed into a pipeline via an interchange, which transfers/transforms the output from one into the input to its successor. Acquiring input, processing throughput and providing output (input–process–output) are the primary function(s) of each viable system.

In the figure, each of the three systems is a viable system: each has its own mission management, behavior management, resource management and, of course, viability management. Each system organization could, in principle, operate on its own. However, for the three systems to behave as a unified whole, i.e., as a system of systems, without prejudicing their viability, some degree of harmonization will be needed:

- *Mission harmonization.* The mission of the whole system will 'guide' the missions of the three viable subsystems. Directors of each viable subsystem will pursue a mission, or missions, consistent with, and possibly approved by, the director of the whole. They may be free to pursue their separate missions without supervision, in a manner of their choosing, but are likely to be held to account should they fail.
- *Behavior harmonization.* The containing system may seek to impose 'ways of doing things,' including a 'corporate image,' and a doctrine, or company procedure, on the three viable subsystems, such that they all separately and together appear and behave as a single, unified whole. An integrated transport system may see the various operatives from their various viable subsystems all wearing the same uniform, for example. Publicity and public relations may be provided only for the new SoS, and no longer for individual contained systems — see Behavior Management on page 130 and The GRM and the Systems Approach on page 135
- *Interchange harmonization.* The smooth flow of e.g., people, materials and information through the whole system, and therefore through the three subsystems in the example, requires, inter alia, that the interchanges between the viable subsystems – which may be miles, or even oceans, apart — should be swift and effective. In a lean volume supply systems, parts and subassemblies flow from part suppliers through subassembly manufacturers, and on to final assembly, before going out to the market. Any parts, subassemblies or assemblies in the system represent a cost, including any waiting for, or transit between, factories. In some situations, interchange may involve some transformation of the throughput, e.g., information may be reformatted, separate parts configured into sets, etc.
- *Primary function coordination.* It can be shown that the minimum amount of WIP in a lean volume supply pipeline is consistent with a steady manufacture and assembly rate, rather than one that works in batches, or where demand rises and falls. Similarly, transfer between sites (viable subsystems) should be of individual units, rather than batches. Primary function coordination seeks to initiate and regulate activity such that outputs are available from one unit just in time to be used as an input to the next unit, allowing for transit time. This primary function coordination may benefit from a dedicated information system, as shown in Figure 17.1.

Harmonizing and coordinating the various flows and interchanges between the subsystem is not unlike 'spinning the plates,' except that it is potentially able to address all the 'plates' at once and in parallel. The objective is to promote synergy, improve responsiveness and hence increase overall effectiveness for less effort. In the process, the various viable subsystems may change; they may also resist change, as people systems are wont to do. Achieving harmonization turns out, in practice, to be a management and leadership task, and may either take time, or invoke a 'bloodbath,' with transfers, redundancies, strikes, etc. Much depends on the suitability of the original viable systems that were selected to form the SoS . . . .

SoS complementary architecture

An alternative architecture, SoS complementary architecture, is shown at Figure 17.2. Three viable subsystems are shown: the actual number would depend on situation, but there should be sufficient to complete the complement. For a naval task force, for instance, there would be various ships and aircraft to provide air defense, surface defense and subsurface defense: the three defensive arenas are complementary since, given those three, there should be no more. In a distributed enterprise, there might be separate viable facilities for research & development, design, manufacture, marketing and sales; these four might constitute a complementary set.

The name of the departments in the President's cabinet are indicative of a complementary set: State, Treasury, Defense, Justice, Interior, Agriculture, Commerce, Labor, Health and Human Services, Housing and Urban Development, Transportation, Energy and Education. The test of complementarity is to consider which of the set could be left out without leaving an evident gap, one that would result in an incomplete and probably dysfunctional system.

The human body can be viewed as a complementary set of subsystems, although none of the subsystems is viable: instead, they are all mutually dependant. There are subsystems for ingesting

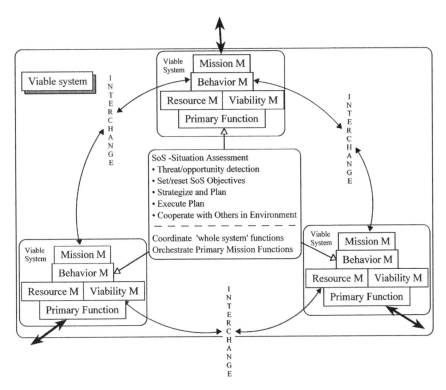

Figure 17.2 SoS complementary architecture. A requisite variety of complementary systems (hence primary mission functions) are brought together: their behaviors are cooperative, coordinated and complementary, encouraging synergy. Additionally, whole system functions may be established, perhaps with functional parts distributed amongst the viable subsystems — see text. Potentially, the central focus may be vestigial, or even conceptual only . . . .

food and energy, breathing and exhaling, circulating blood, sensing, moving, thinking. . . the list goes on. With such a complementary set, the question becomes — which of these could we do without and still remain viable? (Our answers may change with advancing years!)

In a not unrelated vein (sorry!), the twelve cranial nerves form a complementary set of nerves emerging from the human brain. As trainee doctors will have learned, using various mnemonics, the nerves are: olfactory, optic, oculomotor, trochlear, trigeminal, abducens, facial, acoustic, glossopharyngeal, vagus, accessory and hypoglossal; these nerves comprise a complementary set. It would be apparent if any one of these were missing or inoperative. All would be needed for a fully operative system . . . .

In the general case of Figure 17.2, it may not be simple to establish when the set of different, complementary viable subsystem is complete. For the naval task force, might there be a space-borne threat? And to a significant degree, variety in complementary system sets is driven by risk and threat — see Threats and Strategies, part of SM3, on page 226.

Bringing together a complementary set of subsystems to form a SoS whole may provide the opportunity to develop and introduce whole system functions, i.e., functions of the SoS that cannot be exclusively attributed to any of the viable subsystems.

For an integrated transport system as an SoS, a whole system function might be dynamic capacity management, in which the capacity for travel on particular routes is adjusted by putting on more, or larger, vehicles as routes start to get crowded, at the same time as reducing the capacity on routes that are not so busy. This would be a complex arrangement to execute 'on the fly,' involving traffic flow sensors, dynamic resource allocation, rerouting, etc., etc. But, with an integrated transport system, it may be both feasible and beneficial — especially during rush hour times.

For an integrated air defense system, with complementary systems including ground radar and airborne early warning radar, fighter interceptors, surface to air missiles, etc., a subset of viable subsystems, perhaps the airborne radar and a fleet of interceptor fighters, might suffice. Suppose they were all fitted with track-while-scan radars — see Case F: Fighter Avionics System Design — and could automatically exchange target track information via some data link. Then each aircraft could display a set of all identified targets being tracked on a map. This would be a dynamic map, one that moved with the fleet, and would operate even when the fleet moved out of ground radar range; indeed, it might improve as the fleet moved towards the target area. The formation of this so-called recognized air picture (RAP) would be a whole SoS (fleet?) function.

Suppose further that each aircraft could see potential targets on the map display that were within the capability of that aircraft to intercept. The aircraft, under crew control, could transmit details of their preferred targets to other interceptors. These others may also have seen targets, and some may prefer the same target(s) as the first interceptor, so there would follow an automatic negotiation in which particular targets were automatically allocated between interceptors in the way most likely to achieve some objective: maximum interceptions in a set period; maximum interceptions before the targets crossed some boundary line; etc., (Hitchins, 1987.) This would be automatic target allocation: it would be a second whole SoS function.

Complementary architectures of SoS, then, may exhibit emergent properties, capabilities and behaviors, even without a central coordinating capability. Indeed, the point of bringing a set of complementary systems under a single SoS umbrella may be to generate synergy, and, in so doing to generate emergent properties, capabilities and behaviors of the whole SoS, i.e., not exclusively attributable to the separate parts. In some instances, as in the air defense example, survivability of the whole may be enhanced by the elimination of a vulnerable node, such as a central, coordinating focus.

SoS — Unified Whole, or Dissociated Set?

From the foregoing SoS models, it seems that there may be 'degrees of system-ness;' i.e., there may be SoSs, with viable interacting (sub) systems where the mission and behavior management of each subsystem is less than, or entirely, congruent with that of the whole.

For full congruency, the SoS could be represented by an instantiation of the layered generic reference model as shown in Figure 17.3 — (see The GRM and the Systems Approach on page 135.) The figure is a standard version of the layered GRM, with the Form layer highlighting the viable subsystems. Each subsystem is shown as having discrete resource management and viability management, as well as performing primary functions of the SoS as a whole. The mission management and behavior management elements of each viable subsystem have been subsumed into the two upper layers of the diagram, indicating either that:

■ the management of the subsystems is entirely congruent with that of the SoS as a whole, or that
■ there is no discrete management of the whole: management of the viable subsystems is by, and of, the subsystems — by mutual agreement, by negotiation, etc. This would be consistent with Figure 17.2, and air defense example above, where whole system functions had been set up, enabling cooperation, coordination, and negotiation to take place between the parties, without the need for a central, nodal figurehead, commander, CEO, chairman, president, etc . . . .

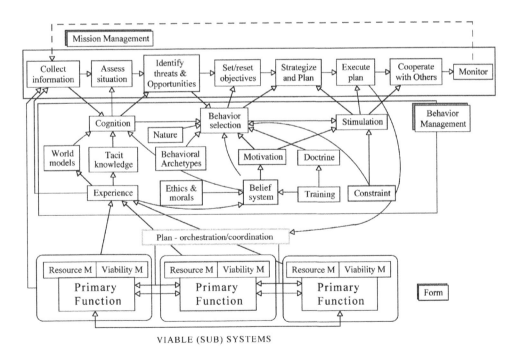

Figure 17.3 Layered generic reference model for a system of systems — see text.

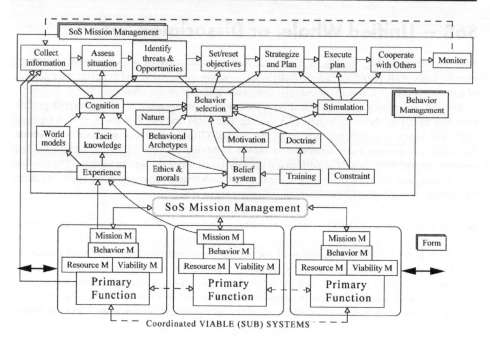

Figure 17.4 Layered generic reference model of a SoS, where each of the systems has discrete mission and behavior management. The GRM is presented in three layers as usual. SoS Mission Management is shown at the top, and is repeated in the center, in abbreviated form. See text.

On the other hand, some viable subsystems may have a 'mind of their own;' i.e., each of the viable subsystems has its independent approach to mission management and or, has its own culture, doctrine, and 'way of doing things.'

An alternative instantiation is presented in Figure 17.4. This considers the situation where there is a central, containing system element, itself a system. As the figure shows, each of the viable (sub)systems has its discrete mission and behavior management, in addition to resource and viability management (resource and viability management for the central, directing element are omitted for clarity).

This might suggest that each system is independent and may retain, or create, its own culture, belief system, way of doing things, etc. However, that need not be the case. The three systems are shown 'tied into' the SoS Mission Management. That indicates that the 'directors' of each of the viable systems contribute to SoS Mission management, providing information, assessing situation, identifying threats and opportunities, setting and resetting objectives, strategizing and planning, etc. It also indicates that the mission of each of the viable systems is harmonized within the plan and that the viable systems coordinate and cooperate intelligently to prosecute the plan. As illustrated, the SoS is potentially capable of exhibiting intelligent behavior, particularly since it is using experience of what has gone before in assessing situations and formulating strategies. Note that Experience also feeds Belief System, such that Beliefs may be grounded in reality, and may evolve in consequence.

Managing Change in a SoS

Provided subsystems in a relatively dissociated SoS perform their primary functions in a coordinated, cooperative, complementary manner, it might be that there is no immediate issue — the SoS, in effect, is operating as it should. Looked at another way, an outside observer would not be aware that the parts of the whole were culturally and managerially independent. Some global volume supply systems appear to operate in this manner, with different organizations/enterprises in different countries forming links in supply and manufacturing chains. So long as the links continue to form strong, coordinated and cooperative connections, then so long will the SoS maintain the status quo.

However, few SoS are in a steady state for any appreciable duration — SoSs are open, interacting systems and, as such, are continually adapting to their environment and to their systems in that environment — or not, in which case they may soon become terminal. The ability and manner in which dissociated and fully integrated SoS accommodate change may differ: see Figure 17.5.

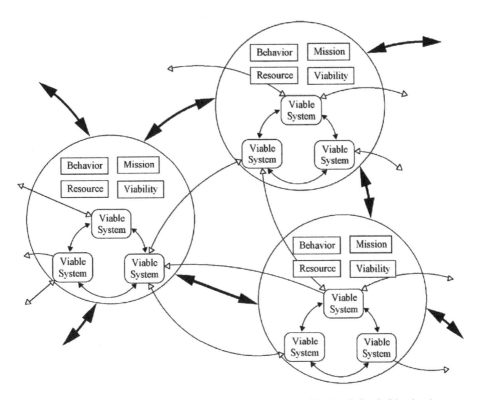

Figure 17.5 A SoS, left, containing three viable systems, interacting with other SoS at SoS level and at system level. The environment for any SoS is constantly changing, perhaps even weakly chaotic. From a contained system perspective, it appears to be a node within an ever-changing network of changeable systems. To survive and flourish, and SoS will need to adapt and evolve. For a unified SoS, adaptation may invoke change within and between the integrated systems composing the SoS. For a more dissociated SoS, adaptation may invoke discarding some contained systems and replacing them with others.

For the dissociated SoS, rapid environmental change may present problems: the state of dissociation militates against the SoS as a whole being able to adapt quickly and in a coordinated manner, without losing what little coherence exists in the process. Instead, one way of accommodating change would be to replace one or more viable subsystems with others: effectively, to reconfigure the SoS. This approach may seem attractive and quick, but a settling down period will follow as new cultures adapt, cooperation develops, and coordination synchronizes.

For the fully integrated system, speed of adaptation to changing circumstances may conceptually be greater; provided, that is, that the viable subsystems — either HASs or sociotechnical systems — are themselves able to change. It is possible for a large, integrated SoS to become a monolith. Alternatively, an SoS might be comprised of large numbers of small units, each able to form and reform in response to e.g., market change, new opportunities, availability of new technology, etc., etc. Essentially, speed of adaptation in this context can be looked upon as an architectural characteristic. Architecture with few, monumental blocks, is potentially slower out of the starting blocks than architecture of many small, relatively independent parts.

By analogy, the rate at which an organism adapts and evolves to changes in its environment is also dependent on size and structure. A large dinosaur may have taken 25–30 years to grow to maturity and have offspring, which it may then need to nurture for 10–15 years or more. A small mammal, by comparison, may mature in perhaps two years and may have offspring in even less time. The mammal can create many more generations within, say 100 years than can the dinosaur, giving the mammal much more opportunity to adapt and mutate in ways that improve its prospects. A rapidly changing environment, then, favors the survival of a species that has a relatively short lifespan and which can generate variety in its offspring.

What is true in ecology may also be true in economics. It is not always the large, monolithic company that survives and overcomes the opposition. IBM, Big Blue, found that out to their cost, and had to regroup, reorganize and reconfigure such that they could adapt to changes in the market that had previously passed them by.

System of Systems Engineering

Is system of systems engineering (SoSE) any different from conventional systems engineering? Opinion on that score will rather depend on where one is coming from. For those steeped in ideas of systems engineering as being concerned with creating artifacts, SoSE will seem alien: systems engineering as propagated by some defense agencies does not overly concern itself with systems in general, but almost exclusively with engineered artifacts, and their continuing support. However, those for whom 'system' includes organizations, teams of people, sociotechnical systems, enterprises, industries, economies, etc., there will be little or no distance to travel. To them, SoS engineering will be virtually indistinguishable from classic systems engineering — which has not been confined to artifacts.

For SoSE — if indeed there is such a discrete subject — the best approach would seem to be to employ the systems methodology as it stands, since it is designed to address systems of all kinds, remembering that within the systems methodology, the Generic Reference Model serves as a keystone for bridging from problem to the solution. In addressing SoS, then, the key will be to instantiate the layered GRM in the form of the SoS, and to set the resulting whole in the context of a dynamic, interactive network of systems as in Figure 17.5.

Instead, however, of trying to establish a variety of different functional/physical partitions, (see SM5: Architecting/Designing System Solutions on page 253) it may be appropriate to accept the inherent partition formed by the choice of viable systems that are to be brought together. Coupled with Continual Redesign, the whole SoS will then be harmonized, adapted and evolved as it seeks to flourish and survive in its dynamic environment.

Summary

Systems of systems (SoS) may be formed by bringing together extant, complementary, viable systems under a single umbrella. The systems in question are human activity systems (HASs) and sociotechnical systems. SoS fields of endeavor include Levels 3, 4 and 5: see the The 5-layer systems model on on page 113, i.e., enterprise/business, industry, economy; they do not appear to include artifacts-comprised-of-artifacts (AoA?), such as an avionics system, a tank, or a racing car, although the term SoS is so inadequately defined that it is not possible to be definite. Another term, family of systems, for example, does not appear to mean the same as 'SoS,' since the latter must sensibly exhibit the characteristics of a system. 'Association of systems,' on the other hand, is unclear — it could, but need not, refer to a SoS.

Instead, SoS appears to refer to commercial enterprises or military systems formed by bringing together other, complementary commercial enterprises or military systems. The whole will also be a system, and will become, more or less, a unified whole.

Why 'more or less'? Because the enterprises/systems composing a SoS were viable systems prior to 'incorporation,' they will have had their own purpose, concept of operations, etc. In bringing several such, previously independent, systems together, it would be fortuitous if they fitted together perfectly. Instead, it is likely that there will be a need to harmonize their previously independent missions and behaviors; similarly, interchanges between the viable systems may need to be made easier, swifter and more transparent. The functions performed by each viable system will now become (some of) the primary (mission) functions of the whole SoS, harmonized to be cooperative and coordinated, such that the SoS can develop synergy between its constituent parts.

Harmonization may prove less than straightforward: the systems to be harmonized are HASs and sociotechnical systems (which may exhibit the behavioral aspects of their human elements), and as such are nonlinear. There are different strategies that might be adopted to bring about a degree of harmonization within a complex of non-linear systems. One strategy is the so-called spinning plates approach, where imbalances, dysfunctions and disharmonies are tackled one at a time and resolved before moving on to the next.

An alternative harmonization strategy is Continual Redesign, which seeks to balance the whole nonlinear structure. Using the systems approach and the systems methodology, or some other way, the problem space is continually explored to detect shifts and changes in the problem. Remedial solution concepts are generated, purpose is focused, etc., and a functional design for the whole is developed, taking into account threats, risks and strategies to address them. A functional design and architecture are developed, and mapped on to the existing partitions formed by the extant viable systems. The resulting functional/physical design is compared with the real world: if the difference between the two exceeds a sensible threshold, then further harmonization, reconfiguration, etc., are required. The method adopts the systems approach with dynamic simulation using open systems models to address nonlinearity, environment, adaptation, evolution, survival, etc.

SoS physical architectures, comprised of contained systems, may vary: two archetypes are presented; the pipeline architecture and the complementary systems architecture. Additionally, a SoS may be hierarchically organized, with some central system serving as overall mission and behavior management. This might be the case where a company had a number of subsidiaries.

On the other hand, it is both feasible and practicable to form SoS architecture from complementary systems, but without central focus or coordination — not unlike a committee without a chairman.

For such a SoS to function and behave as a whole, there must exist a basis for the systems to cooperate and coordinate, implying a means of negotiation that will operate successfully without independent arbitration. Modern digital communication systems are making such SoS practicable, as in military nodeless, network centric operations. Invariably, such nodeless SoS, like the committee without a chairman, will need methods for negotiating agreements and deriving corporate decisions. These methods and capabilities are 'functions of the whole,' i.e., of the SoS, and are not attributable exclusively to any of its logically separable parts: essentially, they are emergent capabilities.

A variety of Layered Generic Reference Models (LGRMs) is presented for differing SoS configurations. They suggest that the ability of a SoS to adapt swiftly and effectively to changing environments may depend in part on the degree of integration/dissociation of and between the various viable systems comprising the whole.

Dissociated viable subsystems are those that, although performing the appropriate primary functions, say, in a pipeline SoS, may retain their original culture prior to incorporation, and still possess their own, independent mission and behavior management. Changes in the environment for the SoS as a whole may see such dissociated systems either unwilling or unable to adapt; instead, the SoS may adapt by replacing the system. In effect, some SoSs may comprise a 'floating population' of largely dissociated, viable subsystems.

Integrated viable subsystems of a SoS are those that participate in and share the same, overall mission and behavior management, although perhaps retaining discrete resource and viability management. As is more general in HASs, integration involves less control, more leadership and influence. In particular, fully integrated subsystems would tend to share a common Belief System.

For those steeped in the more recent defense version of systems engineering, which restricts itself to creating and supporting linear technological artifacts, SoSE may seem to be novel. However, SoSE appears from the assessment and analysis above to be in the mold of classic systems engineering, i.e., appropriate for application of the system methodology presented in this book. It should be noted, however, that the prospects for optimizing a SoS may be limited: potential synergies and emergence may be similarly limited by the original choice of systems to be brought together, and by the degree to which they can be harmonized and coordinated. With SoS, it may be more a case of satisficing than of solving the original problem — see The Systems Approach on page 16.

Assignment

You are the Business Development executive for a large manufacturing company in the automobile industry. The company is seeking to introduce a new product, unlike those currently manufactured. The company will need to form a supply chain of parts from which to assemble the new product. You have identified five companies that supply similar parts to the industry.

Your initial idea is that your manufacturing company should purchase these supply companies so that they may be incorporated and dedicated to supplying parts exclusively for the new product. On second thoughts, you realize that buying these potential suppliers may not be the best plan.

Develop a conceptual strategy for setting up and establishing the supply chain for the new product that does not involve buying the supply companies, but which, instead, seduces them to supply high-quality parts predominantly to your company. You should consider harmonization and interchange factors, and how requisite quality can not only be achieved, but also improved upon. You should also consider how supply of parts might be guaranteed. . . .

Case G: Defense Procurement in the 21st Century

The Problem Space

Defense is an expensive business, not only in terms of the finance, but also in terms of lives lost, dreams shattered, and communities disrupted. Defense is also a sensitive issue in a democracy. The University of Michigan's Correlates of War Project showed that democracies seldom make war on other democracies — with no exceptions since 1815. This suggests — at least — that voters in a democracy are reluctant to vote for war against another democracy, and that war is likely to be a vote loser.

There are counteracting influences. Should a democracy be attacked, history shows that the people will band together against the common enemy: they will vote for war; they will go to war; and, they will count the cost. In times of increasing human rights and freedoms, the cost of losing young lives is becoming increasingly painful and less acceptable. The people would, it seems, want to respond to attack with a clean war, one in which their young men and women did their duty, but survived unscathed both physically and mentally.

Which means that the armed forces should be equipped with the best weapons, the best armor, the best everything to enable them to defeat opponents with minimal risk to themselves. And, since technology is advancing at breakneck speed, our armed forces should have the latest, up to date, innovative technology. Or, so one argument goes . . . .

Difficulties in predicting the need

It is difficult to forecast what weapon systems a particular force will need. It depends so much on the situation, the opposition and the rules of engagement. Changing sensitivities have made it unacceptable to engage an enemy who is embedded in civilian enclaves — a favorite hiding place of the terrorist or insurgent just because he believes that he won't be attacked there — raising the profile of the so-called surgical strike: a precise hit with no collateral damage. Is there such a thing

Systems Engineering: A 21st Century Systems Methodology   Derek K. Hitchins
© 2007 John Wiley & Sons, Ltd

as a genuine surgical strike? It would require perfect, timely intelligence: precise target detection, location and guidance; and a weapon payload that was powerful enough to fulfill its purpose, but tightly circumscribed in its effect.

The foregoing is predicated on the notion that technology is the answer — what was the question? In the real world, warfare is not quite like that. For a start, conflicts are rarely one-on-one; instead, they are many-on-many. And the rules keep changing, too. At the present time, for instance, it is unacceptable to simply overpower a foe with superior weapons and numbers. 'Proportional response' is the order of the day, so that a watching world does not see the more powerful nation as 'bullying.' In such an environment, deploying some advanced technologies may be 'beyond the pale.'

Military men, too, have argued interminably over the need for more, simpler, cheaper, weapon systems, versus the need for fewer, more sophisticated, more expensive weapon systems. Generally, the argument has gone in favor of the fewer, more sophisticated and more expensive . . . and that has led in turn to the use of advanced technologies, long, protracted procurements, and weapons systems that may be rather too precisely directed towards a threat that – by the time the weapon is available — may have changed or disappeared.

It has been the practice in the past for the military to determine what they need, and to issue requirements for weapons, platforms (tanks, ships, plane, etc.) This has not worked well for several reasons:

- The need for particular weapons has often been predicated on intelligence claims about a potential enemy's developing capability. Such intelligence has not always been accurate, tending to represent the enemy as more powerful, more technologically advanced, and even more aggressive, than events proved to be the case
- Military staffs formulate the requirements for new weapons systems based on their own operational experience. This would seem to be sensible and appropriate: who could know better than a 'fighting man'? Unfortunately, this turns out to be neither sensible nor appropriate. The one-time 'fighting man,' now a senior 'desk jockey' in some government department, may not have fought for twenty years, and then under entirely different conditions and situations. Moreover, when he did fight, he may have been using platforms and weapons systems that were themselves ten years old, having been designed at least twenty years earlier still, using the then technology. So, he formulates his requirement based on his understanding of fifty-year-old technological capabilities.
- New technologies may afford the opportunity to engage an enemy in an entirely different way: experienced military men may find it difficult to conceive or accept such changes — see Case F: Fighter Avionics System Design on page 425.
- In consequence, there is a marked tendency for military requirements to seek replacements for that that has worked previously, only 'up-gunned,' faster, more survivable, more reliable etc. Often, such requirements can tend to be rather specific to particular situations and circumstances that have applied in the past, rendering the new system unattractive for export.

Governments of developed nations tend to invest heavily in advanced technology weapon systems, and seek to defray the cost in export sales to nations of all persuasions, taking care to moderate exported weapon systems capabilities.

Meanwhile, history shows that force capability is dependent on much more than the technology of weapons and transport systems. It is also to do with training, with balanced forces, with synergy between force elements, with motivation, *esprit– de -corps*, and discipline, with organization and command, and many more. Capability, it seems, may be emergent.

Cutting edge of technology — was defense, now commerce

With all the talk of advanced technology for defense, it used to be thought that defense was the breeding ground for new technologies, which would later 'trickle down' into commercial use. With rare exceptions, the reverse has been true for some time. Advanced, sophisticated systems that were once the preserve of the military are now appearing first in general commercial use. This is particularly true in electronics, electro-optics, software, imaging, graphics, communications, and many more. (It may not be true in materials technology.)

One reason is that research in these areas is driven by sales, and commercial applications afford much greater sales potential than military applications. Commercial products have become as robust as their militarized counterpart — indeed, there is often no need to militarize commercial products. Commercial off-the-shelf (COTS) products are now available to the defense system designer and procurer that are capable of being fitted unaltered into military platforms.

The use of COTS products presents the defense industry with a dilemma: whereas conventionally procured defense systems might have a potential operational (and support) lifetime of perhaps 20–25 years, COTS products and systems are intended to make use of the latest commercial technology, and are viewed as consumables. In other words, they have a life in the marketplace of perhaps 18 months, after which they are overtaken by the next wave of technological innovation. Moreover, instead of repairing — which conventionally requires a vast, expensive, logistic support capability — COTS products are generally thrown away or recycled when they fail, and are replaced by a new device — it is cheaper, quicker and much easier.

This short life presents both the military and the defense industry with difficulties. How are they to maintain the capability of, say, an aircraft or a ship over a lifespan of 20–50 years if they are employing COTS systems that change every 18 months? As the defense industry would see it, there is more to it than saving money. What about maintaining compatibility between the various subsystems if they are currently being superceded: are using different intercommunication systems and protocols; are able to do new things/have new capabilities that were not anticipated in the initial design for the whole?

Bureaucracy blunting the cutting edge for defense

In the mid-1990s, the US administration of the day realized that there was an alternative to conventional defense procurement — one that did not place such a heavy burden on government research and development, and could instead make significant use of COTS equipments and systems. Not only would COTS save on time and money, it could in principle mean that military platforms deployed the latest technology; that, although the platform might age, its carried systems would not.

The phenomenal global success of Japanese lean volume supply (LVS) systems brought about this realization. In the lean volume supply system, the US administration saw a way of procuring advance technology weapon systems in much the same way that they might buy a car from a showroom.

Would-be buyers of a new car do not impose a detailed requirement for a new car on some car manufacturer, and then wait some 20 years for the car to appear. Instead, they go along to a number of showrooms with some general ideas of what they would like in mind. On seeing what is available, they may expand their wishes, or shop around for good deal.

For the car manufacturer to succeed, he has to provide an attractive range of vehicles at competitive prices, with good availability, innovative features, etc. The manufacturer carries out research and development, funded out of profits. And, the Japanese had shown, are still showing, the world how to do this on a grand scale, producing innovative, high-quality, reliable goods at affordable prices (see Case A: Japanese Lean Volume Supply Systems on page 145.)

The prospect was exciting. The administration 'cleared the decks' by eliminating much of their long-winded, reductionist procurement management procedures (including MIL-STD-499A, Engineering Management — widely, but perhaps unadvisedly, regarded as the manual on defense systems engineering). A number of major defense organizations would reorganize into two or more, competing, large-scale volume supply systems. The capability would exist to create and supply advanced, up-to-date weapons systems in volume, inexpensively; and, saving hugely on tax dollars, expensive defense research and development would be funded by the commercial manufacturer. The stage was set, seemingly, for a revolution in defense procurement. It did not happen.

The administration had reckoned without the defense bureaucracy, with its vested interests. Bureaucracies, like aircraft carriers, exist to defend themselves: the defense bureaucracy was no different. If the plans were to go ahead unchecked, many thousands of civil servants working in the many and various defense project offices, laboratories, etc., the length and breadth of the nation, would be redundant.

Essentially, the proposed reforms were culturally unacceptable. The idea that industry could work out for itself, and provide, what the military needed was anathema. Government had a pathological distrust of the defense industry, matched only by the defense industry's distrust of government. At all costs (literally), the defense industry had to be controlled, regulated and hedged in with defense standards, and who was going to do that in this new era of commercial procurement? Instead of thousands of bureaucrats being made redundant and turning to the defense industry for employment, the defense bureaucracy would back itself up and reimposed itself as the arbiter of requirements, specifications, architectures, integrated project teams, etc., etc.

Some of this metamorphosis paid lip service to the Japanese lean volume supply concepts. In the automobile business, the concept of integrated product teams (IPTs) had emerged, and proved invaluable. The idea was simple, but radical. When a design change was proposed, it would be examined, assessed and approved by a small, multidisciplinary group of people who worked on the assembly shop floor — not, as previously by senior managers, administrators and designers sitting in conference.

Decisions would be swift, made by people who understood the problem, and would not involve communications and approvals up and down some hierarchy. The IPT would have the authority to make changes on the spot. An IPT was an *ad hoc* grouping appropriate to the problem. To approve a design change proposed for a car seat, for example, would need the seat designer, a commercial man to evaluate costs, a sales and marketing man to assess any impact on advertising, publicity materials, etc., and a technician familiar with automobile manufacturing regulations in various countries. IPTs typically comprised four or five people.

US attempts to shift towards lean volume supply were mirrored in the UK in 1997 in a program called Smart Procurement, which was similarly dedicated to shortening procurement timescales, introducing the use of COTS, etc. UK government defense bureaucracy strangled the attempted change at birth, by adopting the role of organizer and controller of Smart Procurement. The bureaucrats even went so far as to introduce IPTs — not Integrated Product Teams, but Integrated Project Teams, which might typically involve between 250 and 500 people. Instead of facilitating fast, on the spot decisions, these parodies of Japanese IPTs were, of course, unsuited to making any decisions at all.

Smart Procurement did not live up to expectations. Now re-badged as Smart Acquisition, it seems to be having similar problems of curbing mounting cost overruns and delays for major projects... (National Audit Office: Ministry of Defence Major Projects Report, 2004.)

It still takes a very long time to introduce a new platform or major weapon system. Typically, the time from conception to being in service might be some 20 years. Smart Acquisition promises to reduce that time, perhaps to 13–15 years, but has yet to deliver on that promise. Meanwhile, industry on its own, without government's multi-layer defense administration overlays, controls, regulations, etc., can create a new fighter aircraft from scratch in under four years — and at a fraction of the cost.

Security

One cause for concern expressed by bureaucrats was security: expressing concern for security is intended to scatter the chickens and send them dodging for cover in the henhouse; it is a vague threat with which to beat the weak and defenseless. How could anyone argue about the need for security? How could anyone even dare to discuss it? And was it not obvious that commercial lean volume supply would inevitably leak national defense secrets like a sieve?

Security is an issue, of course, but not in the way that the defense bureaucrats might like to pretend. It is important to protect and conceal some aspects of defense such that a potential enemy or attacker is unable to detect any weak points or develop successful adversarial strategies.... Moreover, there is a need for commercial secrecy, too, where through extensive/expensive R&D, or perhaps through serendipity, major advances have been made in materials, processes, techniques, algorithms, etc. Such advantages can provide an edge for a nation in conflict, at least until other nations catch up.

And that is important to understand — other nations are always 'catching up:' technology and advanced engineering are no longer the preserves of rich advanced Western nations. On the contrary, in many spheres the burden of excellence is shifting/has shifted towards India, China, Japan and the so-called Asian Tiger economies. If the West has secrets, they won't be secret for long — secrets have a short shelf-life. Only continual R&D to create new secrets will keep an industry ahead of the game.

But what has security, *per se*, got to do with defense lean volume manufacture and procurement, Japanese style? Consider, for instance, a new military aircraft. What could be secret about it: its operational capability? That can easily be estimated simply by looking at the aircraft, assessing its all-up weight, looking at the wing sweep, the shock diamonds in the jet efflux, etc. etc. It might be more difficult to determine what advanced composite material the wing is made from, and how it was formed. So, performance of the whole might be relatively accessible, while the materials from which something is made could be more sensitive.

Consider, too, something as seemingly sensitive as electronic surveillance measures (ESM) equipment. It might be thought that the equipments for 'listening in' were security sensitive: generally not, however. The equipments are radio receivers of various kinds, and are usually nothing out of the ordinary. However, the digital signal processing (DSP) that is applied to the signals, once received, might be highly sophisticated, and may employ algorithms that might confer advantage on whoever possessed them. So, does that mean the electronic surveillance systems should not be manufactured commercially? Hardly. The signal processing algorithms may be held in secure software — dummy versions of which can be installed during manufacture and test, such that no one can access any sensitive information. The real algorithms would be employed only when on military operations, and even then it would be simple to include an 'auto-wipe' arrangement into the system, say, on power-down.

Security, though something always to consider carefully, is not a showstopper for defense lean volume procurement and supply.

Conceptual Remedial Solutions

So, is there a straightforward solution to the issue of defense procurement? Well, there certainly ought to be. The main obstacles standing in the way are deeply entrenched bureaucracies, defense customer conservatism and distrust, and some major defense support reorganizational issues.

The Japanese Lean Volume Procurement and Supply model is undoubtedly the right one for future procurements. However, the model may not extend far enough. In the case of national defense, the model needs perhaps to include the customer/operator within an overall closed loop.

Figure G.1 shows the concept. At top left is a notional lean volume supply system (LVSS), with a lead contractor, first-tier suppliers, second-tier suppliers, and raw materials suppliers. At right is a defense market, into which are presented products, services and capabilities from competing LVSSs. The military, represented bottom right in an operational loop, identify a potential product, service or capability and will try before they buy, just as they would if buying a new car, or as they might like to if buying a new house. If, after this test and evaluation phase, they like one of the products, services or capabilities on offer, then the new facility will be installed, incorporated,

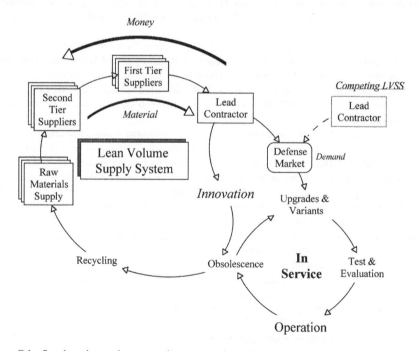

Figure G.1 Supply and operations as contiguous open, interacting subsystems within a closed-loop whole.

or whatever and will go into virtually immediate operation, integrated, backed up and supported by the LVSS.

Meanwhile, the LVSS is continuing to innovate, and will present new upgrades and variants for Test and Evaluation. If these are attractive to the military operators, then they will incorporate the upgrade, or the variant as appropriate, and carry on operating with enhanced capability. If these are not attractive to the military, then they will not buy, but will carry on as before, waiting for a better opportunity to upgrade . . . . The onus, then, is on the LVSS to come up with upgrades and variants that are so attractive to the military operators that the upgrade/variant renders the existing facility obsolescent. An upgrade will then proceed, and any redundant equipment will be recycled to the LVSS, so closing the loop.

It is possible in this scheme to incorporate COTS quite freely. The LVSS that supplies the system, the upgrade, or the variant is also responsible for ensuring it integrates correctly with other on-board systems and for supporting it during its operational life. Indeed, it is even possible to consider that the whole platform is provided and maintained by the LVSS, but owned and operated by the military. In some scenarios, the military may effectively lease the platform, rather than buy it: at the termination of the lease, the platform, like any other artifact, reverts to the LVSS for recycling.

The procurement–supply–operation–recycle loop is subject to competition, for three reasons:

- To keep costs down — if a competing LVSS can offer an equivalent capability, then it is incumbent upon the current supplier to keep costs down.
- To afford variety of contribution. Competing LVSS may offer different features, some of which may be preferred in the situation facing the military.
- Security of supply: in the unlikely event that one LVSS might be unable to supply for whatever reason, the competitor should be able to fill the void.

Lean volume supply systems depend on throughput to stay in existence: it is their lifeblood. If a LVSS were dedicated exclusively to the conception and creation of defense artifacts, products, capabilities, etc., the question must arise: would there be enough regular throughput to maintain a Defense LVSS? If there were not, the closed loop of Figure G.1 would break open, and the military would find themselves without support. Having an alternative, competing LVSS might be little comfort in such circumstances if it, too, were to experience inadequate throughput to sustain itself.

This situation has not arisen in the commercial world, operating Japanese style: instead, they perceive a steadily expanding and more diverse throughput over time. Might it be possible, then, to combine commercial and defense LVSSs in one?

The situation is illustrated in Figure G.2. A single, agile LVSS is shown serving both the commercial and defense markets. This would afford great benefits to both parties, since the accumulated wealth would support R&D into new products and capabilities for both, with defense benefiting from the rapid advances being made in commercial technology.

Although the notion of a common LVSS for both commercial and defense products may seem *avant-garde*, it becomes less so when considering, e.g., avionics COTS products which might be fitted equally in a military transport and in a commercial airliner. And it is not too great a stretch of the imagination to think of defense ground radars being assembled and tested on the same assembly line as air traffic management radars, *en route* control and reporting radars, area surveillance radars, etc. It is not difficult to envisage a factory making ship navigation, steering and engine control systems for both military and commercial ships. And, with the advent of commercial spread spectrum, frequency hopping radios, combined assembly lines for defense and commercial communication products are readily conceivable.

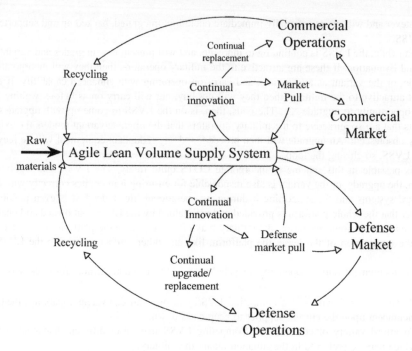

Figure G.2   For some defense systems, it may be appropriate to operate using a common, agile LVSS, as opposed to one dedicated to defense products. This offers the advantage of greatly increased throughput, allowing the LVSS to acquire more wealth and to afford more R&D. National security considerations might militate against this concept for some facilities and equipments . . . .

The main objection to a combined defense/commercial LVSS, such as that illustrated in Figure G.2, is likely to be security. Not so much, perhaps, security in the sense of revealing secrets, but security of supply in the sense of guaranteed support and supply over the life of a weapon system . . . although, there is no guarantee that a defense contractor will not go out of business under the current procurement regimes.

CONOPS

Defense Acquisition has evolved: instead of seeking to procure weapon systems, the objective is to procure defense capabilities. A capability statement describes what is to be achieved, rather than how. So, there might be a defense capability requirement to establish a defended sea–land bridgehead on third world territory in the face of enemy ground and air attack.

The implications are that a ship-borne military force may sail to a designated area, to put ashore with a complement of men, machines, equipment, etc., under an umbrella defensive screen. There is no mention of the numbers of men, the type of equipment, what the enemy threat might be, how the defensive screen is to be supplied, etc, etc: so, the 'what' without the 'how.' A defense capability statement might indicate that this capability should be replicated, such that defense forces could conduct such operations at several locations in different parts of the world at the same time. This suggests the need for capacity and redundancy in the system.

It is, in principle, up to the military to decide how they will provide such capabilities. This puts the onus on the military to decide if they already have said capabilities, and if they need to expand and diversify. It also gives them the opportunity to view each capability as an open system, with interacting parts, adapting to its changing environment as it interacts with other systems in the environment. Military operators, designers and planners understand about complementary systems, about cooperation and coordination between systems to create synergies.

They know, too, what technological facilities they will need to support them — but how would they get what they needed in a world where defense products were procured according to Japanese-style lean volume supply rules?

One way of approaching that issue is to consider how defense procurement could be controlled. Table G.1 shows a potential national defense capability procurement procedure. The left-hand column shows inputs to each of the processes, which are shown in sequence from top to bottom in the center; outputs are shown in the right hand column. The whole is systems engineering at Level 5.

The first process is to set Goal Defense Capabilities: these are dependent upon Foreign Policy and National Security Issues, top left; these are the stuff of high politics and government. On the basis of these Goal Defense Capabilities, once established, there follow Doctrine, Strategies and Concepts of Operations: how, in principle, should we, as nation, go about achieving these Goals, what are the risks and how might we mitigate them? The output from this second process will be Formal Policies.

Table G.1 Defense capability procurement model (Level 5 Socioeconomic Systems Engineering.)

Stage	Input	Process	Output
1	Foreign policy National security	Set goal defense capabilities	Defense capability goal
2	Sociopolitical analysis	Establish doctrine, strategy and CONOPS	Formal policies

3	Force representation:
	    current
	    potential
		Synthesize alternative force structures	Structure options

4	Effectiveness criteria	Select optimal force structure	Target force structure
5	Market intelligence	Identify suitable components in the market	Component options
6	Component shortlist	Test and evaluate contribution to capability	Component contribution
7	Contribution criteria	Select Optimal Components	Target contribution
8	Funding	Buy 'Off the Shelf'	Acquisition
9		Integrate into Structure	Achieved capability

The third process is to Synthesize Alternative Force Structures: how can the military best configure itself to achieve the Defense Capability Goals within the constraints of national policies, doctrines, strategies and CONOPS — which might, for instance, include cooperation with other nations.

Alternative Force Structures can be tested, compared and evaluated '*in vitro*' using large-scale simulations, national and international joint defense exercises, etc., where 'joint' indicates tri-service involvement. Simulations, in particular, can envisage and test the employment of different weapons systems, and can assess their likely effect on speed and effectiveness of operations, losses and casualties, etc. — see Case D: Architecting a Defense Capability on page 313.

From simulations, exercises, previous campaigns and military experience, etc., it is possible to identify an optimal force structure, including platforms, weapons systems, transport facilities, communication systems, etc. Simulations in particular will show the ideal characteristics of such systems, in terms such as range, kill potential, ability to operate in prevailing conditions, need for, and availability of support, interoperability, bandwidth, jam resistance, exploitability, survivability, etc., etc.

Instead of then placing requirements on a number of defense contractors to conceive, design and create these various defense artifacts, to be received in 15–20 years time, it should be possible to go directly into the market place and see what is available that meets the bill, nearly meets the bill, or perhaps exceeds the bill. In the commercial procurement scheme of things, the commercial supplier will have anticipated what is going to be needed, particularly if the suppliers have been privy to the first three or four outputs shown at the right on Table G.1.

Competing Defense LVSSs will, if enabled, produce competing components of each defense capability for defense forces to 'try before buying,' in the commercially approved fashion. They may select what they prefer and then buy off the shelf, with the LVSS willing, and more than able, to integrate the various technological offerings into the force structure. If the LVSS does not come up with the goods, then 'no sale.'

Can it be that simple? Note that no public monies are at risk throughout. One of the major causes of long, protracted and difficult defense procurements under the present schemes is the great care taken by bureaucrats when spending public money. Bureaucrats are inclined to take so much care that they will spend billions to save millions. Defense industries, on the other hand, have on occasion been known to charge startlingly high prices, often blaming high costs on unnecessarily high standards, and on restrictive government regulations, processes and procedures. In this alternative approach, responsibilities are more sensibly allotted:

- Politicians are responsible for deciding what capability defense forces should have and for determining — or at least understanding and approving — doctrine, high-level strategies and CONOPS, on the basis of essential national security and avowed foreign policy
- The military are responsible for determining how to achieve the various capabilities, and therefore what resources of men, machines, weapons will be needed, and how they should best be configured to promote synergy and optimize performance. They are also directly responsible for buying (leasing?) what they need.
- Industry is responsible for conceiving and creating ranges of innovative products to support, enable and empower military manpower in achieving its capability.

There is no extensive defense government bureaucracy. But then, there is no reason for any of the three parties, government, military and industry, to distrust the other. This alternative approach encourages cooperation and synergy between the parties, all of whom stand to gain.

System Design

Accepting, for the moment at least, that this radical alternative to defense capability procurement is acceptable (and there will always be exceptions to the rule, such that not all particularly sensitive items may be procured in this relatively open fashion), how might it look on a more global scale?

Figure G.3 illustrates a bird's-eye view of the globe, with the US under the left-hand lean volume supply circle, and Europe under the right-hand lean volume supply circle (see Industry Circle, Figure A.5 on page 157). In this instance, the US LVSSs — there may be more than one — involve US sources, US companies and US market-directed products, services and capabilities. In contrast, the European LVSSs — there would be more than one — are open to a wider range of organizations, particularly at second, third and subsequent tiers. The approach here would be to encourage both existing and potential defense customer countries to participate in the LVSSs.

Note that the two LVSS circles are in competition and both are competing to supply military customers across the board, i.e., armies, navies, air forces, marines, coastguards, emergency services, and so on.

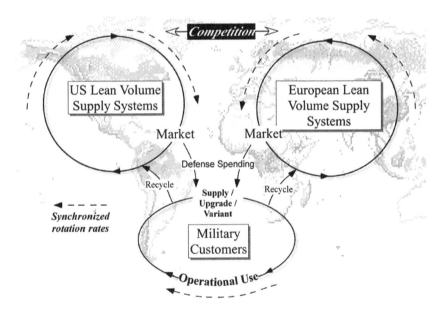

Figure G.3 Global LVSS competition/cooperation. Top left is a (stylized) US LVSS. Top right is a European LVSS. Both LVSS's compete to supply military capability, including COTS to military customers. Note that the whole system of supply–operation–recycle will develop a synchronized, circulatory rhythm, not unlike that of blood circulating in the body, but with the pulse rate determined by the operational life of individual upgrades, variants and replacements. The shorter the life, the faster the rhythm, and the more frequently money will flow around the LVSS loops. By time-shifting different upgrades, variants and replacements for different customers, the LVSS loops can operate in a dynamically steady state, with a continuous flow of products in one direction and contraflow of money in the other.

Conclusions

Western countries, observing the phenomenal success of Japanese lean volume supply systems, sought a decade ago to emulate that success in national defense procurement/acquisition. One of the keys to the proposed revolution was the wide-scale introduction of COTS — commercial off-the-shelf-equipments and systems. Such systems are, as the title infers, available immediately as consumables, so that in-service military systems can be equipped and upgraded with the latest, innovative technology.

National defense bureaucracies hijacked attempts to reform acquisition organization, regulations, methods and procedures, stifling them at birth: the proposed reforms were, simply, culturally unacceptable. As a result, it is still taking far too long to procure defense capabilities, weapon systems, products and services. A new platform (aircraft, ship, tank) is likely to take 15–20 years from start to first in-service date: left to industry, it could take as little as three or four years without the continuous bureaucratic regulation, intervention, distrust and 'politicking.'

There is, as yet, little COTS to be found in operational defense systems: unsurprisingly, the term 'COTS' has all but disappeared from current defense acquisition glossaries . . . . Meanwhile, COTS systems have a lifetime of only two or three years before being replaced by newer versions with technology that is more recent, has greater capability and offers better performance.

The original proposal, to emulate Japanese best procurement practices, was sound — as far as it went. It is better observed by widening the system from a procurement system to a combined procurement and consumption system, i.e., to couple the supplier and the user/customer into one system with two mutually interdependent (sub)systems. Defense operators need lean volume supply of the highest quality innovative goods, and the lean volume suppliers need customers who use high-quality innovative systems and services: it could/should be a marriage made in heaven.

18

Systems Engineering: Intelligent Systems

Tax not the royal Saint with vain expense,
With ill-matched aims the Architect who planned —
Albeit labouring for a scanty band
Of white robed Scholars only — this immense
And glorious work of fine intelligence!
Give all thou canst; high Heaven rejects the lore
Of nicely-calculated less or more.

(Of King's College Chapel, Cambridge)
William Wordsworth – 1770–1850

Introduction

There is considerable business management interest in businesses and enterprise as entities with the potential to learn and to behave intelligently. Concepts such as the Learning Organization (Senge, 1990) and the Intelligent Enterprise (Quinn, 1992) have entered the business management psyche. The ideas, broadly, were that the people working within an enterprise could learn together, develop intellectual capital, could manage their knowledge in such a way as to enhance their achievement of objectives.

Peter Senge wrote about the Fifth Discipline, which he identified as Systems Thinking (see page 17), where the system was the enterprise, and the thinkers were the employees who came together to understand and address problems, and develop ways of overcoming them. Warfield's interpretive structural modeling (ISM) (on page 191) and interactive management (IM) (Warfield, 1973, 1989) were significant methods for exploring issues and developing understanding, cooperation and consensus. Major international organizations rose to the challenge and converted themselves into learning organizations, often with surprisingly swift and effective results. In some instances, group learning became the norm; in others, the experiment made a significant difference in the short term,

Systems Engineering: A 21st Century Systems Methodology   Derek K. Hitchins
© 2007 John Wiley & Sons, Ltd

but there was subsequent reversion to previous practices. Some managers and accountants saw little merit in employees 'spending time learning about the business, which was really none of their concern,' when they could be 'getting on with their proper work.' Essentially, too, accountants saw little return in the short term — their principal interest — from group learning and group problem solving.

Although Peter Senge in particular popularized the notion of corporate systems thinking, it was not new. The systems houses of the 1950s, 1960s and 1970s had been centers of systems thinking (see Preface). In at least one well known systems house of the period, the systems engineering paradigm (see Systems Methodology: conceptual model on page 172) was used extensively to address virtually any problem that arose. An *ad hoc* group would form: members would outline a problem, identify the criteria for a good solution, and generate a range of possible solutions. The group would then trade off the potential solutions against the criteria, arriving at a 'best fit' solution. The result was a solution to which all who had participated were committed: it not only worked, it worked well.

The result of this extensive use of a systems thinking/problem solving approach was a vibrant, creative organization populated with intelligent, articulate, highly motivated people: yet, the systems houses were largely driven out of business by the large aerospace and defense companies. Systems thinking, it seems, was not at a premium

What is an Intelligent System?

Rather than situate the idea of learning and intelligence in enterprise and business, it is useful initially to abstract the concept, and to consider any open system existing, flourishing and trying to survive in its dynamically changing environment, where it interacts with and adapts to, other systems.

Employing the classic systems approach, then, the image is of an entity (open system) that experiences inflow and outflow of energy, materials and information, yet maintains dynamic stability, seeking to survive and flourish in a complex of other, open, interacting, adapting entities, all similarly engaged in trying to survive and flourish. The entity does work and expends energy in processing the flowthrough: the entity may also extract material and energy from the flow through. The entity, although interacting with, and adapting to, other entities, maintains a level of dynamic stability (homeostasis), such that inflow rates and outflow rates are balanced and, if disturbed, a new point of balance may be reached (in accordance with Le Chatelier's Principle). The entity constantly maintains its internal processing facilities. The entity, as an open system, stabilizes dynamically at a high energy level, reinforcing the appropriateness of the organic, as opposed to the mechanistic metaphor — where stability would be consistent with a low energy state — (see The Concept of the Open System on page 11: this book addresses open systems; although some manmade systems may be more open than others, all are essentially open. If a system were truly closed, we would be unable to detect it and would be unaware of its existence.)

Examples of open systems that fit this description abound, of course. A cell in the human body is an archetypal open system. So, too, are individuals, families, communities, societies, gangs, teams, businesses, enterprises, platoons, sociotechnical systems, civilizations, economies, governments, organizations/organisms, etc., etc.

So, are some open systems intelligent, and how would we know? In posing the question, we are asking a wider question than that posed and answered by the pundits mentioned in the opening paragraph: they presumed that any prospective intelligence was to be found in the people of the organization; our question allows that intelligence could be emergent, i.e., a property of the whole not exclusively attributable to any of its logically separable parts.

Defining Intelligence

Kinds of intelligence

If we consider the open system as (being like) an organism, we may reconsider the question: are any organisms intelligent, and how would we know?

It is not easy to define intelligence. Some like to define it as an ability to learn facts and skills, and to apply them. For others, intelligence is the capacities to reason, plan, solve problems, think abstractly, comprehend ideas and language, and learn. Yet again, the terms 'smart,' and 'quick-witted' often indicate situational and behavioral intelligence.

There seem to be two contradictory threads:

1. Intelligence is seen as 'deep think;' reasoning, learning, planning, problem solving, abstract thinking, etc.
2. Intelligence is the ability to 'think on one's feet,' to be aware of what is going on around you, and to respond in a decisive, effective manner.

Intelligence seems to be related to the ability to unravel problems and make decisions. Jacques Cousteau, the French naturalist, filmed an octopus in a water-filled glass tank on a table, on the deck of his yacht, the Calypso. The octopus climbed out of the tank, slithered to the corner of the table, felt with its tentacles for a table leg that it could not see, slid down the table leg, dragged itself across the deck and dropped to the safety of the sea below. In so doing, it put itself at great risk: it has no skeleton to sustain its weight when out of water, nor could it breathe; yet it was prepared to go to great risks to escape.

Was this intelligent behavior? Most people might think so: the animal faced a problem, saw an opportunity, formulated an escape strategy and a plan, executed the plan at great risk to life, and succeeded. So, the octopus was brave, as well as smart and decisive. This might be classified as situational and behavioral intelligence. Of course, as it dropped over the side of the yacht, the octopus may have fallen into the open jaws of some waiting shark — we will never know: what price intelligence then?

Intelligence and survival

Are intelligent organisms better suited to survive than nonintelligent ones? There does not seem to be much evidence for such a hypothesis. Starfish (echinoderms) may not be classified as intelligent — they have no central brain — yet they has survived in harsh environments for hundreds of millions of years: longer than the dinosaurs, apparently. But, is survival in this context related to intelligence? The starfish survives, not as an individual, but as a species by reproducing (sexually and asexually) many offspring — a strategy that recognizes that most offspring will fall victim to predation. (Recently, time-lapse photography has revealed echinoderms as having unexpectedly complex interaction behavior: such behavior occurs so slowly in relation to that of other creatures that we humans have been unaware of it, making the phylum's survival all the more remarkable.

Is this very slow behavior something to do, perhaps, with the echinoderms' apparent absence of lifecycle: starfish apparently do not age? We have much to learn, it seems, about organisms — natural as well as man-made.)

Some organizations operate in a not-dissimilar manner, comprising a variety of relatively small, viable companies each of which is aimed directly at different market targets within the same market segment (e.g. consumer electronics, defense avionics). Market dynamics favor growth in some areas, shrinkage in others, and the constant introduction of innovative products. As the market shifts, some companies expand; some shrink, some fold and new ones start up. The organization as a whole, however, is dynamically stable, maintaining broadly the same number of companies, while there is a continual turnover of those companies, with the resources of those that fold being transferred to start-up replacements.

Is this intelligent behavior: not at individual company level, it seems, but at organizational level? Certainly, the organization as a whole is open and continually adapting to its environment and certainly, the strategy promotes survival and longevity of the organization, if not of its parts. Perhaps this is intelligence in the sense of the whole being smart in situational and behavioral terms.

Consider too the Japanese technique of Heijunka, flow smoothing — see The Open System Viewpoint on page 149, part of Case A: Japanese Lean Volume Supply Systems. It can be shown that WIP can be minimized if the flow of a product through a lean volume supply system is steady, as opposed to rising and falling, or batching.

Heijunka operates by sensing the sales rate for the product and either increasing marketing and introducing 'special offers' if the rate starts to fall, or conversely reducing marketing if the sales rate starts to rise. When the rate looks set to fall irretrievably, Heijunka triggers a switchover from the current product, soon to be obsolescent, to a new product in such a way that the flow through rate is maintained.

To an observer, this might appear as intelligent organic behavior: the organism/organization is continually manipulating the market to smooth out demand peaks and troughs, which in turn minimizes WIP, which in turn keeps unit production costs down so that products remain affordable while profits rise. This might be seen as intelligence more in the mould of reasoning, learning, planning and problem solving. And Heijunka is a philosophy, leading to a strategy, leading to an execution of a plan by the organism, as much as by any individuals. So, is Heijunka 'institutional intelligence?'

Once the system for Heijunka is in place, the individual actors have no particular need to show intelligence — the 'smarts' are in the way the system is conceived, set up and organized. . . . So, is this intelligence promoting survival? Or is this simply a human-powered, rate-feedback servomechanism? In either event, Heijunka adapts the environment to the open system, rather than the other way around — and that is neat.

Predicting the future

There is an interesting dilemma here: the variability of the environment, both business and natural world, militates against the idea of making 'intelligent decisions' about the future. For an organism to make intelligent decisions concerning future actions presupposes that it has some knowledge of how things are going to turn out in the future: in that way, the organism can pursue advantageous situations as they develop and opportunities as they arise, while at the same time it can avoid, absorb, or protect itself against risks, threats and attacks. But, can it know the future?

Our environment, and the various interacting systems within it, seem to vary chaotically — not in a deterministically chaotic sense, but more in the way of *Weak chaos*: see page 27 *et seq*. Weakly chaotic behavior, unlike random or chaotic behavior, is consequential, in the sense that conditions in following time periods are dependent on conditions in previous time periods. The weather is weakly chaotic: tomorrow's weather stands a (slightly) better than even chance of being the same as today's, which would not be true of weather were random. However, the weather ten days from now is not knowable, at least not in any detail. Ask any meteorologist.

This means that for most things there is a Horizon of Knowability — a period in our immediate future during which we have a reasonable prospect of knowing what is going to happen. Beyond that time-horizon, things become increasing unknown and unknowable.

Humans, and possibly other organisms too, seem to ignore this fact. Humans predict the weather, the outcome of some future situation, the way the stock market is going to move, the fashions for next summer, etc. The predictions are invariably, wrong. We ignore the fact that they are wrong and we make more predictions — and they are wrong too. Humans are hooked on making predictions and basing long-term decisions on them. And humans are in full denial, clinging firmly to the belief that they can, after all, predict the future. It seems that we humans are driven by our intelligence to believe that we can predict the future, and no amount of contrary evidence will convince us otherwise.

(A classic example of this curious propensity is our formulation of project and program plans, stretching five, ten or even twenty years into the future. Such plans are often out of date before the ink dries, yet we continue to make them, consider them sagely, and base major investment decisions upon them. We even go so far a to create project plans, using Program Evaluation and Review Technique (PERT), that are designed to change from week to week, from month to month. So, we know in our hearts that we lack the ability to predict what is going to happen in the weeks and months ahead, yet we still insist on making predictions about outturn, outcome, end costs, timescales, performance, cost effectiveness, etc., etc., years ahead of the events.)

So, does that mean that intelligence is of little value to an organism? Not really, but it does limit the prospects for intelligent decision-making to result in favorable outcomes.

About Making Decisions

Part of intelligent behavior seems to be about acting decisively. Which involves making decisions. One way to understand decision-making is to study it in the military context, where two opponents face each other. Each wishes to impose his or her will on the other. Each has only imperfect, incomplete knowledge (Intel) about the other. Moreover, when either party acts, the results of that action are uncertain since the opposing party may parry, defend, evade, or be acting simultaneously, so changing the assumptions on which plans were made, as well as the outcome.

Interactions between protagonists are therefore generally nonlinear, and the results of actions taken by either party are seldom going to turn out as planned. In other words, cause and effect do not correspond as in a simple linear model. Commanders may adopt 'empathic reasoning': 'if I do this, then he is likely to respond by doing that: but if I do something else, instead, he will be obliged to respond in a different way.' Sometimes, as in a game of chess, protagonists may think several moves ahead. Empathic reasoning only works, of course, if commanders have some understanding of situation and opponent: and then, there is always the 'double bluff. . . .'

Opposing commanders have to make decisions. Circumstances dictate that sometimes those decisions must be virtually instantaneous: at other times, there may be an opportunity to strategize, plan and coordinate action. So, the military become interested in 'what is a good decision.'

Figure 18.1 is an intent structure showing aspects of a good decision in an adversarial military

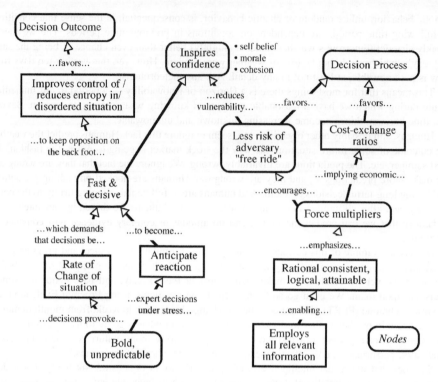

Figure 18.1 Decision-making in command and control. What makes a good decision? At left, factors favor decision outcome. At right, factors favor decision process. Both strands inspire confidence. Force multiplier is a concept, similar to 'leverage,' in which some forces may be seen as having more effect than might be expected, often due to agility and the ability to adopt more than one role. A 'free ride' may occur when all aspects of an adversary's offensive capabilities are not parried.

situation. As with intelligence above, there seem to be two contradictory threads: one thread favors the comprehensive, systematic decision-making process as defining a good decision — this is the process view. The second thread is concerned with outcome — never mind how the decision was reached, did it result in the right outcome? This is the outcome view. Both kinds of decision inspire confidence in others, although probably under different circumstances. The process view is acceptable when time is not pressing, and where the decision maker may be inexperienced. The outcome view is acceptable when under pressure, but generally requires an experienced leader to take the initiative.

Conflict within a military command and control HQ may arise when the commander's staff systematically develop a strategy and a plan and present it to the commander, who then uses his experience and 'instinct' to do something quite different. Similarly, conflict can arise in a boardroom when the chairman overrides the well planned, well structured, carefully thought out proposals of board members . . . .

Are military decision models relevant to enterprises? Military formations (platoons, brigades, etc.) are enterprises, of course, but are often considered, as in this instance, to be in direct confrontation with an adversary. In the west, we tend not to think of enterprises as being at war. Some

Japanese commercial enterprises, on the other hand, follow bushido ('the way of the warrior') and the teachings of Sun Tzu (Wee Chow Hou, 1991): for them, business is warfare; and, they seem to be doing quite well . . . .

What Characterizes a Learning Organization/Intelligent Enterprise ...?

We may deduce that an organism/organization may be deemed intelligent:

■   by virtue of its behavior in context, and that. . .
■   . . . said behavior may be an emergent property of the whole, or . . .
■   . . . it may be vested in the ability of parts within the organization to make 'good decisions,' and for . . .
■   . . . other parts to act cooperatively and cohesively upon those decisions,
■   so making the whole decisive,
■   such that intelligence may again be perceived as an emergent property of the whole.

All of which amounts to intelligence being perceived as a property of the whole: one that may emerge, like any other emergent property, from coherent, cooperative, coordinated and complementary behavior of the parts.

What does this suggest about enterprises, learning, intelligence, adaptation and evolution?

■   An enterprise may be deemed intelligent in context by virtue of its organization and configuration, if it adapts (evolves?) to changes and perturbations in its environment such that it can both survive and flourish: this without any apparent discrete focus of 'intelligence' within the enterprise
■   Individuals, teams and groups comprising an enterprise can learn, be aware, of changes and perturbations in their environment such that they can anticipate and respond to change quickly and coherently as a social group or groups. Such group learning enables swift adaptation to change and opportunity, which will favor survival and sustained flow through.' Such an enterprise would appear intelligent by virtue of its ability to adapt swiftly to changing circumstances
■   An enterprise may set out to be intelligent, may seek to understand its context, to anticipate change and perturbation, and to act decisively to take advantage of opportunities/avoid threats. Such an enterprise would be deemed intelligent if it made good decisions, ones in which the outcome favored the enterprise both surviving and flourishing. In a turbulent environment, 'good decisions' would be bold and innovative, to accommodate foreshortened timescales, rather than process driven.

It is reasonable to conclude that the learning organization gains survival advantage not from what it learns, *per se*, but from the shared group motivation, social cohesion and coordinated behavior that such learning encourages. People in a learning organization become aware of their respective positions, roles and tasks, but also of the 'bigger picture:' of their situation and objectives as a group, rather than as individuals; and, of how they may cooperate to achieve those objectives. Through the continual exercise of learning in groups, teams, etc., the various parts/groups, etc., become more closely coupled, more responsive, and better able to react together as a unified whole. A key property of the learning organization, then, is that it is reactive.

The intelligent enterprise seeks survival advantage through anticipating changes in situation, environment, etc., and through perception of opportunities and avoidance of risks. As such, it attempts to 'look ahead' to determine the best course of action to take. Making such intelligent choices depends on the ability to perceive the future, which is not practicable beyond a time horizon, the distance ahead being a function of situational and environmental turbulence: the intelligent enterprise will gather Intel to help it identify threats, and opportunities, and to anticipate trends. A key property of the intelligent enterprise, then, is that it is proactive.

Situation Facing Intelligent Enterprises

A commercial enterprise may be visualized as a node within a complex network of interconnected nodes — see Figure 18.2, which shows a notional Intelligent Enterprise (IE) existing, like any other, in a complex multidimensional network of systems, exchanges, markets, etc. In a free market economy, the network is continually shifting, with enterprises coming and going: linkages, shown as arrows in the figure, may be transitory, and may flip from one enterprise to another. Enterprises will have many inflow sources, and many outflow sinks: none need be permanent. The figure, then, represents a snapshot: another snapshot, taken earlier or later, may look quite different. Despite this, the whole may be dynamically stable. . .

In this turbulent environment, the goal of the enterprise in the first instance is to survive — not, as some would have it, to make a profit. It will prove necessary to make a profit in order to survive, but profitability is a means to survival, not the goal. As with any organism in a hostile, competitive environment, the enterprise needs sustenance. It extracts sustenance from 'flowthrough:' that proportion of the flow of energy, materials and information through the network that can be 'persuaded' to flow through the enterprise. Flowthrough can be processed to create products and services for sale to downstream enterprises and markets: flow-through depends, of course, on downstream wants, needs and ability to pay. Similarly, flow-through depends upon sufficient and appropriate inflow: an enterprise's sources are as important as its markets. (There is an obvious analogy here with anabolism, catabolism and metabolism.)

Figure 18.3 shows a 'homeostatic' view of an (intelligent) enterprise, with resource levels of manpower, machinery, materials, money and information/intelligence (M^4I) being kept in dynamic balance so that together they may support flow-through. Changes in flowthrough rate may be 'followed' by changes in any or all of M^4I. Ideally, as the figure suggests, a steady flowthrough enables M^4I stability, which potentially enhances economy, efficiency and effectiveness. Where flowthrough either dips or rises, there may be a concomitant need for an IE to adjust M^4I inflow/outflow rates to achieve a new point of dynamic balance. Some IEs achieve this adjustment by outsourcing, subcontracting, etc.

Flowthrough can be encouraged by stimulating downstream demand, through publicity, advertising, marketing and selling practices. Creating innovative products and services, ones that do not exist in the marketplace at present, and which afford benefits to the owner, also stimulate demand.

One way to go about innovating is to undertake market analyses, often examining trends, to suggest what kind of product would sell well. An alternative approach, favored by some Japanese entrepreneurs, is to make a variety of innovative products 'on spec,' not knowing whether they will sell well or not, and finding out by putting them on the market. Those that do not sell are discontinued, while those that do sell spark off new products ranges.

So, is it intelligent to analyze the market and to predict what will sell? Or, is it intelligent to 'throw' a variety of innovative products at the market and 'see which ones stick?' This 'suck it and

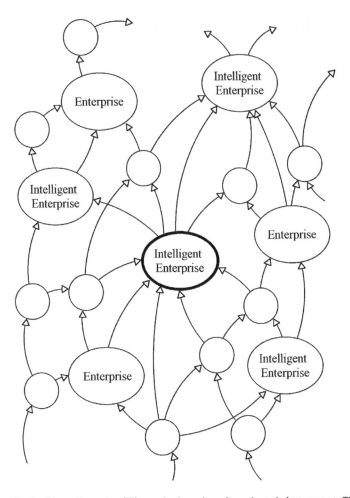

Figure 18.2 The Intelligent Enterprise (IE) as a business in a dynamic, turbulent context. The IE will be one of many enterprises, operating in an extensive, multi-dimensional market economy. Its primary goal is to survive. Prospects for survival are enhanced by establishing, sustaining and enhancing 'flowthrough:' the proportion of material, energy and information flowing through the net that passes through the IE. The IE flourishes if it able to process flowthrough to produce goods and services that downstream enterprises need, want and are willing and able to pay for. Intelligence can be seen as manipulating the network in such a way as to sustain and enhance flowthrough as opportunities and threats appear and develop in the network, and as variations occur in the total flowthrough.

see' approach has revealed holes in the ability to analyze the market and to predict what innovative products will be needed and will sell. 'Suck it and see' has come up with many products that have appealed to the public, but which sober analysis would have rejected out of hand as unworthy of investment. On the other hand, not to analyze the market would seem foolish. . . so, perhaps taking both approaches in parallel might be the intelligent thing to do!

Making intelligent choices — intelligent enterprise model

It may be possible to create a model of an intelligent enterprise — see Figure 18.4. The model is based on the layered Generic Reference Model (see The Generic Reference Model (GRM) Concept on page 125), but with specific features highlighted:

- Mission Management shows Intel being fed from sensors to enable Situation Assessment. Good Intel may allow anticipation of threats and exploitation of opportunities.
- Mission Management shows two different kinds of decision making: naïve decision-making (methodical/systematic) and Recognition-primed Decision-making (RPD); see Instantiated layered GRM on page 138.
- Mission Management is moderated throughout by Behavior, which is founded on Cognition such that situations are interpreted on the basis of Belief: perceptions of threats, opportunities and decisions are colored by Belief.

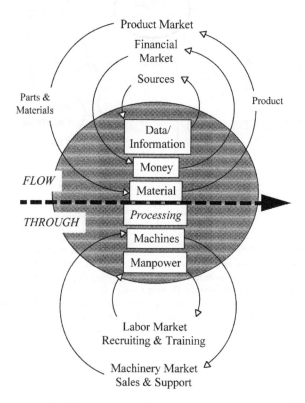

Figure 18.3 Despite existing in a dynamic environment, the intelligent enterprise will seek to maintain homeostasis — local dynamic stability. The figure shows 'homeostatic loops' for M^4I: Men, Machines, Material, Money and Information/Intelligence. So, although there will be a variable turnover of manpower, for instance, recruiting and training rates are adjusted to keep the manpower level and skills as needed to support flow-through.

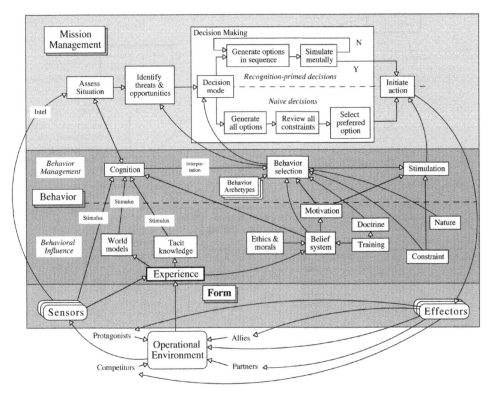

Figure 18.4   Intelligent enterprise: layered generic reference model (LGRM). Mission Management shows alternative models of decision-making: Naïve and Recognition Primed, for methodical and expert decisions respectively. Note how Experience of what has happened in the operational environment can modify World Models, Tacit Knowledge and, importantly, Belief System, which can in turn moderate Cognition, Behavior Selection, Motivation, and Stimulation. This is situational learning — learning what works and what does not — leading to Behavior modification. Operational Environment in this context will depend upon the kind of enterprise. For a commercial business, it would encompass the market, conferences, trade associations, etc., where intelligence might be picked up. For a gang, it might be the street, or a gang hideout.

■   The whole appears to represent a control loop, with Mission Management decisions being made about what actions to take, the actions occurring, and sensors feeding the results of the actions back to Mission Management. However, competitors, protagonists and even allies and partners may confound this simple control notion by taking contrary actions such that cause does not result in the expected effect.

Innovative decision-making model

A typical decision-making process, outlined in the Intelligent Enterprise LGRM, is shown in more detail in Figure 18.5. A situation, indicated at the top of the diagram, generates complex input stimuli to those seeking to understand the situation and make decisions relating to it.

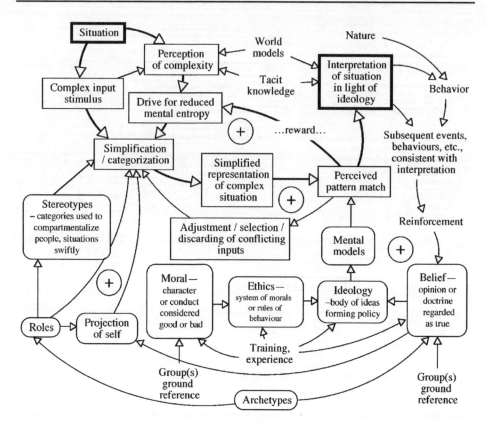

Figure 18.5 Making innovative decisions — process model for continual situation assessment and decision-making. N.B. Behavior occurs as a result of making decisions.

People categorize entities and features perceived in a situation as a means of reducing and managing the otherwise overwhelming degree of disorder/perceived complexity: this allows them to develop a simplified representation of the complex situation, one that is simple enough to accommodate mentally. In driving towards this simplified representation, they may adjust, select, or discard conflicting pieces of information. They employ stereotypes of people, places and 'things,' perhaps subconsciously, to categorize and to reduce mental entropy. This may be unavoidable: humans may have developed this capability as a means of survival, where it was vital to identify friend from foe, safety from danger, etc., and where instantaneous reaction meant the difference between life or death.

Successive representations of the situation are compared with stored mental models to find a fit that is deemed 'close enough,' i.e., one that reduces mental entropy — giving that familiar sense of the 'penny dropping.' As the process model shows, stored mental models are largely founded in belief systems, which have been developed over time through training, experience, and through individual psyches.

A perceived pattern match will provide an interpretation of the situation based on belief, ideology, training, etc. Using this interpretation of the real world situation, people will make decisions and take actions (Behavior in the figure) that, if they turn out as expected, will reinforce their belief system which underlay the decision. If not, then, they will either put it down to 'bad luck,' or they might, just might, question the basis for their judgment. Rational decision-making may not be quite as rational as we might like to think. . . and one person's rationale may be quite different to another's.

If this model of decision-making is accepted, then where does that leave innovative decision-making? Interestingly, it suggests that innovation is less likely to emerge from a group of people who share the same culture, beliefs and training, project similar mental models, and possess similar world models and tacit knowledge — all of which are functions of nurture. Variety may be the basis for innovative decisions; variety in terms of culture, education, training, experience, character, personality, etc., brought together in an unconstrained, accommodating and supportive environment.

Learning and Intelligent Behavior

The foregoing sections have suggested that intelligent enterprises (IEs) and learning organizations may not be the same thing. A graphic comparison between the two concepts is drawn in Figure 18.6. The two concepts are represented left and right respectively. Their common goal, survival, is shown in the center in the context of a dynamic environment.

At left, the proactive IE is seen as gathering intelligence about markets, suppliers, competitors, allies, with a view to anticipating some impending change, opportunity or threat. Perceiving some future change stimulates the IE to adapt in some way so as to be able to take advantage of the opportunity as soon as it arises, or even, perhaps, to assist in its arising. . . . The IE is hampered by system inertia, the resistance to change that systems/organizations all exhibit to a greater or lesser degree. So, the IE invests in changing the enterprise, allowing for the time it may take to effect the necessary change. In the ideal world, the IE has, then, adapted to the impending change just as it arises, enabling the IE to capitalize on the new opportunity for flow-through.

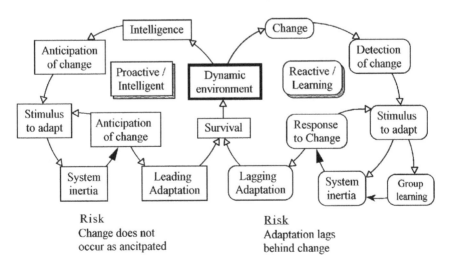

Figure 18.6 Proactive intelligent behavior vs reactive learning behavior.

As the figure suggests, there is some degree of exposure in this. The greater the period of anticipation, the greater the exposure: the greater the corporate inertia, the longer to achieve the necessary change, and, again, the greater the exposure. However, the potential rewards of being first are enticing: first in the market often dominates that market.

At right, the learning organization is reactive: observing and detecting changes stimulates it to adapt. The learning organization has inertia, too, but the process of learning may reduce this such that all the players are 'singing from the same hymn sheet,' all are motivated, all pursue the same — or at least consistent — goals. Inevitably, the learning organization's adaptation to change will lag behind that change, but the learning process will minimize that lag, and hence reduce the risk that adaptation comes too late to save the organization. Essentially, the learning organization invests up front in its people to reduce inertia and resistance to change; instead, they may learn to embrace change.

Should the concepts combine: should an IE also be a learning organization? If the analysis above is valid, then the issue is one of speed of adaptation to change, with corporate inertia being the crucial factor. There may be several ways to address corporate inertia other than group learning. Group inertia rises with the size of the group. (Informal experiments suggest that the time taken to reach a consensus rises broadly as the square of the number of participants — based on using Warfield's interactive management and ISM to develop consensus within different sized groups of military officers, across a range of subjects.)

This suggests that one strategy might be to organize in many small groups, rather than few large groups, although that then raises the issue of coordinating the many small groups. Another strategy employs competition between teams, coupled with immediate reward for success. Competition motivates and creates its own dynamic.

A third, intelligence-led strategy is to establish a separate small team to find and address prospective opportunities. If the search team becomes the nucleus of a group dedicated to seizing the new opportunity, then inertia is principally associated with recruiting additional team members. This inertia may be limited, since each new member is imprinted, as he or she joins, with the culture already existing in the team. This third strategy is similar to earlier examples of institutional intelligence, rather than of intelligent individuals, per se.

Keeping the Enterprise Intelligent

There is a tendency for any enterprise to mature and become set in its ways — see the Social Genotype on page 50. In many ways this is a good thing, indicating as it does that social bonds have been formed, that routines have developed, processes established, etc. It also suggests that groups have formed (nascent organizations often lack any formal group structure), and that group barriers are likely to be rising.

Group barriers, which impede group-to-group interaction and cooperation, are a potential source of social inertia within an enterprise. However, groups will form naturally, if not by design, probably because we humans are anthropologically more comfortable working in family-size groups; we will form such groups willy-nilly, at parties, dances, meetings, etc. Teams of people greater than about seven and less than about fourteen in number will divide into two groups, perhaps three. Psychologists suggest that there are different conceptual group sizes: 'natural' group size populations might be of some 50 people, and there is another natural group size of about 300 people, or so. These may have to do with instincts related to nuclear family, extended family and tribal populations. . . .

Be that as it may, groups proffer psychological comfort of belonging, and they develop their own in-group cultures — all of which can lead to social cohesion and social inertia — two sides, perhaps, of the same coin. The way in which an intelligent enterprise is organized into groups is of importance in this context, since social cohesion is desirable for cooperative and coordinated behavior, while social inertia resists change.

Continual redesign — reprise

One approach to keeping the enterprise intelligent is continual redesign — see page 436. Sensibly, for the IE, this would be supported by a dynamic simulation of the environment, of interacting systems, of markets, etc., essentially of Figures 18.2 and 18.3. The usual approach (see The GRM and the Systems Approach on page 135) would be augmented by the inclusion of markets, sources, sinks and many competitors, to create the necessary complex dynamic simulation – see Evolving Adaptive Systems on page 25; see also Create an intelligent, auto-adaptive, evolving solution system? on page 179.

The simulation would be of the behavior of each of the many interacting entities: developing such a simulation is not as daunting as it might appear, since the behavior of an enterprise is less diverse, more predictable and much slower than, say, of individuals comprising that enterprise. For many enterprises competing and cooperating in a complex market simulation, it would be necessary only to represent their emergent behavior. . . for the IE, however, more detail would be appropriate, as indicated in Figure 18.4. Additionally, the IE simulacrum would show interacting functional groups and would represent their dynamics, synergies, homeostatic influences, etc.

Such a simulation would enable the IE to experiment with different organizational configurations, pose situations and different strategies to exploit them, all without risk. Of course, the simulation could not predict the future either, but it would be possible to posit a variety of futures, and hence to make a well-educated guess (SWAG) at the best course of action. (In marketing jargon, SWAG means a scientific wild-ass guess — a deliberate oxymoron to indicate limited credibility: cf., WAG, which has no credibility.)

Summary

The concept and identification of intelligent behavior is considered, in both the natural and manmade systems worlds. The relationship, if any, between intelligence and survival is also considered, where survival is seen as the ultimate goal of both organism and organization. No firm connection is identified: there are examples of intelligence aiding survival, but then again there are examples of intelligence defeating itself.

For IEs, intelligence is seen as either an emergent property of the whole, or as a property of people who form (part of) the enterprise, and who condition its behavior — making intelligence again an emergent property. It may be that some organizations and enterprises may behave intelligently by virtue of their adaptable configurations.

An attempt is made to define intelligence, but it seems to mean different things to different people. Similarly, making 'good decisions' means different things to different people For some, a good decision is one with a favorable outcome: for others, a good decision is one that was reached carefully and comprehensively, taking all relevant factors into account. One characteristic of a good decision is that it inspires confidence in others.

Decision-making is explored, both naïve and recognition-primed decision-making: they are seen as alternatives rather than complementary, with RPD being employed by experts of long-standing, such that they believe their decisions to be 'instinctive.'

A distinction is drawn between learning organizations and intelligent enterprises. The learning organization is seen as reacting to changes in its environment, with the speed of response being enhanced by group and corporate learning. In contrast, intelligent enterprises may be proactive, seeking to look ahead for opportunities and threats, and to anticipate them by preparing and configuring itself in advance. If the intelligence is accurate, the intelligent enterprise is potentially able to avoid delay and exploit the opportunity/avoid the threat directly. By anticipating events, situations or changes, the IE may increase its exposure. Taking the risk may reap the benefit; on the other hand, it may not.

Ways in which an IE might be designed and structured are considered, together with how such organizations can be kept intelligent, in view of the general tendencies of many enterprises to mature to the point of petrifaction. Since intelligent behavior consists, not only of making good decisions, but also of acting swiftly upon those decisions, if follows that system inertia should be minimized. The learning organization achieves this through group learning, developing shared motivation and consistent goals. Another factor seems to be group size within the enterprise, where smaller groups can be more cohesive and can reach consensus views and act more quickly. On the other hand, reorganization of an existing enterprise into small groups implies more groups, all of which may then need to be coordinated. It is likely that there is an optimum group size, between the large and few, and the small and many. . . that favors best motivation, cooperation, coordination and synergy.

Assignment

You are the project systems architect for a human-like robot that is being developed for prospective use as a Peace Officer — a policeman whose role is the maintenance of peace, as opposed to crime fighting — see Case E: The Police Command and Control System on page 405, for background. You are required to endow the robot with 'intelligence,' although the 'specification of concept' is far from clear about what robot intelligence is supposed to mean.

1.	Consider and outline the emergent properties, capabilities and behaviors of the proposed police robot that you would consider identified it as behaving intelligently. You should consider not only one robot Peace Officer operating, but also squads of Police Officers, some robot, some human, some mixed robot and human.

Police operating together as a team in potentially hostile conditions would constitute an intelligent enterprise. Confrontational policing has developed a standard tactic in recent years of 'tit-for-tat.' In other words, responses in crowd situations are always proportionate, never 'over the top (OTT), and hopefully never 'undercooked,' either: OTT draws others into the conflict in support of the underdog; undercooked risks the police being outnumbered and overpowered.

2.	How, if at all, would you consider that the police might employ the intelligent police robot, supposing it to have been successfully developed, tested and employed as a Peace Officer, in a tit-for-tat, many-on-many conflict situation. . . (e.g. a riot at a football game, or similar). You should consider the psychological impact of the robot on the crowd, as well as on associated human policemen.

3.   Can you envisage a situation where such an intelligent, robotic Peace Officer might be equipped with: (a) nonlethal weapons to temporarily incapacitate miscreants; or (b) lethal weapons for robot self-defense; or (c) lethal weapons for the defense of human police, other robotic police, or innocent people caught up in a situation? Discuss the pros and cons in each case. Justify your viewpoints.

Case H: Global Warming, Climate Change and Energy

Global warming and global energy supplies are subjects of such complexity and importance that to consider the following as a case study would be inappropriate. Instead, it may be more apposite to view it as an essay prompted by despair at our human inability to look problems squarely in the face.

Energy, Demands, Resources and Reserves

Energy demands in an industrialized society rise inexorably. Societies that meet energy demands from fossil reserves must inevitably consume those reserves. Since demand is continually accelerating, the time horizon at which reserves run out is continually foreshortening. Discovery of more fossil reserves merely postpones the evil day.

Burning fossil reserves pollutes the atmosphere. The fossil reserves stored up carbon as hydrocarbons, carbonates, carbohydrates, etc., during their periods of growth as living flora and fauna millions of years ago: burning the fossil fuels releases carbon back into the atmosphere as carbon dioxide (CO_2), methane (CH_4), and other gases. Some CO_2 is reabsorbed by photosynthesis into present day flora, but not enough to prevent an abundance of CO_2 in the atmosphere from burning fossil fuels, which deposited their carbon content progressively over hundreds of millions of years, only to have it released in hundreds of years.

Rising CO_2 levels have been observed over the last 5000 years. It may be that the levels have risen partly as a result of widespread agriculture, since that also started some 5000 year ago. The rate of rising has accelerated alarmingly in recent years, suggesting that natural mechanisms for reabsorbing CO_2 are being swamped.

Systems Engineering: A 21st Century Systems Methodology   Derek K. Hitchins
© 2007 John Wiley & Sons, Ltd

Global Warming and Climate Change

The mean temperature of Earth appears to be rising. This is not unusual: on the contrary, it fluctuates continually, due to a number of different effects, including predictable variation in the Earth's relationship with the Sun's radiation. The eccentricity, axial tilt, and precession of the Earth's orbit vary. The Earth's axis completes one full cycle of precession approximately every 26 000 years. At the same time, the elliptical orbit rotates, more slowly, leading to a 22 000-year cycle in the equinoxes. In addition, the angle between Earth's rotational axis and the normal to the plane of its orbit changes from 21.5 to 24.5° and back again on a 41 000-year cycle.

Moreover, there is reason to believe that, like other stars in its class, the Sun's output is variable, sometimes increasing and sometimes decreasing. Indicators of this variability are not well understood, but may be associated with sunspot activity. The solar system also passes from time to time through interstellar dust clouds — spirals of dust from the center of our galaxy — that can reduce the solar radiation reaching the Earth.

Taking the predictable variations, the less well-understood variability in the Sun's output and the interstellar dust clouds into consideration should lead us to the conclusion that we do not really know if the Earth is heating up through insolation, or for some other reason. Indeed, for all we know, the Earth may be on the cusp of a period of Earth cooling. Or so various climatologists were agreeing some 20 years ago.

Of course, this does not take account of the so-called greenhouse effect, in which re-radiation from the Earth's surface is trapped inside the Earth's atmosphere. As every schoolchild will be aware, CO_2 is a so-called 'greenhouse gas,' and has been identified as one of the culprits in global warming.

Latest estimates suggest that the mean temperature of the globe will rise by some 3 °C over the next 100 years. Some climatologists insist that it will be much more — perhaps 5 °C, or even 7 °C. Ice caps will melt, and the climate will change: just how it will change is difficult to predict.

Climatologists have created sophisticated models that purport to show the likely outcomes, none of which looks comforting. But even without such plausible simulations, it seems reasonable to suppose that global warming will put more energy into the global weather systems, i.e., will 'drive' an already chaotic system. In general terms, the results of that are likely to be greater extremes of weather: fiercer storms; longer periods of calm; torrential rains; severe droughts; areas of high temperature; areas of very low temperatures; rising sea levels; flooding; submerging of low-lying islands; etc., etc. But which of these is going to happen where and when is more difficult: global weather is a complex, nonlinear chaotic system of systems — see Generating Chaos on page 35.

Choices

Meanwhile, armed with forecasts of this doomsday scenario, what can we do? There seem to be at least three approaches:

1. Try to prevent the continuing rise in atmospheric CO_2 levels with a view to curbing the greenhouse effect
2. Take measures to ameliorate the effects of climate change: e.g., build flood defenses, move populations to higher ground . . . .
3. Manage the climate and limit global warming — or cooling.

Controlling rising greenhouse gas levels

Concerned politicians in some countries are trying to rally international support for a concerted, international effort at curbing emissions of carbon dioxide, methane, nitrous oxide, sulfur hexafluoride, hydroflourocarbons, and perfluorocarbons. The Kyoto Accord was the first such effort, at which participating nations undertook to meet emission reduction targets. Many signatory nations are trying to meet agreed targets, reducing domestic energy consumption, reducing fossil fuel dependency, introducing alternative energy sources, etc.

Not every nation participated: notable exceptions were Australia and the US. Other countries, including India and China, ratified the Accord but were not required to cut their emissions because of their status as developing nations: it was agreed that developed nations had emitted most greenhouse gases, and that developing nations, with relatively low emissions per capita, should not be unfairly inhibited. With India and China being the most populated nations on Earth, they may soon overtake the USA, currently the world's biggest emitter. China is currently building a coal-fired power plant every week, and intends doing so for years to come. Some predict that China will overtake the US in 2–3 years as the principal emitter of greenhouse gases. China points out that their emissions *per capita* are still relatively low on account of their large population....

Which brings to mind the Tragedy of the Commons (TOTC) (Hardin, 1968). Originally believed to have occurred in medieval times near Oxford in England, the Tragedy of the Commons came about through herdsmen sharing common land upon which they grazed their cattle. They could either graze on private land, for a price, or on common land, for nothing. Individual farmers found it beneficial to graze more of their stock on common land, free of charge: it was also beneficial to increase stock numbers since, with free grazing, there was more profit to be gained. Since all farmers, as individuals, came to the same conclusion, the common land became overgrazed and useless, leaving cattle to die. Hence, the tragedy: by acting separately, the herdsmen ruined each other and themselves.

TOTC applies to pollution, in a reverse sense. If several parties pollute their common environment, and if it would cost each of them to clean up waste rather than pollute, then it is not in their individual interest to clean up. Hence, none will clear up — unless, of course, some binding agreement can be reached in which all parties clean up. And, it has to be *all* parties, since any party left outside the agreement will see advantage — reduced cost — in not cleaning up their waste — and so will in effect have a polluter's license, granted by those who cleaned up their respective acts.

So, in greenhouse gas emissions, do we see an unfolding Tragedy of the Commons? It seems likely. In spite of Herculean efforts by the United Nations Framework Convention on Climate Change, and the setting of specific targets for signatories to the Kyoto Protocol, greenhouse gases are still rising alarmingly, the output from developing nations is rising dramatically, and developed nations with the best of intentions are experiencing difficulties in reaching their agreed targets.

Even supposing that further increases in atmospheric CO_2 levels were curbed over the next decade or two, it is possible that the damage has already been done. Kyoto and its possible successors may be too little, too late.

Measures to ameliorate the effects of climate change

Not everyone accepts that global warming is caused by human agricultural and industrial activity. However, to some extent disagreements over the causes are irrelevant: the symptoms of climate change are presenting around the globe. Not sufficiently, perhaps, to cause wide-scale panic, but noticeably nonetheless. Ice caps are receding. Ocean currents are changing temperature and salinity: fish stocks are following the changing patterns. On average, winters are milder. Spring comes a little earlier; flowers may bloom out of season. Birds are laying eggs earlier. Severe storms may occur more frequently; so may severe droughts.

Accepting, for the moment at least, that climate change is inevitable: what can be done to protect the environment, the flora, fauna and, in particular, humanity? The answer seems to be: precious little.

- As climate shifts, the changes will affect food production: occasionally change may be beneficial; generally, it will mean that plants that grew well before the change will need to be replaced with different varieties or even different species altogether.
- As food production is disrupted, even if temporarily as old crops are replaced by new, major food shortages are predictable.
- Current practices in which developed countries import low-cost foods over vast distances may have to change, with a return to more homegrown foodstuffs.
- Climate change will be accompanied by local variations in rainfall, as well as temperature, so improved means will be needed for collecting and storing fresh water — already a commodity in short supply around the world.
- Plants for making fresh water from seawater by reverse osmosis may become more popular, increasing energy demand
- Domestic energy demand is likely to rise as people depend on cooling and heating systems to combat both rising and falling temperatures
- Mass migration may occur as people find their littoral environments threatened. Countries with Mediterranean coastlines would be greatly affected, for instance. Greece, with 160 permanently inhabited islands, and hundreds more uninhabited islands, would be particularly affected: Indonesia even more so.
- Major flood defenses would be raised around the world as people resisted rising waters. Major disasters would be inevitable as floodwaters found their way around, underneath, or over the top of such defenses

Climate change may also affect the viability of a number of so-called alternative energy sources, reliance on which is presently the cornerstone of some national strategies....

Rational and Irrational Alternative Energy Sources

Much reliance is being placed in some quarters on alternative energy sources, where 'alternative' implies alternative to fossil fuel burning — hence 'green.' Hydroelectric schemes have a history as alternative sources of energy, using as they do the Earth's gravitation attraction as their driving force. Not all schemes have been entirely without their drawbacks.

Hydroelectricity

The Aswan High Dam in Egypt, for instance, certainly generates a huge amount of electricity, but at the expense of submerging a nation (Nubia) and altering the climate and ecology of central and east Africa. Moreover, its life is limited. The dam traps silt being washed down the Nile which previously would have been carried into southern and northern Egypt and out into the Mediterranean. The dam basin is filling up, not with water, but with silt. It will not be many years before the dam is unusable . . . .

Some countries have capitalized on their mountainous terrain by developing hydroelectric schemes without such drawbacks: the Nordic countries are prime examples. However, with the advent of climate change, and the movement of weather systems, will such countries continue to have the abundance of snow and rain which fills the rivers and drives the turbines? It certainly cannot be guaranteed.

Wind power

Wind power is popular with alternative energy buffs: it comes in several forms. Generally, they raise the issue of energy density: energy extracted from wind machines varies as the cube of the wind speed.

This means that windmill generators do best in windy locales — which is where people do not want to live. Electrical energy generated at a distance from its point of application will incur transmission costs and energy losses — as heat. Current estimates of the efficiency of propeller generators sited near centers of population are reputedly in the region of 23%. On the other hand, siting windmills in very windy locales not only distances them from centers of population, but can also result in them being overdriven by very strong winds, and damaged. And, with climate change, future strong winds could be much stronger than at present. Then again, places that are windy now may be becalmed as climate patterns shift . . .

Objections to siting windmill 'farms' inland on aesthetic grounds have resulted in many being sited offshore, in shallow waters. Their durability and reliability in the face of tornados, storm surges and rising tides has yet to be established: dipping blades in the water is, presumably, a showstopper.

Tidal flow

Tidal flow schemes have been in use for some times, and they seem to be successful: suitable sites may prove difficult to locate, however, as a considerable flow of water is needed from which to extract energy — essentially Moon energy. And, of course, any site that is suitable now may not be suitable if and when the waters rise . . . .

A tidal barrier scheme is proposed for the Bristol Channel in England, which is famous as having a very high tidal reach, second in the world only to Newfoundland. A large barrier is proposed across the estuary, with a road running along the top between England and Wales. The barrier would serve as a flood defense, resisting rising tides and storm surges, and would protect a large, highly populated estuary coastal area. Additionally, it would also serve to generate large amounts of electricity through inset turbines. These would generate electricity both on the incoming and outgoing tides: enough, it has been estimated, to meet 6% of the UK's electricity demand. So, combined flood defense, road and power generation: win–win–win. Environmentalists, concerned that the barrier would affect shore flora and fauna, and upriver wetlands, have raised serious objections, however . . . .

Wave power

Wave power seems attractive to some, with the short-term choppy motion of waves providing relatively rapid movements of generating elements, including Salter's Duck — a 300-tonne floating canister designed to drive a generator from the motion of bobbing up and down on waves like a duck. Named for Professor Salter of Edinburgh University, the Duck (project) was killed off in a dispute with proponents of wind power at the British Atomic Energy Authority!

Moon power

Moon power can be extracted in a different way. A large floating platform tethered off shore would rise and fall with the tides: that motion could be used to turn electrical generators, effectively generating electricity from the Moon's gravitational attraction. Rising tides would not affect the capability, and may even enhance it.

If the large floating platform were to double as a floating airfield, for instance, then international flights could set down there and takeoff without polluting the land environment with noise and noxious emissions. Waves could be prevented from breaking over the edge of the platform/airfield by surrounding it with Salter's (dead) ducks, which would take energy out of the waves and prevent them from swamping the platform. The whole could contribute significant amounts of energy to the national grid . . .

Biomass solar power

Biomass sources of alternative energy have a history.

> The fuel of the future is going to come from fruit like that sumac out by the road, or from apples, weeds, sawdust — almost anything. There is fuel in every bit of vegetable matter that can be fermented. There's enough alcohol in an acre of potatoes to drive the machinery necessary to cultivate the field for a hundred years. Henry Ford, 1906.

Prophetic, perhaps, in other ways, Henry Ford got this one wrong, although Cleveland Discol, for instance, a British subsidiary of Standard Oil, promoted alcohol blends from the 1930s to the 1960s for clean, cool and efficient combustion. It was particularly popular with riders of single-engine motorcycles as it made engines run 'smooth and cool.' However, the US oil industry has generally been opposed to synthetic oils, as they were called.

Recent interest in alternatives to imported petroleum products has focused on ethanol production, from corn (maize), sugar cane, etc. Although politicians in the US are pushing aggressively for ethanol from homegrown corn as a substitute for foreign oil, the conversion requires copious amounts of fossil fuels to achieve. Producing ethanol from corn creates almost the same amount of greenhouse gas as gasoline production does. Burning ethanol in vehicles offers little, if any, pollution reduction. (Wald, 2007)

Biomass, essentially solar energy, can be used in a quite different way. It may be practicable to run a power plant on wood grown for the purpose (Odum, 1971). A power plant might be sited at the center of a circular area divided into sectors, and planted with, e.g., fast-growing willow saplings. Each sector is coppiced (cut back) in turn to fuel the power plant. By the time that harvesting has gone right around the area and returned to the first sector, each coppice has grown a fresh stock of wood, such that power generation is continuous.

Such power generation is, of course, carbon neutral: burning the wood releases as much CO_2 as was absorbed into the growing trees from the atmosphere by photosynthesis. The amount of energy generated depends on the area under cultivation and the growth rate of the wood concerned. This is green, low technology solar power. Like other alternative sources, it is subject to the vagaries of sun, wind and weather. In the event that such a system grew large volumes of identical plants, there would also be a risk of systemic disease: crop rotation would be necessary. Large amounts of continuous power would require large estates to be put under cultivation.

Alternative summary

In this brief overview of some of the principal alterative energy sources, none has been found to be ideal on its own: each has drawbacks when climate change is taken into account. Conventionally, however, things are looked at the other way around: if we convert to alternative energy sources, and if we 'go green,' then climate change will not occur.

The cat is out of that bag, however. Climate change has been occurring, and is occurring, even at current gas emission levels — and they are still rising. There is no evidence that an immediate switch to totally green power sources — even if such a switch were feasible — would make a difference.

In the circumstances, it might seem rather late in the day to switch to alternative energy sources, particularly since they seem to be threatened right now by the very climate change they seek to prevent in the future. It might, on the other hand, be socially and psychologically comforting to think that we are 'doing something.'

Nuclear Energy

Nuclear energy is not alternative energy, of course, but it is nonetheless green: very green. The mainstay for energy needs in the future is unlikely to be anything other than nuclear energy: this, despite the risks and costs that undoubtedly exist in creating, running and decommissioning nuclear power plants.

The general public is taught to fear nuclear energy by the media. Much is made of the problems of disposing of highly radioactive waste, yet the total amount of such waste in the UK, for instance, since nuclear energy generation started half a century ago, is less than the volume of the average small cottage!

In practice, there are remarkably few nuclear power plant incidents around the world. France has 59 nuclear plants producing 78% of the entire country's electricity, and is the largest exporter of nuclear electricity in the European Union. France is second in the world (behind the United States) in terms of total nuclear power generation, contributing 15.9% of the world's nuclear electricity. Their CO_2 emissions per capita are correspondingly low.

However, the goal for mass power generation seems to be nuclear fusion, which promises cheap, plentiful energy with no radioactive side effects. Nuclear fission energy is sometimes regarded as a stopgap, waiting for nuclear fusion to emerge from its protracted research and development program. It is taking a long time . . . .

Meantime, the rise in international terrorism has highlighted other risks from nuclear fission power plants, i.e., that they may be attacked, and that fissionable materials may be stolen to make nuclear bombs. It is a risk that we may have to guard against, but there seems to be little sensible alternative to nuclear fission energy in the short to mid term. Besides, there are other ways for terrorists to extract fissionable materials besides stealing them . . . In any event, it hardly seems sensible to address the risk from terrorism by degrading our energy generation capabilities and running down our economies in the process. If there is a threat, then either it must be faced, or civilization will rollover and submit to chaos.

Future Imperfect...

Taking a step back, and looking in the longer term, the current global warming/fossil fuel debacle will, hopefully, appear as a 'blip' on the continuing development of human civilization on Planet Earth. Why? Well, the trading environment as it stands is one of competitive international economies, each trying to expand at a sustainable rate: competition makes for vibrant and energetic interchange between economies.

Sustainable growth has been founded, so far, on relatively stable climates and environmental conditions; and, sustained growth has meant, and will continue to mean, expanding energy demands. Alternative energy sources, unable to fully shoulder the increasing burden of energy supply in developed countries, will provide a reducing fraction of national energy supplies as demands continue to rise.

Faced with the prospects of switching entirely to alternative energy sources, with the concomitant inability to sustain economic growth, most nations are likely to opt for mixed energy supplies: fossil fuels will continue to be used by developing nations in the short to medium term, and with nuclear energy — fission first, then fusion - supplanting fossil fuels in the longer term, as dwindling supplies become unaffordable. (At this point, of not before, global energy conflict — World War III — must be a distinct possibility . . . .)

Despite valiant efforts to curb their growth, greenhouse gas levels will therefore continue to rise — perhaps not so quickly, but that remains to be seen. In any event, global warming will continue and — since humanity is 'forcing' an already-chaotic system — the climate is likely to change both drastically and unpredictably. At some point, politicians will realize that they will have to take radical action if their economies are not to be substantially shattered, and that will mean either:

- a means must be found of preventing more heat energy from entering the atmosphere, such that the earth's mean temperature can stabilize and perhaps drop a notch, or . . .
- An alternative means must be found of removing energy from the atmosphere and similarly curbing temperature rises and restoring climatic status quo.
- At the same time, sustained economic growth will necessitate continuing growth in global energy supplies.

So, a conundrum presents itself: Earth will, as things go, contain too much energy in the atmosphere, earth and seas, yet at the same time it will be demanding more and more energy to sustain its burgeoning human population and their economies. Using that increasing energy to do work will generate heat, which — in the circumstances — will be like throwing gasoline on a burning fire!

Dyson spheres

In the 1950s, mathematician Freeman Dyson observed that every human technological civilization had constantly increased its demand for energy. If human civilization were to survive for long enough, he reasoned, it would eventually reach a point where it required the total output from the Sun as its energy supply.

Dyson reasoned that an advanced civilization would create a system of orbiting structures that would surround their sun, so capturing all its emitted radiation. These orbiting structures would constitute a sphere, centered on their sun. The concept became known as the Dyson sphere. (Dyson, 1959). He went on to reason that the energy receptors in the sphere would re-radiate energy in such a way as to alter the stellar spectrum perceived by others at a distance: human-like entities would probably re-radiate energy predominantly in the infrared. Hence, one way to search for advanced alien civilizations was to look for atypical radiation spectra in space.

Dyson did not concern himself with the physical form that such a sphere might take, nor with issues such as global warming. However, the concept of the Dyson sphere captured the minds of scientists and of the general public at the time. The technological capability for constructing such a sphere was thought to be beyond humanity at the time, but perhaps in the future? And, interestingly, the energy captured by such a sphere would be clean, green, solar fusion energy.

Nuclear winter

Research into the possible effects of global nuclear war in the second half of the 20th Century suggested that one outcome would be a so-called nuclear winter (Turco *et al.*, 1983). This would arise from fallout dust clouds obscuring the Sun for many months, possibly many years, with a consequent cooling of the Earth. Later estimates by the same authors (Turco *et al.*, 1990) suggested temperature drops of several degrees in land temperatures and 2–6 °C drop in ocean temperatures, together with an ozone depletion of 50% leading to 200% increase in solar UV radiation incident on the ground.

Horrifying though the prospect of a global nuclear war might be, nonetheless the studies did show that it was possible to reverse some of the effects of global warming by reducing the amount of solar energy penetrating the atmosphere and reaching the ground.

Volcanic effects on climate

The climatic effects of volcanic eruptions had been apparent long before the possibilities of nuclear fallout on climate had been considered. However research focused on the Mount Pinatubo, Philippines, eruption in June, 1991, indicated that: 'the direct radiative effect of volcanic aerosols causes general stratospheric heating and tropospheric cooling, with a tropospheric warming pattern in the winter.' (Kirchner *et al.*, 1999). Evidently, volcanic eruptions can affect climate in unpredictable ways . . . .

Industrial pollution

There are controversial suggestions that a drought from 1970 to 1985 in the Sahel region of Africa, stretching from Senegal to Ethiopia, may have been 'assisted' by European industrial pollution.

Sulfur dioxide aerosols from coal burning seem to contribute to drought by altering cloud formations. Researchers said the particles remain suspended in the clouds instead of falling as rain, and the heavier clouds reflected more of the sun's energy back to space. The aerosols add to conventional causes of drought, such as the overuse of land and natural atmospheric changes.

Taking these unconfirmed suggestions at face value suggests that it is possible to effectively alter the Earth's albedo at high altitude: clearly, in this instance, that would be a risky thing to do; on the other hand, if it were possible to alter the albedo uniformly, and to a very small degree, the net effect might be simply to reduce the solar energy reaching Earth, and hence reduce global warming.

The suspicion that drought in the Sahel was assisted by industrial pollution as far away as Europe points to the degree of coupling between complex weather systems in the environment.

The Case for Active Climate Control

Foregoing paragraphs have indicated that global warming is probably inevitable: that current attempts to limit the emission of greenhouse gases, presuming them to be the causative warming agents, are unlikely to be successful; that, even if they were successful, it may already be too late; and that the prospects from global warming are potentially damaging to life on the planet.

Current political considerations focus on a short time horizon of less than 100 years. If human civilization is to continue, much longer time horizons must be considered, many thousands of years into the future. From that perspective it is possible to see that energy demands will continue to rise, far beyond the meager capabilities of today's alternative energy sources, even beyond nuclear energy sources. If they do not continue to rise, it will because economies and societies will have failed, and technological civilizations along with them.

So, on the one hand we have too much low-grade energy in the environment while on the other hand, we will demand ever more high-grade energy to sustain our technological, economy-based civilizations. Meanwhile, the basis for economic stability, a stable environment, is threatened by climate change.

If, as seems not unlikely, Man is responsible for global warming and climate change, then Man is surely responsible for doing something to either prevent or ameliorate the damaging prospect. Unpleasant though the conclusion may be, it is nonetheless unavoidable: at some point, sooner rather than later, Homo sapiens will attempt to control the climate, to combat global warming in such a way as to restore the stable environment of the last 5000 years of developing human civilization. Unless, that is, Homo sapiens leaves it too late . . . .

Risk from 'doing too little, too late'

The dangers of attempting to control Earth's climate are evident. But the dangers from not controlling the situation may be even greater.

Graph H.1 shows the varying output from a simple system model that can be set to be near the edge of, but not in, a chaotic state: see page 35. The initial ordered state is evident in the regular behavior between zero and $x = 0.25$ on the graph. The simple model is then stimulated by an impulse at time $x = 0.25$, after which it recovers to its former, ordered state. A second impulse at time $x = 0.75$ drives the system into chaotic behavior again, from which it starts to recover, only to be stimulated once more at time $x = 0.9$ before the system has had sufficient time to 'recover' from the previous impulse. This last impulse drives the system into catastrophic instability, from which there is no recovery.

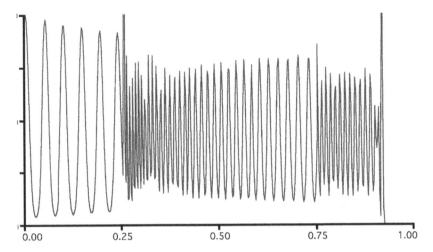

Graph H.1 Chaotic stability. A nonlinear, orderly system is stimulated to behave chaotically at time $x = 0.25$, and $x = 0.75$. The first stimulation promotes the system into a chaotic state, from which it recovers towards its former ordered state. The second stimulation, at time $x = 0.75$, promotes the system into a chaotic state again, from which it starts to recover. However, it is stimulated a third time at time $x = 0.9$, before it has had time to recover. The consequence of this last stimulation, occurring before recovery has taken place, is catastrophic system instability from which there is no recovery . . .

Of course, Graph H.1 does not represent the complex behavior of Earth's climate: but it may be an allegory. The model indicates that chaotically stable systems can be driven too far, beyond a point of no return. Earth's climate is already in a chaotic state: adding energy to the system will undoubtedly make the chaotic behavior more pronounced: but, will it drive the already chaotic system unstable? That is probably impossible to predict; but there is a distinct risk that all life on Earth might be extinguished — Planet Earth could end up like our sister planet, Venus, where:

- the greenhouse effect has resulted in surface temperatures of up to 400 °C;
- water and water vapor are rare due to the high temperature;
- the atmosphere is mostly CO_2 with a small percentage of nitrogen;
- the atmospheric pressure at the surface is 90 times that on Earth; and
- the clouds are composed of high concentrations of sulfuric acid.

Exaggeration? Could not happen here on Earth? Who knows? Predicting the behavior of complex chaotic systems such as Earth's climate system is extremely difficult. And, as Graph H.1 suggests, different stresses on the system may be cumulative . . . .

Remedial Solution Concept

Appreciating the problem

Treating earth as a whole, the problem is seen as one of failing homeostasis — see The Concept of the Open System on page 11, and Equation (1.1) in particular. As an open system, earth receives radiated energy from the sun, on the side of earth facing the sun by day, and earth reradiates (largely infrared) energy into space predominantly on the side facing away from the sun at night. For homeostasis, the energy received should equal the energy re-radiated. Essentially, the earth is less able to reradiate infrared energy because of greenhouse gases. So, there are two potential remedies:

■ Reduce the greenhouse gases, allowing reradiated energy levels to rise back to their former, steady-state levels, or . . .
■ . . . reduce radiated energy received from the Sun.

The first bullet seems to represent an insurmountable obstacle, at least in the short term. Earth is presently having great difficulties in trying to stop the rise in greenhouse gases, let alone reduce the levels significantly.

But what about the second bullet? Is that feasible? If it were, if the energy inflow to earth's systems could be reduced, even by a small amount, and if that reduction were uniform, then the net effect would be to simply 'take the heat out of the system:' literally. In principle, that would not so much change climate, as limit its more extreme manifestations — weather systems would revert slowly to what they had always been over recent millennia. And, if at the same time greenhouse gas emissions and atmospheric levels could be brought under control, and even perhaps reversed; then the status quo would have been restored with minimal risk.

Regulating the solar constant

The solar constant is the solar energy incident on unit area normal to the Sun's rays at the Earth's mean distance per unit time. It is currently some $1366\,J\ m^{-2}\ s^{-1}$, although it varies over the course of a day and is affected by sunspots. The energy received at ground level is affected by latitude. It is largely this energy, with its latitude variation, together with the Earth's rotation, that generates Earth's chaotic weather systems.

One approach to countering global warming would be to cut down the energy reaching earth, i.e., effectively to reduce the solar constant. Even modest reductions, say of 1–2%, might be expected to have a significant effect over time. It would be important to make any such reduction entirely uniform — as would be the case if the sun's output were to reduce slightly. Uniformity would be vital to prevent disproportionate effects upon different areas of the Earth, which would not so much 'take the heat of out of the situation' as perturb the interacting chaotic weather systems, inevitably making things worse — at least locally.

Two complementary concepts come to mind: one that could be implemented in the near term, hopefully before things get out of hand; and a second for the medium to long term. Both involve reducing the solar energy received on earth by a small amount. To appreciate how this might be sensibly achieved, consider the schematic at Figure H.1. The figure shows five Lagrange points, L_1–L_5 that will exist in theory between any two orbiting bodies. In the diagram, assuming that the massive central body is the sun and the small orbiting body is earth, the Lagrange point of immediate interest is L_1. Lagrange point L_1 is located directly between the Earth and the Sun, and at 1 500 000 km (1.5×10^9 m) from Earth, is outside the Moon's orbit (mean distance from earth $= 3.844 \times 10^8$ m).

A small object placed at L_1 would tend to remain in position relative to the other rotating bodies: Although L_1 is closer to the Sun than Earth, and would be expected to orbit more quickly due to increased pull from the Sun, Earth's gravity acts to reduce the effect of the Sun's gravity, thereby reducing the object's orbital velocity to coincide with that of Earth. Hence, L_1, and the other Lagrange points, which similarly balance out opposing centripetal and gravitational forces,

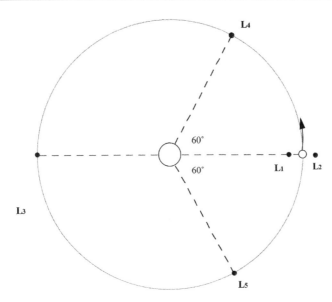

Figure H.1 Lagrange points. Where two orbiting bodies exist, one much more massive than the other, five Lagrange, or libration, points exist. At these points, a small object placed there will feel no net force, and so will remain in position relative to the other bodies. The Earth and Moon generate five Lagrange points: so do the Sun and Earth.

are stationary with respect to each other; i.e., they rotate together with the orbiting body. L_4 and L_5 are sometimes described as 'gravity wells,' for this reason. L_2 is on the other side of Earth from L_1, away from the Sun, at about the same distance from the Earth as L_1. L_3 is opposite Earth, on the other side of the Sun, and is never visible from Earth: for many years, science fiction writers conceived of a ghost planet Earth at L_3. . . .

The Lagrange points are well known to space scientists: The Sun— Earth L_1 is ideal for observing the Sun: neither Earth nor Moon ever occults satellites there. The Solar and Heliospheric Observatory (SOHO) and the Advanced Composition Explorer (ACE) are both in orbits at the L_1 point.

The L_1 cloud concept

It is known that the solar system occasionally passes through galactic spiral arms of interstellar dust, and that these have the effect of reducing the energy received on earth from the Sun. The L_1 cloud concept consists of emulating the effects of the interstellar dust, by injecting a cloud of particles at, or near the L_1 Lagrange point, to scatter a small proportion of solar radiation. The concept is illustrated in Figure H.2.

The figure shows a particle cloud, which has been injected into the path of the Sun's rays just to the Sunward side of the L_1 point. A small proportion of the Sun's radiation is scattered by the particles in the cloud and does not reach Earth: the bulk of the radiation does reach Earth, otherwise unaffected.

As explained above, the L_1 is the balance point between the gravitational attractions of the Sun and Earth and the centripetal force that arises because the earth is orbiting the Sun. The forces at L_1

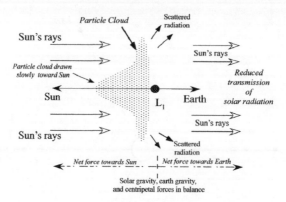

Figure H.2 Particle cloud screen. To reduce solar energy arriving at Earth, a cloud of particles is injected just to the Sunward side of the L_1 Lagrange point, where it is held near and about the Sun–Earth axis. The cloud scatters a small proportion of the Sun's energy, principally infrared emissions, with the bulk passing through unaffected. The cloud disperses gradually under the Sun's gravitational attraction and the solar wind.

are such that a small object to either side of the Sun–Earth line would be moved back towards that line. However, a small object placed nearer to the sun than the L_1 would be attracted differentially toward the Sun. So, the L_1 is stable only in a plane at right angles to the Sun–Earth axis. By placing the cloud just to the sunward side of that axis, the particles will be slowly drawn toward the sun over time, and the cloud will disperse.

L_1 cloud CONOPS

So, the concept of operations would be to position a satellite in halo orbit at the L_1 point: to inject a plume of particles outwards around the halo orbit; to create a disc of particles about the diameter of the Earth just to the Sunward side of the L_1; to observe the effect of the cloud on earth; to observe the dispersion of the cloud toward the Sun; and to re-inject the cloud as needed, using the results of test and observations on earth as a guide to frequency of injection, particle cloud composition, etc.

Further, by carefully choosing the particle size in the cloud, it may be possible to scatter a small proportion of incoming infrared radiation only, leaving light and ultraviolet radiation unaffected: this would involve particles to scatter radiation of wavelength 700 nm – 1mm. Leaving visible light and ultraviolet unchanged would minimize even minor effects on photosynthesis, for instance. And, scattering only incoming infrared radiation would directly balance the adverse effects of greenhouse gases, which inhibit largely outgoing infrared radiation . . .

The L_1 cloud concept is one that could be implemented with today's technology: it could be controlled remotely; injections could be very limited initially, to anticipate any counterintuitive effects; in the scheme of things, it would be inexpensive; and it could be done now.

The L_1 sunshield concept

L_1 would also be an ideal spot to locate a sunshield. L_1 is readily accessible from Earth and Moon, facilitating the construction and subsequent control/adjustment of the device, which would be large in area terms, but made of light, expendable materials. A large sunshield raised over the Earth, between the Earth and the Sun, could reduce the effective solar constant by a small amount across the full cross-sectional area of the Earth — some $127, 400, 000 \, km^2$.

Consider, too, that any proposed solution will be uncertain in its effect. To mitigate risks, a viable solution will be one where any shielding effect can be controlled — reduced, increased, even removed altogether — in the event that the outcome is not as predicted.

Sunshield construction

How would the sunshield be constructed, and from what would it be made?

It is not the intention here to solve all the engineering problems that deploying a shield at the Sun–Earth L_1 would incur. The L_1 is stable only in the plane perpendicular to the line joining Sun and Earth. Moving an object either nearer to the Sun or to the Earth would result in a corresponding increase in the respective gravitational attraction. Nonetheless, stable orbits can be formed around L_1.

Positioning a stationary structure, then, is not going to be simple. Another factor to consider is the solar wind — exposed to the solar wind, a sunshield would be pushed from L_1 towards Earth. On the other hand, it might be possible to take advantage of the solar wind. Major projects have been proposed in which the solar wind is used to sail a satellite in space much as a sailing yacht uses the Earth's winds. The Advanced Composition Explorer provides continual real-time measurements of varying solar wind velocity, temperature, density and pressure at the L_1 point. Pressure, for example is typically about 1nPa (1 nanoPascal), but variable.

Figure H.3 shows an outline schematic of the sunshield in elevation and plan. It is formed of a number of large louver slats, arranged concentrically around a central point L_1. The louver angles may be rotated in such a way that the solar wind (velocity $433 \, km \, s^{-1}$, density 2.6 particles cm^{-3} at the time of writing, but varying) impinges on each louvre and creates a thrust towards the center of the sunshield, which is rotated slowly to maintain the structure's integrity. By adjusting the louvers, the amplitudes and direction of the net thrust can be used to maintain the sunshield at the L_1 position and with its plane normal to the Sun's rays. As with sailing yachts, this might best be achieved using automatic steering to accommodate wind variability (Pressure has dropped 20% to 0.8 nPa in the time taken to write this paragraph.)

The solar wind would push the sunshield towards the Earth. This force towards the Earth may be countered by locating the sunshield on the sunwards side of the L_1, where the Sun's gravitational attraction is greater than that of the Earth, It may be possible for auto-navigation, using the louvres, to maintain the sunshield in this dynamically stable position, effectively 'resting' on the solar wind.

Louvre material will be an important consideration. The panels have to be

- thin enough to filter only a small proportion of the incident radiation;
- light enough to be easily transported and manipulated;
- firm enough to be rotated and to withstand the solar wind pressure;
- ideally, self-repairing to accommodate cosmic particles which will tear holes in them; and
- ideally able to conserve the filtered proportion of solar energy.

N.B. Although louvers are shown in the conceptual diagrams, the panels could be formed in many different ways, including geodetic panels for example. They could be hollow, and contain a working fluid or gas.

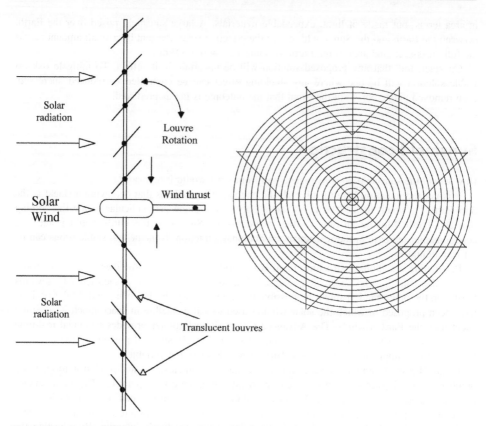

Figure H.3 Sunshield concept. At right is a face-on view of the shield, showing a web-like structure supporting concentric 'louvre' panels of transparent material. The sunshield would be rotated around the central axis. At left, the structure can be seen side on. Louvres overlap, so that all radiation can be filtered in the fully closed position, or pass through unfiltered in the open position. Louvres are angled so that thrust from solar wind is directed towards the center of the web: louvre angles may be altered automatically to maintain the structure near the L_1 point, as with automatic steering on a sailing yacht, and may also be angled to rotate the whole sunshield structure. Locating the sunshield slightly nearer to the sun than the L_1 point will counter the axial thrust from the solar wind, and balance out solar tidal drag.

The last bullet is, of course, a tribute to the Dyson sphere: although different in form, and on a much smaller scale than Dyson envisaged, the sunshield could, in principle, gather a considerable amount of energy. And similar, even larger structures could be sited at L_4 and L_5 specifically for capturing solar energy and converting it for use on Earth, or on the Moon. They would thus have no direct impact on climate, but would affect it indirectly by reducing the generation of energy on the ground, with its concomitant heat wastage.

Solar energy arriving at the earth amounts to some 1.74×10^{17} J s^{-1} (i.e., solar constant multiplied by the cross-sectional area of the earth.) If only 1% of that energy were filtered off by the sunshield, that would amount to 1.74×10^{16} W. It would be spread over a vast web of louver panels, so that power transmission might incur losses, suggesting a case for superconducting elements in the sunshield structure.

If it were possible to focus the captured energy, it might be transmitted separately to Earth or Moon, providing a much-needed alternative to dwindling fossil fuels, alternative energy sources and even, perhaps, nuclear fusion plants.

L₁ sunshield concept of operations

The construction of the sunshield lends itself to flexible operational introduction and control. The complete structure could be constructed and assembled in space, and put into position at the L₁ point. Automatic navigation using the louvers as sails may render orbiting unnecessary. In any event, the structure would be set up with the louvers 'feathered,' i.e., edge into wind so that there would be no solar filtering, no energy generation, and no effect on the earth's solar constant.

Testing would then commence to establish control of the structure to establish the reliability of the auto-navigation system, and to test the physical stability of the structure in the face of varying solar winds, solar sunspot perturbations, etc.

Operational tests would then commence with the louvers ('sails') being closed for short periods, sufficient to observe the impact on Earth, in terms of incident energy, ground, sea and air temperatures, and counterintuitive climatic reactions. Assuming tests to be satisfactory, the sails could be 'unfurled' for longer periods, with further tests, until finally — if all went according to plan — the sails could be left open, navigation could be set to automatic, and the whole would be subject to continual maintenance.

If at any time there was a suspicion of undesirable longer-term effects, then the sails could be partially or fully closed, so regulating effective solar constant, earth temperature and inhibiting climate change — but, essentially, without trying to control climate directly. As with the L₁ cloud concept, the sunshield could be designed to filter a proportion of infrared radiation only, if greenhouse gasses were still an issue on earth.

For the future, it might be advantageous to have a second shield positioned, but at L₂. Looking ahead to future times when greenhouse gases are reduced, and the possibility of a returning ice age arise, a shield at L₂ could be used to moderate the loss of infrared energy from the Earth at night, so providing a counterbalance to the sunshield at L₁.

Comparative timescales

Unlike the L₁ cloud, the L₁ sunshield is presently beyond our technological capability — not by much perhaps, but we are unlikely to be able to mount such a project in less than a few decades — which may be too late. The two concepts are seen as complementary: the L₁ cloud, which we could tackle now, with today's technology and know-how, would buy earth enough time to get its act together, permit undeveloped nations to develop, exhaust our fossil fuel reserves, and get our atmosphere back on track with minimal greenhouse gases.

The L₁ cloud is not a long-term solution, however: the L₁ sunshield would be more appropriate for the mid-to-long term and it would take some considerable time to build, test and put into operation: even supposing we had the technology. However, were we to undertake a full feasibility and systems design study, it would point to the necessary goals for future research. All of which would seem to be in the best interests of the planet, the environment, the flora, fauna — oh; and yes, humanity, too!

Risks

Of course, there are risks: risks are unavoidable; both concepts address risk. The L_1 cloud can be injected progressively, starting with a very 'thin' cloud to measure the effect. This can then be allowed to disperse, both to observe any after effects on earth and to observe the pattern of dispersion, which could be complex in view of the solar wind, which will flow against the solar gravitational 'tide.' If things go according to plan, a slightly denser cloud can be injected, and so on. If, on the other hand, things do not go according to plan, then there will be time to revise the plan and start again.

In any event, the L_1 cloud concept involves making only a slight change in the effective solar constant, and then only in the infrared region of the solar spectrum: although sensitive instruments may detect the effect on earth immediately, it would be decades before enough heat had been 'taken out of the system' to counteract the effects of global warming. Only when extreme weather events became rarer would the impact be realized.

The L_1 sunshield similarly addresses risk, particularly in terms of impact on earth environment and climate. The effects would be progressive, with the rate of energy flow reduction being very small. The L_1 sunshield is technologically challenging, however, particularly if we wish to 'harvest' the filtered solar energy for use on the moon or earth.

Critique

The case for doing something to combat global warming, whether it be manmade or not, is mounting. Just what can be done to combat climate change is not really clear — the climate is, after all, a vast system of many, close-coupled, opens systems interacting chaotically. Any direct attempt to control climate in any particular region is unlikely to work. On the contrary, it is likely to make matters worse: if not in the region of interest, then counterintuitively in other regions.

Reducing the effective solar constant, as proposed above, is something to be concerned about. The solar constant appears to have varied naturally in the past, but then, we have had several ice ages in the past. Could we jump out of the frying pan into the refrigerator? It has to a possibility. However, if the concept worked, it would effectively take the heat, gently, smoothly and uniformly out of the chaotic climate, with a view to restoring the chaotic climate to its more moderate self — still chaotic, but without some of the recent, more frequent and forecast extremes.

Is the risk of reducing the effective solar constant worth taking? That is difficult. But it has to be weighed against a host of other risks, and possibilities. Could the planet be on the cusp of a new ice age, as some climatologists argue: if so, could global warming actually be deferring a serious problem?

What is the probability that the politicians will really get together and reduce global greenhouse gas emissions? Is the Tragedy of the Commons scenario on the cards? To believe that the politicians will succeed, and that none of the big players will opt out, in their own, selfish national interests, is to ignore both history and the current evidence.

Supposing that greenhouse gas emissions are eventually brought under control: will their control solve the problem? We are dealing with systems with enormously long time constants, sometimes in the millions of years. There has to be a significant prospect that immediate control of greenhouse gases would be accompanied by an overrun, such that the effect of the high gas levels took some time to wear off, just as they took some 5000 years to build.

Many of these factors are imponderables. Global warming is upon us, climate change is happening, and we seem, as an uncoordinated, antagonistic planet, helpless to do much about it.

Reducing the effective solar constant is doing something. The concept would, of course, create widespread opposition, even supposing the physics to be sound, the systems science to be solid, the engineering to be feasible, and the climatological impact to be as expected. Is it the right thing to do? It is a moral dilemma.

Consider as a comparison: is it right, having detecting a meteor on course to strike the Earth, to try to divert it, break it up, etc? Most people would say yes, although the outcome from attempts to divert such a large body, with so much momentum, is somewhat unpredictable. We might create a rain of intermediate-size meteors, which would hit the Earth, for instance. Should we still try? Most people would still say yes. So, perhaps we should try to save our planet. Else, we may behave like the boiled frog, which sits in water as it is gently warmed on the stove, unable to sense the slow rate of heating, until finally the frog expires — boiled alive without moving a muscle.

References

Ackoff, R.L. (1981) *Creating the Corporate Future*, Wiley, New York.

Ashby, W.R. (1964) in Mesarovic, Mihajlo D. (ed.), *Views on General Systems Theory*, Wiley, New York.

Bak, P & Chen, K. (1991) Self-Organized Criticality, *Scientific American*, **264**(1).

Beers, S. (1972) *Brain of the Firm*; Allen Lane, The Penguin Press, London.

Bertalanffy, L. (1950) The Theory of Open Systems in Physics and Biology, Science **III**.

Bertalanffy, L. (1968) *General Systems Theory*, New York, NY: George Braziller Inc., pp. 140–141.

Briggs, I.M. (1990) *Introduction to Type*, Consulting Psychologists Press, Palo Alto, CA.

Checkland, P.B. (1972) *Towards a System Based Method for Real World Problem Solving, Systems Behaviour*, Harper.

Checkland, P.B. (1981) *Systems Thinking, Systems Practice*, Wiley, Chichester.

Dyson, F. (1959) Search for Artificial Stellar Sources of Infra-Red Radiation, *Science*.

Forrester J.W. (1971) *Counterintuitive Behavior of Social Systems*, Cambridge Press, MA.

Espejo, R. & Harnden, R. Eds. (1989) *The Viable Systems Model — Interpretations and Applications of Stafford Beers' VSM*, Wiley, Chichester.

Global Security Organization (2006), http://www.globalsecurity.org/military/systems/ground/fcs.htm

Jones, J. and Wilson W. (1987), *An Incomplete Education*, Unwin Hyman Ltd., London.

Jung, C.G., (1917) *On the Psychology of the Unconscious*, in Joseph Campbell (ed.), *The Portable Jung*, Penguin.

Hall, A.D. III (1989) *Metasystems Methodology*, Pergamon Press, Oxford, England.

Hardin, G. (1968) The Tragedy of the Commons, *Science* **162**.

Haywood, J. (2005) *Historical Atlas of Ancient Civilizations*, Penguin Books, London.

Hitchins, D. K. (1987) MOSAIC — Future Deployment of Air Power in European NATO: *Advances in Command, Control and Communications Systems*, Peter Perigrinus, London.

Hitchins, D.K. (1992) *Putting Systems to Work*, Wiley, Chichester.

Hitchins, D.K. (2000) Getting to Grips with Complexity, http://www.hitchins.net/ASE_Book.html.

Hitchins, D.K. (2003) *Advanced Systems Thinking, Engineering and Management*, Artech House, Boston MA.

Institute for Defense Analysis (2003) A Transformational Approach to Capabilities-Based Planning, http://www.dtic.mil/jointvision/ideas_concepts/ujtl_cap.ppt.

INCOSE (2006) A Consensus of the INCOSE Fellows, http://www.incose.org/practice/fellowsconsensus.aspx.

Janes. F.R. (1988) Interpretive Structural Modeling: a Methodology for Structuring Complex Issues, *Trans. Inst. Measurement and Control*, **10**(3), London.

Jones, J. and Wilson W. (1987) *An Incomplete Education*, Unwin Hyman, London.

Kast, F.E., and Rosenzweig, J.E. (1970) *Organization and Management: a Systems Approach*, McGraw-Hill.

Katz, D. and Kahn, R.L. (1966) *The Social Psychology of Organizations*, Wiley, New York.

Kelling, G.L. and Coles, C.M. (1996) *Fixing Broken Windows*, Free Press, New York.

Kirchner, I., Stenchikov, G., Graf, H.-F., Robock. A., Antuna, J. (1999) Climate model simulation of winter warming and summer cooling following the 1991 Mount Pinatubo volcanic eruption, *J. Geophys. Res.*, **104**, 19,039–19,055.

Klein, G.A. (1989) Recognition-Primed Decisions, *Advances in Man-Machine Research*, **5**, JAI Press.

Kotov, V., Systems of Systems as Communicating Structures, Hewlett Packard Computer Systems Laboratory Paper HPL-97–124, (1997), pp. 1–15.

Manthorpe, W.H. (1996) The Emerging Joint System of Systems: A Systems Engineering Challenge and Opportunity for APL, *John Hopkins APL Technical Digest*, **17**(3) 305–310.

Marx, Karl (1887) Capital Vol. 1: The Process of Production of Capital, Progress Publishers, Moscow, USSR.

M'Pherson, P.K. (1981) *The Radio and Electronic Engineer*, **51**(2), Paper 1971/ACS 82.

Odum, H.T. (1971) *Environment, Power and Society*, Wiley Interscience, New York.

Peak, K.J. and Glensor, R.W. (1996) *Community Policing and Problem Solving*, Prentice Hall, New Jersey.

Popper, K. (1972) *Conjectures and Refutations: the Growth of Scientific Knowledge*, Routledge & Kegan Paul, London.

Putnam, R. (2000) *Bowling Alone: The Collapse and Revival of American community*, New York: Simon and Schuster: 288–290.

Quinn, J.B. (1992) *The Intelligent Enterprise*, Free Press

Ramachandran, V.S. and Oberman, L.M. (2006) Broken Mirrors: a Theory of Autism, *Scientific American* **295**(5).

Rapoport, A. and Horvath, W.J. (1959) Thoughts on organization theory and a review of two conferences, *General Systems IV*, pp. 87–93.

Rechtin, E. and Meier, M. (2000) *The Art of System Architecting*, CRC Press, New York.

Richmond, B. (1992) *An Introduction to Systems Thinking*, High performance Systems Inc., Hanover, New Hampshire.

Richardson, L.F. (1960) *The Statistics of Deadly Quarrels*, Boxwood Press, Pittsburgh, Pennsylvania.

Roberts, N., et al (1983) *Introduction to Computer Simulation: A Systems Dynamics Modeling Approach*, Addison-Wesley, Reading, Massachusetts.

Romme, W.H. and Despain, D.G. (1989) *The Yellowstone Fires*, Scientific American, **261**(5).

Sage, A.P. and C.D. Cuppan (2001) On the Systems Engineering and Management of Systems of Systems and Federations of Systems, *Information, Knowledge, Systems Management*, **2**(4) 325–345.

Senge, Peter M. (1990) *The Fifth Discipline: The Art and Practice of the Learning Organization*, Random House Business Books; 2nd revised edn (2006).

Smith, Adam. (1904) ed. Edwin Cannan, *An Inquiry into the Nature of the Wealth of Nations* (5th edn), Methuen, London. Fifth edition

Tuckman, Bruce W. (1965) Developmental sequence in small groups, *Psychological Bulletin*, **63**, 384–399.

Turco, R.P., Toon, O.B., Ackerman, T.P., Pollack, J.B., Sagan, C. (1983) Nuclear Winter: Global Consequences of Multiple Nuclear Explosions, *Science*, **222**, 4630.

Turco, R.P., Toon, A.B., Ackerman, T.P., Pollack, J.B., Sagan, C. (TTAPS) (1990) Climate and Smoke: An Appraisal of Nuclear Winter, *Science*, **247**, 167–168.

von Bertalanffy, L. (1968) *General Systems Theory*, George Braziller, *NY*.

Wald, M.L. (2007) Is Ethanol for the Long Haul? *Scientific American*, **296** (1).

Warfield, J.N. (1973) Intent Structures, *IEEE Transaction: Systems Man and Cybernetics*, SMC **3**(2), 133–140.

Warfield, J.N. (1989) *Societal Systems*, Intersystems Publications, Salinas, California.

Wee Chow Hou (1991) *Sun Tzu: War and Management*, Addison Wesley, Singapore.

Wilson, B. (1984) *Systems: Concepts Methodologies and Applications*, Wiley, Chichester.

Womack, J.P., Jones, D.T. & Roos, D (1990) *The Machine that Changed the World*, Rawson Associates, New York.

Index

YEFIM FASSER and DONALD BRETTNER
Management for Quality in High-Technology Enterprises

THOMAS B. SHERIDAN
Humans and Automation: System Design and Research Issues

ALEXANDER KOSSIAKOFF and WILLIAM N. SWEET
Systems Engineering Principles and Practice

HAROLD R. BOOHER
Handbook of Human Systems Integration

JEFFREY T. POLLOCK AND RALPH HODGSON
Adaptive Information: Improving Business Through Semantic Interoperability, Grid Computing, and Enterprise Integration

ALAN L. PORTER AND SCOTT W. CUNNINGHAM
Tech Mining: Exploiting New Technologies for Competitive Advantage

REX BROWN
Rational Choice and Judgment: Decision Analysis for the Decider

WILLIAM B. ROUSE AND KENNETH R. BOFF (editors)
Organizational Simulation

HOWARD EISNER
Managing Complex Systems: Thinking Outside the Box

STEVE BELL
Lean Enterprise Systems: Using IT for Continuous Improvement

J. JERRY KAUFMAN AND ROY WOODHEAD
Stimulating Innovation in Products and Services: With Function Analysis and Mapping

WILLIAM B. ROUSE
Enterprise Tranformation: Understanding and Enabling Fundamental Change

JOHN E. GIBSON, WILLIAM T. SCHERER, AND WILLAM F. GIBSON
How to Do Systems Analysis

WILLIAM F. CHRISTOPHER
Holistic Management: Managing What Matters for Company Success

WILLIAM B. ROUSE
People and Organizations: Explorations of Human-Centered Design

DEREK K. HITCHINS
Systems Engineering: A 21st Century Systems Methodology

Printed and bound in the UK by
CPI Antony Rowe, Eastbourne

Printed and bound by CPI Group (UK) Ltd, Croydon, CR0 4YY

16/04/2025

14658474-0003